Phytoplasmas: Plant Pathogenic Bacteria - I

Govind Pratap Rao • Assunta Bertaccini
Nicola Fiore • Lia W. Liefting
Editors

Phytoplasmas: Plant Pathogenic Bacteria - I

Characterisation and Epidemiology of Phytoplasma - Associated Diseases

Springer

Editors
Govind Pratap Rao
Division of Plant Pathology
Advance Center for Plant Virology
Indian Agricultural Research Institute
New Delhi, India

Assunta Bertaccini
Department of Agricultural and Food Sciences
Alma Mater Studiorum – University of Bologna
Bologna, Italy

Nicola Fiore
Departamento de Sanidad Vegetal
Universidad de Chile
La Pintana, RM – Santiago, Chile

Lia W. Liefting
Ministry for Primary Industries
Plant Health and Environment Laboratory
Auckland, Auckland, New Zealand

ISBN 978-981-13-4332-2 ISBN 978-981-13-0119-3 (eBook)
https://doi.org/10.1007/978-981-13-0119-3

This Springer imprint is published by the registered company Springer Nature Singapore Pte Ltd.
The registered company address is: 152 Beach Road, #21-01/04 Gateway East, Singapore 189721, Singapore

Contents

About the Editors

Govind Pratap Rao is working as a Principal Scientist (Plant Pathology) at Indian Agricultural Research Institute, New Delhi. He did his M.Sc (Botany) and Ph.D. in Plant Virology from Gorakhpur University. He did Post Doc at University of Urbana-Champaign, Illinois, USA, with Prof. R.E. Ford on Characterization of Sugarcane mosaic and maize dwarf mosaic viruses in 1994. Dr. Rao has 30 years of research experience on plant pathology especially on plant virology and phytoplasmas. He did significant contributions in characterization of viruses infecting cucurbits, sugarcane, maize and sorghum. He has published over 130 research publications and authored and edited nearly 17 books. He has also guided 3 M.Sc. and 12 Ph.D. students on different aspects of plant pathology. He has worked in different capacities as Scientific Officer, Sr. Scientific Officer (Plant Pathology), Head, Div. Plant Pathology and Officer-in-Charge at Research Stations of UP Council of Sugarcane Research centres at Seorahi, Gorakhpur and Shahjahanpur from 1987 to 2010. He has been awarded several prestigious awards. The most important ones are: National Biotechnology Associateship Award (1991–1992), DBT, Govt. of India; Young Scientist Award (1994–1995) from DST, Govt. of India; Overseas BOYSCAST Award (1996) from DST, Govt. of India; President Award, Society for General Microbiology, UK, 1998; Best U.P. Agriculture Scientist Award (UPCAR), Govt. of Uttar Pradesh in 2002; Vigyan Ratna Award by CST, Govt. of UP for the year 2003–2004; Jin Xiu Qiu Award in 2006 by People's Govt. of Guangxi Province, Nanning, China;

Global Award of Excellence, IS 2008, Al-Ahrish, Egypt; Dr. Ram Badan Singh Vishisht Krishi Vaigyanik Puraskar – 2014 by UPCAR, Lucknow, India and Leadership Excellence Award in Sugarcane Crop Protection by Thailand Society of Sugar Cane Technologists, Bangkok.

Dr. Rao is Editor-in-Chief of *Sugar Tech*, an international journal on sugar crops and related industries, and *Phytopathogenic Mollicutes*, an international journal on phloem limited micrmedian organisms. Dr. Rao is also Secretary General of Indian Virological Society, New Delhi, and member of several prestigious scientific societies and organizations like, APS; ASM; ISSCT; IPWG; SSRP and IPS. Dr. Rao has visited over 30 countries as visiting scientists, for invited talk, post doc fellow, research training, panel discussion and for attending workshop and conferences. At present Dr. Rao is working on characterization, epidemiology and management of virus infected cereal crops, millets and maize and phytoplasmas infecting important agricultural and horticultural crops in India.

Assunta Bertaccini is a Plant Pathology Professor at the University of Bologna, invited speaker at many national and international meetings and seminars, member of scientific committees of several international meetings on biotechnology and virology/phytoplasmology, and referee for numerous scientific international journals. Among others, she was awarded with the Emmy Klienenberger-Nobel award for distinguished research in mycoplasmology. She is responsible for the phytobacteriology laboratories where she is leading a team mainly working on several aspects of plant bacteriology with emphasis on the plant diseases associated with phytoplasmas. Recently in her laboratory, phytoplasma cultivation in complex media was eventually achieved. She mentored a number of Ph.D. students, Plant Pathology specialization students, and undergraduate and master thesis students. She is author or coauthor of more than 800 publications, books and book chapters, Editor-in-Chief of *Phytopathogenic Mollicutes*, Senior Editor of *Phytopathologia Mediterranea*, and since 2007, she founded and is leading the International Phytoplasmologist Working Group (IPWG) (http://www.ipwgnet.org/).

Nicola Fiore is a Plant Pathology Associate Professor at the University of Chile, Faculty of Agronomical Sciences, Department of Plant Health, Santiago, Chile. He is a member of several phytopathological societies. He is an invited speaker at many national and international meetings and seminars, member of scientific committees of several international meetings on virology/phytoplasmology, and referee for several scientific international journals. Award received: best poster in 22nd ICVF, Rome, 3–8 June, 2012. He is responsible for the Phytovirology Laboratory focusing on the diagnosis, epidemiology, and control of plant diseases caused by viruses, viroids, and bacteria (including phytoplasmas), applying biological approaches and molecular tools. He has elaborated and participated as a researcher or director in 20 research projects. He has mentored several students: 3 Ph.D., 6 Master's, and 6 undergraduate theses. He maintains constant professional contact with wine, table grape, and fruit producers, advising them on the prevention of diseases caused by viruses, viroids, and bacteria.

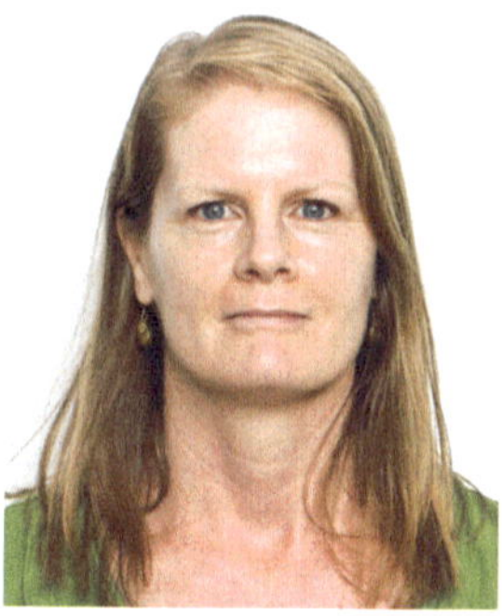

Lia W. Liefting is a Principal Scientist in the Virology team at the Plant Health and Environment Laboratory at the Ministry for Primary Industries (MPI) in New Zealand. The overall role of the team involves diagnosing virus and virus-like diseases including phytoplasmas and liberibacters of a wide range of host plants as well as providing technical advice on these diseases. Prior to her role at MPI, Lia completed a Ph.D. and two post-doctoral positions on phytoplasmas. The first post-doc at the University of California at Davis, USA, involved sequencing the genome of Western-X disease phytoplasma. On return to New Zealand, Lia's second post-doc was to sequence the genome of '*Candidatus* Phytoplasma australiense'. Both phytoplasma genome sequencing projects were performed by Sanger sequencing prior to the advent of next-generation sequencing. Lia was lead author on the IPPC diagnostic protocols for phytoplasmas and '*Candidatus* Liberibacter solanacearum' and is an Editor for *Plant Pathology*.

Chapter 1
Phytoplasmas: An Update

Assunta Bertaccini and Ing-Ming Lee

Abstract A summary of the research carried out on phytoplasma-associated diseases 50 years after their discovery is presented. The great majority of this research was devoted to classification and differentiation of these prokaryotes by molecular and bioinformatic tools applied to specific phytoplasma genes. The availability of a robust classification system has greatly facilitated phytoplasma identification leading to an increased knowledge of plant diseases worldwide. Phytoplasma biology study still needs to be improved to allow better management solutions to reduce the impact of these diseases in both agricultural and natural environments.

Keywords Taxonomy · Biology · Insect vector · Plant disease · Epidemiology

1.1 Introduction

Many yellows-type diseases, including aster yellows and paulownia witches' broom, were believed to be caused by viruses until 1967 when a group of Japanese scientists observed microorganisms resembling animal mycoplasmas in the phloem sieve tube elements of diseased plants by electron microscopy (Doi et al. 1967). These pleomorphic cell wall lacking bacteria were then named mycoplasma-like organisms (MLOs) (Fig. 1.1). They possess a unique lifestyle that allows them to live across plant and insect kingdoms. In the following decades, their detection was primarily based on electron microscopy images of diseased phloem tissue and biological properties, such as unique disease symptoms, specific insect vector, and plant host range. In subsequent years, the development of serological and molecular

A. Bertaccini (✉)
Department of Agricultural and Food Sciences, *Alma Mater Studiorum* – University of Bologna, Bologna, Italy
e-mail: assunta.bertaccini@unibo.it

I-M. Lee
Molecular Plant Pathology Laboratory, United States Department of Agriculture, Agriculture Research Service, Beltsville, MD, USA

G. P. Rao et al. (eds.), *Phytoplasmas: Plant Pathogenic Bacteria - I*,
https://doi.org/10.1007/978-981-13-0119-3_1

Fig. 1.1 On the right electron microscopy picture of cross section of sieve tubes with phytoplasmas (X 6,000), and on the left symptoms associated with phytoplasma presence in aster (*Callistephus chinensis*)

tools such as monoclonal antibodies and cloned DNA probes greatly improved their detection.

DNA-specific amplification and sequencing provided evidence that MLOs constitute a large monophyletic group within the class *Mollicutes*, and the trivial name of "phytoplasma" was adopted followed by designation of the '*Candidatus* Phytoplasma' genus (IRPCM 2004). Phytoplasmas have variable sizes and shapes (Fig. 1.1) and survive and multiply in the isotonic environments provided by plant phloem and insect hemolymph. The full genome sequence has been completed for two strains of aster yellows ('*Candidatus* Phytoplasma asteris'), two strains of '*Ca.* P. australiense', and a strain of '*Ca.* P. mali' (Oshima et al. 2004; Bai et al. 2006; Kube et al. 2008; Tran-Nguyen et al. 2008; Andersen et al. 2013). A number of draft genome sequences are also reported (Saccardo et al. 2012; Mitrović et al. 2014; Zamorano and Fiore 2016). Phytoplasmas possess one of the smallest genomes among living organisms (Marcone et al. 1999) yet code complex metabolic pathways that allow them to interact with both their plant and insect hosts (Hogenhout et al. 2008). Phytoplasmas are quite often associated with severe and rapidly spreading plant diseases, they are also able to increase the metabolic activity of their hosts, modify insect fitness, increase plant shoot production, and change flower shape and color (Bertaccini et al. 2014); in other cases they are associated with severe decline and death of the infected plants. It is however not uncommon to detect phytoplasmas in asymptomatic plants, and this leaves open questions related to their pathogenicity and biology as relevant characteristics that must be provided together with phytoplasma identification when describing a new phytoplasma-associated disease.

1.2 Biological and Molecular Basis for Classification

Fifty years after their discovery, the role of phytoplasmas as plant pathogens is still only based on indirect biological proof, such as electron microscopy observation, specific DNA amplification, and symptom elimination after tetracycline treatments (Ishiie et al. 1967). Insect and dodder transmission are the main tools available to confirm phytoplasma association with plant disease. Plants infected by phytoplasmas very often exhibit symptoms indicative of profound unbalance of growth regulators. Symptoms include virescence and phyllody of flowers, sterility, loss of apical dominance generating the proliferation of axillary buds with witches' broom formation (Fig. 1.1), abnormal internode elongation, and generalized stunting. The characteristic symptomatology is very useful for preliminary indication of the possible phytoplasma involvement in a disease. Some phytoplasmas also confer desirable features such as for poinsettia, for which the presence of a specific strain allowed it to be grown as an ornamental pot plant (Bertaccini et al. 1996; Lee et al. 1997).

Serological diagnostic techniques for detection of phytoplasmas began to emerge in the 1980s when polyclonal and monoclonal antisera were produced and tested (Lee et al. 1993b; Chen et al. 1993, 1994; Saeed et al. 1994) in plant tissues as well as in leafhopper vectors or potential vectors using immunofluorescence (Lherminier et al. 1990) and immunosorbent electron microscopy (Sinha 1979; Sinha and Benhamou 1983), dot blot, or ELISA (Boudon-Padieu et al. 1989). In other approaches, tissue blotting with direct or indirect antigen detection has been used for specific phytoplasma detection (Lin and Chen 1985). In more recent years, antibodies have been prepared to partial sequences of the major immunodominant proteins of some phytoplasmas (Berg et al. 1999; Blomquist et al. 2001; Hong et al. 2001; Mergenthaler et al. 2001; Kakizawa et al. 2001; Barbara et al. 2002; Wei et al. 2004; Arashida et al. 2008; Siampour et al. 2012). Starting from the 1990s, the application of molecular probes (Kirkpatrick et al. 1987; Lee and Davis 1988; Lee et al. 1992; Bertaccini et al. 1993), PCR (Ahrens and Seemüller 1992; Namba et al. 1993; Lee et al. 1993a; Schneider et al. 1993), and nested-PCR (Lee et al. 1994, 1995) together with restriction fragment length polymorphism (RFLP) analyses or sequencing allowed the broad detection of phytoplasmas (Lee et al. 1998a). The introduction of quantitative PCR assays (qPCR) has shown these assays to be sensitive with reduced risk of contamination making this technique a reliable alternative to nested PCR assays in routine testing (Bianco et al. 2004; Torres et al. 2005; Crosslin et al. 2006; Angelini et al. 2007; Baric et al. 2006; Hren et al. 2007; Hodgetts et al. 2009; Aldaghi et al. 2009; Berger et al. 2009; Pelletier et al. 2009; Nejat et al. 2010; Manimekalai et al. 2011; Monti et al. 2013; Córdova et al. 2014; Ikten et al. 2016; Satta et al. 2017; Linck et al. 2017). Microarray (Nicolaisen and Bertaccini 2007; Lenz et al. 2015), deep amplicon sequencing (Nicolaisen et al. 2011), and LAMP (Tomlinson et al. 2010; Bekele et al. 2011; Obura et al. 2011; Sugawara et al. 2012; Kogovšek et al. 2015; Vu et al. 2016) are other techniques used for phytoplasma detection, but not yet fully exploited and not always adequate due to lack of specificity or sensitivity.

1.3 16S Ribosomal DNA-Based Classification

The development of a robust and quite exhaustive classification system based on 16S ribosomal gene sequence (Lee et al. 1998a) was followed by the multilocus typing on other genes, which are usually different according to the diverse '*Candidatus* Phytoplasma' species (Bertaccini 2015). A consensus for naming novel phytoplasmas was reached in 2004 (IRPCM) for which "a '*Candidatus* (*Ca.*) Phytoplasma' species description should refer to a single, unique 16S rRNA gene sequence (>1200 bp)," and "a strain can be recognized as a novel '*Ca.* Phytoplasma' species if its 16S rRNA gene sequence has <97.5% similarity to that of any previously described '*Ca.* Phytoplasma' species." Phytoplasma taxonomy still relies on these rules due to the lack of phenotypic characteristics of these bacteria that are needed to classify them in formal genus and species. Some phytoplasma strains which may warrant designation of a new taxon, but fail to meet the requirement of sharing <97.5% sequence similarity with existing '*Ca.* Phytoplasma', can be differentiated and classified using additional unique biological properties such as antibody specificity, host range, and vector transmission specificity (Seemüller and Schneider 2004).

Phytoplasma 16S rRNA genes are highly conserved; however, they possess sufficient diversity to be used for their classification. Phytoplasmas are classified either into a '*Candidatus* Phytoplasma' species based on percent sequence identity (IRPCM 2004) or into ribosomal groups and subgroups based on the presence of restriction sites (Lee et al. 1998a; Wei et al. 2007; Zhao et al. 2009b). The number of '*Candidatus* Phytoplasma' species and ribosomal groups and subgroups is continually growing due to increased awareness of the importance of these pathogens in agriculture and environment worldwide (Bertaccini et al. 2014; Maejima et al. 2014). A formal description remains to be published for phytoplasmas associated with some well-known diseases such as grapevine "flavescence dorée" and some of the lethal yellowing associated agents for which still the grouping and subgrouping are the only official taxonomy (Table 1.1). A number of new '*Candidatus*' species were proposed in the last few years indicating the great biodiversity in these microorganisms intensified by the frequent report of interoperon heterogeneity and/or mixed phytoplasma infection in both insects and plants (Schneider and Seemüller 1994; Lee et al. 1995, 1998b; Ho et al. 2001). Very recently two other '*Candidatus*' species were officially proposed '*Ca.* P. wodyetiae' and '*Ca.* P. noviguineense' (Naderali et al. 2017; Miyazaki et al. 2018). There are thousands of 16S rRNA gene sequences of phytoplasmas deposited in the public databases as well as sequences to other conserved genomic regions used as supplemental tools for finer phytoplasma differentiation (Duduk and Bertaccini 2011). A barcode screening system was also developed in agreement with the 16S rDNA classification for fast phytoplasma detection and identification (Makarova et al. 2012). The *rp*, *tuf*, and *secY* genes show more variation than the 16S rRNA gene (Gundersen et al. 1996; Schneider et al. 1997; Cimerman et al. 2009; Marcone et al. 2000; Lee et al. 2006b, 2010, 2012; Danet et al. 2007; Arnaud et al. 2007; Martini et al. 2002, 2007;

Table 1.1 Phytoplasma classification based on RFLP analyses and/or sequencing of 16S rDNA

16Sr grouping	Strain (acronym)	'*Candidatus*' sp.	GenBank accession number	References
16SrI – aster yellows				
I-A	Aster yellows witches' broom (AY-WB)		NC_007716	Bai et al. (2006)
I-B	Aster yellows (MAY)	**'*Ca.* P. asteris'**	M30790	Lee et al. (2004a)
I-C	Clover phyllody (CPh)		AF222065	Lee et al. (2004a)
I-D	Paulownia witches' broom (PaWB)		AY265206	Lee et al. (2004a)
I-E	Blueberry stunt (BBS3)		AY265213	Lee et al. 2004a
I-F	Aster yellows apricot Spain (A-AY)		AY265211	Lee et al. (2004a)
I-I	Strawberry witches' broom (STRAWB1)		U96614	Jomantiene et al. (1998)
I-K	Strawberry witches' broom (STRAWB2)		U96616	Jomantiene et al. (1998)
I-L	Aster yellows (AV2192)		AY180957	Lee et al. (2003)
I-M	Aster yellows (AVUT)		AY265209	Lee et al. (2004a)
I-N	Aster yellows (IoWB)		AY265205	Lee et al. (2004a)
I-O	Soybean purple stem (SPS)		AF268405	Lee et al. (2002)
I-P	Aster yellows from *Populus* (PopAY)		AF503568	Šeruga et al. (2003)
I-Q	Cherry little leaf (ChLL)		AY034089	Valiunas et al. (2005)
I-R	Strawberry phylloid fruit (StrawbPhF)		AY102275	Jomantiene et al. (2002)
I-S	Pepper little leaf (PeLL)		DQ092321	Santos-Cervantes et al. (2008)
I-T	Tomato little leaf (ToLL)		DQ375238	Santos-Cervantes et al. (2008)
I-U	Mexican potato purple top (JAL8)		FJ914650	Santos-Cervantes et al. (2010)
I-V	Mexican potato purple top (SON18)		FJ914642	Santos-Cervantes et al. (2010)
I-W	Peach rosette-like disease (PRU0382)		HQ450211	Arocha-Rosete et al. (2011)
I-X	Papaya bunchy top (BTS)		JF781308	Acosta et al. (2013)
I-Y	Tomato "brote grande"	**'*Ca.* P. lycopersici'**	EF199549	Arocha et al. (2007)
I-Z	Papaya bunchy top (BTS)		JF781311	Acosta-Pérez et al. (2017)

(continued)

Table 1.1 (continued)

16Sr grouping	Strain (acronym)	'*Candidatus*' sp.	GenBank accession number	References
16SrII – peanut witches' broom				
II-A	Peanut witches' broom (PnWB)		L33765	Gundersen et al. (1994)
II-B	Lime witches' broom (WBDL)	**'*Ca.* P. aurantifolia'**	U15442	Zreik et al. (1995)
II-C	Faba bean phyllody (FBP)		X83432	Seemüller et al. (1998)
II-D	Papaya mosaic (PpM)	**'*Ca.* P. australasia'**	Y10096	White et al. (1998)
II-E	Picris echioides phyllody (PEY)		Y16393	Seemüller et al. (1998)
II-F	Cotton phyllody (CoP)		EF186827	Martini et al. (2007)
II-G	Cactus witches' broom (CWB)		EU099568	Cai et al. (2008)
II-J	Cactus witches' broom (CWB)		EU099552	Cai et al. (2008)
II-H	Cactus witches' broom (CWB)		EU099569	Cai et al. (2008)
II-K	Cactus witches' broom (CWB)		EU099572	Cai et al. (2008)
II-I	Cactus witches' broom (CWB)		EU099551	Cai et al. (2008)
II-L	Cactus witches' broom (CWB)		EU099546	Cai et al. (2008)
II-M	Potato purple top		FJ914643	Yadav et al. (2014)
II-N	Papaya BTSp		JF781309	Acosta et al. (2013)
II-O	Tabebuia witches' broom		EF647744	Mafia et al. (2007)
II-P	Cuban papaya phytoplasma		DQ286948	Perez-López et al. (2016)
II-Q	Papaya bunchy top (TSpHav02-IIA)		JF78131	Perez-López et al. (2016)
II-R	*Echinopsis* yellow patch		DQ535900	Perez-López et al. (2016)
II-S	*Amaranthus hypochondriacus* strain 52A		FJ357164	Perez-López et al. (2016)
II-T	Tomatillo witches' broom		U125185	Perez-López et al. (2016)
II-U	Papaya little leaf		KP057205	Yang et al. (2016)

(continued)

Table 1.1 (continued)

16Sr grouping	Strain (acronym)	'*Candidatus*' sp.	GenBank accession number	References
16SrIII – X-disease				
III-A	Peach X-disease (PX11CT1)	**'*Ca.* P. pruni'**	JQ044393	Davis et al. (2013)
III-B	Clover yellow edge (CYE)		AF173558	Davis et al. (2013)
III-C	Pecan bunch (PB)		GU004371	Davis et al. (2013)
III-D	Goldenrod yellows (GR1)		GU004372	Davis et al. (2013)
III-E	Spiraea stunt (SP1)		AF190228	Davis et al. (2013)
III-F	Milkweed yellows (MW1)		AF510724	Davis et al. (2013)
III-G	Walnut witches' broom (WWB)		AF190226/ AF190227	Davis et al. (2013)
III-J	Chayote witches' broom (ChWBIII)		AF147706	Montano et al. (2000)
III-K	Strawberry leafy fruit (SLF)		AF274876	Jomantiene et al. (1998)
III-H	Poinsettia branch-inducing (PoiBI)		AF190223	Davis et al. (2013)
III-I	Virginia grapevine yellows (VGYIII)		AF060875	Davis et al. 1998
III-L	Cassava frog skin disease (CFSD)		EU346761	Alvarez et al. (2009)
III-M	Potato purple top (MT117)		FJ226074	Davis et al. (2013)
III-N	Potato purple top (AKpot6)		GU004365	Davis et al. (2013)
III-O	Dandelion virescence (DanVir)		AF370120	Jomantiene et al. (2002)
III-P	Dandelion virescence (DanVir)		AF370119/ AF370120	Jomantiene et al. (2002)
III-Q	Black raspberry witches' broom (BRWB7)		AF302841	Davis et al. (2001)
III-R	Cirsium white leaf (CirWL)		AF373105	Zhao et al. (2009b)
III-S	Western peach X-disease (WX)		L04682	Zhao et al. (2009b)
III-T	Sweet and sour cherry (ChD)		FJ231728	Valiunas et al. (2009)
III-U	Cirsium white leaf (CWL)		AF373105/ AF373106	Jomantiene et al. (2002)

(continued)

Table 1.1 (continued)

16Sr grouping	Strain (acronym)	'*Candidatus*' sp.	GenBank accession number	References
III-V	Passion fruit phytoplasma (PassWB-Br4)		GU292082	Davis et al. (2012)
III-W	*Heterothalamus* little leaf (HetLL)		KC412029	Galdeano et al. (2013)
III-X	*Conyza* witches' broom		KC412026	Galdeano et al. (2013)
III-Y	Cranberry false blossom		KF62652	Lee et al. (2014)
III-Z	Broccoli stunt strain BSP-21		JX626327	Perez-López et al. (2016)
16SrIV – coconut lethal yellowing				
IV-A	Coconut lethal yellowing (LYJ-C8)		AF498307	Harrison et al. (2002)
IV-B	Yucatan coconut lethal decline (LDY)		U18753	Harrison et al. (1994)
IV-C	Tanzanian coconut lethal decline (LDT)		X80117	Harrison et al. (1994)
IV-D	Texas phoenix decline (TPD)		AF434969	Harrison et al. (2008)
IV-E	Coconut lethal yellowing (LYDR-B5)		DQ631639	Martinez et al. (2008)
IV-F	*Washingtonia robusta* decline		EU241512	Harrison et al. (2008)
16SrV – elm yellows				
V-A	Elm yellows (EY)	**'*Ca.* P. ulmi'**	AY197655	Lee et al. (2004b)
V-B	Jujube witches' broom (JWB-G1)	**'*Ca.* P. ziziphi'**	AB052876	Jung et al. (2003a)
V-C	"Flavescence dorée" (FD-C)		X76560	Martini et al. (1999)
V-D	"Flavescence dorée" (FD-D)		AJ548787	Martini et al. 1999
V-E	Rubus stunt (RuS)	**'*Ca.* P. rubi'**	AY197648	Malembic-Maher et al. (2011)
V-F	Balanite witches' broom (BltWB)	**'*Ca.* P. balanitae'**	AB689678	Win et al. (2013)
V-G	Korean jujube witches' broom		AB052879	Jung et al. (2003a)
V-H	*Bischofia polycarpa* witches' broom		KJ452547	Lai et al. (2014)
V-I	Blackberry witches' broom		KR233473	Fránová et al. (2016)

(continued)

Table 1.1 (continued)

16Sr grouping	Strain (acronym)	'*Candidatus*' sp.	GenBank accession number	References
16SrVI – clover proliferation				
VI-A	Clover proliferation (CP)	**'*Ca.* P. trifolii'**	AY390261	Hiruki and Wang (2004)
VI-B	Strawberry multiplier disease (MD)		AF190224	Jomantiene et al. (1998)
VI-C	Illinois elm yellows (EY-IL1)		AF409069	Jacobs et al. (2003)
VI-D	Periwinkle little leaf (PLL-Bd)		AF228053	Siddique et al. (2001)
VI-E	*Centaurea solstitialis* virescence (CSVI)		AY270156	Faggioli et al. (2004)
VI-F	Catharanthus phyllody (CPS)		EF186819	Martini et al. (2007)
VI-H	Portulaca little leaf phytoplasma (PLL-Ind)		EF651786	Samad et al. (2008)
VI-I	Passionfruit (WB-Br4)	**'*Ca.* P. sudamericanum'**	GU292081	Davis et al. (2012)
16SrVII – ash yellows				
VII-A	Ash yellows (AshY)	**'*Ca.* P. fraxini'**	AF092209	Griffiths et al. (1999)
VII-B	Erigeron witches' broom (ErWB)		AY034608	Barros et al. (2002)
VII-C	Argentinian alfalfa witches' broom (ArAWB)		AY147038	Conci et al. (2005)
VII-D	Erigeron witches' broom (EboWB-Br0)		KJ831066	Flôres et al. (2015)
16SrVIII – loofah witches' broom				
VIII-A	Loofah witches' broom (LufWB)	**'*Ca.* P. luffae'**	AF086621	Davis et al. (2017)
16SrIX – pigeon pea witches' broom				
IX-A	Pigeon pea witches' broom (PPWB)		AF248957	Gundersen et al. (1996)
IX-B	Almond witches' broom (AlWB)	**'*Ca.* P. phoenicium'**	AF515636	Verdin et al. (2003)
IX-C	Naxos periwinkle virescence (NAXOS)		HQ589191	Duduk et al. (2008)
IX-D	Almond witches' broom (AlWB)		AF515637	Verdin et al. (2003)
IX-E	*Juniperus* witches' broom		GQ925918	Davis et al. (2010)

(continued)

Table 1.1 (continued)

16Sr grouping	Strain (acronym)	'*Candidatus*' sp.	GenBank accession number	References
IX-F	Almond and stone fruit witches' broom (N27-2)		HQ407532	Molino Lova et al. (2011)
IX-G	Almond and stone fruit witches' broom (A1-1)		HQ407514	Molino Lova et al. (2011)
IX-H	Sarson phyllody		KU892213	Ahmad et al. (2017)
16SrX – apple proliferation				
X-A	Apple proliferation (AP)	**'*Ca*. P. mali'**	AJ542541	Seemüller and Schneider (2004)
X-B	European stone fruit yellows (ESFY)	**'*Ca*. P. prunorum'**	AJ542544	Seemüller and Schneider (2004)
X-C	Pear decline (PD)	**'*Ca*. P. pyri'**	AJ542543	Seemüller and Schneider (2004)
X-D	Spartium witches' broom (SpaWB)	**'*Ca*. P. spartii'**	X92869	Marcone et al. (2003a)
X-E	Black alder witches' broom (BAWB(BWB)		X76431	Seemüller et al. (1994)
16SrXI – rice yellow dwarf				
XI-A	Rice yellow dwarf (RYD)	**'*Ca*. P. oryzae'**	AB052873	Jung et al. (2003b)
XI-B	Sugarcane white leaf (SCWL)		X76432	Lee et al. (1998a, b)
XI-C	Leafhopper-borne (BVK)		X76429	Lee et al. (1998a, b)
XI-D	Sugarcane white leaf (SCWL)		KR020685	Zhang et al. (2016)
XI-E	Cirsium phytoplasma	**'*Ca*. P. cirsii'**	KR869146	Šafářová et al. (2016)
XI-F	Sugarcane grassy shoot (SCGS)		HF586636	Yadav et al. (2017)
16SrXII – "stolbur"				
XII-A	"Stolbur" (STOL11)	**'*Ca*. P. solani'**	AF248959	Quaglino et al. (2013)
XII-B	Australian grapevine yellows (AUSGY)	**'*Ca*. P. australiense'**	L76865	Davis et al. (1997)
XII-C	Strawberry lethal yellows (StrawLY)		AJ243045	Padovan et al. (2000)
XII-D	Japanese hydrangea phyllody	**'*Ca*. P. japonicum'**	AB010425	Sawayanagi et al. (1999)
XII-E	Yellows diseased strawberry (StrawY)	**'*Ca*. P. fragariae'**	DQ086423	Valiunas et al. (2006)
XII-F	"Bois noir" (BN-Op30)		EU836630	Quaglino et al. (2009)

(continued)

Table 1.1 (continued)

16Sr grouping	Strain (acronym)	'*Candidatus*' sp.	GenBank accession number	References
XII-G	"Bois noir" (BN-Fc3)		47EU8366	Quaglino et al. (2009)
XII-H	Bindweed yellows (BY-S57/11)	**'*Ca.* P. convolvuli'**	JN833705	Martini et al. (2012)
XII-I	Potato strain from China (169/Hezuo 88)		EU338445	Cheng et al. (2015)
16SrXIII – Mexican periwinkle virescence				
XIII-A	Mexican periwinkle virescence (MPV)	**'*Ca.* P. hispanicum'**	AF248960	Davis et al. (2016)
XIII-B	Strawberry green petal (STRAWB2)		U96616	Jomantiene et al. (1998)
XIII-C	Chinaberry yellows (CBY1)		AF495882	Harrison et al. (2002)
XIII-D	Mexican potato purple top (SINPV)		FJ914647	Santos-Cervantes et al. (2010)
XIII-E	Papaya apical curl necrosis (PACN)		EU719111	Melo et al. (2013)
XIII-F	Strawberry red leaf		KJ921641	Fernández et al. 2015
XIII-G	Chinaberry yellowing (ChTY)	**'*Ca.* P. meliae'**	KU850940	Fernández et al. (2016)
16SrXIV – Bermuda grass white leaf				
XIV-A	Bermuda grass white leaf (BGWL)	**'*Ca.* P. cynodontis'**	AJ550984	Marcone et al. (2003b)
XIV-B	Bermuda grass white leaf Iran strain		EF444485	Salehi et al. (2009)
XIV-C	Bermuda grass white leaf (RS304/13)		KP019339	Mitrovic et al. (2015)
16SrXV – hibiscus witches' broom				
XV-A	Hibiscus witches' broom (HibWB)	**'*Ca.* P. brasiliense'**	AF147708	Montano et al. (2001)
XV-B	Guazuma witches' broom (GWB)		HQ258882	Villalobos et al. (2011)
16SrXVI – sugarcane yellow leaf				
XVI-A	Sugarcane yellow leaf	**'*Ca.* P. graminis'**	AY725228	Arocha et al. (2005)
16SrXVII – papaya bunchy top				
XVII-A	Papaya bunchy top	**'*Ca.* P. caricae'**	AY725234	Arocha et al. (2005)
16SrXVIII – American potato purple top wilt				
XVIII-A	American potato purple top wilt	**'*Ca.* P. americanum'**	DQ174122	Lee et al. (2006a, b)

(continued)

Table 1.1 (continued)

16Sr grouping	Strain (acronym)	'*Candidatus*' sp.	GenBank accession number	References
16SrXIX – chestnut witches' broom				
XIX-A	Chestnut witches' broom	**'*Ca.* P. castaneae'**	AB054986	Jung et al. (2002)
16SrXX – Rhamnus witches' broom				
XX-A	Rhamnus witches' broom	**'*Ca.* P. rhamni'**	AJ583009	Marcone et al. (2003a)
16SrXXI – Pinus phytoplasma				
XXI-A	Pinus phytoplasma (PinP)	**'*Ca.* P. pini'**	AJ310849	Schneider et al. (2005)
16SrXXII – lethal yellowing-type				
XXII-A	Lethal yellowing Mozambique (LYDM 178)	**'*Ca.* P. palmicola'**	KF751387	Harrison et al. (2014)
XXII-B	Cape Saint Paul Wilt disease Ghana (LDG)		Y13912	Tymon et al. (1998)
16SrXXIII[a]				
XXIII-A	Buckland valley grapevine yellows		AY083605	Wei et al. (2007)
16SrXXIV[a]				
XXIV-A	Sorghum bunchy shoot		AF509322	Wei et al. (2007)
16SrXXV[a]				
XXV-A	Weeping tea witches' broom		AF521672	Wei et al. (2007)
16SrXXVI[a]				
XXVI-A	Sugarcane phytoplasma D3T1		AJ539179	Wei et al. (2007)
16SrXXVII[a]				
XXVII-A	Sugarcane phytoplasma D3T2		AY539180	Wei et al. (2007)
16SrXXVIII[a]				
XXVIII-A	Derbid phytoplasma		AY744945	Wei et al. (2007)
16SrXXIX – cassia witches' broom				
XXIX-A	Cassia witches' broom (CaWB)	**'*Ca.* P. omanense'**	EF666051	Al-Saady et al. (2008)
XXIX-B	Bindweed witches' broom (RBiWB)		KY047493	Esmailzadeh Hosseini et al. (2016)
16SrXXX – salt cedar witches' broom				
XXX-A	Salt cedar witches' broom	**'*Ca.* P. tamaricis'**	FJ432664	Zhao et al. (2009a)
16SrXXXI – soybean stunt				
XXXI-A	Soybean stunt (SoyST1c1)	**'*Ca.* P. costaricanum'**	HQ225630	Lee et al. (2011)

(continued)

Table 1.1 (continued)

16Sr grouping	Strain (acronym)	'*Candidatus*' sp.	GenBank accession number	References
16SrXXXII – Malaysian periwinkle virescence and phyllody				
XXXII-A	Malaysian periwinkle virescence (MaPV)	**'*Ca.* P. malaysianum'**	EU371934	Nejat et al. (2013)
XXXII-B	Malayan yellow dwarf (MYD)		EU498727	Nejat et al. (2013)
XXXII-C	Malayan oil palm (MOP)		EU498728	Nejat et al. (2013)
16SrXXXIII – *Allocasuarina muelleriana* phytoplasma				
XXXIII-A	Allocasuarina phytoplasma	**'*Ca.* P. allocasuarinae'**	AY135523	Marcone et al. (2003a)

[a]Groups designed only on the basis of GenBank deposited sequences

Hodgetts et al. 2008; Mitrović et al. 2011, 2015) and are useful for epidemiological studies. However using only genetic differentiation to characterize phytoplasmas could end in producing just lists of genotypes if the knowledge of phytoplasma biology is not accompanying the appropriate taxonomy. Biological characteristics of phytoplasmas are needed in particular to help finding management solutions to reduce phytoplasma disease impact on worldwide agriculture.

1.4 Transmission and Epidemiology

In nature phytoplasmas infect numerous plant species and as equally numerous insects that serve as their vectors in a successful three-way interaction. Phytoplasmas are mainly spread between plants by insects in the families Cicadellidae (leafhoppers) and Psyllidae (psyllids) and superfamily Fulgoroidea (plant hoppers), which feed on the phloem sap of infected plants; therefore their host range is dependent upon feeding habits of their insect vectors (Bertaccini 2007). Phytoplasmas overwinter in insect vectors or in perennial plant hosts and interact with insect hosts also reducing or enhancing their fitness (Sugio et al. 2011a). Transovarial transmission (Alma et al. 1997; Kawakita et al. 2000; Hanboonsong et al. 2002; Tedeschi et al. 2006) and seed transmission (Khan et al. 2002; Botti and Bertaccini 2006; Calari et al. 2011; Chung and Jeong 2014; Satta et al. 2016) were also demonstrated in some plant species/insect-phytoplasma combinations. Phytoplasmas are also efficiently spread via vegetative propagation such as cuttings, grafting, and micropropagation practices (Bertaccini et al. 1992; Jarausch et al. 1996; Bertaccini 2007).

1.5 Genomic Sequencing and Metabolic Features

A lot of information has been achieved by full genome sequencing especially related to putative biochemical pathways showing that phytoplasmas are very special microorganisms because they lack a lot of relevant features such as cell wall, mobility, key enzymes, and pathways. Phytoplasmas possess the smallest genome among bacteria; however, gene duplication and redundancy are well represented (Oshima et al. 2004). Moreover, extrachromosomal DNA or plasmids of various sizes have also been found in several phytoplasma groups together with sequences of variable mosaic (SVM) termed potential mobile units or PMUs (Schneider et al. 1992; Kuboyama et al. 1998; Rekab et al. 1999; Oshima et al. 2001; Jomantiene and Davis 2006; Jomantiene et al. 2007; Wei et al. 2008; Ishii et al. 2009; Toruno et al. 2010). Microarray analysis of '*Ca.* P. asteris' strain OY-M revealed that expression of approximately 33% of the genes change during host switching between plant and insect (Oshima et al. 2011) and the phytoplasma may use transporters, secreted proteins, and metabolic enzymes in a host-specific manner. Several factors, namely, tengu, SAP11, SAP54, and P38, that could modulate possible phytoplasma pathogenicity were reported to induce symptoms similar to those observed in phytoplasma-infected plants when inserted into transgenic plants (Hoshi et al. 2009; Sugio et al. 2011a, b; MacLean et al. 2011; Neriya et al. 2014). Phytoplasmas possess two secretion systems, YidC and Sec, the latter seems to be common to most or all phytoplasmas (Kakizawa et al. 2004, 2009). It was also demonstrated that phytoplasmas lack ATP synthase genes suggesting a metabolism strongly dependent on glycolysis. While two sets of glycolytic enzymes were encoded in a duplicated genomic region of a strongly pathogenic strain of '*Ca.* P. asteris', mild strains do not possess this duplication, suggesting this as a possible pathogenicity mechanism (Oshima et al. 2007). The glycolysis genes are completely absent in the full genome of '*Ca.* P. mali' in which a gene-encoding 2-dehydro-3-deoxy-phosphogluconate aldolase was retrieved leading to the hypothesis that in this phytoplasma pyruvate is formed independently from glycolysis (Kube et al. 2012). It is possible that pathogenic mechanisms may differ according to the strain genomic content and/or the diverse environmental conditions such as different host species. The occurrence of major surface epitopes that are unique to each phytoplasma group or '*Candidatus* species', suggests their role in specific interactions with host/insect cells. The Amp protein of the OY phytoplasma forms a complex with insect microfilaments correlated with their phytoplasma-transmitting capacity (Suzuki et al. 2006; Galetto et al. 2011).

1.6 Cultivation in Artificial Media

After preliminary evidence that phytoplasmas can be grown in cell-free media (Bertaccini et al. 2010; Contaldo et al. 2012, 2013), a recent description of a suitable and flexible medium was published (Contaldo et al. 2016). This is important information to enable the study of their biology: substantially pure cultures can be

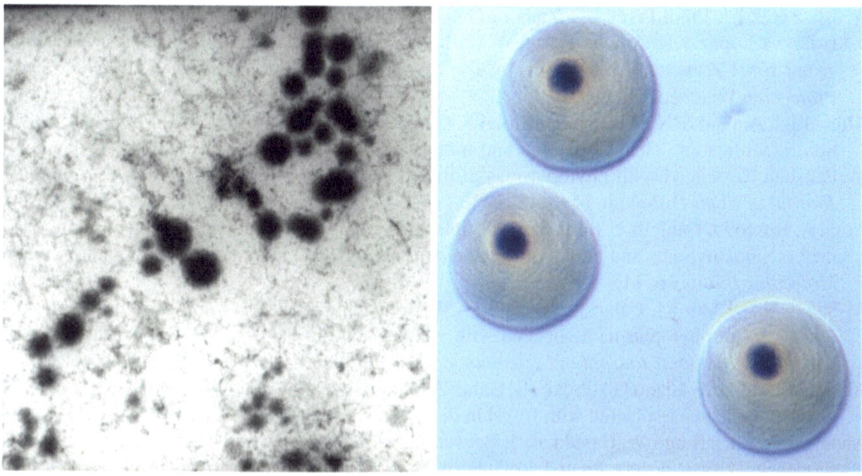

Fig. 1.2 Left. Ultrathin section of phytoplasma-like cells in agar medium embedded in Spurr resin and observed under transmission electron microscope at 5,000 X magnification (A. Bertaccini and A. Calzolari, 1984, unpublished). Right. Phytoplasma colonies growing in CB medium photographed under bifocal microscope 40 X (courtesy of N. Contaldo)

obtained to verify predicted metabolisms and biological properties of phytoplasmas outside their hosts. Colonies with identical morphology and positive for phytoplasma DNA presence (Fig. 1.2) confirm that despite the reduced genome size, phytoplasmas retain an independent metabolism that allows them to survive as parasites in environments as diverse as plant phloem and insect hemolymph and also in cell-free media.

Although further research is needed in order to optimize the culture system, the prospect of phytoplasma cultivation is now a real option. It will facilitate genome sequencing of further phytoplasma species and strains to allow comparative genomics which has been hampered by their intimate association with plant and insect hosts. Moreover, along with genomics, biochemical and physiological studies, phytoplasma cultivation will define their taxonomy. Selection and screening of plants resistant to phytoplasma infection as well as the study of the modes of colonization by phytoplasmas of plant and insect vectors will also be possible. As a consequence, strategies to manage and/or prevent phytoplasma-associated plant diseases more efficiently will be prepared and employed.

References

Acosta KI, Zamora L, Piñol BE, Fernández A, Chávez A, Flores G, Méndez J, Santos M, Leyva N, Arocha Y (2013) Identification and molecular characterization of phytoplasmas and rickettsia pathogens associated with Bunchy Top Symptom (BTS) and Papaya Bunchy Top (PBT) of papaya in Cuba. *Crop Protection* **45**, 49–56.

Acosta-Pérez KI, Piñol-Pérez BE, Zamora-Gutierrez L, Quiñones-Pantoja ML, Miranda-Cabrera I, Leyva-López NE, Arocha-Rosete Y (2017) A phytoplasma representative of a new subgroup 16SrI-Z associated with Bunchy Top Symptoms (BTS) on papaya in Cuba. *Revista de Protección Vegetal* **32**, 52–59.

Ahmad JN, Ahmad SJN, Irfan M, Paltrinieri S, Contaldo N, Bertaccini A (2017) Molecular detection, identification, characterization and transmission study of sarsoon phyllody in Punjab – Pakistan associated with phytoplasmas affiliated to the new subgroup 16SrIX-H. *European Journal of Plant Pathology* **149**, 117–125.

Alma A, Bosco D, Danielli A, Bertaccini A, Vibio M, Arzone A (1997) Identification of phytoplasmas in eggs, nymphs and adults of *Scaphoideus titanus* Ball reared on healthy plants. *Insect Molecular Biology* **6**, 115–121.

Al-Saady NA, Khan AJ, Calari A, Al-Subhi AM, Bertaccini A (2008) '*Candidatus* Phytoplasma omanense', a phytoplasma associated with witches' broom of *Cassia italica* (Mill.) Lam. in Oman. *International Journal of Systematic and Evolutionary Microbiology* **58**, 461–466.

Alvarez E, Mejía JF, Llano GA, Loke JB, Calari A, Duduk B, Bertaccini A (2009) Characterization of a phytoplasma associated with frogskin disease in cassava. *Plant Disease* **93**,1139–1145.

Andersen MT, Liefting LW, Havukkala I, Beever RE (2013) Comparison of the complete genome sequence of two closely related isolates of '*Candidatus* Phytoplasma australiense' reveals genome plasticity. *BMC Genomics* **14**, 529.

Angelini E, Bianchi GL, Filippin L, Morassutti C, Borgo M (2007) A new TaqMan method for the identification of phytoplasmas associated with grapevine yellows by a real-time PCR assay. *Journal of Microbiological Methods* **68**, 613–622.

Arashida R, Kakizawa S, Ishii Y, Hoshi A, Jung HY, Kagiwada S, Yamaji Y, Oshima K, Namba S (2008) Cloning and characterization of the antigenic membrane protein (Amp) gene and *in situ* detection of Amp from malformed flowers infected with Japanese hydrangea phyllody phytoplasma. *Phytopathology* **98**, 769–775.

Ahrens U, Seemüller E (1992) Detection of plant pathogenic mycoplasmalike organisms by a polymerase chain reaction that amplifies a sequence of the 16S rRNA gene. *Phytopathology* **82**, 828–832.

Aldaghi M, Massart S, Dutrecq O, Bertaccini A, Jijakli MH, Lepoivre P (2009) A simple and rapid protocol of crude DNA extraction from apple trees for PCR and real-time PCR detection of '*Candidatus* Phytoplasma mali'. *Journal of Virological Methods* **156**, 96–101.

Arnaud G, Malembic-Maher S, Salar P, Maixner M, Marcone C, Boudon-Padieu E, Foissac X (2007) Multilocus sequence typing confirms the close genetic inter-relatedness between three distinct "flavescence dorée" phytoplasma strain clusters and group 16SrV phytoplasmas infecting grapevine and alder in Europe. *Applied and Environmental Microbiology* **73**, 4001–4010.

Arocha Y, Lopez M, Pinol B, Fernandez M, Picornell B, Almeida R, Palenzuela I, Wilson MR, Jones P (2005) '*Candidatus* Phytoplasma graminis' and '*Candidatus* Phytoplasma caricae', two novel phytoplasmas associated with diseases of sugarcane, weeds and papaya in Cuba. *International Journal Systematic and Evolutionary Microbiology* **55**, 2451–2463.

Arocha Y, Antesana O, Montellano E, Franco P, Plata G, Jones P (2007) '*Candidatus* Phytoplasma lycopersici', a phytoplasma associated with "hoja de perejil" disease in Bolivia. *International Journal Systematic and Evolutionary Microbiology* **57**, 1704–1710.

Arocha-Rosete Y, Zunnoon-Khan S, Krukovets I, Crosby W, Scott J, Bertaccini A, Michelutti R (2011) Identification and molecular characterization of the phytoplasma associated with peach rosette-like disease at the Canadian clonal Genebank based on the 16S rRNA gene analysis. *Canadian Journal of Plant Pathology* **33**, 127–134.

Bai X, Zhang J, Ewing A, Miller SA, Radek AJ, Shevchenko DV, Tsukerman K, Walunas T, Lapidus A, Campbell JW, Hogenhout SA 2006. Living with genome instability: the adaptation of phytoplasmas to diverse environments of their insect and plant hosts. *Journal of Bacteriology* **188**, 3682–3696.

Barbara DJ, Morton A, Clark MF, Davies DL (2002) Immunodominant membrane proteins from two phytoplasmas in the aster yellows clade (chlorante aster yellows and clover phyllody) are highly divergent in the major hydrophilic region. *Microbiology* **148**, 157–167.

Baric S, Kerschbamer C, Dalla-Via J (2006) TaqMan real-time PCR *versus* four conventional PCR assays for detection of apple proliferation phytoplasma. *Plant Molecular Biology Reporter* **24**, 169–184.

Barros TSL, Davis RE, Resende RO, Dally EL (2002) Erigeron witches' broom phytoplasma in Brazil represents new subgroup VII-B in 16S rRNA gene group VII, the ash yellows phytoplasma group. *Plant Disease* **86**, 1142–1148.

Bekele B, Hodgetts J, Tomlinson J, Boonham N, Nikolić P, Swarbrick P, Dickinson M (2011) Use of a real-time LAMP isothermal assay for detecting 16SrII and -XII phytoplasmas in fruit and we of the Ethiopian Rift Valley. *Plant Pathology* **60**, 345–355.

Berg M, Davies DL, Clark MF, Vetten J, Maier G, Seemüller E (1999) Isolation of a gene encoding an immunodominant membrane protein gene in the apple proliferation phytoplasma and expression and characterization of the gene product. *Microbiology* **145**, 1937–1943.

Berger J, Dalla Via J, Baric S (2009) Development of a TaqMan allelic discrimination assay for the distinction of two major subtypes of the grapevine yellows phytoplasma "bois noir". *European Journal of Plant Pathology* **124**, 521–526.

Bertaccini A (2007) Phytoplasmas: diversity, taxonomy, and epidemiology. *Frontiers in Bioscience* **12**, 673–689.

Bertaccini A (2015) Phytoplasma research between past and future: what directions? *Phytopathogenic Mollicutes* **5**(1-Supplement), S1–S4.

Bertaccini A, Davis RE, Lee I-M (1992) *In vitro* micropropagation for maintenance of mycoplasmalike organisms in infected plant tissues. *Horticultural Science* **27**, 1041–1043.

Bertaccini A, Arzone A, Alma A, Bosco D, Vibio M (1993) Detection of mycoplasmalike organisms in *Scaphoideus titanus* Ball reared on "flavescence dorée" infected grapevine by dot hybridizations using DNA probes. *Phytopathologia Mediterranea* **32**, 20–24.

Bertaccini A, Bellardi MG, Vibio M (1996) Virus diseases of ornamental shrubs. X. *Euphorbia pulcherrima* Willd. Infected by viruses and phytoplasmas. *Phytopathologia Mediterranea* **35**, 129–132.

Bertaccini A, Contaldo N, Calari A, Paltrinieri S, Windsor HM, Windsor D (2010) Preliminary results of axenic growth of phytoplasmas from micropropagated infected periwinkle shoots. 18th Congress Iternational Oorganization for Mycoplasmology, Chianciano Terme, Italy **147**, 153.

Bertaccini A, Duduk B, Paltrinieri S, Contaldo N (2014) Phytoplasmas and phytoplasma diseases: a severe threat to agriculture. *American Journal of Plant Sciences* **5**, 1763.

Bianco PA, Casati P, Marziliano N (2004) Detection of phytoplasmas associated with grapevine "flavescence dorée" disease using real-time PCR. *Journal of Plant Pathology* **86**, 257–261.

Blomquist CL, Barbara DJ, Davies DL, Clark MF, Kirkpatrick BC (2001) An immunodominant membrane protein gene from the western X-disease phytoplasma is distinct from those of other phytoplasmas. *Microbiology* **147**, 571–580.

Bojan Duduk, Assunta Bertaccini, (2011) Phytoplasma classification: taxonomy based on 16S ribosomal gene, is it enough?. *Phytopathogenic Mollicutes* **1**, 1–13.

Botti S, Bertaccini A (2006) Phytoplasma infection trough seed transmission: further observations. 16th Congress International Organization for Mycoplasmology, Cambridge, United Kingdom, **76**, 113.

Boudon-Padieu E, Larrue J, Caudwell A (1989) ELISA and dot blot detection of "flavescence dorée" MLO in individual leafhopper vector during latency and inoculative state. *Current Microbiology* **19**, 357–364.

Cai H, Wei W, Davis RE, Chen H, Zhao Y (2008). Genetic diversity among phytoplasmas infecting *Opuntia* species: virtual RFLP analysis identifies new subgroups in the peanut witches' broom phytoplasma group. *International Journal Systematic and Evolutionary Microbiology* **58**, 1448–1457.

Calari A, Paltrinieri S, Contaldo N, Sakalieva D, Mori N, Duduk B, Bertaccini A (2011) Molecular evidence of phytoplasmas in winter oilseed rape, tomato and corn seedlings. *Bulletin of Insectology* **64**(Supplement), S157–S158.

Chen KH, Guo JR, Wu XJ, Loi N, Carraro L, Guo HJ, Chen YD, Osler R, Pearson R, Chen TA (1993) Comparison of monoclonal antibodies, DNA probes, and PCR for detection of the grapevine yellows disease agent. *Phytopathology* **83**, 915–922.

Chen KH, Credi R, Loi N, Maixner M, Chen TA (1994) Identification and grouping of mycoplasmalike organisms associated with grapevine yellows and clover phyllody diseases based on immunological and molecular analyses. *Applied and Environmental Microbiology* **60**, 1905–1913.

Cheng M, Dong J, Lee I-M, Bottner-Parker KD, Zhao Y, Davis RE, Laski PJ, Zhang Z, McBeath JH (2015) Group 16SrXII phytoplasma strains, including subgroup 16SrXII-E ('*Candidatus* Phytoplasma fragariae') and a new subgroup, 16SrXII-I, are associated with diseased potatoes (*Solanum tuberosum*) in the Yunnan and Inner Mongolia regions of China. *European Journal of Plant Pathology* **142**, 305–318.

Chung BN, Jeong MI (2014) Identification of "stolbur" phytoplasmas in *Petunia hybdrida* seedlings. *Phytopathogenic Mollicutes* **4**, 5–8.

Cimerman A, Pacifico D, Salar P, Marzachì C, Foissac X (2009) Striking diversity of vmp1, a variable gene encoding a putative membrane protein of the "stolbur" phytoplasma. *Applied Environmental Microbiology* **75**, 2951–2957.

Conci L, Meneguzzi N, Galdeano E, Torres L, Nome C, Nome S (2005) Detection and molecular characterisation of an alfalfa phytoplasma in Argentina that represents a new subgroup in the 16S rDNA ash yellows group ('*Candidatus* Phytoplasma fraxini'). *European Journal of Plant Pathology* **113**, 255–265.

Contaldo N, Bertaccini A, Paltrinieri S, Windsor HM, Windsor GD (2012) Axenic culture of plant pathogenic phytoplasmas. *Phytopathologia Mediterranea* **51**, 607–617.

Contaldo N, Bertaccini A, Paltrinieri S, Windsor GD, Windsor HM (2013) Cultivation of several phytoplasmas from a micropropagated plant collection. *Petria* **23**, 13–18.

Contaldo N., Satta E, Zambon Y, Paltrinieri S, Bertaccini A (2016) Development and evaluation of different complex media for phytoplasma isolation and growth. *Journal of Microbiological Methods* **127**, 105–110.

Córdova I, Oropeza C, Puch-Hau C, Harrison N, Collí-Rodríguez A, Narvaez M, Nic-Matos G, Reyes C, Sáenz L (2014) A real-time PCR assay for detection of coconut lethal yellowing phytoplasmas of group 16SrIV subgroups -A, -D and -E found in the Americas. *Journal of Plant Pathology* **96**, 343–352.

Crosslin J, Vandemark G, Munyaneza JE (2006) Development of a real-time quantitative PCR for detection of the Columbia Basin potato purple top phytoplasma in plants and beet leafhoppers. *Plant Disease* **90**, 663–667.

Danet J-L, Bonnet P, Jarausch W, Carraro L, Skoric D, Labonne G, Foissac X (2007) Imp and secY, two new markers for MLST (multilocus sequence typing) in the 16SrX phytoplasma taxonomic group. *Bulletin of Insectology* **60**, 339–340.

Davis RE, Dally EL, Gundersen DE, Lee I-M, Habili N (1997) '*Candidatus* Phytoplasma australiense', a new phytoplasma taxon associated with Australian grapevine yellows. *International Journal of Systematic Bacteriology* **47**, 262–269.

Davis RE, Jomantiene R, Dally EL, Wolf TK (1998) Phytoplasmas associated with grapevine yellows in Virginia belong to group 16SrI, subgroup A (tomato big bud phytoplasma subgroup), and group 16SrIII, new subgroup I. *Vitis* **37**, 131–137.

Davis RE, Dally EL, Converse RH (2001) Molecular identification of a phytoplasma associated with witches' broom disease of black raspberry in Oregon and its classification in group 16SrIII, new subgroup Q. *Plant Disease* **85**, 1121.

Davis RE, Dally E, Zhao Y, Lee I-M, Jomantiene R, Detweiler AJ, Putnam ML (2010) First report of a new subgroup 16SrIX-E ('*Candidatus* Phytoplasma phoenicium'-related) phytoplasma associated with juniper witches' broom disease in Oregon, USA. *Plant Pathology* **59**, 1161.

Davis RE, Zhao Y, Dally EL, Jomantiene R, Lee I-M, Wei W, Kitajima EW (2012) '*Candidatus* Phytoplasma sudamericanum', a novel taxon, and strain PassWB-Br4, a new subgroup 16SrIII-V phytoplasma, from diseased passion fruit (*Passiflora edulis* f. *flavicarpa* Deg.). *International Journal Systematic and Evolutionary Microbiology* **62**, 984–989.

Davis RE, Zhao Y, Dally EL, Lee I-M, Jomantiene R, Douglas SM (2013) '*Candidatus* Phytoplasma pruni', a novel taxon associated with Xdisease of stone fruits, *Prunus* spp.: multilocus characterization based on 16S rRNA, *secY*, and ribosomal protein genes. *International Journal Systematic and Evolutionary Microbiology* **63**, 766–776.

Davis RE, Harrison NA, Zhao Y, Wei W, Dally EL (2016) '*Candidatus* Phytoplasma hispanicum', a novel taxon associated with Mexican periwinkle virescence disease of *Catharanthus roseus*. *International Journal Systematic and Evolutionary Microbiology* **66**, 3463–3467.

Davis RE, Zhao Y, Wei W, Dally EL, Lee I-M (2017) '*Candidatus* Phytoplasma luffae', a novel taxon associated with witches' broom disease of loofah, *Luffa aegyptica* Mill. *International Journal of Systematic and Evolutionary Microbiology* **67**, 3127–3133.

Doi Y, Teranaka M, Yora K, Asuyama H (1967) Mycoplasma or PLT grouplike microorganisms found in the phloem elements of plants infected with mulberry dwarf, potato witches' broom, aster yellows or paulownia witches' broom. *Annals Phytopathological Society of Japan* **33**, 259–266.

Duduk B, Mejia JF, Calari A, Bertaccini A (2008) Identification of 16SrIX group phytoplasmas infecting Colombian periwinkles and molecular characterization on several genes. 17th Congress Intaernational Organization for Mycoplasmology, Tienjin, China, **112**, 83.

Esmailzadeh Hosseini SA, Salehi M, Mirchenari SM, Contaldo N, Paltrinieri S, Bertaccini A (2016) Occurrence of a '*Candidatus* Phytoplasma omanense'-related strain in bindweed showing a witches' broom disease in Iran. *Phytopathogenic Mollicutes* **6**, 63–68.

Faggioli F, Pasquini G, Lumia V, Campobasso G, Widmer TL, Quimby PC (2004) Molecular identification of a new member of the clover proliferation phytoplasma group (16SrVI) associated with yellow starthistle virescence in Italy. *European Journal of Plant Pathology* **110**, 353–360.

Fernández FD, Meneguzzi NG, Guzmán FA, Kirschbaum DS, Conci VC, Nome CF, Conci LR (2015) Detection and identification of a novel 16SrXIII subgroup phytoplasma associated with strawberry red leaf disease in Argentina. *International Journal of Systematic and Evolutionary Microbiology* **65**, 2741–2747.

Fernández FD, Galdeano E, Kornowski MV, Arneodo JD, Conci LR (2016) Description of '*Candidatus* Phytoplasma meliae', a phytoplasma associated with Chinaberry (*Melia azedarach* L.) yellowing in South America. *International Journal of Systematic and Evolutionary Microbiology* **66**, 5244–5251.

Flôres D, Amaral Mello AO, Pereira TBC, Rezende JAM, Bedendo IP (2015) A novel subgroup 16SrVII-D phytoplasma identified in association with erigeron witches' broom. *International Journal of Systematic and Evolutionary Microbiology* **65**, 2761–2765.

Fránová J, de Sousa E, Mimoso C, Cardoso F, Contaldo N, Paltrinieri S, Bertaccini A (2016) Multigene characterization of a new '*Candidatus* Phytoplasma rubi'-related strain associated with blackberry witches' broom in Portugal. *International Journal of Systematic and Evolutionary Microbiology* **66**, 1438–1446.

Galdeano E, Guzmán FA, Fernández F, Conci RG (2013) Genetic diversity of 16SrIII group phytoplasmas in Argentina. Predominance of subgroups 16SrIII-J and B and two new subgroups 16SrIII-W and X. *European Journal of Plant Pathology* **137**, 753–764.

Galetto L, Bosco D, Balestrini R, Genre A, Fletcher J, Marzachì C (2011) The major antigenic membrane protein of 'Candidatus Phytoplasma asteris' selectively interacts with ATP synthase and actin of leafhopper vectors. *PLoS One* **6**, e22571.

Griffiths HM, Sinclair WA, Smart CD, Davis RE (1999) The phytoplasma associated with ash yellows and lilac witches' broom: '*Candidatus* Phytoplasma fraxini'. *International Journal of Systematic Bacteriology* **49**, 1605–1614.

Gundersen DE, Lee I-M, Rehner SA, Davis RE, Kingsbury DT (1994) Phylogeny of mycoplasma-like organisms (Phytoplasmas): a basis for their classification. *Journal of Bacteriology* **176**, 5244–5254.

Gundersen DE, Lee I-M, Schaff DA, Harrison NA, Chang CJ, Davis RE, Kingsbury DT (1996) Genomic diversity among phytoplasma strains in 16S rRNA group I (aster yellows and related phytoplasmas) and III (X-disease and related phytoplasmas). *International Journal of Systematic Bacteriology* **46**, 64–75.

Hanboonsong Y, Choosai C, Panyim S, Damak S (2002) Transovarial transmission of sugarcane white leaf phytoplasma in the insect vector *Matsumuratettix hiroglyphicus* (Matsumura). *Insect Molecular Biology* **11**, 97–103.

Harrison NA, Richardson PA, Kramer JB, Tsai JH (1994) Detection of the mycoplasma-like organism associated with lethal yellowing disease of palms in Florida by polymerase chain reaction. *Plant Pathology* **43**, 998–1008.

Harrison NA, Womack M, Carpio ML (2002) Detection and characterization of a lethal yellowing (16SrIV) group phytoplasma in Canary Island date palms affected by lethal decline in Texas. *Plant Disease* **86**, 676–681.

Harrison NA, Helmick EE, Elliott ML (2008) Lethal yellowing-type diseases of palms associated with phytoplasmas newly identified in Florida, USA. *Annals of Applied Biology* **153**, 85–94.

Harrison NA, Davis RE, Oropeza C, Helmick EE, Narváez M, Eden-Green S, Dollet M, Dickinson M (2014) '*Candidatus* Phytoplasma palmicola', associated with a lethal yellowing-type disease of coconut (*Cocos nucifera* L.) in Mozambique. *International Journal of Systematic and Evolutionary Microbiology* **64**, 1890–1899.

Hiruki C, Wang KR (2004) Clover proliferation phytoplasma: '*Candidatus* Phytoplasma trifolii'. *International Journal of Systematic and Evolutionary Microbiology* **54**, 1349–1353.

Ho K, Tsai C, Chung T (2001) Organization of ribosomal RNA genes from a loofah witches' broom phytoplasma. *DNA and Cell Biology* **20**, 115–122.

Hodgetts J, Boonham N, Mumford R, Harrison N, Dickinson M (2008) Phytoplasma phylogenetics based on analysis of secA and 23S rRNA gene sequences for improved resolution of candidate species of '*Candidatus* Phytoplasma'. *International Journal of Systematic and Evolutionary Microbiology* **58**, 1826–1837.

Hodgetts J, Boonham N, Mumford R, Dickinson M (2009) Panel of 23S rRNA gene-based real-time PCR assays for improved universal and group specific detection of phytoplasmas. *Applied and Environmental Microbiology* **75,** 2945–2950.

Hong Y, Davies DL, Wezel RV, Ellerker BE, Morton A, Barbara D (2001) Expression of the immunodominant membrane protein of chlorantie-aster yellows phytoplasma in *Nicotiana benthamiana* from a potato virus X-based vector. *Acta Horticulturae* **550**, 409–415.

Hogenhout SA, Oshima K, Ammar E-D, Kakizawa S, Kingdom HN, Namba S (2008) Phytoplasmas: bacteria that manipulate plants and insects. *Molecular Plant Pathology* **9**, 403–423.

Hoshi A, Oshima K, Kakizawa S, Ishii Y, Ozeki J, Hashimoto M, Komatsu K, Kagiwada S, Yamaji Y, Namba S (2009) A unique virulence factor for proliferation and dwarfism in plants identified from a phytopathogenic bacterium. *Proceedings of the National Academy of Sciences USA* **106**, 6416–6421.

Hren M, Boben J, Rotter A, Kralj P, Gruden K, Ravnikar M (2007) Real-time PCR detection systems for "flavescence dorée" and "bois noir" phytoplasmas in grapevine: comparison with conventional PCR detection and application in diagnostics. *Plant Pathology* **56**, 785–796.

Ikten C, Ustun R, Catal M, Yol E, Uzun B (2016) Multiplex real-time qPCR assay for simultaneous and sensitive detection of phytoplasmas in sesame plants and insect vectors. *PLOS One* **11**, e0155891.

IRPCM (2004) '*Candidatus* Phytoplasma', a taxon for the wall-less, non-helical prokaryotes that colonise plant phloem and insects. *International Journal of Systematic and Evolutionary Microbiology* **54**, 1243–1255.

Ishii Y, Kakizawa S, Hoshi A, Maejima K, Kagiwada S, Yamaji Y, Oshima K, Namba S (2009) In the non-insect-transmissible line of onion yellows phytoplasma (OY-NIM), the plasmid-encoded transmembrane protein ORF3 lacks the major promoter region. *Microbiology* **155**, 2058–2067.

Ishiie T, Doi Y, Yora K, Asuyama H (1967) Suppressive effects of antibiotics of tetracycline group on symptom developement of mulberry dwarf disease. *Annals Phytopathological Society of Japan* **33**, 267–275.

Jacobs KA, Lee I-M, Griffiths HM, Miller FD Jr, Bottner KD (2003) A new member of the clover proliferation phytoplasma group (16SrVI) associated with elm yellows in Illinois. *Plant Disease* **87**, 241–246.

Jarausch W, Lansac M, Dosba F (1996) Long-term maintenance of nonculturable apple-proliferation phytoplasmas in their micropropagated natural host plant. *Plant Pathology* **45**, 778–786.

Jomantiene R, Davis RE (2006) Clusters of diverse genes existing as multiple, sequence-variable mosaics in a phytoplasma genome. *FEMS Microbiology Letters* **255**, 59–65.

Jomantiene R, Davis RE, Maas J, Dally EL (1998) Classification of new phytoplasmas associated with diseases of strawberry in Florida, based on analysis of 16S rRNA and ribosomal protein gene operon sequences. *International Journal of Systematic Bacteriology* **48**, 269–277.

Jomantiene R, Maas JL, Takeda F, Davis RE (2002) Molecular identification and classification of strawberry phylloid fruit phytoplasma in group 16SrI, new subgroup. *Plant Disease* **86**, 920.

Jomantiene R, Zhao Y, Davis RE (2007) Sequence-variable mosaics: composites of recurrent transposition characterizing the genomes of phylogenetically diverse phytoplasmas. *DNA Cell Biology* **26**, 557–564. *Erratum* **26**, 695.

Jung H-Y, Sawayanagi T, Kakizawa S, Nishigawa H, Miyata S, Oshima K, Ugaki M, Joon-Tak L, Namba S (2002) '*Candidatus* Phytoplasma castaneae', a novel phytoplasma taxon associated with chestnut witches' broom disease. *International Journal of Systematic and Evolutionary Microbiology* **52**, 1543–1549.

Jung H-Y, Sawayanagi T, Kakizawa S, Nishigawa H, Wei W, Oshima K, Miyata S, Ugaki M, Hibi T, Namba S (2003a) '*Candidatus* Phytoplasma ziziphi', a novel phytoplasma taxon associated with jujube witches' broom disease. *International Journal of Systematic and Evolutionary Microbiology* **53**, 1037–1041.

Jung H-Y, Sawayanagi T, Wongkaew P, Kakizawa S, Nishigawa H, Wei W, Oshima K, Miyata S, Ugaki M, Hibi T, Namba S (2003b) '*Candidatus* Phytoplasma oryzae', a novel phytoplasma taxon associated with rice yellow dwarf disease. *International Journal of Systematic and Evolutionary Microbiology* **53**, 1925–1929.

Kakizawa S, Oshima K, Kuboyama T, Nishigawa H, Jung H-Y, Sawayanagi T, Tsuchizaki T, Miyata S, Ugaki M, Namba S (2001) Cloning and expression analysis of phytoplasma protein translocation genes. *Molecular Plant-Microbe Interactions* **14**, 1043–1050.

Kakizawa S, Oshima K, Nishigawa H, Jung H-Y, Wei W, Suzuki S, Tanaka M, Miyata S, Ugaki M, Namba S (2004) Secretion of immunodominant membrane protein from onion yellows phytoplasma through the Sec protein-translocation system in *Escherichia coli*. *Microbiology* **150**, 135–142.

Kakizawa S, Oshima K, Ishii Y, Hoshi A, Maejima K, Jung H-Y, Yamaji Y, Namba S (2009) Cloning of immunodominant membrane protein genes of phytoplasmas and their in planta expression. *FEMS Microbiology Letters* **293**, 91–101.

Kawakita H, Saiki T, Wei W, Mitsuhashi W, Watanabe K, Sato M (2000) Identification of mulberry dwarf phytoplasmas in the genital organs and eggs of leafhopper *Hishimonoides sellatiformis*. *Phytopathology* **90**, 909–914.

Khan AJ, Botti S, Paltrinieri S, Al-Subhi AM, Bertaccini A (2002) Phytoplasmas in alfalfa seedlings: infected or contaminated seeds? 13th Congress International Organization for Mycoplasmoloy, Vienna, Austria, 6.

Kirkpatrick BC, Stenger BC, Morris TJ, Purcell AH (1987) Cloning and detection of DNA from a nonculturable plant pathogenic mycoplasma-like organism. *Science* **238**, 197–200.

Kogovšek P, Hodgetts J, Hall J, Prezelj N, Nikolić P, Mehle N, Lenarčič R, Rotter A, Dickinson M, Boonham N, Dermastia M, Ravnikar M (2015) LAMP assay and rapid sample preparation method for on-site detection of "flavescence dorée" phytoplasma in grapevine. *Plant Pathology* **64**, 286–296.

Kube M, Schneider B, Kuhl H, Dandekar T, Heitmann K, Migdo AM, Reinhardt R, Seemüller E (2008) The linear chromosome of the plant-pathogenic mycoplasma '*Candidatus* Phytoplasma mali'. *BMC Genomics* **9**, 306.

Kube M, Mitrovic J, Duduk B, Rabus R, Seemüller E (2012) Current view on phytoplasma genomes and encoded metabolism. *The Scientific World Journal* [dx.doi.org/https://doi.org/10.1100/2012/185942]

Kuboyama T, Huang C, Lu X, Sawayanagi T, Kanazawa T, Kagami T, Matsuda I, Tsuchizaki T, Namba S (1998) A plasmid isolated from phytopathogenic onion yellows phytoplasma and its heterogeneity in the pathogenic phytoplasma mutant. *Molecular Plant Microbe Interactions* **11**, 1031–1037.

Lai F, Song CS, Ren ZG, Lin CL, Xu QC, Li Y, Piao CG, Yu SS, Guo MW, Tian GZ (2014) Molecular characterization of a new member of the 16SrV group of phytoplasma associated with *Bischofia polycarpa* (Levl.) Airy Shaw witches' broom disease in China by a multiple gene-based analysis. *Australasian Plant Pathology* **43**, 557–569.

Lee I-M, Davis RE, Chen T-A, Chiykowski LN, Fletcher J, Hiruki C, Schaff DA (1992) A genotype-base system for identification and classification of mycoplasmalike organisms (MLOs) in the aster yellows MLO strain cluster. *Phytopathology* **82**, 977–986.

Lee I-M, Hammond RW, Davis RE, Gundersen DE (1993a) Universal amplification and analysis of pathogen 16S rDNA for classification and identification of mycoplasmalike organisms. *Phytopathology* **83**, 834–842.

Lee I-M, Davis RE, Hsu H-T (1993b) Differentiation of strains in the aster yellows mycoplasma-like organisms strain cluster by serological assay with monoclonal antibodies. *Plant Disease* **77**, 815–817.

Lee I-M, Gundersen DE, Hammond RW, Davis RE (1994) Use of mycoplasmalike organism (MLO) group-specific oligonucleotide primers for nested-PCR assays to detect mixed-MLO infections in a single host plant. *Phytopathology* **84**, 559–566.

Lee I-M, Bertaccini A, Vibio M, Gundersen DE (1995) Detection of multiple phytoplasmas in perennial fruit trees with decline symptoms in Italy. *Phytopathology* **85**, 728–735.

Lee I-M, Klopmeyer M, Bartoszyk IM, Gundersen-Rindal DE, Chou T, Thomson KL, Eisenreich R (1997) Phytoplasma induced free-branching in commercial poinsettia cultivars. *Nature Biotechnology* **15**, 178–182.

Lee I-M, Gundersen-Rindal DE, Davis RE, Bartoszyk IM (1998a) Revised classification scheme of phytoplasmas based on RFLP analyses of 16S rRNA and ribosomal protein gene sequences. *International Journal of Systematic and Evolutionary Microbiology* **48**, 1153–1169.

Lee I-M, Gundersen-Rindal DE, Bertaccini A (1998b) Phytoplasma: ecology and genomic diversity. *Phytopathology* **88**, 1359–1366.

Lee I-M, Martini M, Bottner KD, Dane RA, Black MC, Troxclair N (2003) Ecological implications from a molecular analysis of phytoplasmas involved in an aster yellows epidemic in various crops in Texas. *Phytopathology* **93**, 1368–1377.

Lee I-M, Martini M, Marcone C, Zhu SF (2004a) Classification of phytoplasma strains in the elm yellows group (16SrV) and proposal of '*Candidatus* Phytoplasma ulmi' for the phytoplasma associated with elm yellows. *International Journal of Systematic and Evolutionary Microbiology* **54**, 337–347.

Lee I-M, Gundersen-Rindal D, Davis RE, Bottner KD, Marcone C, Seemüller E (2004b) '*Candidatus* Phytoplasma asteris', a novel taxon associated with aster yellows and related diseases. *International Journal of Systematic and Evolutionary Microbiology* **54**, 1037–1048.

Lee I-M, Bottner KD, Secor G, Rivera-Varas V (2006a) '*Candidatus* Phytoplasma americanum', a phytoplasma associated with a potato purple top wilt disease complex. *International Journal of Systematic and Evolutionary Microbiology* **56**, 1593–1597.

Lee I-M, Zhao Y, Bottner KD (2006b) *SecY* gene sequence analysis for finer differentiation of diverse strains in the aster yellows phytoplasma group. *Molecular and Cellular Probes* **20**, 87–91.

Lee I-M, Bottner-Parker KD, Zhao Y, Davis RE, Harrison NA (2010) Phylogenetic analysis and delineation of phytoplasmas based on *secY* gene sequences. *International Journal of Systematic and Evolutionary Microbiology* **60**, 2887–2897.

Lee I-M, Bottner-Parker KD, Zhao Y, Villalobos W, Moreira L (2011) '*Candidatus* Phytoplasma costaricanum' a novel phytoplasma associated with an emerging disease in soybean (*Glycine max*). *International Journal of Systematic and Evolutionary Microbiology* **61**, 2822–2826.

Lee I-M, Bottner-Parker KD, Zhao Y, Bertaccini A, Davis RE (2012) Differentiation and classification of phytoplasmas in the pigeon pea witches' broom group (16SrIX): an update based on multiple gene sequence analysis. *International Journal of Systematic and Evolutionary Microbiology* **62**, 2279–2285.

Lee I-M, Polashock J, Bottner-Parker KD, Bagadia PG, Rodriguez-Saona C, Zhao Y, Davis RE (2014) New subgroup 16SrIII-Y phytoplasmas associated with false-blossom diseased cranberry (*Vaccinium macrocarpon*) plants and with known and potential insect vectors in New Jersey. *European Journal of Plant Pathology* **139**, 393–400.

Lee ME, Grau CR, Lukaesko LA, Lee I-M (2002) Identification of aster yellows phytoplasmas in soybean in Wisconsin based on RFLP analysis of PCR-amplified products (16S rDNAs). *Canadian Journal of Plant Pathology* **24**, 125–130.

Lenz O, Markov J, Sarkisov T, Franova J, Pribylova J (2015) Discrimination of phytoplasmas using an oligonucleotide microarray targeting *rps*3, *rpl*22, and *rps*19 genes. *Crop Protection* **70**, 47–52.

Lherminier J, Prensier G, Boudon-Padieu E, Caudwell A (1990) Immunolabeling of grapevine "flavescence dorée" MLO in salivary glands of *Euscelidius variegatus*: a light and electron microscopy study. *Journal of Histochemistry and Cytochemistry* **38**, 79–85.

Lin CP, Chen TA (1985) Monoclonal antibodies against the aster yellows agent. *Science* **227**, 1233–1235.

Linck O, Krüger E, Reineke A (2017) A multiplex TaqMan qPCR assay for sensitive and rapid detection of phytoplasmas infecting *Rubus* species. *PLoS One* **12**, e0177808.

MacLean AM, Sugio A, Makarova OV, Findlay KC, Grieve VM, Toth R, Nicolaisen M, Hogenhout S (2011) Phytoplasma effector SAP54 induces indeterminate leaf-like flower development in Arabidopsis plants. *Plant Physiology* **157**, 831–841.

Maejima K, Oshima K, Namba S (2014) Exploring the phytoplasmas, plant pathogenic bacteria. *Journal of General Plant Pathology* **80**, 210–221.

Mafia RG, Barreto RW, Vanetti CA, Hodgetts J, Dickinson M, Alfenas AC (2007) A phytoplasma is associated with witches' broom disease of *Tabebuia pentaphylla* in Brazil. *New Disease Reports* **15**, 49.

Makarova OV, Contaldo N, Paltrinieri S, Kawube G, Bertaccini A, Nicolaisen M (2012) DNA barcoding for universal identification of '*Candidatus* Phytoplasmas' using a fragment of the elongation factor *Tu* gene. *PLoS One* **7**, e52092.

Malembic-Maher S, Salar P, Filippin L, Carle P, Angelini E, Foissac X (2011) Genetic diversity of European phytoplasmas of the 16SrV taxonomic group and proposal of '*Candidatus*

Phytoplasma rubi'. *International Journal of Systematic and Evolutionary Microbiology* **61**, 2129–2134.

Manimekalai R, Nair S, Soumya VP, Roshna OM, Thomas GV (2011) Real-time PCR technique-based detection of coconut root (wilt) phytoplasma. *Current Science* **101**, 1209–1213.

Marcone C, Neimark H, Ragozzino A, Lauer U, Seemüller E (1999) Chromosome sizes of phytoplasmas composing major phylogenetic groups and subgroups. *Phytopathology* **89**, 805–810.

Marcone C, Lee I-M, Davis RE, Ragozzino A, Seemüller E (2000) Classification of aster yellows-group phytoplasmas based on combined analyses of rRNA and *tuf* gene sequences. *International Journal of Systematic and Evolutionary Microbiology* **50**, 1703–1713.

Marcone C, Gibb KG, Streten C, Schneider B (2003a) '*Candidatus* Phytoplasma spartii', '*Candidatus* Phytoplasma rhamni' and '*Candidatus* Phytoplasma allocasuarinae', respectively associated with spartium witches' broom, buckthorn witches' broom and allocasuarina yellows diseases. *International Journal of Systematic and Evolutionary Microbiology* **54**, 1025–1029.

Marcone C, Schneider B, Seemüller E (2003b) '*Candidatus* Phytoplasma cynodontis', the phytoplasma associated with Bermuda grass white leaf disease. *International Journal of Systematic and Evolutionary Microbiology* **54**, 1077–1082.

Martinez RT, Narvaez M, Fabre S, Harrison NA, Oropeza C, Dollet M, Hichez E (2008) Coconut lethal yellowing on the southern coast of the Dominican Republic is associated with a new 16SrIV group phytoplasma. *Plant Pathology* **57**, 366–376.

Martini M, Murari E, Mori N, Bertaccini A (1999) Identification and epidemic distribution of two "flavescence dorée"-related phytoplasmas in Veneto (Italy). *Plant Disease* **83**, 925–930.

Martini M, Botti S, Marcone C, Marzachì C, Casati P, Bianco PA, Benedetti R, Bertaccini A (2002) Genetic variability among "flavescence dorée" phytoplasmas from different origins in Italy and France. *Molecular and Cellular Probes* **16**, 197–208.

Martini M, Lee I-M, Bottner KD, Zhao Y, Botti S, Bertaccini A, Harrison NA, Carraro L, Marcone C, Khan J, Osler R (2007) Ribosomal protein gene-based phylogeny for finer differentiation and classification of phytoplasmas. *International Journal of Systematic and Evolutionary Microbiology* **57**, 2037–2051.

Martini M, Marcone C, Mitrović J, Maixner M, Delić D, Myrta A, Ermacora P, Bertaccini A, Duduk B (2012) '*Candidatus* Phytoplasma convolvuli', a new phytoplasma taxon associated with bindweed yellows in four European countries. *International Journal of Systematic and Evolutionary Microbiology* **62**, 2910–2915.

Melo L, Silva E, Flôres D, Ventura J, Costa H, Bedendo I (2013) A phytoplasma representative of a new subgroup, 16SrXIII-E, associated with papaya apical curl necrosis. *European Journal of Plant Pathology* **137**, 445–450.

Mergenthaler E, Viczian O, Fodor M, Sule S (2001) Isolation and expression of an immunodominant membrane protein gene of the ESFY phytoplasma for antiserum production. *Acta Horticulturae* **550**, 355–360.

Mitrović J, Kakizawa S, Duduk B, Oshima K, Namba S, Bertaccini A (2011) The cpn60 gene as an additional marker for finer differentiation of '*Candidatus* Phytoplasma asteris'-related strains. *Annals of Applied Biology* **159**, 41–48.

Mitrović J, Siewert C, Duduk B, Hecht J, Mölling K, Broecker F, Beyerlein P, Büttner C, Bertaccini A, Kube M (2014) Generation and analysis of draft sequences of "stolbur" phytoplasma from multiple displacement amplification templates. *Journal of Molecular Microbiology and Biotechnology* **24**, 1–11.

Mitrovic J, Smiljković M, Seemüller E, Reinhardt R, Hüttel B, Büttner C, Bertaccini A, Kube M, Duduk B (2015) Differentiation of '*Candidatus* Phytoplasma cynodontis' based on 16S rRNA and groEL genes and identification of a new subgroup, 16SrXIV-C. *Plant Disease* **99**, 1578–1583.

Miyazaki A, Shigaki T, Koinuma H, Iwabuchi N, Rauka GB, Kembu A, Saul J, Watanabe K, Nijo T, Maejima K, Yamaji Y, Namba S (2018) '*Candidatus* Phytoplasma noviguineense', a novel taxon associated with bogia coconut syndrome and banana wilt disease on the island of New Guinea. *International Journal of Systematic and Evolutionary Microbiology* **68**, 170–175.

Molino Lova M, Quaglino F, Abou-Jawdah Y, Choueiri E, Sobh H, Casati P, Tedeschi R, Alma A, Bianco PA (2011) Identification of new 16SrIX subgroups, -F and -G, among '*Candidatus* Phytoplasma phoenicium' strains infecting almond, peach and nectarine in Lebanon. *Phytopathologia Mediterranea* **50**, 273–282.

Montano HG, Davis RE, Dally EL, Pimentel JP, Brioso PST (2000) Identification and phylogenetic analysis of a new phytoplasma from diseased chayote in Brazil. *Plant Disease* **84**, 429–436.

Montano HG, Davis RE, Dally EL, Hogenhout S, Pimentel PP, Brioso PST (2001) '*Candidatus* Phytoplasma brasiliense', a new phytoplasma taxon associated with hibiscus witches' broom disease. *International Journal of Systematic and Evolutionary Microbiology* **51**, 1109–1118.

Monti M, Martini M, Tedeschi R (2013) EvaGreen real-time PCR protocol for specific '*Candidatus* Phytoplasma mali' detection and quantification in insects. *Molecular and Cellular Probes* **27**, 129–136.

Naderali N, Nejat N, Vadamalai G, Davis R, Wei W, Harrison N, Kong L, Kadir J, Tan Y, Zhao Y (2017) '*Candidatus* Phytoplasma wodyetiae', a new taxon associated with yellow decline disease of foxtail palm (*Wodyetia bifurcata*) in Malaysia. *International Journal of Systematic and Evolutionary Microbiology* **67**, 3765–3772.

Namba S, Kato S, Iwanami S, Oyaizu H, Shiozawa H, Tsuchizaki T (1993) Detection and differentiation of plant-pathogenic mycoplasmalike organisms using polymerase chain reaction. *Phytopathology* **83**, 786–791.

Nejat N, Sijam K, Abdullah SNA, Vadamalai G, Sidek Z, Dickinson M (2010) Development of a TaqMan real-time PCR for sensitive detection of the novel phytoplasma associated with coconut yellow decline in Malaysia. *Journal of Plant Pathology* **92**, 769–773.

Nejat N, Vadamalai G, Davis RE, Harrison NA, Sijam K, Dickinson M, Abdullah SN, Zhao Y (2013) '*Candidatus* Phytoplasma malaysianum', a novel taxon associated with virescence and phyllody of Madagascar periwinkle (*Catharanthus roseus*). *International Journal of Systematic and Evolutionary Microbiology* **63**, 540–548.

Neriya Y, Maejima K, Nijo T, Tomomitsu T, Yusa A, Himeno M, Netsu O, Hamamoto H, Oshima K, Namba S (2014) Onion yellow phytoplasma P38 protein plays a role in adhesion to the hosts. *FEMS Microbiology Letters* **361**, 115–122.

Nicolaisen M, Bertaccini A (2007) An oligonucleotide microarray based assay for identification of phytoplasma 16S ribosomal groups. *Plant Pathology* **56**, 332–336.

Nicolaisen M, Contaldo N, Makarova O, Paltrinieri S, Bertaccini A (2011) Deep amplicon sequencing reveals mixed phytoplasma infection within single grapevine plants. *Bulletin of Insectology* **64**(Supplement), S35–S36.

Obura E, Masiga D, Wachira F, Gurja B, Khan ZR (2011) Detection of phytoplasma by loop-mediated isothermal amplification of DNA (LAMP). *Journal of Microbiological Methods* **84**, 312–316.

Oshima K, Kakizawa S, Nishigawa H, Kuboyama T, Miyata S, Ugaki M, Namba A (2001) A plasmid of phytoplasma encodes a unique replication protein having both plasmid and virus-like domains: clue to viral ancestry or result of virus/plasmid recombination. *Virology* **285**, 270–277.

Oshima K, Kakizawa S, Nishigawa H, Jung H-Y, Wei W, Suzuki S, Arashida R, Nakata D, Miyata S, Ugaki M, Namba S (2004) Reductive evolution suggested from the complete genome sequence of a plant-pathogenic phytoplasma. *Nature Genetics* **36**, 27–29.

Oshima K, Kakizawa S, Arashida R, Ishii Y, Hoshi A, Hayashi Y, Kawagida S, Namba S (2007) Presence of two glycolityc gene clusters in a severe pathogenic line of '*Candidatus* Phytoplasma asteris'. *Molecular Plant Pathology* **8**, 481–489.

Oshima K, Ishii Y, Kakizawa S, Sugawara K, Neriya Y, Himeno M, Minato N, Miura C, Shiraishi T, Yamaji Y, Namba S (2011) Dramatic transcriptional changes in an intracellular parasite enable host switching between plant and insect. *PLoS One* **6**, e23242.

Padovan A, Gibb K, Persley D (2000) Association of '*Candidatus* Phytoplasma australiense' 38 with green petal and lethal yellows diseases in strawberry. *Plant Pathology* **49**, 362–369.

Pelletier C, Salar P, Gillet J, Cloquemin G, Very P, Foissac X, Malembic-Maher S (2009) Triplex real-time PCR assay for sensitive and simultaneous detection of grapevine phytoplasmas of the 16SrV and 16SrXII-A groups with an endogenous analytical control. *Vitis* **48**, 87–95.

Perez-López E, Luna-Rodríguez M, Olivier CY, Dumonceaux TJ (2016) The underestimated diversity of phytoplasmasin Latin America. *International Journal of Systematic and Evolutionary Microbiology* **66**, 492–513.

Quaglino F, Zhao Y, Bianco PA, Wei W, Casati P, Durante G, Davis RE (2009) New 16Sr subgroups and distinct single nucleotide polymorphism lineages among grapevine "bois noir" phytoplasma populations. *Annals of Applied Biology* **154**, 279–289.

Quaglino F, Zhao Y, Casati P, Bulgari D, Bianco PA, Wei W, Davis RE (2013) '*Candidatus* Phytoplasma solani', a novel taxon associated with "stolbur"- and "bois noir"-related diseases of plants. *International Journal of Systematic and Evolutionary Microbiology* **63**, 2879–2894.

Rekab D, Carraro L, Schneider B, Seemüller E, Chen J, Chang CJ, Locci R, Firrao G (1999) Geminivirus-related extrachromosomal DNAs of the X-clade phytoplasmas share high sequence similarity. *Microbiology* **145**, 1453–1459.

Saccardo F, Martini M, Palmano S, Ermacora P, Scortichini M, Loi N, Firrao G (2012) Genome drafts of four phytoplasma strains of the ribosomal group 16SrIII. *Microbiology* **158**, 2805–2814.

Saeed E, Sarindu N, Davies DL, Clark MF, Roux J, Cousin MT (1994) Use of polyclonal antibodies to identify mycoplasma-like organisms (MLOs) from the Sudan and from Thailand. *Journal of Phytopathology* **142**, 345–349.

Šafářová D, Zemánek T, Válová P, Navrátil M (2016) '*Candidatus* Phytoplasma cirsii', a novel taxon from creeping thistle [*Cirsium arvense* (L.) Scop]. *International Journal of Systematic and Evolutionary Microbiology* **66**, 1745–1753.

Salehi M, Izadpanah K, Siampour M, Taghizadeh M (2009) Molecular characterization and transmission of Bermuda grass white leaf phytoplasma in Iran. *Journal of Plant Pathology* **91**, 655–661.

Samad A, Ajayakumar PV, Shasany AK, Gupta MK, Alam M, Rastogi S (2008) Occurrence of a clover proliferation (16SrVI) group phytoplasma associated with little leaf disease of *Portulaca grandiflora* in India. *Plant Disease* **92**, 832.

Santos-Cervantes ME, Chávez-Medina JA, Méndez-Lozano J, Leyva-López NE (2008) Detection and molecular characterization of two little leaf phytoplasma strains associated with pepper and tomato diseases in Guanajuato and Sinaloa, Mexico. *Plant Disease* **92**, 1007–1011.

Santos-Cervantes ME, Chávez-Medina JA, Acosta-Pardini J, Flores-Zamora GL, Méndez-Lozano J, Leyva-López NE (2010) Genetic diversity and geographical distribution of phytoplasmas associated with potato purple top disease in Mexico. *Plant Disease* **94**, 388–395.

Satta E, Contaldo N, Paltrinieri S, Bertaccini A (2016) Biological and molecular proof of phytoplasma seed transmission in corn. 21th Congress of the International Organization for Mycoplasmology, Brisbane, Australia **61**, 65–66.

Satta E, Nanni IM, Contaldo N, Collina M, Poveda JB, Ramírez AS, Bertaccini A (2017) General phytoplasma detection by a q-PCR method using mycoplasma primers. *Molecular and Cellular Probes* **35**, 1–7.

Sawayanagi T, Horikoshi N, Kanehira T, Shinohara M, Bertaccini A, Cousin M-T, Hiruki C, Namba S (1999) '*Candidatus* Phytoplasma japonicum', a new phytoplasma taxon associated with Japanese Hydrangea phyllody. *International Journal of Systematic Bacteriology* **49**, 1275–1285.

Schneider B, Seemüller E (1994) Presence of two set of ribosomal genes in phytopathogenic mollicutes. *Applied and Environmental Microbiology* **60**, 3409–3412.

Schneider B, Maurer R, Saillard C, Kirkpatrick BC, Seemüller E (1992) Occurrence and relatedness of extrachromosomal DNAs in plant pathogenic mycoplasmalike organisms. *Molecular Plant-Microbe Interactions* **5**, 489–495.

Schneider B, Ahrens U, Kirkpatrick BC, Seemüller E (1993) Classification of plant-pathogenic mycoplasma-like organisms using restriction-site analysis of PCR-amplified 16S rDNA. *Journal of General Microbiology* **139**, 519–527.

Schneider B, Gibb KS, Seemüller E (1997) Sequence and RFLP analysis of the elongation factor Tu gene used in differentiation and classification of phytoplasmas. *Microbiology* **143**, 3381–3389.

Schneider B, Torres E, Martìn MP, Schroder M, Behnke HD, Seemüller E (2005) '*Candidatus* Phytoplasma pini', a novel taxon from *Pinus silvestris* and *Pinus halepensis*. *International Journal of Systematic and Evolutionary Microbiology* **55**, 303–307.

Seemüller E, Schneider B, Maurer R, Ahrens U, Daire X, Kison H, Lorenz K, Firrao G, Avinent L, Sears BB, Stackebrandt E (1994) Phylogenetic classification of phytopathogenic mollicutes by sequence analysis of 16S ribosomal DNA. *International Journal of Systematic Bacteriology* **44**, 440–446.

Seemüller E, Marcone C, Lauer U, Ragozzino A, Göschl M (1998) Current status of molecular classification of the phytoplasmas. *Journal of Plant Pathology* **80**, 3–26.

Seemüller E, Schneider B (2004) Taxonomic description of '*Candidatus* Phytoplasma mali' sp. nov., '*Candidatus* Phytoplasma pyri' sp. nov. and '*Candidatus* Phytoplasma prunorum' sp. nov., the causal agents of apple proliferation, pear decline and European stone fruit yellows, respectively. *International Journal of Systematic and Evolutionary Microbiology* **54**, 1217–1226.

Šeruga M, Škorić D, Botti S, Paltrinieri S, Juretić N, Bertaccini A (2003) Molecular characterization of a phytoplasma from the aster yellows (16SrI) group naturally infecting *Populus nigra* L. "Italica" trees in Croatia. *Forest Pathology* **33**, 113–125.

Siampour M, Izadpanah K, Galetto L, Salehi M, Marzachì C (2012) Molecular characterization, phylogenetic comparison and serological relationship of the Imp protein of several '*Candidatus* Phytoplasma aurantifolia' strains. *Plant Pathology* **62**, 452–459.

Siddique ABM, Agrawal GK, Alam N, Krishina Reddy M (2001) Electron microscopy and molecular characterization of phytoplasmas associated with little leaf disease of brinjal (*Solanum melongena*) and periwinkle (*Catharanthus roseus*) in Bangladesh. *Journal of Phytopathology* **149**, 237–244.

Sinha RC (1979) Purification and serology of mycoplasmalike organisms antigens from aster yellows-diseased plants by two serological procedures. *Canadian Journal of Plant Pathology* **1**, 65–70.

Sinha RC, Benhamou N (1983) Detection of mycoplasma-like organisms antigens from aster yellows-diseased plants by two serological procedures. *Phytopathology* **73**, 1199–1202.

Sugawara K, Himeno M, Keima T, Kitazawa Y, Maejima K, Oshima K, Namba S (2012) Rapid and reliable detection of phytoplasma by loop-mediated isothermal amplification targeting a housekeeping gene. *Journal of General Plant Pathology* **78**, 389–397.

Sugio A, Kingdom HN, MacLean AM, Grieve VM, Hogenhout SA (2011a) Phytoplasma protein effector SAP11 enhances insect vector reproduction by manipulating plant development and defense hormone biosynthesis. *Proceedings of the National Academy of Sciences USA* **108**, E1254–E1263.

Sugio A, MacLean AM, Kingdom HN, Grieve VM, Manimekalai R, Hogenhout SA (2011b) Diverse targets of phytoplasma effectors: from plant development to defense against insects. *Annual Revue of Phytopathology* **49**, 175–195.

Suzuki S, Oshima K, Kakizawa S, Arashida R, Jung H-Y, Yamaji Y, Nishigawa H, Ugaki M and Namba S (2006) Interaction between the membrane protein of a pathogen and insect microfilament complex determines insect-vector specificity. *Proceedings of the National Academy of Sciences USA* **103**, 4252–4257.

Tedeschi R, Ferrato V, Rossi J, Alma A (2006) Possible phytoplasma transovarial transmission in the psyllids *Cacopsylla melanoneura* and *Cacopsylla pruni*. *Plant Pathology* **55**, 18–24.

Tomlinson JA, Boonham N, Dickinson M (2010) Development and evaluation of a one-hour DNA extraction and loop-mediated isothermal amplification assay for rapid detection of phytoplasmas. *Plant Pathology* **59**, 465–471.

Torres E, Bertolini E, Cambra M, Montón C, Martín MP (2005) Real-time PCR for simultaneous and quantitative detection of quarantine phytoplasmas from apple proliferation (16SrX) group. *Molecular and Cellular Probes* **19**, 334–340.

Toruno TY, Seruga Music M, Simi S, Nicolaisen M, Hogenhout SA (2010) Phytoplasma PMU1 exists as linear chromosomal and circular extrachromosomal elements and has enhanced expression in insect vectors compared with plant hosts. *Molecular Microbiology* **77**, 1406–1415.

Tran-Nguyen LT, Kube M, Schneider B, Reinhardt R, Gibb KS (2008) Comparative genome analysis of '*Candidatus* Phytoplasma australiense' (subgroup tuf-Australia I; rp-A) and '*Ca.* Phytoplasma asteris' strains OY-M and AY-WB. *Journal of Bacteriology* **190**, 3979–3991.

Tymon AM, Jones P, Harrison NA (1998) Phylogenetic relationships of coconut phytoplasmas and the development of specific oligonucleotide PCR primers. *Annals of Applied Biology* **132**, 437–452.

Valiunas D, Jomantiene R, Davis RE (2005) A '*Candidatus* Phytoplasma asteris'-related phytoplasma associated with cherry little leaf disease represents a new subgroup, 16SrI-Q. *Phytopathology* **95**, S106.

Valiunas D, Staniulis J, Davis RE (2006) '*Candidatus* Phytoplasma fragariae', a novel phytoplasma taxon discovered in yellows diseased strawberry, *Fragaria x ananassa*. *International Journal of Systematic and Evolutionary Microbiology* **56**, 277–281.

Valiunas D, Jomantiene R, Ivanauskas A, Abraitis R, Staniene G, Zhao Y, Davis RE (2009) First report of a new phytoplasma subgroup, 16SrIII-T, associated with decline disease affecting sweet and sour cherry trees in Lithuania. *Plant Disease* **93**, 550.

Verdin E, Salar P, Danet J-L, Choueiri E, Jreijiri F, El Zammar S, Gèlie B, Bové J, Garnier M (2003) '*Candidatus* Phytoplasma phoenicium', a new phytoplasma associated with an emerging lethal disease of almond trees in Lebanon and Iran. *International Journal of Systematic and Evolutionary Microbiology* **53**, 833–838.

Villalobos W, Martini M, Garita L, Muñoz M, Osler R, Moreira L (2011) *Guazuma ulmifolia* (Sterculiaceae), a new natural host of 16SrXV phytoplasma in Costa Rica. *Tropical Plant Pathology* **36**, 2.

Vu NT, Pardo JM, Alvarez E, LeKris HH, Kim-Lien W, Le NDT (2016) Establishment of a loop-mediated isothermal amplification (LAMP) assay for the detection of phytoplasma-associated cassava witches' broom disease. *Applied Biological Chemistry* **59**, 151–156.

Wei W, Kakizawa S, Jung H-Y, Suzuki S, Tanaka M, Nishigawa H, Miyata S, Oshima K, Ugaki M, Hibi T, Namba S (2004) An antibody against the SecA membrane protein of one phytoplasma reacts with those of phylogenetically different phytoplasmas. *Phytopathology* **94**, 683–686.

Wei W, Davis RE, Lee I-M, Zhao Y (2007) Computer-simulated RFLP analysis of 16S rRNA genes: identification of ten new phytoplasma groups. *International Journal of Systematic and Evolutionary Microbiology* **57**, 1855–1867.

Wei W, Davis RE, Jomantiene R, Zhao Y (2008) Ancient, recurrent phage attacks and recombination shaped dynamic sequence-variable mosaics at the root of phytoplasma genome evolution. *Proceedings of the National Academy of Sciences USA* **105**, 11827–11832.

White DT, Blackall LL, Scott PT, Walsh KB (1998) Phylogenetic positions of phytoplasmas associated with dieback, yellow crinkle and mosaic diseases of papaya, and their proposed inclusion in '*Candidatus* Phytoplasma australiense' and a new taxon, '*Candidatus* Phytoplasma australasia'. *International Journal of Systematic Bacteriology* **48**, 941–951.

Win NKK, Lee S-Y, Bertaccini A, Namba S, Jung H-Y (2013) '*Candidatus* Phytoplasma balanitae' associated with witches' broom disease of *Balanites triflora*. *International Journal of Systematic and Evolutionary Microbiology* **63**, 636–640.

Yadav A, Bhale U, Thorat V, Shouche Y (2014). First report of a new subgroup 16SrII-M '*Candidatus* Phytoplasma aurantifolia' associated with witches' broom disease of *Tephrosia purpurea* in India. *Plant Disease* **98**, 990.

Yadav A, Thorat V, Deokule S, Shouche Y, Prasad DT (2017) New subgroup 16SrXI-F phytoplasma strain associated with sugarcane grassy shoot (SCGS) disease in India. *International Journal of Systematic and Evolutionary Microbiology* **67**, 374–378.

Yang Y, Jiang L, Che H, Cao X, Luo D (2016) Identification of a novel subgroup 16SrII-U phytoplasma associated with papaya little leaf disease. *International Journal of Systematic and Evolutionary Microbiology* **66**, 3485–3491.

Zamorano A, Fiore N (2016) Draft genome sequence of 16SrIII-J phytoplasma, a plant pathogenic bacterium with a broad spectrum of hosts. *Genome Announcements* **4**, e00602–e00616.

Zhang RY, Li WF, Huang YK, Wang XY, Shan HL, Luo Z-M, Yin J (2016) Group 16SrXI phytoplasma strains, including subgroup 16SrXI-B and a new subgroup, 16SrXI-D, are associated with sugar cane white leaf. *International Journal of Systematic and Evolutionary Microbiology* **66**, 487–491.

Zhao Y, Sun Q, Wei W, Davis RE, Wu W, Liu Q (2009a) '*Candidatus* Phytoplasma tamaricis', a novel taxon discovered in witches' broom-diseased salt cedar (*Tamarix chinensis* Lour.). *International Journal of Systematic and Evolutionary Microbiology* **59**, 2496–2504.

Zhao Y, Wei W, Lee I-M, Shao J, Suo X, Davis RE (2009b) Construction of an interactive online phytoplasma classification tool, *i*PhyClassifier, and its application in analysis of the peach X-disease phytoplasma group (16SrIII). *International Journal of Systematic and Evolutionary Microbiology* **59**, 2582–2593.

Zreik L, Carle P, Bové JM, Garnier M (1995) Characterization of the mycoplasmalike organism associated with witches' broom disease of lime and proposition of a '*Candidatus*' taxon for the organism, '*Candidatus* Phytoplasma aurantifolia'. *International Journal of Systematic Bacteriology* **45**, 449–453.

Chapter 2
Phytoplasmas Infecting Vegetable, Pulse and Oil Crops

Marta Martini, Duška Delić, Lia Liefting, and Helena Montano

Abstract The present chapter provides an overview of phytoplasma diseases affecting vegetable, pulse and oil crops, with an emphasis on symptoms, geographic distribution, associated phytoplasma taxa and insect vectors. Phytoplasma diseases of these crops occur worldwide; however, the majority of reports are from North American, European and Asian countries. These diseases affect plant species belonging mostly to Apiaceae, Asteraceae, Cucurbitaceae, Fabaceae and Solanaceae. They differ considerably in geographic distribution and size of the various taxonomic groups and subgroups of the associated phytoplasmas. Subgroup 16SrI-B phytoplasmas are the prevalent agents among the main infected countries. Phytoplasmas of subgroup 16SrXII-A are widely distributed throughout Europe, whereas phytoplasmas of 16SrII group are mainly distributed in Asia and Australia and those of 16SrIII group in South America. A number of diseases are associated with genetically different phytoplasmas that induce similar symptoms in the host plants.

Keywords Symptoms · Geographic distribution · Phytoplasma taxa associated · Insect vector

M. Martini (✉)
Department of Agricultural, Food, Environmental and Animal Sciences, University of Udine, Udine, Italy
e-mail: marta.martini@uniud.it

D. Delić
Faculty of Agriculture Bulevar vojvode Petra Bojovića, University of Banjaluka, Banja Luka, Bosnia and Herzegovina

L. Liefting
Plant Health and Environment Laboratory, Ministry for Primary Industries, Auckland, New Zealand

H. Montano
Department of Entomology and Plant Pathology, Federal Rural University of Rio de Janeiro, Seropédica, Brasil

G. P. Rao et al. (eds.), *Phytoplasmas: Plant Pathogenic Bacteria - I*,
https://doi.org/10.1007/978-981-13-0119-3_2

2.1 Introduction

Phytoplasma diseases of vegetable, pulse and oil crops occur worldwide; however, the majority of the reports are from North American, European and Asian countries, affecting plant species belonging mostly to Apiaceae, Asteraceae, Cucurbitaceae, Fabaceae and Solanaceae (Fig. 2.1). They differ considerably in geographic distribution and size of the taxonomic groups and subgroups of the associated phytoplasmas. It appears that subgroup 16SrI-B phytoplasmas are the prevalent agents occurring mainly in Europe, North America and Asia. Phytoplasmas of 16SrXII-A subgroup are restricted to Europe, whereas phytoplasmas of the 16SrII group are prevalent in Asia and Australia. Also, phytoplasmas of the 16SrIII group have been mainly identified in diseased vegetable, pulse and oil crops plants in South America. Like other phytoplasma diseases, a number of these diseases are associated with genetically different phytoplasmas, up to nine in some instances, which induce

Fig. 2.1. Symptoms of phytoplasma diseases. Upper row from left to right: pepper and celery infected with 16SrXII-A phytoplasmas from Bosnia and Herzegovina, chicory infected with 16SrIX-C phytoplasmas from north-east Italy. Lower row: tomato infected with 16SrXII-A phytoplasmas from north-east Italy, sesame infected with 16SrVI-A phytoplasmas from Turkey (Courtesy of G. Sertkaya) and carrot infected with aster yellows phytoplasmas from Serbia (Courtesy of B. Duduk)

similar symptoms in a given plant host species and occur either within the same geographic areas or different continents. In many instances, economically important diseases of vegetable crops have wild plants as alternative hosts.

2.2 Broad Bean (Faba Bean, *Vicia faba* L.)

Symptoms associated with phytoplasma diseases of broad bean (Fabaceae) include leaf yellowing, shoe-stringed leaves, phyllody, virescence, shoot proliferation, witches' broom, short internodes, replacement of ovary by a pod-like structure, flower abortion, plant sterility and stunting. Molecular characterization of the detected phytoplasmas identified 16SrI group from Cuba (Arocha et al. 2007b); 16SrII group from Sudan (Alfaro-Fernandez et al. 2012), Saudi Arabia (Al Saleh and Amer 2014) and Iran (Salehi et al. 2016b); and 16SrIII group from Spain (Castro and Romero 2004). A phytoplasma has also been detected from broad bean in Egypt by Hamed et al. (2014), but sequence analysis was not performed. The only insect vector identified to date is the leafhopper, *Orosius albicinctus*, that has been shown to transmit faba bean phyllody phytoplasma in Iran (Salehi et al. 2016b).

2.3 Cabbage (*Brassica oleracea* L. var. *capitata*)

Cabbage (Brassicaceae) plants are affected by phytoplasma diseases named as cabbage proliferation in the USA (Lee et al. 2003), cabbage stunting in Greece (Gkavaleka et al. 2012), cabbage witches' broom in Italy and China (Bertaccini et al. 1992; Mou et al. 2012) and cabbage yellows in Iran (Salehi et al. 2007a). The main symptoms of the diseases are similar and include virescence and phyllody, yellowing, stunting with shortened internodes, proliferation of axillary shoots and deformation of the leaves (Marcone et al. 1997; Mou et al. 2012). In Hungary, diseased cabbage plants were reported with symptoms of shorter and thicker leaves, a long thick shoot and root malformation (Fodor et al. 1999). In cabbage proliferation, diseased plants exhibited purple discoloration of leaves on cabbage heads. Proliferation of sprouts occurred at the base of the stem and between leaf layers, and sprouts sometimes continued to proliferate along extended stems (Lee et al. 2003). Symptoms of aster yellows have been reported in naturally infected cabbage plants displaying dwarfing and leaf malformation in the USA (Severin and Frazier 1945). Phytoplasmas belonging to the aster yellows (16SrI) group have been found associated with cabbage diseases in different countries; in particular, Italy, Hungary, Greece, the USA and China demonstrated the presence of 16SrI-B phytoplasmas (Bertaccini et al. 1992; Marcone et al. 1997; Fodor et al. 1999; Lee et al. 2003; Arocha et al. 2009a; Gkavaleka et al. 2012; Mou et al. 2012). Additionally, phytoplasmas affecting cabbages in the USA were demonstrated to belong to the rpI-B subgroup, and a comprehensive survey in diseased cabbage fields identified

Macrosteles quadrilineatus (=*fascifrons*), *S. irroratus* and *Ceratagallia* (*Aceratagallia*) *abrupta* as potential vectors of the phytoplasma (Lee et al. 2003). In Iran, phytoplasmas affiliated to 16SrVI-A subgroup were identified in plants affected by cabbage yellows exhibiting symptoms of yellowing, leaf malformation (little leaves and opening of the head), stem proliferation and plant stunting. *Circulifer haematoceps* is the vector responsible for transmitting the Iranian cabbage yellows (ICY) agent. Periwinkle, cauliflower and rape were identified as experimental hosts of this phytoplasma (Salehi et al. 2007a).

2.4 Carrot (*Daucus carota* L.)

Carrot (Apiaceae) plants are affected by carrot yellows, witches' broom and proliferation diseases. Phytoplasma infections associated with carrot yellows disease have been reported mainly in North America, Europe and Asia. Main symptoms are yellowing of leaf midribs, which progresses to chlorosis of entire leaves; formation of upright, chlorotic adventitious shoots; mild bronzing, reddening, or purpling of older leaves; and reduction in size and quality of taproots with proliferation of secondary roots (Fig. 2.1). Late in the growing season, flowers of infected plants display phyllody and virescence, with stunting and chlorosis of the plants. The severity of this disease became a limiting factor for carrot production in several regions of the USA (Gabelman et al. 1994), where it was primarily associated with an aster yellows phytoplasma (Lee et al. 2003, 2006b). In particular, in Texas, phytoplasmas of subgroups 16SrI-B/rpI-B and 16SrI-A/rpI-A have been reported, whereas in Oklahoma phytoplasmas of subgroup 16SrI-B/rpI-B have been reported (Lee et al. 2003, 2004b). Phytoplasmas of the aster yellows group were reported also in Canada (16SrI-A subgroup) (Wang and Hiruki 2005), Israel (Orenstein et al. 1999), Serbia (16SrI-A/rpI-A and 16SrI-B/rpI-B subgroups) (Duduk et al. 2007, 2008, 2009), Spain (subgroup 16SrI-A in mixed infection with *Spiroplasma citri* (Cebrián et al. 2010) and 16SrI-B (Font et al. 1999)), United Kingdom (Nisbet et al. 2014), Finland (16SrI-A) (Munyaneza et al. 2011), Lithuania (16SrI-A subgroup) (Valiunas et al. 2001) and Iran (16SrI-B/rpI-L subgroup) (Vali-Sichani et al. 2014). In Lithuania 16SrI-A subgroup phytoplasmas were detected in plants with proliferation of the crown, chlorosis of young leaves and reddening of mature leaves (Valiunas et al. 2001).

Phytoplasmas from other 16S ribosomal groups have been reported associated with carrots with symptoms resembling yellows disease: 16SrIII group in Israel (Orenstein et al. 1999) and Japan (Maejima et al. 2014); 16SrV group in Israel (Weintraub and Orenstein 2004); 16SrVI-A subgroup in the western USA (Lee et al. 2003, 2006b) and Russia (Kastal'eva et al. 2016); 16SrIX group in Italy (Marchi et al. 2015); and 16SrXII-A subgroup in Serbia (Duduk et al. 2008), Spain (Font et al. 1999), Hungary (Viczián et al. 1998), Italy (Marchi et al. 2015) and Russia (Kastal'eva et al. 2016). Carrot witches' broom (CarWB) disease was observed in Iran; the main symptoms were little leaf, yellowing, proliferation of shoots from taproot, stunting of taproot, virescence, phyllody, leaf reddening and witches' broom. The phytoplasmas associated with CarWB disease in Yazd province belonged to 16SrII-C subgroup. The disease agent was transmitted by graft and dodder to periwinkle inducing phytoplasma-type symptoms. *Orosius albicinctus* leafhopper was identified as a natural vector of CarWB disease since it was able to transmit the agent from witches' broom-infected carrot to healthy alfalfa and carrot plants (Salehi et al. 2016a). Phytoplasmas of the 16SrII group were also identified in carrot plants in India with symptoms of leaf yellowing, chlorosis and little leaf (Arocha et al. 2009b) and in Saudi Arabia (16SrII-D) in plants with symptoms of fasciation, phyllody, hairy roots proliferation, yellow and purple leaves (Omar 2017). *M. quadrilineatus* (=*fascifrons*) is the principal vector that transmits phytoplasmas of subgroup 16SrI-A in the northeastern region of the USA and in Canada, whereas *Scaphytopius irroratus* is a principal vector of phytoplasmas of subgroup 16SrI-B in the western coastal regions, the Midwest and the southern USA (Lee et al. 2003). In Japan phytoplasmas of the 16SrIII group are transmitted by *Ophiola flavopicta* (Maejima et al. 2014). In the western coastal USA, 16SrVI-A phytoplasmas are transmitted by the beet leafhopper, *Circulifer tenellus* (Lee et al. 2006b). In Serbia, *Macrosteles quadripunctulatus* and *M. sexnotatus* are potential vector species of aster yellows phytoplasmas (16SrI-A and 16SrI-B subgroups) and *M. laevis* of "stolbur" phytoplasmas (Duduk et al. 2008). Moreover, since some aster yellows- and "stolbur"-infected species of the genera *Psammotettix* and *Anaceratagallia* (especially *P. confinis* and *A. laevis*) were regularly and commonly found in the infected carrot fields during the whole vegetative period, Drobnjaković et al. (2010) suggested that they could play a significant role in transmitting and spreading these pathogens.

2.5 Cauliflower (*Brassica oleracea* L. var. *botrytis*)

In cauliflower (Brassicaceae) plants, at least three diseases have been attributed to phytoplasmas. Cauliflower aster yellows were reported in association with naturally diseased plants with symptoms of flower malformation in the USA (Severin and Frazier 1945). Cauliflower virescence was described on the basis of transmission electron microscopy, in plants from commercial fields in north-central Italy. Diseased plants exhibited dwarfing, shortened internodes, apex rosetting, inflorescence malformation (chlorotic little leaves) and phyllody (Bertaccini et al. 1983). In Brazil the cauliflower stunt disease reported in the states of São Paulo and Rio Grande do Sul (Rappussi et al. 2012; Canale and Bedendo 2013; Pereira et al. 2016) was also associated with phytoplasma presence. Diseased plants were stunted and exhibited reduced plant size, malformation of the inflorescence, reddening of leaves and vessel necrosis. In some cases, small heads arose at the base of stem. In symptomatic cauliflower plants from commercial fields close to the city of São Paulo, the associated phytoplasmas belonged to the 16SrIII-J subgroup (Rappussi et al. 2012). More recently, cauliflower samples collected in the state of São Paulo were found infected by a phytoplasma of 16SrVII-B subgroup (Pereira et al. 2016). Interestingly, phytoplasmas closely related to the 16SrXV-A subgroup were found associated with the disease in the state of Rio Grande do Sul (Canale and Bedendo 2013). In 2016, a phytoplasma of the 16SrII-A subgroup was found associated with cauliflower phyllody (CauPh) in China. The CauPh disease is characterized by the formation of green leaf-like inflorescences with malformed flowers, production of very few seed pods and rare production of seeds that, however, were able to germinate and develop into healthy-looking plants (Cai et al. 2016).

2.6 Celery (*Apium graveolens* L.)

Celery (Apiaceae) plants are affected mainly by yellows and by stunting and phyllody diseases. Celery yellows disease has been reported in North America, Europe, Japan, Australia and New Zealand. Main symptoms consisted of severe stunting and yellowing of celery plants (Fig. 2.1). Inner petioles were short, yellow to white in colour and moderately to severely curved and twisted. Petioles were reported to become very brittle, with peeling back of the epidermis and cracking in older plants. In later stages of the disease, the inner heart of the plant was reported to turn brown and sometimes decay (Severin 1929). Celery yellows has been associated with phytoplasmas in the 16SrI, 16SrII, 16SrIII and 16SrXII groups. In Europe '*Ca.* P. solani' (16SrXII-A subgroup) was reported in France (Seemüller et al. 1998), Spain (Alfaro-Fernandez et al. 2011), Hungary (Viczián et al. 1998), Bulgaria (Sakalieva 2016), Romania (Chireceanu et al. 2016), Czech Republic (Navrátil et al. 2009), Italy (Carraro et al. 2008), Serbia (Ivanović et al. 2011) and Bosnia and Herzegovina (Delić et al. 2016). This phytoplasma was also characterized as tuf-type B. In Northeast Italy, where a severe epidemic of '*Ca.* P. solani' occurred in celery, *H. obsoletus* has been demonstrated to transmit the pathogen to celery plants under controlled conditions (Carraro et al. 2008). In North America, phytoplasmas of the aster yellows group (16SrI) have been reported to be associated with celery diseases, in particular phytoplasmas of the 16SrI-B subgroup in California, USA (Lee et al. 1993), and phytoplasmas of the 16SrI-A subgroup in Canada as the agent of the celery proliferation disease (Wang and Hiruki 2005). Phytoplasmas in the 16SrI-B subgroup were described also in Europe in Alicante, Spain (Alfaro-Fernandez et al. 2011). In Australia, a phytoplasma of the 16SrII-D subgroup was detected in a celery plant from southern Queensland showing stunting, chlorosis and reddening of the leaf tips which is transmitted by the leafhopper vector *Orosius argentatus* (Tran-Nguyen et al. 2003). In New Zealand '*Ca.* P. australiense' (16SrXII-B) has been identified in celery plants with symptoms of pink and yellow foliage and leaf deformation (Liefting et al. 2011); the polyphagous-feeding behaviour of the insect vector *Zeolarius oppositus* suggests that it may also be moving the phytoplasma into celery. In Japan a phytoplasma from the 16SrIII group has been found associated with celery yellows syndrome transmitted by *O. flavopicta* (Maejima et al. 2014). A new disease named celery stunting and phyllody has been described in South Bohemia (Czech Republic) showing purplish/whitening of leaves on newly growing leaves only in the early stages of infection. Other symptoms revealed in the course of 2 months were small leaves, premature flowering, severely bushy tops, virescence and phyllody of inflorescences. Phytoplasmas of 16SrI-C/rpI-C subgroups have been associated with this disease (Fránová and Špak 2013).

2.7 Chayote [*Sechium edule* (Jacq.) Sw.]

Chayote (Cucurbitaceae) witches' broom (ChWB) disease was first observed in the 1960s in the state of Rio de Janeiro, Brazil. In the following decade, transmission electron microscopy (TEM) studies revealed the presence of phytoplasmas, and the disease was further reported in other states of Brazil (Robbs and Kitajima 1977; Kitajima et al. 1981). ChWB disease has been reported in Mexico and Taiwan, on the basis of TEM observations (Chou et al. 1976; Olivas 1978), and in Costa Rica (Villalobos et al. 2002). The main symptoms are plant basal proliferation, witches' broom, generalized stunting and yellowing, leaf and fruit malformation and flower and fruit drop (Robbs and Kitajima 1977; Kitajima et al. 1981; Montano et al. 2000; Villalobos et al. 2002). The phytoplasma associated with ChWB in Brazil is affiliated to the subgroup 16SrIII-J (Montano et al. 2000), whereas in Costa Rica, it is a member of the 16SrI-B subgroup (Villalobos et al. 2002). The epidemiology of ChWB disease is still poorly understood in Brazil, except for the fact that bitter gourd (*Momordica charantia*) is considered an alternative host of the phytoplasma. Vegetating nearby chayote fields, symptomatic bitter gourd showing witches' broom were infected by subgroup 16SrIII-J phytoplasma. Although no transmission of the ChWB from chayote to bitter gourd and vice versa was accomplished, *M. charantia* is implicated as a potential inoculum source for infection of commercial chayote (Montano et al. 2000). The involvement of insect vectors in the dissemination of 16SrIII-J phytoplasmas among species of the family Cucurbitaceae is probable, since strains close to 16SIII-J phytoplasmas were also found associated with pumpkin yellows in fields next to chayote crops (Montano et al. 2006). In Costa Rica, two 16SrI-B phytoplasma-infected cucurbits were found growing near chayote fields: tacaco (*Sechium tacaco*) showing severe size reduction of leaves and fruits and *Rytidostylis carthaginensis* exhibiting abnormal tendril proliferation. These findings suggest that they may act as a reservoir of the ChWB phytoplasma in that epidemic (Villalobos et al. 2002).

2.8 Chickpeas (*Cicer arietinum* L.)

The only phytoplasma recorded in chickpea (Fabaceae) is from the 16SrII group, and the associated disease is commonly referred to as chickpea phyllody phytoplasma (Fig. 2.2). This disease has been recorded from India (Pallavi et al. 2012), Pakistan (Akhtar et al. 2009a), Sudan (Alfaro-Fernandez et al. 2012), Australia (Saqib et al. 2005) and Oman (Al-Saady et al. 2006). The symptoms of chickpea

Fig. 2.2 Symptoms of phytoplasma diseases in Indian crops. From left to right: upper row, tomato big bud and brinjal little leaf (Courtesy of G.P. Rao); middle row, toria phyllody (Courtesy of M. Azdavar), sesame phyllody and lettuce witches' broom; lower row, brassica flat stem and shoot proliferation and chickpea stunt (Courtesy of G.P. Rao)

phyllody differ between countries. The common and most characteristic symptom is phyllody that may only appear on part of the plant, and the rest of the plant bears normal flowers. Other symptoms include shoot proliferation, stunting, leaf yellowing and reddening and the production of small leaves. Seeds developed on partially infected plants are undersized, shriveled and discoloured and have a bitter taste. In Pakistan, the vector of chickpea phyllody phytoplasma has been identified as the leafhopper *Orosius orientalis* (Akhtar et al. 2009a).

2.9 Common Chicory (*Cichorium intybus* L.)

Common chicory (Asteraceae) plants are affected by the chicory phyllody (ChiP) and fasciation diseases. The most characteristic symptoms of chicory phyllody are yellowing, proliferation of slender secondary shoots (proliferation of axillary buds), small leaves, virescence and phyllody (Fig. 2.1) (Ermacora et al. 2013; Marcone 2011). The disease has been reported in Italy and Australia. In Italy, ChiP is associated with two phytoplasmas which induce the same symptoms, one belongs to the pigeon pea witches' broom group (subgroups 16SrIX-C, rp(IX)-C1 and secY(IX)-C1) (Ermacora et al. 2013; Martini et al. 2013) and the other to the peanut witches' broom group (subgroup 16SrII-E) (Marcone 2011). Field surveys and molecular analyses determined that alternative plant hosts of ChiP phytoplasma are wild herbaceous plant species in the Asteraceae family, such as *Erigeron annuus* (annual fleabane) and *Picris echioides* (bristly oxtongue) for the phytoplasma of the subgroup 16SrIX-C (Ermacora et al. 2013; Martini et al. 2013) and *Taraxacum officinale* (dandelion) and *Picris echioides* (bristly oxtongue) for the phytoplasmas of the subgroup 16SrII-E (Marcone 2011). Molecular analyses of the insects captured in the field and transmission trials demonstrated that the vector of ChiP phytoplasma of the 16SrIX-C subgroup is the leafhopper *Neoaliturus fenestratus* (Ermacora et al. 2013). In Australia a phytoplasma of the 16SrII-D subgroup was detected in a chicory plant from southern Queensland with little leaf and phyllody symptoms which is transmitted by *O. argentatus* (Tran-Nguyen et al. 2003). A phytoplasma-associated disease was also detected in Serbia in 2012 showing characteristic symptoms of fasciation in wild plants, proliferation of chicory shoots and flowers, flattening of the stem with a large number of filamentous leaves, contortion and abnormal growth of flowers on the stem (typical fasciation symptoms) and sterility of flowers. Molecular analysis demonstrated that phytoplasmas infecting these chicory plants belong to '*Ca.* P. solani', 16SrXII-A subgroup (Pavlovic et al. 2014). Puna (or Grasslands Puna) chicory (*Cichorium intybus* L.) plants with flat stem (PCFS) disease were observed in an experimental field in Shaanxi Province, China. Infected plants exhibited abnormal leaf, shoot distortion, flat main stem and a cluster of inflorescence (Li et al. 2012). Molecular analysis demonstrated that the PCFS phytoplasma belongs to subgroup 16SrV-B; thus PCFS may represent an alternative host of jujube witches' broom phytoplasma.

2.10 Eggplant (*Solanum melongena* L.)

Eggplant (Solanaceae) is affected by brinjal little leaf (BLL) (Fig. 2.2), one of the most important diseases in this plant (Rao and Kumar 2017). The main symptoms include little leaf, witches' broom, phyllody, proliferated shoots, internode shortening, severe growth stunting, giant calyx and fruits of small and abnormal size. Excessive root branching is also reported together with flower malformation and sterility. The yield loss reaches 100% in the severely diseased plants. The presence of phytoplasmas in the phloem cells was confirmed by TEM (Shantha and Lakshmanan 1984). Phytoplasma diseases in eggplant are prevalent in Asian, American and African continents, in particular Italy (Martelli et al. 1969; Paltrinieri et al. 2010), India (Kumar et al. 2012), Bangladesh (Siddique et al. 2001; Kelly et al. 2009), Japan (Okuda et al. 1997), Egypt (Omar and Foissac 2012), Iran (Tohidi et al. 2015), Oman (Al-Subhi et al. 2011), Turkey (Sertkaya et al. 2007), Russia (Ember et al. 2011), Australia (Davis et al. 1997) and Brazil (Amaral-Mello et al. 2011).

To date, phytoplasmas belonging to six ribosomal groups including 16SrI, 16SrII, 16SrIII, 16SrVI, 16SrIX and 16SrXII groups have been identified in eggplant (Rao and Kumar 2017). In particular, phytoplasmas of 16SrI group have been reported from Japan (16SrI-B), Bangladesh and India (Okuda et al. 1997; Kelly et al. 2009; Kumar et al. 2012); 16SrII-D subgroup from Australia, Oman, Egypt, Iran and India (Davis et al. 1997; Al-Subhi et al. 2011; Omar and Foissac 2012; Siampour et al. 2012; Yadav et al. 2016); 16SrIII-B, 16SrIII-J and 16SrIII-U subgroups from Brazil (Barros et al. 1998; Amaral-Mello et al. 2011); 16SrVI-A subgroup from Turkey (Sertkaya et al. 2007); 16SrVI-D subgroup from Bangladesh and India (Siddique et al. 2001; Azadvar and Baranwal 2012); 16SrIX-C subgroup from Iran (Tohidi et al. 2015); and 16SrXII-A from Russia and Turkey (Ember et al. 2011; Rao and Kumar 2017). *Datura inoxia*, *D. stramonium*, *Cannabis sativa* subsp *sativa*, *Portulaca oleracea* and *Portulaca grandiflora* were reported as alternative hosts for eggplant phytoplasma in India that was also dodder transmitted to tomato, potato and tobacco (Rao and Kumar 2017). In India, the leafhopper *Hishimonus phycitis* was demonstrated as the vector of BLL phytoplasma (16SrVI-D) (Azadvar and Baranwal 2012). It is able to infect eggplant at any stage of growth with an acquisition access feeding of 72 hours followed by an inoculation access period of 7 days (Rao and Kumar 2017). In Japan *Macrosteles striifrons* was proven to transmit 16SrI-B phytoplasma associated with eggplant dwarf (Okuda et al. 1997).

2.11 Groundnuts (Peanuts, *Arachis hypogaea* L.)

Peanut (Fabaceae) plants are affected by peanut witches' broom (PnWB) disease first reported in Java. Since then PnWB was reported with a high incidence in a geographically isolated area of Taiwan, the Penghu Islands (Yang 1975), Indonesia and the Philippines. The disease is also known to occur in India, Thailand, Japan, China, Papua New Guinea (Reddy 1984) and Australia (Davis et al. 1997). The main symptoms are proliferation of axillary buds and reduction of internodes leading to bushy plants, small pale yellow leaves and upward growth of the pod stalks due to loss of positive geotropism resulting in the loss of pod formation. The phytoplasma associated with PnWB from Taiwan belongs to the 16SrII-A subgroup (Lee et al. 1998). The disease agents are transmitted by the insect vector *O. orientalis*. The minimum periods of acquisition and inoculation feeding are 6 hours and 1 hour, respectively. The minimum incubation period in the vector is 11 days (Yang and Wu 1990). Host range studies by insect vector transmission confirmed that peanut, tomato, potato, alyce clover, purple bean, globe amaranth, periwinkle, *Ipomea obscura* and *Ipomea triloba* were susceptible to PnWB (Yang 1985). In Taiwan, the phytoplasma associated with PnWB was proven to be serologically and genetically related to sweet potato witches' broom (SPWB) phytoplasma (Shen and Lin 1993; Lee et al. 1998). In Australia, TBB (16SrII-D) and SPLL-V4 phytoplasmas have been found associated with peanut plants showing witches' broom in Queensland (Davis et al. 1997) and little leaf symptoms in the Northern Territory (Wilson et al. 2001).

2.12 Lettuce (*Lactuca sativa* L.)

Lettuce (Asteraceae) plants are affected by phytoplasma-associated diseases including yellows and phyllody (Fig. 2.2). Lettuce yellows has been reported from various countries including the USA (Errampalli et al. 1991), Canada (Lee et al. 1993), Italy (Vibio et al. 1994; Marcone et al. 1997) and Czech Republic (Navrátil et al. 1999). The main symptoms of lettuce yellows are chlorosis and the production of latex-like ooze droplets on leaf and stem surfaces at the late stage of infection. The ooze is initially milky white and darkens with time. Strain AY-WB (aster yellows witches' broom) is associated with vein clearing followed by chlorosis, stunting, necrosis, bolting and finally plant death (Zhang et al. 2004). Phytoplasmas of the aster yellows group (16SrI) are commonly associated with the disease. In Canada, New Jersey and Oklahoma (USA), a subgroup 16SrI-A/rpI-A/tufI-A (NJAY) phytoplasma was identified in lettuce, while in Italy and New York state (USA), the phytoplasma detected in similar epidemics belonged to subgroup 16SrI-B, tufI-B (for the Italian strain) (Lee et al. 2004b; Seemüller et al. 1998). In Ohio (USA) phytoplasmas of both ribosomal subgroups have been detected (Zhang et al. 2004). In the Great Lakes region of the USA, *M. quadrilineatus* is the predominant vector for aster yellows (Hoy et al. 1992), which affects many different plant species, but can have particular economic importance on lettuce and other vegetables (e.g. carrots and celery) affecting large acreage of crops in certain years with disease incidence up to 100% in some fields. Leafhoppers, predominately female, move into the Midwest and from areas as far south as Texas, Oklahoma, Arkansas and Louisiana on storm fronts (Hoy et al. 1992). Migrants feed on winter wheat and oats and move into lettuce and other crops once they emerge. Colonizing leafhoppers can be infected with phytoplasmas, acquired before or during migration, and can infect crops immediately or soon after arrival. Lettuce phyllody (LP) have been reported from Australia (Seemüller et al. 1998) and Iran (Salehi et al. 2007b). In Iran, LP is an economically important disease; in surveys from 2002 to 2004, LP was observed in many lettuce fields, especially lettuce-seed fields in Fars province with an incidence rate of up to 70%. The main symptoms of lettuce phyllody in Iran were flower virescence, phyllody, proliferation and sterility, proliferation of shoots with small and deformed leaves from stem and crown and witches' broom. In Australia lettuce phyllody was associated with a phytoplasma related to 16SrII group (Seemüller et al. 1998) whereas in Iran with a phytoplasma of the 16SrIX-D subgroup (Salehi et al. 2007b). In Iran the same phytoplasma was also associated with wild lettuce (*L. serriola*) phyllody (WLP) which occurs in epidemic proportions during the fall in central and southern provinces, especially in Fars. Agents of both diseases, LP and WLP, were successfully transmitted using *N. fenestratus* from infected lettuce or wild lettuce to lettuce, wild lettuce, periwinkle and sowthistle (*Sonchus arvensis*), but not to safflower, sunflower, calendula and sesame (Salehi et al. 2007b). A 16SrIX-related phytoplasma was reported previously as being associated with sowthistle phyllody in Iran (Salehi et al. 2005). Sowthistle as a spring weed, and wild lettuce as a common biennial weed, can therefore serve as a reservoir of LP and as a source of primary inoculum in the spring (Salehi et al. 2007b).

2.13 Loofah (*Luffa* spp.)

Loofah (Cucurbitaceae) plants are affected by loofah witches' broom (LfWB) and luffa little leaf diseases. LfWB has been reported for the first time on the basis of symptomatology in Ping-tung district, Taiwan (Yang et al. 1974). The association of phytoplasmas with diseased loofah was demonstrated by TEM (Chung et al. 1975). The LfWB has been reported also in Brazil, in the State of Rio de Janeiro, where it is characterized by witches' broom, generalized stunting and yellowing, fruit malformation and aborted seeds (Montano et al. 2007). Phytoplasmas affiliated to distinct 16Sr groups have been identified as agents of LfWB disease from Taiwan and Brazil (Lee et al. 1993; Montano et al. 2007). The LfWB phytoplasma from Taiwan is '*Ca.* P. luffae' assigned to subgroup 16SrVIII-A (loofah witches' broom) (Lee et al. 1998), whereas the phytoplasma associated with the same disease from Brazil was identified as belonging to subgroup 16SrIII-J, as reported from other cucurbit diseases in this country. Epidemiological studies concerning field transmission of LfWB phytoplasma revealed that *Hishimonus concavus* can be considered the vector for the disease (Chang 1975). A survey searching for possible field vectors in epidemic areas confirmed that among *H. concavus*, *Tartessinae* sp., *Eupelicinae* sp. and *Coelidiinae* sp. captured in southern Taiwan from nine locations near infected loofah plants, only *H. concavus* was positive for LfWB phytoplasma (Kuan et al. 2008). The disease luffa little leaf has been reported in India.

Main symptoms are the development of leaves of minute size, recorded as one-fifth the size of a normal leaf. The phytoplasma associated was identified as belonging to group 16SrI (Kumar et al. 2010).

2.14 Mustard (*Brassica rapa* L.)

Sarson and toria (Brassicaceae) are important oilseed crops used for cooking oil and fodder in Asia. The molecular characterization of the toria phyllody phytoplasma confirmed the presence of a phytoplasma from the pigeon pea witches' broom group (Azadvar et al. 2009). Symptoms include phyllody, virescence, witches' broom, extensive malformation of the floral parts and flower sterility. The phytoplasma was detected in different plant parts including midribs, flowers, siliques, stems, roots and seeds. Toria phyllody disease was reported to be transmitted by grafting and by dodder, and additional plant hosts are yellow sarson (*B. rapa* L. subsp. *trilocularis*), brown sarson (*B. rapa* subsp. *sarson*) and rapeseed (*B. napus* subsp. *oleifera*). This phytoplasma was also detected in *Laodelpax striatellus*, an abundant planthopper in toria fields at IARI, New Delhi, indicating this as a potential vector (Azadvar et al. 2011). Clover proliferation (16SrVI) and rice yellow dwarf group (16SrXI) groups were reported in mustard (*B. juncea*) in India associated with flat stem, phyllody, witches' broom and virescence symptoms (Rao et al. 2016) (Fig. 2.2). In Pakistan, sarson samples showing symptoms similar to those observed in Indian samples were determined to belong to a new ribosomal subgroup, 16SrIX-H, and transmitted by the leafhopper, *O. albicinctus* (Ahmad et al. 2017).

2.15 Onion (*Allium cepa* L.)

Onion (Amarillidaceae) plants are predominantly affected by onion yellows disease which has been reported in the USA, Canada, Japan, Italy, Czech Republic, Lithuania, Pakistan, Korea, Mauritius, Saudi Arabia and Egypt. In bulb crops, foliar symptoms begin as yellow and green streaks at the base of young leaves. Affected leaves will flatten and occasionally twist and intertwine. Eventually, entire leaves become yellow. In seed crops, the umbel will have a star-burst appearance with elongated pedicels and distorted flowers. The most striking symptom is the production of bulbils or bulblets instead of seeds in the umbel. This important vegetable grown worldwide was found to be infected with phytoplasmas of the 16SrI, 16SrII, 16SrVI and 16SrXII groups. There are many reports of group 16SrI phytoplasmas infecting onion around the world: 16SrI-A/rpI-A subgroup in Texas (USA) (Lee et al. 2003), 16SrI-A subgroup in Canada (Khadhair et al. 2002), 16SrI-B subgroup in Japan and Italy (Vibio et al. 1995; Lee et al. 2004b), 16SrI-A/rpI-A and 16SrI-L/rpl-B/secYI-B subgroups in Lithuania (Jomantiene et al. 2010), 16SrI in Pakistan

(Ahmad et al. 2015) and Korea (Jung et al. 2012) and 16SrI-B subgroup in Mauritius (Gungoosingh-Bunwaree et al. 2010). Infection of phytoplasmas from other ribosomal groups have also been reported in Saudi Arabia and Egypt where onion plants with yellowing, streaks and twisting of leaves were found infected with phytoplasmas of 16SrII-D subgroup (Omar 2017; El-Sisi et al. 2017). A phytoplasma of clover proliferation group, subgroup 16SrVI-A, was identified from symptomatic onion collected in Texas (USA) (Lee et al. 2003), whereas phytoplasmas of 16SrXII-A subgroup were identified in onion from Mauritius (Gungoosingh-Bunwaree et al. 2010). In North America the main vector of the disease is *M. quadrilineatus*, but it was also transmitted to onion plants by *Colladonus montanus*, *C. geminatus* and *Acinopterus angulatus* (Severin and Frazier 1945). In Japan, onion yellows is transmitted by *Macrosteles striifrons*, *Hishimonus sellatus,* and *Hishimonoides sellatiformis* (Maejima et al. 2014).

2.16 Pepper (*Capsicum annuum* L.)

Pepper (Solanaceae) is affected by "stolbur" and other phytoplasma-associated diseases. Typical symptoms of "stolbur" in infected pepper plants are leaf yellowing followed by leaf cupping, shortening of internodes, stunting, wilting (Fig. 2.1), fruit deformation and plant decline. The "stolbur" phytoplasma, '*Ca.* P. solani', was reported in pepper in Hungary (Viczián et al. 1998), France (Cimerman et al. 2009), Italy (Murolo et al. 2010), Czech Republic (Fialová et al. 2009), Serbia (Mitrović et al. 2013), Bulgaria (Vlahova 2015), Russia in Volga and North Caucasian regions (Kastal'eva et al. 2016) and Bosnia and Herzegovina (BiH) in Semberia region (Delić et al. 2016). "Stolbur"-like symptoms were reported in sweet and chilli pepper varieties associated with phytoplasmas from the clover proliferation (16SrVI) group in Spain, West Azerbaijan, Turkey, Russia, Lebanon, New Mexico and Mexico (Castro and Romero 2002; Zibadoost et al. 2016; Sertkaya et al. 2007; Kastal'eva et al. 2016; Choueiri et al. 2007; Randall et al. 2009; Mauricio-Castillo et al. 2015). In Turkey the associated phytoplasma belongs to the 16SrVI-A subgroup (Sertkaya et al. 2007), whereas in Mexico the phytoplasma strains were placed into two subgroups, 16SrVI-A and 16SrVI-J (Mauricio-Castillo et al. 2015; Swisher et al. 2017). Potential insect vectors found infected by PCR with 16SrVI-A phytoplasmas are *O. orientalis* in Turkey (Sertkaya et al. 2007) and *C. tenellus* in Mexico (Swisher et al. 2017). Other phytoplasma diseases in pepper characterized by little leaves, witches' broom, crowding of leaves and stunting were associated with phytoplasmas of the aster yellows group (16SrI) in India, Mexico, Cuba and Central region of Russia and China (Khan and Raj 2006; Santos-Cervantes et al. 2008; Arocha et al. 2007b; Kastal'eva et al. 2016; Zheng-Nan et al. 2013). In China the phytoplasma associated with pepper witches' broom belongs to subgroup 16SrI-B (Zheng-Nan et al. 2013), whereas in Mexico the phytoplasma associated with pepper little leaf belongs to subgroup 16SrI-S (Santos-Cervantes et al. 2008). In China *Cicadella viridis* was found infected with 16SrI-B phytoplasma in cultivated pepper (Zheng-Nan et al. 2013). In Iran and Indonesia, phytoplasmas of the 16SrII group were identified in pepper with symptoms of yellowing, big bud, little leaf and virescence (Faghihi et al. 2016; Harling et al. 2009). In Australia phytoplasmas of 16SrII-D subgroup were detected in pepper from southern Queensland (Tran-Nguyen et al. 2003). In Bolivia, Arocha et al. (2010) detected a phytoplasma from the 16SrIII group associated with leaf size reduction and shortening of internodes in sweet pepper. Pepper plants infected with the same phytoplasma group were reported also in Volga and Central regions of Russia where potential vectors are *Euscelis incisus* and *Aphrodes bicinctus* that were found positive for 16SrIII group phytoplasmas in the Moscow region and *Psammotettix striatus* in the Samara region (Girsova et al. 2016; Kastal'eva et al. 2016). In Costa Rica, a purple vein syndrome characterized by dark green and rugose leaves, a zigzag pattern to the midvein and purple vein discoloration, was recorded on sweet pepper infected with '*Ca.* P. costaricanum' (16SrXXXI group) (Lee et al. 2011).

2.17 Pigeon peas (*Cajanus cajan* L.)

The best known phytoplasma disease of pigeon peas (Fabaceae) is associated with pigeon pea witches' broom phytoplasma classified in the 16SrIX group. This phytoplasma has been recorded from China, Mexico, Myanmar, India and Puerto Rico (Caicedo et al. 2015). Symptoms include excessive proliferation and clustering of branches with small pale green leaves. Infected plants rarely produce flowers and pods; if flowers are produced, they appear in clusters with elongated pedicels. In Puerto Rico, the potential vectors of pigeon pea witches' broom phytoplasma are *Empoasca kraemeri*, *Melormenis antillarum* and *Colpoptera maculifrons*. Other phytoplasmas known to infect pigeon peas are from the 16SrI and 16SrII groups. The 16SrI group and 16SrIX-C subgroup phytoplasmas were reported from India associated with symptoms of little leaves, shortening of internodes and petioles resulting in a bunchy appearance and whole plant stunting (Raj et al. 2006; Rao et al. 2017b). In Australia, the 16SrII group phytoplasma induces stunting of apical stems, conversion of petals to green leaf-like structures (phyllody) and sterility of plants (Yang et al. 2013).

2.18 Potato (*Solanum tuberosum* L.)

Potatoes (Solanaceae) are affected by several phytoplasma diseases including "stolbur", witches' broom, purple top and others (Klein 2001), which have increasing importance in many potato-producing areas around the world. Symptoms of "stolbur" on potato plants include upward rolling and purplish or red discoloration of the top leaves, shortened internodes, aerial tubers, early senescence and, finally, plant wilt and death. In addition, "stolbur" infection increases the sucrose content of tubers by three- to six-fold severely affecting the suitability of tubers for fried potato processing (Lindner et al. 2011). Phytoplasmas from nine 16Sr groups including 16SrI, 16SrII, 16SrIII, 16SrV, 16SrVI, 16SrVIII, 16SrXII, 16SrXIII and 16SrXVIII have been identified in potato. In Europe (Belgium, Greece, Serbia, Montenegro, Romania) and Russia, phytoplasmas from subgroup 16SrXII-A ('*Ca.* P. solani') are the most frequently identified in potato plants. Symptoms resembling "stolbur" are prevalent, but round leaf disease, potato witches' broom and potato purple top wilt also occurred (Tahzima et al. 2013; Holeva et al. 2014; Girsova et al. 2016; Ember et al. 2011; Jović et al. 2011; Kastal'eva et al. 2016; Mitrović et al. 2016; Radonjić et al. 2016). In Romania and southern Russia, the phytoplasma was also detected in other solanaceous crops and weeds (*Convolvulus arvensis*, *Cuscuta* sp. and *Euphorbia falcata*) in the vicinity of the surveyed potato fields. In addition, tuf genotyping revealed that all "stolbur" strains detected in potato and other plants were of tuf-type B, a genotype known to be associated with *C. arvensis* (Ember et al. 2011), suggesting that this weed could constitute a source of inoculum for the strains infecting potatoes in these countries. In the Samara region of Russia, phytoplasmas from 16SrXII-A were identified in *Hyalesthes obsoletus* and *Pentastiridius leporinus* (Kastal'eva et al. 2016). In Serbia *H. obsoletus*, *Reptalus panzeri* and *R. quinquecostatus* tested positive for '*Ca.* P. solani'; moreover semi-field experiments with naturally '*Ca.* P. solani'-infected *H. obsoletus* and *R. panzeri* confirmed the ability of both species to successfully transmit the pathogen to potato plants and induce symptoms characteristic of "stolbur" disease (Mitrović et al. 2016). Girsova et al. (2016) and Kastal'eva et al. (2016) also identified phytoplasmas of subgroups 16SrI-B, 16SrI-C and 16SrI-P, 16SrVI-A and 16SrVI-C (West Siberian and Central regions), 16SrIII-B (Northern region, Central and Black earth regions) and 16SrII-A (North Caucasian and Central region) in potato plants from Russia. In a survey for potential vectors in the central region, *Euscelis incisus* and *Macrosteles laevis* were the prevalent species detected as phytoplasma carriers for group 16SrIII and group 16SrI (Girsova et al. 2016), respectively. In Asia, surveys for potato phytoplasmas were carried out in Iran, China and Korea. In the production areas of central, western and northwestern regions of Iran, potato plants with symptoms of potato purple top disease (PPTD) were infected by '*Ca.* P. solani', '*Ca.* P. asteris' and '*Ca.* P. trifolii' (Hosseini et al. 2011). In Yunnan and Inner Mongolia regions of China, potato plants exhibiting symptoms including rosette and upright growth, upward rolling, yellowing and purpling of leaves, shortened and thickened internodes and formation of aerial tubers revealed the presence of phytoplasmas from '*Ca.* P. fragariae' (16SrXII-E subgroup) and several strain variants designated into subgroup 16SrXII-I (Cheng et al. 2015). In Korea, '*Ca.* P. luffae'-related phytoplasmas (16SrVIII group) have been associated with potatoes exhibiting symptoms of potato witches' broom disease (Jung et al. 2003). Little leaf disease symptoms were observed in potato crops grown in Uttar Pradesh (India) and found infected with a phytoplasma isolate assigned to 16SrI-B subgroup (Tiwari et al. 2013). In North America and Mexico, phytoplasmas have been associated with potato witches' broom (PWB) and potato purple top wilt (PPT) diseases. PWB disease is attributed to phytoplasmas belonging to subgroup 16SrVI-A, whereas PPT disease is associated with group 16SrI subgroups 16SrI-A, 16SrI-B, 16SrI-S, 16SrI-T, 16SrI-U and 16SrI-V; group 16SrII subgroups 16SrII-C and 16SrII-M; group 16SrIII subgroups 16SrIII-S and 16SrIII-M; subgroup 16SrVI-A; subgroup 16SrXIII-D; and '*Ca*. P. americanum' from 16SrXVIII-A and 16SrXVIII-B subgroups (Gutiérrez-Ibáñez et al. 2012; Khadhair et al. 2003; Lee et al. 2004a, 2004b, 2006a, 2009; Leyva-López et al. 2002; Santos-Cervantes et al. 2010; Secor et al. 2006). The beet leafhopper *C. tenellus* transmits the PPT phytoplasma (subgroup 16SrVI-A) to potato in the Columbia Basin of Washington, USA (Munyaneza et al. 2006), while *Macrosteles* is the major vector of aster yellows phytoplasma in the north central USA. In Bolivia a disease locally known as "brotes grandes", was prevalent in potato fields in the valleys of Chilon, Saipina, Pulquina and Comarapa and found associated with a phytoplasma of the aster yellows (16SrI) group. In addition, the mora-mora vine (*Serjania perulacea*) with little leaf symptoms in hedgerows surrounding potato fields in La Tranca, Santa Cruz Province, may act as a reservoir for this phytoplasma (Jones et al. 2005). In Colombia, phytoplasmas related to 16SrV and 16SrXII groups were identified in the potato variety Criolla Colombiana (Mejia et al. 2011). In New Zealand, potato plants with symptoms of upward rolling and purpling from a commercial crop showed the presence of '*Ca*. P. australiense' (16SrXII-B subgroup) (Liefting et al. 2009). The only other known hosts of '*Ca*. P. australiense' in New Zealand are strawberry and native plants belonging to the genera *Cordyline*, *Coprosma* and *Phormium*. In Australia, '*Ca*. P. australiense' is associated with Australian grapevine yellows and papaya dieback. Potential vector of this pathogen is *Zeoliarus oppositus*, a polyphagous planthopper which is able to vector '*Ca*. P. australiense' to both *Coprosma robusta* and *Cordyline australis* (Winks et al. 2014).

2.19 Sesame (*Sesamum indicum* l.)

Sesame (Pedaliaceae) is an oilseed plant that is one of the most economically important hosts of phytoplasmas. *S. indicum* plants are affected by sesame phyllody (SP) (Figs. 2.1 and 2.2), a very important disease especially in tropical and subtropical areas of the world where it causes significant economic losses (Rao et al. 2015). The major symptoms of the disease are phyllody, flower virescence, witches' broom, shoot-tip fasciation, flattening of the shoot apex, intense leaf and flower bud proliferation and cracking of seed capsules. In addition to this, infected plants exhibit reduction of internode distance and of leaf size. The carpels are transformed into a leaf fusion at the margins (Fig. 2.1) (Rao et al. 2015). The disease was reported for the first time in Burma (Myanmar), and today, it has been described in many parts of the world, mainly in Asian and African countries: Israel (Klein 1977), Thailand (Choopanya 1973; Nakashima et al. 1995), Pakistan (Akhtar et al. 2009b), Oman (Khan et al. 2007a), India (Khan et al. 2007b; Kumar et al. 2011b; Nabi et al. 2015a; Madhupriya et al. 2015), Iraq (Tamimi et al. 1989), Myanmar (Win et al. 2010), Turkey (Sertkaya et al. 2007; Ikten et al. 2014), Sudan, Nigeria, Tanzania, Ethiopia, Venezuela, Mexico (Kolte 1985), Burkina Faso (Desmits and Laboucheix 1974) and

Taiwan (Tseng and Deng 2014). The first evidence of phytoplasma association with the disease was described by Cousin et al. (1971). The four phytoplasma groups 16SrI, 16SrII, 16SrVI and 16SrIX have been reported as associated with sesame phyllody disease: 16SrI-B in Myanmar, India and South Korea (Lee 2004; Win et al. 2010; Nabi et al. 2015a; Madhupriya et al. 2015); 16SrII-A in Taiwan, Thailand and India (Nakashima et al. 1995; Tseng and Deng 2014; Singh et al. 2016); 16SrII-C in India (Madhupriya et al. 2015; Nabi et al. 2015a); 16SrII-D in Pakistan, Turkey, India, Oman and Iran (Akhtar et al. 2009b; Khan et al. 2007a; Ikten et al. 2014; Madhupriya et al. 2015; Salehi et al. 2016c); and 16SrVI-A and 16SrIX-C in Turkey and Iran (Sertkaya et al. 2007; Ikten et al. 2014; Salehi et al. 2016c). Several phytoplasma subgroups have been reported associated with SP within the same country: in Turkey and Iran phytoplasmas of the three ribosomal subgroups 16SrII-D, 16SrVI-A and 16SrIX-C and in India phytoplasmas of the four subgroups 16SrI-B, 16SrII-A, 16SrII-C and 16SrII-D. Two leafhoppers, *Circulifer* (*Neoaliturus*) *haematoceps* (Mulsant and Rey) and *O. orientalis* (Matsumura) *albicinctus* (Distant) were identified and reported as vectors of sesame phyllody disease in Turkey (Baspinar et al. 1993; Kersting 1993; Ikten et al. 2014) and in Iran (Salehi and Izadpanah 1992; Esmailzadeh Hosseini et al. 2007). *C. haematoceps* was also identified as a vector in Sudan (Kolte 1985) and *O. orientalis* in Pakistan, India and Israel (Klein 1977; Kolte 1985; Selvanarayanan and Selvamuthukumaran 2000; Akhtar et al. 2009b; Pathak et al. 2012). Very recently in Iran, it was shown that *C. haematoceps* transmitted 16SrII-D, 16SrVI-A and 16SrIX-C phytoplasmas, while *O. albicinctus* only transmitted 16SrII-D strains. This is in agreement with reports from Turkey demonstrating the transmission of only 16SrII-D phytoplasma by this insect vector (Ikten et al. 2014). It seems that in sesame fields, *O. albicinctus* is the vector of 16SrII phytoplasmas, and this and the other sesame phyllody-associated phytoplasmas are vectored by *C. haematoceps* or other, still not known, insect vectors (Salehi et al. 2016c). *Orosius cellulosus* was identified as vector in Burkina Faso (Upper Volta) (Desmits and Laboucheix 1974), and recently *H. phycitis* was identified in India based on 16SrI-B phytoplasma detection in insects and transmission assays (Nabi et al. 2015b). Phyllody disease is not restricted to the cultivated species of *Sesamum*, but it has been also observed in *S. alatýun*, *S. indicatum*, *S. occidentale* and *S. radiatum* (Rao and Kumar 2017). SP disease could be transmitted to sunn hemp (*Crotalaria juncea*), chickpea (*Cicer arietinum*), berseem clover (*Trifolium alexandrinum*), *Medicago scutellata*, *Brassica campestris* var. *toria* and the ornamental species *Phlox drummondii* and *Petunia violacea.* All these species were affected by phyllody under natural conditions and have been used to infect sesame and sunn hemp through *Orosius* spp. (Vasudeva and Sahambi 1955). Recently *Sclerocarpus africanus* and *Cannabis sativa* were reported as natural plant hosts of SP disease associated with 16SrI-B subgroup phytoplasmas (Nabi et al. 2015b, 2015c). The 16SrVI-A phytoplasma detected in sesame was also identified in pepper and eggplant in Turkey (Sertkaya et al. 2007) and in tomato, cabbage, maize, cucumber and alfalfa in Iran (Salehi et al. 2007a; Jamshidi et al. 2014; Zibadoost et al. 2016; Esmailzadeh Hosseini et al. 2016). The 16SrIX-C subgroup phytoplasmas have been previously detected in eggplant (Tohidi et al. 2015). In

Iran, other reported hosts of 16SrII-D phytoplasmas are squash (Salehi et al. 2015a), tomato (Salehi et al. 2014), sunflower (Salehi et al. 2015b) and alfalfa (Esmailzadeh Hosseini et al. 2015).

2.20 Soybean (*Glycine max* L.)

Soybean (Fabaceae), one of the most widely marketed agricultural commodities, is grown on a very large area on a worldwide basis and is affected by several phytoplasma-associated diseases. First disease reports were associated with unclassified phytoplasma(s) such as "machismo" disease of soybean in tropical regions such as Colombia and Mexico (Fletcher et al. 1984; Granada 1979), soybean witches' broom in Indonesia (Iwaki et al. 1978) and soybean bud proliferation (SBP) in Louisiana (Derrick and Newsom 1984). SBP caused severe bud proliferation and delayed maturity of soybeans and was found to be transmitted by the leafhopper *Scaphytopius acutus* (Derrick and Newsom 1984), whereas "machismo" disease found in Colombia was transmitted by *S. fuliginosus* (Granada 1979). Soybean witches' broom disease agent was transmitted to groundnut, soybean, mung bean (*Vigna radiata*), cowpea and sunn hemp (*Crotalaria juncea*) by *O. argentatus* showing typical symptoms (Iwaki et al. 1978).

Several other soybean diseases have been associated with phytoplasmas around the world such as phyllody and/or witches' broom, stunt and veinal necrosis. Soybean phyllody and/or witches' broom were reported to occur in Thailand (Saeed et al. 1994), Malawi and Mozambique (Kumar et al. 2011a), Tanzania (Murithi et al. 2015) and India (Thorat et al. 2016). Main symptoms consisted of shoot proliferation, reduced leaflets, shortened internodes, proliferated auxiliary shoots producing witches' brooms, virescence and phyllody. The disease is associated with phytoplasmas of the 16SrII-C subgroup in Thailand (Khan et al. 2002), Malawi and Mozambique (Kumar et al. 2011a) and Tanzania (Murithi et al. 2015) and 16SrI-B and 16SrII-D subgroup in India (Kumar et al. 2015; Thorat et al. 2016). Soybean stunt was reported in Costa Rica (Lee et al. 2011) and Cuba (Acosta et al. 2015) in which the major symptoms of the disease are stunting, little leaf, shoot proliferation, chlorosis, crinkle and aborted seed pods. In Costa Rica the disease was associated with phytoplasmas belonging to the 16SrXXXI group (Lee et al. 2011) in the '*Ca.* P. costaricanum' taxon. In Cuba phytoplasmas belonging to the ribosomal subgroup 16SrI-B and ribosomal subgroups 16SrI-W and 16SrI-Y were identified. The 16SrI-B was the prevalent subgroup (45% of positive samples), and mixed infections of different subgroups were observed (10% of the positive samples) (Acosta et al. 2015). In Costa Rica '*Ca.* P. costaricanum', or a very closely related strain, also infected sweet pepper (*C. annuum*) with purple vein syndrome and passion fruit vine (*Passiflora edulis*) with bud proliferation disease in the same region (Lee et al. 2011). Soybean veinal necrosis was reported in Lithuania in plants with a normal growth habit, but exhibiting veinal necrosis where a phytoplasma belonging to the 16SrIII-B subgroup was identified (Jomantiene et al. 2000). In Brazil, phytoplasmas of the 16SrI-B subgroup were found in soybean plants with symptoms of leaf deformation, reduction in the number of sheaths with small seeds and sheath immaturity (Acosta et al. 2015).

2.21 Squash (*Cucurbita maxima* Duchesne - *C. moschata* Duchesne ex Poir) and Summer Squash (*C. pepo* L.)

Cucurbita maxima and *C. moschata* are commonly known as squash, although there are reports of phytoplasma diseases in which "pumpkin" is the common name attributed to these botanic species (Streten et al. 2005; Montano et al. 2006). Phytoplasma diseases of squashes and summer squashes (Cucurbitaceae) were named pumpkin yellows (Seemüller et al. 1998; Montano et al. 2006), pumpkin yellow leaf curl (Streten et al. 2005), squash phyllody (Salehi et al. 2015a) and squash virescence (Omar and Foissac 2012). Pumpkin yellows has been reported in Italy (Seemüller et al. 1998) and Brazil (Montano et al. 2006) affecting *C. pepo* and *C. moschata*, respectively. In certain Brazilian areas, *C. moschata* is grown next to chayote (*S. edule*), and symptomatic plants with chlorosis of the shoots and leaves, reduced leaf size and fruit malformation were found infected by the same phytoplasma associated with chayote witches' broom, belonging to the 16SrIII-J subgroup (Montano et al. 2006). Pumpkin yellow leaf curl (PYLC) disease has been reported in Australia affecting plants showing yellow, curled leaves and stunted growth. Molecular analyses revealed that the phytoplasma from symptomatic pumpkin was indistinguishable from '*Ca.* P. australiense' (16SrXII-B). The role of *C. maxima* and *C. moschata* as a source of phytoplasmas to more valuable crops was considered since pumpkin crops are often grown in close proximity to papaya crops affected by papaya dieback (PpDB) and to strawberry crops affected by strawberry lethal yellows (SLY) (Streten et al. 2005). Squash phyllody (SqP) affected *C. pepo* plants was reported in Yazd province of Iran (Salehi et al. 2015a). Characteristic symptoms of SqP included proliferation of short spindly shoots along the stem, reduced size of leaves, shortening of internodes, fruit cracking, virescence and phyllody, floral proliferation, sterility, witches' broom, branch malformation and failure to fruit especially in case of early infection. Phytoplasmas associated with SqP were identified as members of the subgroup 16SrII-D. *O. albicinctus* leafhoppers successfully transmitted SqP phytoplasmas to healthy squash plants and also from squash to periwinkle, alfalfa, cucumber, carrot, sesame, sunflower, pot marigold, eggplant, tomato and parsley. These results showed that squash fields are sources of 16SrII phytoplasmas and *O. albicinctus* is a possible vector also for phytoplasma infection of other important crops in the Yazd province of Iran. Virescence of cucurbit crops has been reported in Egypt (squash virescence) and Argentina (cucurbita virescence). Squash virescence in Egypt was characterized by symptoms of stunting and virescence on *C. pepo* plants (Omar and Foissac 2012) and by a disease incidence of about 1% among the squash fields surveyed. The associated phytoplasma was assigned to the 16SrII-D phytoplasma subgroup. In the same study, 16SrII-D phytoplasmas were proven to also infect tomato and eggplant plants. Cucurbita virescence (CucVir) is a disease of summer squash (*Cucurbita maxima* var. *zapallito*) reported in Argentina. The plants showed typical symptoms of phytoplasma diseases, such as leaf size reduction, stunting and virescence. The molecular identification of the phytoplasma associated with CucVir demonstrated its affiliation to subgroup 16SrIII-J (Galdeano et al. 2013). Recently witches' broom disease on *C. pepo* was identified to be associated with 16SrI-B subgroup phytoplasmas in India (Rao et al. 2017).

2.22 Tomato (*Solanum lycopersicum* L.)

Several phytoplasma diseases, associated with similar symptoms on tomato (Solanaceae), have been described worldwide under different names: "stolbur," big bud, yellows, proliferation, stunt and little leaf. Major symptoms of these diseases (Figs. 2.1 and 2.2) are as follows: internodes near to the plant apex are shorter with smaller and curled leaves, leaf tissues are often thicker or even brittle, the leaves are yellow and/or purple, adventitious roots sometimes appear on the stems and plants infected at early stages present a bushy appearance by development of numerous axillary shoots. The flowers are abnormally straight and often sterile with various morphological changes: sepals with purple veins remain completely sealed, and the calyx is enlarged (big bud); the flowers showed virescence, phyllody and sometimes malformations (absence of petals, stamens and carpels, overdevelopment of petioles). The few fruits formed are small and dense, develop colour slowly and irregularly and have a rather thick stem (Blancard 2012). Six phytoplasma groups including 16SrI, 16SrII, 16SrIII, 16SrV, 16SrVI and 16SrXII infect tomato plants worldwide (Blancard 2012). In particular, phytoplasmas of the 16SrI group have been reported from the USA (16SrI-A/rpI-A/tufI-A) (Lee et al. 2004b), Japan, Italy (16SrI-B) (Okuda et al. 1997; Del Serrone et al. 2001; Giuliani et al. 2010), Greece (Vellios and Lioliopoulou 2007), Poland, Mauritius (16SrI-C) (Krawczyk et al. 2010; Gungoosingh-Bunwaree et al. 2013), Mexico, Cuba (Santos-Cervantes et al. 2008; Zamora et al. 2014) and Bolivia (16SrI-Y, '*Ca.* P. lycopersici') (Arocha et al. 2007a; Bertaccini et al. 2014); 16SrII group from Australia, Egypt, Iran, Saudi Arabia, India (16SrII-D) (Omar and Foissac 2012; Salehi et al. 2014; Alhudaib and Razq 2011; Singh et al. 2012) and China (16SrII-A) (Dong et al. 2013); 16SrIII group from Brazil, Italy (Amaral-Mello et al. 2006; Del Serrone et al. 2001), Argentina (16SrIII-B and -J) (Galdeano et al. 2013) and Mexico (16SrIII-F) (Tapia-Tussell et al. 2010); 16SrV group from Italy and Mauritius (Del Serrone et al. 2001; Gungoosingh-Bunwaree et al. 2013); 16SrVI group from North America (California), China, West Azerbaijan and Russia (16SrVI-A) (Shaw et al. 1993; Munyaneza et al. 2006; Du et al. 2013; Zibadoost et al. 2016; Kastal'eva et al. 2016), Jordan and Lebanon (Anfoka et al. 2003; Choueiri et al. 2007); 16SrXII-A subgroup ('*Ca.* P. solani') from Italy, France, Greece, Turkey, South Moravia (Czech

Republic) and Russia (Albanese et al. 1998; Del Serrone et al. 2001; Garnier 2000; Vellios and Lioliopoulou 2007; Sertkaya et al. 2007; Navrátil et al. 2009; Kastal'eva et al. 2016). '*Ca.* P. solani' infects a wide range of weeds and cultivated plants in Europe, such as other solanaceous crops (tobacco, potato, pepper, aubergine, *Solanum nigrum* and *Datura stramonium*), several Asteraceae/Apiaceae (carrot, celery, wild chicory and chervil), grapevine, maize, sugar beet, strawberry and lavender. '*Ca.* P. solani' is naturally transmitted by polyphagous planthoppers of the family Cixiidae, mainly *H. obsoletus* and *R. panzeri*. The planthopper *P. leporinus* has been reported to transmit '*Ca.* P. solani' to sugar beet. Some other insect species, the planthopper *Reptalus quinquecostatus* and the leafhoppers *Anaceratagallia ribauti* and *M. quadripunctulatus*, are reported as experimental vectors. In Europe, '*Ca.* P. solani' planthopper vectors are monovoltine; the acquisition stage is achieved by overwintering nymphs feeding on infected roots, and the plant-to-plant transmission by flying adults takes place in summer. Phytoplasmas of the 16SrI group have a wide host range affecting more than 350 different plant species, both cultivated and wild (Lee and Davis 2000), and are transmitted by a large number of leafhopper species including *Macrosteles* spp., *Euscelis* spp., *Scaphytopius* spp. and *Aphrodes* spp. In Australia, the tomato big bud (TBB) phytoplasma (16SrII-D subgroup) is a widely distributed phytoplasma and can be transmitted by the leafhopper vector *O. argentatus*. In Iran, a phytoplasma of 16SrII-D subgroup associated with squash phyllody (SqP) was experimentally transmitted to tomato plants using *O. albicinctus* (Salehi et al. 2015a). In North America, phytoplasmas of the 16SrVI-A subgroup have also been reported to infect potato and to be transmitted by the vector *C. tenellus* (Shaw et al. 1993; Munyaneza et al. 2006) considered as the leafhopper vector of beet leafhopper transmitted virescence agent (16SrVI-A subgroup) in California (Golino et al. 1987).

2.23 Conclusion

Phytoplasmas are widespread in horticultural crops and are associated with economic losses that may reach 100% during epidemics. A wide variety of phytoplasmas are associated with diseases that have identical symptomatology in different geographic areas. Despite the major role played by insect vectors in dissemination of phytoplasma diseases, their transmission is enhanced by cultural practices such as grafting and micropropagation.

References

Acosta K, Silva FN, Zamora L, Quiñones M, Martínez R, Piñol B, Urquiza GPC, Carvalho CM, Zerbini FM (2015) Detection and molecular characterization of an aster yellows group phytoplasma associated with soybean and evidence of two new 16SrI subgroups in Cuba. *Journal of Plant Pathology* **97**, 339–344.

Ahmad SJN, Ahmad JN, Irfan M, Ahmad M, Aslam M (2015) Report on phytoplasma new host plants in Pakistan. *Phytopathogenic Mollicutes* **5**(1–Supplement), S71–S72.
Ahmad JN, Ahmad SJN, Irfan M, Paltrinieri S, Contaldo N, Bertaccini A (2017) Molecular detection, identification, characterization and transmission study of sarsoon phyllody in Punjab – Pakistan associated with phytoplasmas affiliated to the new subgroup 16SrIX-H. *European Journal of Plant Pathology* **149**, 117–125.
Akhtar KP, Shah TM, Atta BM, Dickinson M, Hodgetts J, Khan RA, Haq MA, Hameed S (2009a) Symptomatology, etiology and transmission of chickpea phyllody disease in Pakistan. *Journal of Plant Pathology* **91**, 649–653.
Akhtar KP, Sarwar G, Dickinson M, Ahmad M, Haq MA, Hameed S and Iqbal MJ (2009b) Sesame phyllody disease: symptomatology, etiology and transmission in Pakistan. *Turkish Journal of Agriculture and Forestry* **33**, 477–486.
Albanese G, Polizzi G, Grimaldi V, Collodoro S, Tessitori M, Davis RE (1998) Detection and molecular identification of a phytoplasma in diseased tomato plants in Calabria (Italy). *Phytopathologia Mediterranea* **37**, 83–87.
Alfaro-Fernandez A, del Carmen Cebrian M, Jose Villaescusa F, Font-San-Ambrosio MI (2011) Detection and identification of aster yellows and "stolbur" phytoplasmas in various crops in Spain. *Bulletin of Insectology* **64**(Supplement), S63–S64.
Alfaro-Fernandez A, Ali MA, Abdelraheem FM, Saeed EAE, Font-San-Ambrosio MI (2012) Molecular identification of 16SrII-D subgroup phytoplasmas associated with chickpea and faba bean in Sudan. *European Journal of Plant Pathology* **133**, 791–795.
Alhudaib K, Razq A (2011) First report of witches' broom disease of tomato associated with phytoplasmas in Saudi Arabia. *Bulletin of Insectology* **64** (Supplement), S237–S238.
Al-Saady NA, Al-Subhi AM, Al-Nabhani A, Khan AJ (2006) First report of a group 16SrII phytoplasma infecting chickpea in Oman. *Plant Disease* **90**, 973.
Al Saleh MA, Amer MA (2014) Molecular characterisation of the 16SrII group of phytoplasma associated with faba bean (*Vicia faba* L.) in Saudi Arabia. *Journal of Animal Plant Science* **24**, 221–228.
Al-Subhi AM, Al-Saady NA, Khan AJ, Deadman ML (2011) First report of a group 16SrII phytoplasma associated with witches' broom of eggplant in Oman. *Plant Disease* **95**, 360.
Amaral-Mello APO, Bedendo IP, Kitajima EW, Ribeiro LF, Kobori R (2006) Tomato big bud associated with a phytoplasma belonging to group 16SrIII in Brazil. *International Journal of Pest Management* **52**, 233–237.
Amaral-Mello APO, Eckstein B, Flores D, Kreyci PF, Bedendo IP (2011) Identification by computer-simulated RFLP of phytoplasmas associated with eggplant giant calyx representative of two subgroups, a lineage of 16SrIII-J and the new subgroup 16SrIII-U. *International Journal Systematic and Evolutionary Microbiology* **61**, 1454–1461.
Anfoka GHA, Khalil AB, Fattash I (2003). Detection and molecular characterization of a phytoplasma associated with big bud disease of tomatoes in Jordan. *Journal of Phytopathology* **151**, 223–227.
Arocha Y, Antesana O, Montellano E, Franco P, Plata G, Jones P (2007a). '*Candidatus* Phytoplasma lycopersici', a phytoplasma associated with "hoja de perejil" disease in Bolivia. *International Journal Systematic and Evolutionary Microbiology* **57**, 1704–1710.
Arocha Y, Pinol B, Picornell B, Almeida R, Jones P (2007b) Broad bean and sweet pepper: two new hosts associated with '*Candidatus* Phytoplasma asteris' (16SrI phytoplasma group) in Cuba. *Plant Pathology* **56**, 345.
Arocha Y, Piñol B, Almeida R, Acosta K, Quiñones M, Zayas T, Varela M, Marrero Y, Boa E, Lucas JA (2009a) First report of phytoplasmas affecting organoponic crops in central and eastern Cuba. *Plant Pathology* **58**, 793.
Arocha Y, Singh A, Pandey M, Tripathi AN, Chandra B, Shukla SK, Singh Y, Kumar A, Srivastava RK, Zaidi NW, Arif M, Narwal S, Tiwari AK, Gupta MK, Nath PD, Rabindran R, Khirbat SK, Byadgi AS, Singh G, Boa E (2009b) New plant hosts for group 16SrII, '*Candidatus* Phytoplasma aurantifolia', in India. *Plant Pathology* **58**, 391.

Arocha Y, Plata G, Franco J, Maín G, Veramendi S, Lazcano F, Crespo JL, Lino V, Calderón C, Llerena R, Andrew R, Antezana O, Gutiérrez A, Coca M, Boa E (2010). First report of a 16SrIII phytoplasma (X-disease group) affecting bell pepper, strawberry (frutilla), *Schinus molle* and *Arracacia xanthorrhiza* in Cochabamba, Bolivia. *Plant Pathology* **59**, 395.

Azadvar M, Baranwal VK (2012) Multilocus sequence analysis of phytoplasma associated with brinjal little leaf disease and its detection in *Hishimonus phycitis* in India. *Phytopathogenic Mollicutes* **2**, 15–21.

Azadvar M, Baranwal VK, Yadav DK (2009) First report of a 16SrIX (pigeon pea witches' broom) phytoplasma associated with toria (*Brassica rapa* cv. Toria) phyllody disease in India. *Plant Pathology* **20**, 27.

Azadvar M, Baranwal VK, Yadav DK (2011) Transmission and detection of toria (*Brassica rapa*) phyllody phytoplasma and identification of a potential vector. *Journal of General Plant Pathology* **77**, 194–200.

Barros TSL, Kitajima EW, Resende RO (1998) Diversidade de isolados brasileiros de fitoplasmas atrave's da ana'lise do 16S rDNA. *Fitopatologia Brasileira* **23**, 459–465.

Baspinar H, Korkmaz S, Onelge N, Cinar A, Uygun N and Kersting U (1993) Studies on citrus stubborn disease and sesame phyllody in sesame and their related leafhopper vectors. *Journal of Turkish Phytopathology* **22**, 1–8.

Bertaccini A, Pisi A, Marani F (1983) Virescenza e fillodia del cavolfiore e del broccolo. *Informatore Fitopatologico* **5**, 57–60.

Bertaccini A, Davis RE, Hammond RW, Bellardi MG, Vibio M, Lee I-M (1992) Sensitive detection of mycoplasmalike organisms in field-collected and in vitro propagated plants of *Brassica*, *Hydrangea* and *Chrysanthemum* by polymerase chain reaction. *Annals of Applied Biology* **121**, 593–599.

Bertaccini A, Duduk B, Paltrinieri S, Contaldo N (2014). Phytoplasmas and phytoplasma diseases: a severe threat to agriculture. *American Journal of Plant Sciences* **5**, 1763.

Blancard D (2012) Tomato Diseases. 2nd edition, Academic Press, 556–560 pp.

Cai H, Wang L, Mu W, Wan Q, Wei W, Davis RE, Chen H, Zhao Y (2016) Multilocus genotyping of a '*Candidatus* Phytoplasma aurantifolia'–related strain associated with cauliflower phyllody disease in China. *Annals of Applied Biology* **169**, 64–74.

Caicedo JD, Rivera-Vargas LI, Segarra AE, Davis RE (2015) Detection and molecular characterization of a group 16SrIX phytoplasma infecting citrus (*Citrus sinensis* and *C. limon*), coffee (*Coffea arabica*), periwinkle (*Catharanthus roseus*), and tabebuia (*Tabebuia heterophylla*) in Puerto Rico. *Australasian Plant Disease Notes* **10**, 28.

Canale MC, Bedendo IP (2013) '*Candidatus* Phytoplasma brasiliense' (16SrXV-A subgroup) associated with cauliflower displaying stunt symptoms in Brazil. *Plant Disease* **97**, 419.

Carraro L, Ferrini F, Martini M, Ermacora P, Loi N (2008) A serious epidemic of "stolbur" on celery. *Journal of Plant Pathology* **90**, 131–135.

Castro S, Romero J (2002) The association of clover proliferation phytoplasma with "stolbur" disease of pepper in Spain. *Journal of Phytopathology* **150**, 25–29.

Castro S, Romero J (2004) First detection of a phytoplasma infecting faba bean (*Vicia faba* L.) in Spain. *Spanish Journal of Agricultural Research* **2**, 253–256.

Cebrián MC, Villaescusa FJ, Alfaro-Fernández A, de Mendoza AH, Córdoba-Sellés MC, Jordá C, Ferrándiz JC, Sanjuán S, Font MI (2010). First report of *Spiroplasma citri* in carrot in Europe. *Plant Disease* **94**, 1264.

Chang A (1975) Studies on the vector of loofah witches' broom. 1. Identification and morphology of *Hishimonus concavus* Knight. *Plant Protection Bulletin* **21**, 144–146.

Cheng M, Dong J, Lee I-M, Bottner-Parker KD, Zhao Y, Davis RE, Laski PJ, Zhang Z, McBeath JH (2015) Group 16SrXII phytoplasma strains, including subgroup 16SrXII-E ('*Candidatus* Phytoplasma fragariae') and a new subgroup, 16SrXII-I, are associated with diseased potatoes (*Solanum tuberosum*) in the Yunnan and Inner Mongolia regions of China. *European Journal of Plant Pathology* **142**, 305–318.

Chireceanu C, Chiriloaie A, Teodoru A (2016) Molecular detection of "stolbur" phytoplasma in celery plants in Romania. *Phytopathogenic Mollicutes* **6**, 46–49.

Choopanya D (1973) Mycoplasma like bodies associated with sesamum phyllody in Thailand. *Phytopathology* **63**, 1536–1537.

Chou TG, Yang SJ, Huang PY (1976) Mycoplasmalike bodies observed in the plants of bottle gourd, chayote and balsam-pear with witches' broom in Taiwan. *Plant Disease Reporter* **60**, 378–380.

Choueiri E, Salar P, Jreijiri F, El Zammar S, Massaad A, Abdul-Nour H, Bové J-M, Danet J-L, Foissac X (2007) Occurrence and distribution of '*Candidatus* Phytoplasma trifolii' associated with diseases of solanaceous crops in Lebanon. *European Journal of Plant Pathology* **118**, 411–416.

Chung SJ, Huang PY, Yang SJ, Chou TG (1975) Studies on loofah (*Luffa cylindrica* Roem) witches' broom in Taiwan. II Electron microscopic observation of the diseased tissue. *Plant Protection Bulletin* **13**, 329–332.

Cimerman A, Pacifico D, Salar P, Marzachì C, Foissac X (2009) Striking diversity of vmp1, a variable gene encoding a putative membrane protein of the "stolbur" phytoplasma. *Applied Environmental Microbiology* **75**, 2951–2957.

Cousin MT, Kartha KK, Deliatre R (1971) First evidence of mycoplasma association of disease sesame phyllody. *Revue of Plant Pathology* **50**, 430.

Davis RI, Schneider B, Gibb KS (1997) Detection and differentiation of phytoplasmas in Australia. *Australian Journal of Agricultural Research* **48**, 535–544.

Del Serrone P, Marzachì C, Brangaloni M, Galeffi P (2001) Phytoplasma infection of tomato in central Italy. *Phytopathologia Mediterranea* **40**, 137–142.

Delić D, Contaldo N, Lolić B, Moravcevic D, Bertaccini A (2016) First report of '*Candidatus* Phytoplasma solani' in pepper and celery in Bosnia and Herzegovina. *Journal of Plant Pathology* **98**, 184.

Derrick KS, Newsom LD (1984) Occurrence of a leafhopper-transmitted disease of soybeans in Louisiana. *Plant Disease* **68**, 343–344.

Desmits M, Laboucheix J 1974. Relationship between cotton phyllody and a similar disease of sesame. *FAO Plant Protection Bulletin* **22**, 19–20.

Dong JH, Zhang L, Li WH, McBeath JH, Zhang ZK (2013) '*Candidatus* Phytoplasma aurantifolia'-related strain associated with tomato yellows disease in China. *Journal of General Plant Pathology* **79**, 366–369.

Drobnjaković T, Perić P, Marčić D, Picciau L, Alma A, Mitrović J, Duduk B, Bertaccini A (2010) Leafhoppers and cixiids in phytoplasma-infected carrot fields: species composition and potential phytoplasma vectors. *Pesticides and Phytomedicine* **25**, 311–318.

Du Y, Mou H, Shi B, XiaXu X, Xiang B (2013) Molecular detection and identification of a 16SrVI group phytoplasma associated with tomato big bud disease in Xinjiang, China. *Journal of Phytopathology* **161**, 870–873.

Duduk B, Bulajić A, Duduk N, Calari A, Paltrinieri S, Krstić B, Bertaccini A (2007) Identification of phytoplasmas belonging to aster yellows ribosomal group (16SrI) in vegetables in Serbia. *Bulletin of Insectology* **60**, 341–342.

Duduk B, Perić P, Marčić D, Drobnjaković T, Picciau L, Alma A, Bertaccini A (2008) Phytoplasmas in carrots: disease and potential vectors in Serbia *Bulletin of Insectology* **61**, 327–331.

Duduk B, Calari A, Paltrinieri S, Duduk N, Bertaccini A (2009) Multigene analysis for differentiation of aster yellows phytoplasmas infecting carrots in Serbia. *Annals of Applied Biology* **154**, 219–229.

El-Sisi Y, Omar AF, Sidaros SA, El-Sharkawy MM (2017) Characterization of 16SrII-D subgroup associated phytoplasmas in new host plants in Egypt. *Archives of Phytopathology and Plant Protection* **50**, 504–513.

Ember I, Zoltan A, Munyaneza JE, Crosslin JM, Kölber M (2011) Survey and molecular detection of phytoplasmas associated with potato in Romania and Southern Russia. *European Journal of Plant Pathology* **130**, 367–377.

Ermacora P, Ferrini F, Martini M, Pavan F, Loschi A, Loi N, Moruzzi S, Osler R (2013) Biological and molecular characterization of a phytoplasma associated with chicory phyllody in North-East Italy. *Petria* **23**, 29–32.

Errampalli D, Fletcher J, Claypool PL (1991) Incidence of yellows in carrot and lettuce and characterization of mycoplasma-like organism isolates in Oklahoma. *Plant Disease* **75**, 579–584.

Esmailzadeh Hosseini SA, Mirzaie A, Jafari-Nodooshan A, Rahimian H (2007) The first report of transmission of a phytoplasma associated with sesame phyllody by *Orosius albicinctus* in Iran. *Australasian Plant Disease Notes* **2**, 33–34.

Esmailzadeh Hosseini SA, Khodakaramian G, Salehi M, Mirchenari SM, Bertaccini A (2015) Incidence, distribution and economic importance of alfalfa witches' broom disease in Sistan-Baluchestan (Iran) and characterization of associated phytoplasma. *Phytopathogenic Mollicutes* **5**, 1–7.

Esmailzadeh Hosseini SA, Khodakaramian G, Salehi M, Bertaccini A (2016) First report of 16SrVI-A and 16SrXII-A phytoplasmas associated with alfalfa witches' broom disease in Iran. *Journal of Plant Pathology* **98**, 174.

Faghihi MM, Taghavi SM, Salehi M, Siampour M, Modarres Najafabadi SS (2016) First report of a phytoplasma associated with bell pepper big bud disease in Iran. *New Disease Reports* **33**, 15.

Fialová R, Válová P, Balakishiyeva G, Danet J-L, Safárová D, Foissac X, Navrátil M (2009) Genetic variability of "stolbur" phytoplasma in annual crop and wild plant species in South Moravia. *Journal of Plant Pathology* **91**, 411–416.

Fletcher J, Irwin ME, Bradfute OE, Granada GA (1984) Discovery of a mycoplasmalike organism associated with diseased soybeans in Mexico. *Plant Disease* **68**, 994–996.

Fodor M, Viczián O, Mergenthaler E, Süle S (1999) Cabbage infected with phytoplasma from the aster yellows group in Hungary. *Acta Phytopathologica Entomologica Hungaricae* **34**, 1–6.

Font I, Abad P, Albinana M, Espino AI, Dally EL, Davis RE, Jordía C (1999) Amarilleos y enrojecimientos en zanahoria: una enfermedad a diagnóstico. *Bolletin Sanidad Vegetal Plagas* **25**, 405–415.

Fránová J, Špak J (2013) First report of a 16SrI-C phytoplasma infecting celery (*Apium graveolens*) with stunting, bushy top and phyllody in the Czech Republic. *Journal of Plant Pathology* **161**, 666–670.

Gabelman WH, Goldman IL, Breitbach DN (1994) Field evaluation and selection for resistance to aster yellows in carrot (*Daucus carota* L.). *Journal of American Society for Horticultural Science* **119**, 1293–1297.

Galdeano E, Guzmán FA, Fernández F, Conci RG (2013) Genetic diversity of 16SrIII group phytoplasmas in Argentina. Predominance of subgroups 16SrIII-J and B and two new subgroups 16SrIII-W and X. *European Journal of Plant Pathology* **137**, 753–764.

Garnier M (2000) The "stolbur" phytoplasma: an ubiquitous agent. *Comptes Rendus de l'Academie d'Agriculture de France* **86**, 27–33.

Girsova NV, Bottner-Parker KD, Bogoutdinov DZ, Meshkov YI, Mozhaeva KA, Kastalyeva TB, Lee I-M (2016) Diverse phytoplasmas associated with potato stolbur and other related potato diseases in Russia. *European Journal of Plant Pathology* **145**, 139–153.

Giuliani MM, Contaldo N, Nardella E, Quitadamo M, Vocale E, Bertaccini A (2010) Il giallume dell'astro colpisce i pomodori da industria pugliesi. *L'Informatore Agrario* **11**, 75–76.

Gkavaleka S, Efthimiou K, Lotos L, Katis NI (2012) First report of a '*Candidatus* Phytoplasma asteris' related strain associated with cabbage stunting in Greece. *Journal of Plant Pathology* **94**, S4.86.

Golino DA, Oldfield GN, Gumpf DJ, (1987) Transmission characteristics of the beet leafhopper transmitted virescence agent. *Phytopathology* **77**, 954–957.

Granada GA (1979) Machismo disease of soybeans. I. Symptomatology and transmission. *Plant Disease Reporter* **63**, 47–50.

Gungoosingh-Bunwaree A, Contaldo N, Vally V, Benimadhu SP, Duduk B, Bertaccini A (2010) Detection of phytoplasmas in watercress and onion plants from Mauritius. *Plant Health Progress* **10**, 1094.

Gungoosingh-Bunwaree A, Contaldo N, Mejia J F, Paltrinieri S, Benimadhu SP, Bertaccini A (2013) Phytoplasma and virus diseases on tomato in Mauritius. *Australasian Plant Pathology* **42**, 659–665.

Gutiérrez-Ibáñez AT, Laguna-Cerda A., Rojas-Martínez R., González-Garza R, Salgado-Siclán ML (2012) Molecular detection and classification of the phytoplasma that causes purple top in potatoes (*Solanum tuberosum*) in the State of Mexico. *Ciencias Investigation Agraria* **39**, 339–346.

Hamed AH, Attar AKE, El-Banna OM (2014) First record of a phytoplasma associated with faba bean (*Vicia faba* L.) witches' broom in Egypt. *International Journal of Virology* **10**, 129–135.

Harling R, Arocha Y, Harju V, Tobing C, Boa E, Kelly P, Reeder R (2009) First report of 16SrII '*Candidatus* Phytoplasma aurantifolia' infecting chilli and tamarillo in Indonesia. *New Diseases Reports* **19**, 3.

Holeva MC, Glynos PE, Karafla CD, Koutsioumari EM, Simoglou KB, Eleftheriadis E (2014) First report of '*Candidatus* Phytoplasma solani' associated with potato plants in Greece. *Plant Disease* **98**, 1739.

Hosseini P, Bahar M, Madani G, Zirak L (2011) Molecular characterization of phytoplasmas associated with potato purple top disease in Iran. *Journal of Phytopathology* **159**, 241–246.

Hoy CW, Heady SE, Koch TA (1992) Species composition, phenology and possible origins of leafhoppers (Cicadellidae) in Ohio vegetable crops. *Journal of Economic Entomology* **85**, 2336–2343.

Ikten C, Catal M, Yol E, Ustun R, Furat S, Toker C, Uzun B (2014). Molecular identification, characterization and transmission of phytoplasmas associated with sesame phyllody in Turkey. *European Journal of Plant Pathology* **139**, 217–229.

Ivanović Ž, Trkulja N, Živković S, Pfaf Dolovac E, Dolovac N, Jović J, Mitrović M (2011) First report of "stolbur" phytoplasma infecting celery in Serbia. *Bulletin of Insectology* **64**(Supplement), S239–S240.

Iwaki M, Roechan M, Saleh N, Sugiura M, Hibino H (1978) Identity of mycoplasma-like agents of legume witches' brooms in Indonesia. *Contributions from the Central Research Institute for Agriculture Bogor* **41**, 1–11.

Jamshidi E, Jafarpour B, Rouhani H, Salehi M (2014) Association of members of clover proliferation (16SrVI) and pigeon pea witches' broom (16SrIX) phytoplasma groups with tomato big bud disease in Iran. *Iran Journal of Plant Pathology* **50**, 77–89.

Jomantiene R, Davis RE, Antoniuk L, Stanulius J (2000) first report of phytoplasmas in soybean, alfalfa, and *Lupinus* sp. in Lithuania. *Plant Disease* **84**, 198.

Jomantiene R, Davis RE, Lee I-M, Zhao Y, Bottner-Parker K, Valiunas D, Petkauskaite R (2010). Onion is host for two phytoplasma lineages, subgroups 16SrI-A and 16SrI–(B/L)L, in Lithuania: a *Hinf*I site revealed a snp marking divergent branches of evolution. *Journal of Plant Pathology* **92**, 461–470.

Jones P, Arocha Y, Antesana O, Montellano E, Franco P (2005) "Brotes grandes" (big bud) of potato: a new disease associated with a 16SrI-B subgroup phytoplasma in Bolivia. *Plant Pathology* **54**, 234.

Jović J, Ember I, Mitrović M, Cvrković T, Krstić O, Krnjajić S, Acs Z, Kölber M, Toševski I (2011) Molecular detection of potato "stolbur" phytoplasma in Serbia. *Bulletin of Insectology* **64**(Supplement), S83–S84.

Jung H-Y, Hahm YI, Lee JT, Hibi T, Namba S (2003) Characterization of a phytoplasma associated with witches' broom disease of potatoes in Korea. *Journal of General Plant Pathology* **69**, 87.

Jung H-Y, Win NKK, Kim Y-H (2012) Current status of phytoplasmaa and their related diseases in Korea. *Journal of Plant Pathology* **28**, 239–247.

Kastal'eva TB, Bogoutdinov DZ, Bottner-Parker KD, Girsova NV, Lee I-M (2016) Diverse phytoplasmas associated with diseases in various crops in Russia – pathogens and vectors. *Agricultural Biology* **51**, 367–375.

Kelly P, Arocha Y, Dider SZ (2009) First report of a 16SrI '*Candidatus* Phytoplasma asteris' isolate affecting eggplant and *Mikania* sp. in Bangladesh. *Plant Pathology* **58**, 789.

Kersting U 1993. Symptomatology, etiology and transmission of sesame phyllody in Turkey. *Journal of Turkish Phytopathology* **22**, 47–54.
Khadhair A-H, Evans IR, Choban B (2002). Identification of aster yellows phytoplasma in garlic and green onion by PCR-based methods. *Microbiological Research* **157**, 161–167.
Khadhair A-H, Duplessis McAllister P, Ampong-Nyarko K, Bains P (2003) Transmission and characterization of phytoplasma diseases associated with infected potato cultivars in Alberta. *Acta Horticulturae* **619**, 167–176.
Khan AJ, Botti S, Al-Subhi AM, Gundersen-Rindal DE, Bertaccini A (2002) Molecular identification of a new phytoplasma associated with alfalfa witches' broom in Oman. *Phytopathology* **92**, 1038–1047.
Khan AJ, Bottner K, Al-Saadi N, Al Subhi, Lee I-M (2007a) Identification of phytoplasma associated with witches' broom and virescence diseases of sesame in Oman. *Bulletin of Insectology* **60**, 133–134.
Khan MS, Raj SK (2006) First report of molecular detection of an aster yellows phytoplasma ('*Candidatus* Phytoplasma asteris') isolate infecting chilli (*Capsicum annuum*) in India. Plant Pathology **5**, 822.
Khan MS, Raj SK, Snehi SK (2007b) First report of molecular identification of '*Candidatus* Phytoplasma asteris' affecting sesame (*Sesamum indicum*) cultivation in India. *Journal of Plant Pathology* **89**, 301–305.
Kitajima EW, Robbs CF, Kimura O, Wanderley LJG (1981) O irizado do chuchuzeiro e o superbrotamento do maracujá – duas enfermidades associadas a microorganismos do tipo micoplasma constatadas nos Estados do Rio de Janeiro e Pernambuco. *Fitopatologia Brasileira* **6**, 115–122.
Klein M (1977) Sesame phyllody in Israel. *Phytopathology* **88**, 165–171.
Klein M (2001) Phytoplasma diseases. In: Virus and virus-like disorders of potatoes and production of seed-potatoes. Eds G Loebenstein, PH Berger, AA Brunt, RH Lawson. Kluwer Academic Publishers, 145–158 pp.
Kolte SJ (1985) Diseases of annual edible oil seed crops. Vol. II. CRS Press, 135 pp.
Krawczyk K, Pospieszny H, Kamasa J (2010) Identification of new members of '*Candidatus* Phytoplasma asteris' affecting tomato plants in Poland. *Journal of Phytopathology* **158**, 496–502.
Kuan CP, Kuan MM, Hsu MC, Li ML (2008) Rapid detection by genomic amplification of loofah witches' broom phytoplasma in leafhopper vectors. *Journal of Phytopathology* **156**, 293–296.
Kumar C, Singh V, Lakhanpaul S (2010) First report of '*Candidatus* Phytoplasma asteris' (16SrI) associated with little leaf of cotton and luffa in India. *Australasian Plant Disease Notes* **5**, 117–119.
Kumar J, Gunapati S, Singh SP, Lalit A, Sharma NC, Tuli R (2012) First report of a '*Candidatus* Phytoplasma asteris' (16SrI group) associated with little leaf disease of *Solanum melongena* (brinjal) in India. *New Disease Reports* **26**, 21.
Kumar PL, Sharma K, Boahen S, Tefera H, Tamò M (2011a) First report of soybean witches' broom disease caused by group 16SrII phytoplasma in soybean in Malawi and Mozambique. *Plant Disease* **95**, 492.
Kumar S, Singh V, Lakhanpaul S (2011b) Occurrence of spiroplasma and phytoplasma in sesame affected with yellowing disease in India. *Phytopathogenic Mollicutes* **1**, 47–49.
Kumar S, Sharma P, Sharma S and Rao GP (2015) Mixed infection and natural spread of '*Candidatus* Phytoplasma asteris' and *Mungbean yellow mosaic Indiavirus* affecting soya bean crop in India. *Journal of Phytopathology* **163**, 395–406.
Lee I-M, Davis RE (2000). Aster yellows. In Maloy OC, Murray TD (Ed) Encyclopedia of Plant Pathology. New York: Wiley. 60–63 pp.
Lee I-M, Hammond RW, Davis RE, Gundersen DE (1993) Universal amplification and analysis of pathogen 16S rDNA for classification and identification of mycoplasmalike organisms. *Phytopathology* **83**, 834–842.

Lee I-M, Gundersen-Rindal DE, Davis RE, Bartoszyk IM (1998) Revised classification scheme of phytoplasmas based on RFLP analyses of 16S rRNA and ribosomal protein gene sequences. *International Journal of Systematic and Evolutionary Microbiology* **48**, 1153–1169.

Lee I-M, Martini M, Bottner KD, Dane RA, Black MC, Troxclair N (2003) Ecological implications from a molecular analysis of phytoplasmas involved in an aster yellows epidemic in various crops in Texas. *Phytopathology* **93**, 1368–1377.

Lee I-M, Bottner KD, Munyaneza JE, Secor GA, Gudmestad NC (2004a) Clover proliferation group (16SrVI), subgroup a (16SrVI-A) phytoplasma is probable causal agent of potato purple top disease in Washington and Oregon. *Plant Disease* **88**, 429.

Lee I-M, Gundersen-Rindal D, Davis RE, Bottner KD, Marcone C, Seemüller E (2004b) '*Candidatus* Phytoplasma asteris', a novel taxon associated with aster yellows and related diseases. *International Journal of Systematic and Evolutionary Microbiology* **54**, 1037–1048.

Lee I-M, Bottner KD, Secor G, et al (2006a) '*Candidatus* Phytoplasma americanum', a phytoplasma associated with a potato purple top wilt disease complex. *International Journal of Systematic and Evolutionary Microbiology* **56**, 1593–1597.

Lee I-M, Bottner KD, Munyaneza JE, Robert E. Davis, Crosslin JM, du Toit LJ, Crosby T (2006b) Carrot purple leaf: a new spiroplasmal disease associated with carrots in Washington State. *Plant Disease* **90**, 989–993.

Lee I-M, Bottner KD, Sun M (2009) An emerging potato purple top disease associated with a new 16SrIII group phytoplasma in Montana. *Plant Disease* **93**, 970.

Lee I-M, Bottner-Parker KD, Zhao Y, Villalobos W, Moreira L (2011) '*Candidatus* Phytoplasma costaricanum' a novel phytoplasma associated with an emerging disease in soybean (*Glycine max*). *International Journal of Systematic and Evolutionary Microbiology* **61**, 2822–2826.

Lee JT (2004) Phytoplasmal diseases in Korea. National Institute of Agricultural Science and Technology, Suwan, Korea, 226 pp.

Leyva-López NE, Ochoa-Sánchez JC, Leal-Klevezas DS, Martínez-Soriano JP (2002) Multiple phytoplasmas associated with potato diseases in Mexico. *Canadian Journal of Microbiology* **48**, 1062–1068.

Li ZN, Zhang L, Bai YB, Liu P, Wu Y-F (2012) Detection and identification of the elm yellows group phytoplasma associated with Puna chicory flat stem in China. *Canadian Journal of Plant Pathology* **34**, 34–41.

Liefting LW, Veerakone S, Ward LI, Clover GRG (2009) First report of '*Candidatus* Phytoplasma australiense' in potato. *Plant Disease* **93**, 969.

Liefting LW, Beever RE, Andersen MT, Clover GRG (2011) An update on phytoplasma diseases in New Zealand. *Bulletin of Insectology* **64**(Supplement), S93–S94.

Lindner K, Haase NU, Roman M, Seemüller E (2011) Impact of "stolbur" phytoplasmas on potato tuber texture and sugar content of selected potato cultivars. *Potato Research* **54**, 267–282.

Madhupriya, Rao GP, Kumar A, Baranwal VK (2015) Classification of sesame phytoplasma strain in India at 16Sr subgroup level. *Journal of Plant Pathology* **3**, 523–528.

Maejima K, Oshima K, Namba S (2014) Exploring the phytoplasmas, plant pathogenic bacteria. *Journal of General Plant Pathology* **80**, 210–221.

Marchi G, Cinelli T, Rizzo D, Stefani L, Goti E, Della Bartola M, Luvisi A, Panattoni A, Materazzi A (2015) Occurrence of different phytoplasma infections in wild herbaceous dicots growing in vineyards affected by "bois noir" in Tuscany (Italy). *Phytopathologia Mediterranea* **54**, 504–515.

Marcone C, Ragozzino A, Seemüller E (1997) Detection and identification of phytoplasmas infecting vegetable, ornamental and forage crops in southern Italy. *Journal of Plant Pathology* **79**, 211–217.

Marcone C (2011) Current status of phytoplasma diseases of medicinal and nutraceutical plants in southern Italy. *Bulletin of Insectology* **64**(Supplement), S233–S234.

Martelli GP, Russo M, Citrulli M (1969) Brevi note sullo "stolbur" del pomodoro in Puglia. *Phytopathologia Mediterranea* **8**, 150–154.

Martini M, Moruzzi S, Ermacora P, Ferrini F, Loi N (2013) Multiplex and real-time PCR assays for the specific detection of chicory phyllody phytoplasma (16SrIX-C). In: COST ACTION FA0807, Final meeting. Ed Bertaccini A. Lisbon, Portugal, 46–47 pp.

Mauricio-Castillo JA, Salas-Muñoz S, Velásquez-Valle R, Ambríz-Granados S, Reveles-Torres LR (2015) '*Candidatus* Phytoplasma trifolii' (16SrVI) in mirasol chili pepper (*Capsicum annuum* L.) cultivated in Zacatecas, México. *Revista Fitotecnia Mexicana* **38**, 389–396.

Mejia JF, Contaldo N, Paltrinieri S, Pardo JM, Rios CA, Alvarez E, Bertaccini A (2011) Molecular detection and identification of group 16SrV and 16SrXII phytoplasmas associated with potatoes in Colombia. *Bulletin of Insectology* **64**(Supplement), S97–S98.

Mitrović J, Pavlović S, Duduk B (2013) Survey and multigene characterization of "stolbur" phytoplasmas on various plant species in Serbia. *Phytopathologia Mediterranea* **52**, 434–441.

Mitrović M, Jakovljević M, Jović J, Krstić O, Kosovac O, Trivellone V, Jermini M, Toševski I, Cvrković T (2016) '*Candidatus* Phytoplasma solani' genotypes associated with potato "stolbur" in Serbia and the role of *Hyalesthes obsoletus* and *Reptalus panzeri* (Hemiptera, Cixiidae) as natural vectors. *European Journal of Plant Pathology* **144**, 619–630.

Montano HG, Davis RE, Dally EL, Pimentel JP, Brioso PST (2000) Identification and phylogenetic analysis of a new phytoplasma from diseased chayote in Brazil. *Plant Disease* **84**, 429–36.

Montano HG, Brioso PST, Pimentel JP, Figueiredo DV, Cunha Junior JO (2006) *Cucurbita moschata*, new phytoplasma host in Brazil. *Journal of Plant Pathology* **88**, 226.

Montano HG, Brioso PST, Cunha Junior JO, Figueiredo DV, Pimentel JP (2007) First report of group 16SrIII phytoplasma in loofah (*Luffa cylindrica*). *Bulletin of Insectology* **60**, 277–278.

Mou HQ, Zhang YJ, Li HX, Zhu S-F, Li Z-H, Zhao W-J (2012) Molecular identification of a '*Candidatus* Phytoplasma asteris' associated with cabbage witches' broom in China. *Journal of Phytopathology* **160**, 304–307.

Munyaneza JE, Crosslin LJM, Upton JE (2006) Beet leafhopper (Hemiptera: Cicadellidae) transmits the Columbia Basin potato purple top phytoplasma to potatoes, beets, and weeds. *Journal of Economic Entomology* **99**, 268–271.

Munyaneza JE, Lemmetty A, Nissinen AI, Sengodaand VG, Fisher TW (2011) Molecular detection of aster yellows phytoplasma and '*Candidatus* Liberibacter solanacearum' in carrots affected by the psyllid *Trioza apicalis* (Hemiptera: Triozidae) in Finland. *Journal of Plant Pathology* **93**, 697–700.

Murithi H, Owati A, Madata, CS, Joosten M, Beed F, Lava Kumar P (2015) First report of 16SrII-C subgroup phytoplasma causing phyllody and witches' broom disease in soybean in Tanzania. *Plant Disease* **99**, 886.

Murolo S, Marcone C, Prota V, Garau R, Foissac X, Romanazzi G (2010) Genetic variability of the "stolbur" phytoplasma *vmp*1 gene in grapevines, bindweeds and vegetables. *Journal of Applied Microbiology* **109**, 2049–2059.

Nabi S, Madhupriya, Dubey DK, Rao GP, Baranwal VK, Sharma P (2015a) Characterization of phytoplasmas associated with sesame (*Sesamum indicum*) phyllody disease in North India utilizing multilocus genes and RFLP analysis. *Indian Phytopathology* **68**, 112–119.

Nabi S, Madhupriya, Dubey DK, Rao GP, Baranwal VK, Sharma P (2015b) Molecular characterization of '*Candidatus* Phytoplasma asteris' subgroup I-B associated with sesame phyllody disease and identification of its natural vector and weed reservoir in India. *Australasian Plant Pathology* **44**, 289–297.

Nabi S, Madhupriya, Dubey DK, Rao GP (2015c) Identification of *Cannabis sativa* L. ssp. *sativa* as putative alternate host of sesame phyllody phytoplasma belongs to 16SrI group in India. *Medicinal Plants* **7**, 68–70.

Nakashima K, Hayashi T, Chaleeprom W, Wongkaew P, Sirithorn P (1995) Detection of DNA of phytoplasmas associated with phyllody disease of sesame in Thailand. *Annals of Phytopathological Society of Japan* **61**, 519–528.

Navrátil M, Fialová R, Válová P, Šimková M, Fránová J, Nebesářová J (1999) Lettuce (*Lactuca sativa* L.): a phytoplasma host in the Czech Republic. *Eucarpia Leafy Vegetables* **99**, 165–169.

Navrátil M, Válová P, Fialová R, Lauterer P, Safarova D, Star M (2009) The incidence of "stolbur" disease and associated yield losses in vegetable crops in South Moravia (Czech Republic). *Crop Protection* **28**, 898–904.

Nisbet C, Ross S, Monger WA, Highet F, Jeffries C (2014) First report of '*Candidatus* Phytoplasma asteris' in commercial carrots in the United Kingdom. *New Disease Reports* **30**, 16.

Okuda S, Prince JP, Davis RE, Dally EL, Lee I-M, Mogen B, Kato S (1997). Two groups of phytoplasma from Japan distinguished on the basis of amplification and restriction analysis of 16S rDNA. *Plant Disease* **81**, 301–305.

Olivas E (1978) Estudio de una nueva enfermedad del chayote. *Fitopatologia (Lima)* **13**, 14.

Omar AF, Foissac X (2012) Occurrence and incidence of phytoplasmas of the 16SrII-D subgroup on solanaceous and cucurbit crops in Egypt. *European Journal of Plant Pathology* **133**, 353–360.

Omar AF (2017) Detection and molecular characterization of phytoplasmas associated with vegetable and alfalfa crops in Qassim region. *Journal of Plant Interaction* **12**, 58–66.

Orenstein S, Franck A, Kuznetzova L, Sela I,Tanne E (1999) Association of phytoplasmas with a yellows disease of carrot in Israel. *Journal of Plant Pathology* **81**, 193–199.

Pallavi MS, Ramappa HK, Shankarappa KS, Rangaswamy KT, Wickramaarachchi WART, Maruthi MN (2012) Detection and molecular characterisation of phytoplasma associated with chickpea phyllody disease in south India. *Phytoparasitica* **40**, 279–286.

Paltrinieri S, Contaldo N, Davino M, Davino S, Bertaccini A (2010) Gravi malformazioni in melanzana in Sicilia associate alla presenza di fitoplasmi. *Petria* **20**, 697–698.

Pathak DM, Parakhia AM, Akbari LF (2012) Symptomatology and transmission of sesame phyllody disease caused by phytoplasma. *Journal of Mycological Plant Pathology* **42**, 479–484.

Pavlovic S, Starović M, Stojanovic SD, Josic D (2014) First report of "stolbur" phytoplasma affecting *Cichorium intybus* in Serbia. *Plant Disease* **98**, 839.

Pereira TBC, Dally EL, Davis RE, Banzato TC, Galvão SR, Bedendo IP (2016) Cauliflower is a new host of a subgroup 16SrVII-B phytoplasma associated with stunting disease in Brazil. *Plant Disease* **100**, 1007.

Radonjić S, Hrnčić S, Kosovac A, Krstić O, Mitrović M, Jović J, Toševski I (2016) First report of '*Candidatus* Phytoplasma solani' associated with potato "stolbur" disease in Montenegro. *Plant Disease* **100**, 1775.

Raj SK, Khan MS, Snehi SK, Srivastava S, Singh HB (2006) '*Candidatus* Phytoplasma asteris' isolate associated with a little leaf disease of pigeon pea in India. *Plant Pathology* **55**, 823.

Randall JJ, Bosland PW, Hanson SF (2009). Brote grande, a new phytoplasma-associated disease of chile peppers. *Plant Disease* **93**, 968.

Rao GP, Nabi SU, Madhupriya (2015) Overview on a century progress in research on sesame phyllody disease. *Phytopathogenic Mollicutes* **5**, 74–83.

Rao GP, Kumar M (2017) World status of phytoplasma diseases associated with eggplant. *Crop Protection* **96**, 22–29.

Rao GP, Madhupriya, Singh SK, Dubey D, Rao S (2016) Association of rice yellow dwarf (16SrXI) and clover proliferation (16SrVI) phytoplasmas with flat stem and phyllody disease in mustard. *Phytopathogenic Mollicutes* **6**, 82–86.

Rao GP, Gopala, Goel S, Rao A (2017) First report of a '*Candidatus* Phytoplasma asteris'–related strain (16SrI-B subgroup) associated with witches' broom disease in *Cucurbita pepo* in India. *New Disease Reports* **35**, 33.

Rao GP, Madhupriya, Kumar M, Tomar S, Maya B, Singh SK, Johnson JM (2018) Detection and identification of four 16Sr subgroups of phytoplasmas associated with different legume crops in India. *European Journal of Plant Pathology* **150,** 507–513.

Rappussi MCC, Eckstein B, Flôres D, Haas, ICR, Amorim L, Bedendo IP (2012) Cauliflower stunt associated with a phytoplasma of subgroup 16SrIII-J and the spatial pattern of disease. *European Journal of Plant Pathology* **133**, 829–840.

Reddy DVR (1984) Witches' broom. In: Compendium of Peanut Diseases. Eds Porter DM, Smith DH, Rodriguez-Kabana R. American Phytopathological Society, St Paul, USA, 50 pp.

Robbs CF, Kitajima EW (1977) "Envassouramento" do chuchuzeiro na Baixada Fluminense: moléstia de provável etiologia micoplasmática. *Fitopatologia Brasileira* **2**, 99–100.
Saeed E, Sarindu N, Davies DL, Clark MF, Roux J, Cousin MT (1994) Use of polyclonal antibodies to identify mycoplasma-like organisms (MLOs) from the Sudan and from Thailand. *Journal of Phytopathology* **142**, 345–349.
Sakalieva D (2016) Molecular diagnostics of "stolbur" phytoplasma disease in celery *Apium graveolens* L. *Agrarni Nauki* **8**, 101–106.
Salehi E., Salehi, M., Taghavi SM, Izadpanah Jahromi K (2014) A 16SrII-D phytoplasma strain associated with tomato witches' broom in Bushehr province, Iran. *Journal of Crop Protection* **3**, 377–388.
Salehi M, Izadpanah K (1992) Etiology and transmission of sesame phyllody in Iran. *Journal of Phytopathology* **135**, 37–47.
Salehi M, Heydarnejad J, Izadpanah K (2005) Molecular characterization and grouping of 35 phytoplasmas from central and southern provinces in Iran. *Iran Journal of Plant Pathology* **41**, 62–65.
Salehi M, Izadpanah K, Nejat N, Siampour M (2007a) Partial characterization of phytoplasmas associated with lettuce and wild lettuce phyllodies in Iran. *Plant Pathology* **56**, 669–676.
Salehi M, Izadpanah K, Siampour M (2007b) Characterization of a phytoplasma associated with cabbage yellows in Iran. *Plant Disease* **91**, 625–630.
Salehi M, Esmailzadeh SA, Salehi E (2015a) Characterisation of a phytoplasma associated with sunflower phyllody in Fars, Isfahan and Yazd provinces of Iran. *New Diseases Reports* **31**, 6.
Salehi M, Siampour M, Esmailzadeh Hosseini SA, Bertaccini A (2015b) Characterization and vector identification of phytoplasmas associated with cucumber and squash phyllody in Iran. *Bulletin of Insectology* **68**, 311–319.
Salehi M, Esmailzadeh Hosseini SA, Salehi E, Bertaccini A (2016a) Molecular and biological characterization of a 16SrII phytoplasma associated with carrot witches' broom in Iran. *Journal of Plant Pathology* **98**, 83–90.
Salehi M, Rasoulpour R, Izadpanah K (2016b) Molecular characterization, vector identification and partial host range determination of phytoplasmas associated with faba bean phyllody in Iran. *Crop Protection* **89**, 12–20.
Salehi M, Hosseini SE, Salehi E, Bertaccini A (2016c) Genetic diversity and vector transmission of phytoplasmas associated with sesame phyllody in Iran. *Folia Microbiologica* **62**, 99–109.
Santos-Cervantes ME, Chávez-Medina JA, Méndez-Lozano J, Leyva-López NE (2008) Detection and molecular characterization of two little leaf phytoplasma strains associated with pepper and tomato diseases in Guanajuato and Sinaloa, Mexico. *Plant Disease* **92**, 1007–1011.
Santos-Cervantes ME, Chávez-Medina JA, Acosta-Pardini J, Flores-Zamora GL, Méndez-Lozano J, Leyva-López NE (2010) Genetic diversity and geographical distribution of phytoplasmas associated with potato purple top disease in Mexico. *Plant Disease* **94**, 388–395.
Saqib M, Bayliss KL, Dell B, Hardy GEStJ, Jones MGK (2005) First record of a phytoplasma-associated disease of chickpea (*Cicer arietinum*) in Australia. *Australasian Plant Pathology* **34**, 425–426.
Secor GA, Lee I-M, Bottner KD, Rivera-Varas V, Gudmestad NC (2006) First report of a defect of processing potatoes in Texas and Nebraska associated with a new phytoplasma. *Plant Disease* **90**, 377.
Seemüller E, Marcone C, Lauer U, Ragozzino A, Göschl M (1998) Current status of molecular classification of the phytoplasmas. *Journal of Plant Pathology* **80**, 3–26.
Selvanarayanan V, Selvamuthukumaran T (2000) Field resistance of sesame cultivars against phyllody disease transmitted by *Orosius albicinctus* Distant. *Sesame and Safflower Newsletter* **15**, 71–74.
Sertkaya G, Martini M, Musetti R, Osler R (2007) Detection and molecular characterization of phytoplasmas infecting sesame and solanaceous crops in Turkey. *Bulletin of Insectology* **60**, 141–142.

Severin HHP (1929) Yellows disease of celery, lettuce and other plants, transmitted by *Cicadula sexnotata* (Fall). *Hilgardia* **3**, 543–571.

Severin HHP, Frazier NW (1945) California aster yellows on vegetable and seed crops. *Hilgardia* **16**, 573–596.

Shantha PR, Lakshmanan M (1984) Vector-borne MLOs of brinjal little leaf. *Current Science* **53**, 265–267.

Shaw ME, Kirkpatrick BC, Golino DA (1993) The beet leafhopper-transmitted virescence agent causes tomato big bud disease in California. *Plant Disease* **77**, 290–295.

Shen WC, Lin CP (1993) Production of monoclonal antibodies against a mycoplasmalike organism associated with sweet potato witches' broom. *Phytopathology* **83**, 671–675.

Siampour M, Izadpanah K, Galetto L, Salehi M, Marzachì C (2012) Molecular characterization, phylogenetic comparison and serological relationship of the Imp protein of several '*Candidatus* Phytoplasma aurantifolia' strains. *Plant Pathology* **62**, 452–459.

Siddique ABM, Agrawal GK, Alam N, Krishina Reddy M (2001) Electron microscopy and molecular characterization of phytoplasmas associated with little leaf disease of brinjal (*Solanum melongena*) and periwinkle (*Catharanthus roseus*) in Bangladesh. *Journal of Phytopathology* **149**, 237–244.

Singh J, Rani A, Kumar P, Baranwal VK, Saroj PL, Sirohi A (2012) First report of a 16SrII-D phytoplasma '*Candidatus* Phytoplasma australasia' associated with a tomato disease in India. *New Disease Reports* **26**, 14.

Singh V, Kumar S, Reddy NSB, Naik MK, Bhatc KV, Lakhanpaula S (2016) Devastation of sesame (*Sesamum indicum* L.) crops in different agroclimatic zones of India by genetically diverse subgroups of phytoplasma. *Crop Protection* **86**, 24–30.

Streten C, Conde B, Herrington M, Moulder J, Gibb K (2005) '*Candidatus* Phytoplasma australiense' is associated with pumpkin yellow leaf curl disease in Queensland, Western Australia and the Northern Territory. *Australasian Plant Pathology* **34**, 103–105.

Swisher KD, Munyaneza J, Velásquez-Valle ER, Mena-Covarrubias J (2017) First report of beet leafhopper transmitted virescence agent phytoplasma in *Capsicum annuum* and *Circulifer tenellus* in Mexico. *Plant Disease* **101**, 1032.

Tahzima R, Maes M, De Jonghe K (2013) First occurence of '*Candidatus* Phytoplasma solani' associated with seed potato in Belgium. 65th International Symposium on Crop Protection (ISCP). Ghent, Belgium, 649.

Tamimi KM, Fattah FA, Al-Hamdany MA (1989) Shoot apex fasciation in *Sesamum indicum* associated with mycoplasmalike organisms. *Plant Pathology* **38**, 300–304.

Tapia-Tussell R, Suaste-Dzul A, Cortés-Velázquez A, Quijano-Ramayo A, Martín-Mex R, Nexticapan-Garcez A, Córdova-Lara I, Sáenz-Carbonell L, Pérez-Brito D (2010). First report of a 16SrIII, X-disease phytoplasma affecting tomato plants in Mexico. *Plant Pathology* **59**, 395.

Thorat V, More V, Jadhav P, Mane SS, Nandanwar RS, Surayavanshi M, Shouche Y, Yadav A (2016) First report of a 16SrII-D group phytoplasma associated with witches' broom disease of soybean (*Glycine max*) in Maharashtra, India. *Plant Disease* **100**, 2521.

Tiwari AK, Khan MS, Iqbal A, Chun SC, Priya M (2013) Molecular identification of '*Candidatus* Phytoplasma asteris'(16SrI-B) associated with the little leaf disease of potato in India. *Journal of Plant Pathology* **95**, 659–668.

Tohidi Z, Salehi M, Ghasemi S, Khanchezar A, Shahamiri M (2015) Association of a 16SrIX-C phytoplasma with eggplant phyllody in Iran. *Journal of Crop Protection* **4**, 247–256.

Tran-Nguyen LTT, Persley DM, Gibb KS (2003) First report of phytoplasma disease in capsicum, celery and chicory in Queensland, Australia. *Australasian Plant Pathology* **32**, 559–560.

Tseng YW, Deng WL (2014) First report on the association of a 16SrII-A phytoplasma with sesame (*Sesamum indicum*) exhibiting abnormal stem curling and phyllody in Taiwan. *Plant Disease* **98**, 990.

Vali-Sichani F, Bahar M, Zirak L (2014) Characterization of phytoplasmas related to '*Candidatus* Phytoplasma asteris' subgroup rpI-L in Iran. *Journal of Plant Protection Research* **54**, 199–203.

Valiunas D, Alminaite A, Staniulis J, Jomantiene R, Davis RE (2001) First report of aster yellows-related subgroup I-A phytoplasma strains in carrot, phlox, sea-lavender, aconitum, and hyacinth in Lithuania. *Plant Disease* **85**, 804.

Vasudeva RS, Sahambi HS (1955) Phyllody in sesamum (*Sesamum orientale* L.). *Indian Phytopathology* **8**, 124–129.

Vellios E, Lioliopoulou F (2007) Detection and characterization of phytoplasmas infecting tomato plants in Greece. *Bulletin of Insectology* **60**, 157–158.

Vibio M, Bertaccini A, Lee I-M, Davis RE, Minucci C, Milne RG (1994) Etiology of lettuce yellows in Italy: genetic characterization of associated mycoplasmalike organisms. *Phytopathologia Mediterranea* **33**, 179–186.

Vibio M, Camele I, Bertaccini A, Rana GL (1995) Molecular identification of phytoplasmas infecting onion in Italy. 8th Conference on Virus Diseases of Vegetables. Prague, Czech Republic, 176–179.

Viczián O, Süle S, Gáborjányi R (1998) Detection and identification of "stolbur" phytoplasma in Hungary by PCR and RFLP methods. *Acta Phytopathologica Entomologica Hungaricae* **33**, 255–260.

Villalobos W, Moreira L, Rivera C, Bottner KD, Lee I-M (2002) First report of an aster yellows subgroup 16SrI-B phytoplasma infecting chayote in Costa Rica. *Plant Disease* **86**, 330.

Vlahova V (2015) Research on "stolbur" of pepper (*Capsicum annuum* L.) cultivated under the conditions of organic farming. *New Knowledge Journal of Science* **4**, 32–39.

Wang K, Hiruki C (2005) Distinctions between phytoplasmas at the subgroup level detected by heteroduplex mobility assay. *Plant Pathology* **54**, 625–633.

Weintraub PG, Orenstein S (2004) Potential leafhopper vectors of phytoplasma in carrots. *International Journal of Tropical Insect Science* **24**, 228–235.

Wilson D, Blanche KR, Gibb KS (2001). Phytoplasmas and disease symptoms of crops and weeds in the semi-arid tropics of the Northern Territory, Australia. *Australian Plant Pathology* **30**, 159–163.

Win NKK, Back C, Jung HY (2010) Phyllody phytoplasma infecting sesame in Myanmar belongs to group 16SrI and subgroup 16SrI-B. *Tropical Plant Pathology* **35**, 310–313.

Winks CJ, Andersen MT, Charles JG, Beever RE (2014) Identification of *Zeoliarus oppositus* (Hemiptera: Cixiidae) as a vector of '*Candidatus* Phytoplasma australiense'. *Plant Disease* **98**, 10–15.

Yadav V, Mahadevakumar S, Tejaswini GS, Shilpa N, Amruthavalli C, Janardhana GR (2016) First report of 16SrII-D phytoplasma associated with eggplant big bud (*Solanum melongena* L.) in India. *Plant Disease* **100**, 517.

Yang I-L (1975) Studies on the peanut witches' broom disease in Taiwan. *T'ai-wan-Nung-Yeh-Chi-k'an* **11**, 93–108.

Yang I-L (1985) Host responses of peanut witches' broom disease. *Journal of Agricultural Research in China* **34**, 464–468.

Yang I-L, Wu SY (1990) The latent period of peanut witches' broom agent in the vector *Orosius orientalis*. *Journal of Agricultural Research in China* **39**, 204–207.

Yang SJ, Chou TG, Chung SJ, Huang PY (1974) Studies on loofah (*Luffa cylindrica* Roem) witches' broom in Taiwan. I Occurrence, symptoms and transmission test. *Plant Protection Bulletin* **16**, 162–166.

Yang SY, Habili N, Aoda A, Dundas I, Paull JG, Randles JW (2013) Three group 16SrII phytoplasma variants detected in co-located pigeon pea, lucerne and tree medic in South Australia. *Australasian Plant Disease Notes* **8**, 125–129.

Zamora L, Acosta K, Piñol B, Quiñones M, Bertaccini A (2014) First report of '*Candidatus* Phytoplasma asteris' (16SrI group) causing stunt of tomato in Cuba. *New Disease Reports* **30**, 10.

Zhang J, Hogenhout SA, Nault LR, Hoy CW, Miller SA (2004) Molecular and symptom analyses of phytoplasma strains from lettuce reveal a diverse population. *Journal of Phytopathology* **94**, 842–849.

Zheng-Nan L, Zhang L, Song J-G, Wu Y-F (2013) Molecular detection and identification of phytoplasma associated with pepper witches' broom in China. *Phytoparasitica* **41**, 429–434.
Zibadoost S, Rastgou M, Tazehkand SA (2016) Detection and molecular identification of '*Candidatus* Phytoplasma trifolii' infecting some cultivated crops and vegetables in West Azarbaijan province, Iran. *Australasia Plant Disease Notes* **11**, 1–4.

Chapter 3
Occurrence and Epidemiological Aspects of Phytoplasmas in Cereals

Elizabeth de Oliveira, Deividas Valiūnas, Jelena Jović, Ivan Paulo Bedendo, Laima Urbanavičienė, and Charles Martins de Oliveira

Abstract The symptoms of the main cereal diseases associated with phytoplasma presence such as barley deformation, maize bushy stunt, maize redness, green ear, oat proliferation, oat stunt, oat yellows, rice orange leaf, rice yellow dwarf, triticosecale stunt, wheat blue dwarf and wheat streak, yellowing, and stunting are described. Moreover the phytoplasmas associated with each disease are described together with their geographical distribution. In some cases, factors that affect the occurrence, development, and severity of these diseases are discussed. The insect vectors involved in transmission of some of these diseases are presented, along with the main morphological characteristics adopted for their identification.

Keywords Phytoplasma diseases · Barley · Maize · Rice · Oat · Wheat · Millets · Sorghum · Insect vectors

---

E. de Oliveira (✉)
Embrapa Maize and Sorghum, Sete Lagoas, MG, Brazil
e-mail: elizabeth.o.sabato@embrapa.br

D. Valiūnas · L. Urbanavičienė
Nature Research Centre, Phytovirus Laboratory, Vilnius, Lithuania

J. Jović
Department of Plant Pests, Institute for Plant Protection and Environment, Zemun, Serbia

I. P. Bedendo
University of São Paulo, College of Agriculture, Piracicaba, SP, Brazil

C. M. de Oliveira
Embrapa Cerrados, Brasília, DF, Brazil

G. P. Rao et al. (eds.), *Phytoplasmas: Plant Pathogenic Bacteria - I*,
https://doi.org/10.1007/978-981-13-0119-3_3

3.1 Introduction

Historically, the grains of cereals have been used mainly for human and animal nourishment, in natural and transformed form in all regions of the world, and the straw of these plants has been used for many purposes. These important crops, all species of the Poaceae family, are cultivated from seeds and have been genetically improved for several characteristics such as productivity, nutritional quality, and resistance to diseases caused by a diversity of pathogens. Detailed knowledge about each disease of these plant species is very important and useful for development of resistant cultivars and delineation of strategies for their management, aiming to increase productivity. The occurrence and characteristics of distinct diseases associated with phytoplasmas in wheat, maize, rice, barley, oat, millet, triticale, and sorghum are discussed.

3.2 Barley (*Hordeum vulgare* L.)

Barley Deformation (BaDef). Barley is an economically important cereal crop, grown in many countries of Europe, North America, Eurasia, and Africa. Globally, barley is mainly a livestock feed and in some countries is the major feed grain. Barley plants are susceptible to numerous diseases, many of which cause significant damages. Reports dealing with fungal, viral, and bacterial diseases affecting barley plants are numerous in the literature. However, little is known about phytoplasma diseases of barley in Europe and other areas of the world.

Aster yellows (AY) infection in barley, when the disease was thought to be caused by a virus, was described in North America. Symptoms which include chlorotic blotching, stunting, and backward rolling of leaves were very similar to those expressed by AYV (aster yellows virus) in barley (Banttari and Moore 1960; Chiykowski 1963). Hollingsworth et al. (2008) described aster yellows disease symptoms in wheat, barley, and oat plants as similar to those caused by *Barley yellow dwarf virus* (BYDV). Symptoms exhibited by diseased plants range from chlorotic leaf blotch to complete plant collapse. Many plants exhibited red to purple blotches on leaves that turned necrotic with time.

Numerous diseased plants of barley exhibiting twisted, abnormally thin and yellowed awns, reduced spikelets, general stunting, and yellowing were observed in fields in the Vilnius and Kaunas regions of Lithuania (Fig. 3.1). The possible association of a phytoplasma with the disease, termed barley deformation (BaDef), was assessed using polymerase chain reaction (PCR). The BaDef phytoplasma was identified and classified according to Lee et al. (1998) and Marcone et al. (2000) on the basis of collective RFLP patterns of amplified 16S rRNA and *rp* gene sequences and classified as a member of group 16SrI (aster yellows), subgroup B (16SrI-B), and rp subgroup rpI-B (Urbanavičienė et al. 2004, 2005).

Fig. 3.1 *Hordeum vulgare* plants infected by phytoplasmas, healthy spike on the left

Some data show, for example, that AY phytoplasma may be associated with severe losses in barley and wheat (Chiykowski 1963). Other reports considered AY disease generally of little significance, with an overall incidence of less than 1% per year, despite occasional epidemics (Olivier et al. 2011). Insect vector(s) of the BaDef pathogen have not been identified.

3.3 Maize (*Zea mays* L.)

Maize is cultivated in many regions in the world and in some of the warmest regions, with more than one harvest per year, especially under irrigation conditions. This cereal is widely used as food for humans and animals, like corn on the cob, popcorn, and cooked kernels, or processed into meal, oil, and many other industrial products.

Maize Bushy Stunt (MBS) This disease has been known since the 1950s, and in 1977, Bascopé-Quintanilla demonstrated phytoplasma association by electron microscopy. Phytoplasmas with dimensions of approximately 100 nm × 130 nm were observed in the plant's phloem tissue. MBS is transmitted by the leafhoppers *Dalbulus maidis* (DeLong &Wolcott) (Fig. 3.2) and *D. elimatus* (Ball). The transmission of the agent associated with MBS was confirmed by Nault (1980), who also demonstrated its experimental transmission by the leafhopper *Graminella nigrifrons* (Forbes). Since then, several studies showed association of phytoplasmas with MBS disease (Costa et al. 1971; Kitajima and Costa 1972; Oliveira et al. 2002b, 2007) although, until now, Koch's postulates for the agent of this disease were not fulfilled due to its inability to be cultured. Polymorphisms in 16S rDNA amplified

Fig. 3.2 Corn leafhopper adults (*Dalbulus maidis*)

from maize diseased plants from Brazil classified the phytoplasma in the16S rDNA group 16SrI, subgroup -B (Bedendo et al. 1997, 2000).

This disease has been reported only in maize, in the tropical and subtropical regions of the American continent where this cereal is cultivated (Bascopé-Quintanilla 1977; Oliveira et al. 1998, 2002b; Lee et al. 2000; Pérez-López et al. 2016). However, *Zea mays* sbp. *mexicana* was experimentally infected by the MBS phytoplasma (Nault 1980). Also, guinea grass (*Panicum maximum* Jacq.), alexandergrass [*Brachiaria plantaginea* (Link) Hitch], and *Brachiaria decumbens* Stapf. were experimentally infected by this phytoplasma at low frequency (Haas 2010).

The characteristic symptoms of MBS disease appear at the maize producing stage, and before this plant age, usually when the infected plants are grown in pots. Depending on the maize genotype, MBS symptoms include reddening of leaves and shortening of the internodes. Some genotypes present symptoms such as high severity of stunting and high intensity of the red color, which can be purple in the older leaves. The symptomatic plants also present tearing, twisting, and deformation of leaves and numerous ears and tillers arising from the leaf axils and from the plant base (Figs. 3.3a and b). However, other genotypes do not present drastic reduction in plant height or tillers. The MBS-affected plant can dry prematurely. Some genotypes became chlorotic and dry quickly, and the leaves do not became reddish (Fig. 3.3c). Symptoms of MBS also include lower development of the aerial part and roots of the diseased plants and smaller ears, frequently with few grains, in relation to healthy plants (Costa et al. 1971; Bascopé-Quintanilla 1977; Nault 1980; Oliveira et al. 1998, 2002b, 2015; Massola Júnior et al. 1999; Sabato 2017). MBS development is favored by high temperatures (Nault 1980; Oliveira et al. 2007). Usually, MBS disease occurs in the same areas or in the same maize plant together with corn stunt spiroplasma (CSS) caused by *Spiroplasma kunkelii*, because both pathogens are transmitted by the same insect vectors, mainly by *D. maidis* (Oliveira et al. 2002a, 2015). Outbreaks of these corn stunting diseases had been reported in countries of the Central and South America, like Nicaragua and Brazil (Hruska et al. 1996; Oliveira et al. 1998; Sabato 2017).

Fig. 3.3 Severe symptoms of MBS in different maize genotypes

The corn leafhopper *D. maidis* is the main vector of MBS phytoplasma (Nault 1980; Oliveira et al. 2013), although it is also able to be transmitted by *D. elimatus* and *G. nigrifrons* (Bascopé-Quintanilla 1977; Nault 1980). *D. maidis* is considered the vector of epidemiological importance because it presents high transmission efficiency and the same geographical distribution of the maize bushy stunt phytoplasma in the Americas (Triplehorn and Nault 1985; Oliveira et al. 2013; van Nieuwenhove et al. 2015). Specimens of *D. maidis* are small (about 3.7–4.3 mm in length), with general yellow pale coloration, and females are larger than males. Adults have two well-marked black circular spots on the crown, which allow them to be distinguished from other leafhoppers commonly found in maize (Oliveira et al. 2017; Oliveira and Querino 2017). *D. maidis* occurs in the Americas, being present from the southern USA to the temperate regions of Argentina (Oman 1948; Triplehorn and Nault 1985; Nault 1990; Carloni et al. 2013). The biological cycle of *D. maidis* at temperatures close to 25 °C lasts approximately 25 days (Marín 1987; Tsai 1988; Waquil et al. 1999), and this species can produce from five to six generations during the rainy season in tropical regions (Waquil et al. 1999). It has no diapause and can continuously develop throughout the year (Oliveira et al. 2013). MBS phytoplasma is transmitted by *D. maidis* in a persistent and propagative manner. This phytoplasma can be acquired after 2 h of feeding in an infected plant; the latent period varies from 22 to 28 days; the inoculation occurs from 0.5 h of feed; and the retention in the vector may vary from 29 to 48 days, with no evidence of transovarial transmission (Gonzales and Gámez 1974; Alivizatos and Markham 1986; Nault 1980; Legrand and Power 1994; Moya-Raygoza and Nault 1998). Phytoplasma 16Sr I-B subgroup associated with a maize leaf redness disease in India was recently reported (Rao et al. 2017a) but its vector has not been identified.

Maize Redness (MR) This is a severe disease of maize, also known as corn reddening, "crvenilo kukuruza" (црвенилокукуруза) or покраснениемкукурузы,

Fig. 3.4 Symptoms of maize redness disease (MR) in the field (Courtesy of I. Toševski)

that is induced by '*Ca.* P. solani' (16SrXII-A subgroup), known under the trivial name "stolbur" (Duduk and Bertaccini 2006; Jović et al. 2007; Grigoryevna 2015). The first occurrence of the disease was observed 60 years ago when it was described as maize redness syndrome in the South Banat region of northeastern Serbia (Marić and Kosovac 1959) and has since been recorded annually with cycles of epidemics and epiphytotic appearance (Duduk and Bertaccini 2006; Bekavac et al. 2007; Jović et al. 2009). Symptoms of MR include reddening of midribs, leaves, and stalks followed by whole-plant desiccation. Abnormal ear development, reduced seed numbers, and poor, shriveled grains lead to significant yield reduction (Fig. 3.4). Diseased plants do not show symptoms of stunting or bushy appearance as in the cases of maize bushy stunt phytoplasma or corn stunt spiroplasma. The "stolbur" phytoplasma naturally infects numerous wild plants that act as its reservoir, source of inoculum, and hosts for the vector(s), such as field bindweed (*Convolvulus arvensis* L.), stinging nettle (*Urtica dioica* L.), and monks pepper (*Vitexagnus-castus* L.) (Maixner 2010; Quaglino et al. 2013; Kosovac et al. 2016). However, the most important hosts for the MR epidemiological cycle are those associated with its insect vector: (a) maize itself where the vigorous maize roots remaining after autumn harvest are infected by the vector's parent population, (b) Johnson grass [*Sorghum halepense* (L.) Pers.] which acts as a natural weed reservoir hosting both vector and pathogen populations, and (c) winter wheat (*Triticum aestivum* L.), an asymptomatic alternative host consequently infected due to the traditional maize-wheat rotation in affected fields, thereby influencing the disease intensification through fostering a higher vector population number and a higher vector infection rate (Jović et al. 2009, 2011).

The MR-associated "stolbur" phytoplasma is transmitted to maize by the ciixid planthopper *Reptalus panzeri* (Löw) (Hemiptera: Cixiidae) (Fig. 3.5) (Jović et al. 2007, 2009). *R. panzeri* is the main vector responsible for the transmission of the MR agent mainly occurring in Serbia, Romania, Bulgaria, and more recently north

Fig. 3.5 Cixiid planthopper adult (*R. panzeri*) (Courtesy of I. Toševski)

Italy, Hungary, Bosnia and Herzegovina, and Russia (Šutić et al. 2002; Jović et al. 2009; Calari et al. 2010; Acs et al. 2011; Kovačević et al. 2014; Grigoryevna 2015). Prior to its discovery as an efficient transmitter of "stolbur" phytoplasma and a major vector of MR disease, *R. panzeri* was not considered to be of agronomic importance. It is a polyphagous species generally found on shrubs and herbaceous plants in hot and dry areas. In the Banat region of northeastern Serbia and in surrounding areas affected by MR (Romania, Bulgaria, Hungary), *R. panzeri* seems to associate its life cycle with maize fields and is well adapted to the environmental conditions and agronomic practice of maize-winter wheat rotation (Jović et al. 2009). MR transmission and epidemiology studies in the South Banat region of Serbia (Jović et al. 2009) showed that the disease is strictly linked to this specific insect vector. However, the "stolbur" genotype transmitted by *R. panzeri* is not distinguished from the one carried by *H. obsoletus* and *R. quinquecostatus* (Cvrković et al. 2014; Kosovac et al. 2016). Therefore, further experimentation is required to determine the role of these cixiids in the MR disease emergence and maintenance through intercrossing transmission pathways. Moreover in a single experiment carried out in a South Banat corn field, transmission by *Hyalesthes obsoletus* (Signoret) from nettle to corn was shown, although the MR symptomatology observed under the experimental cages was mainly present in the leaves (Mori et al. 2013).

The specimens of *R. panzeri* measure between 4.5 and 7.0 mm; they have a dark general color and translucent wings with dark venation and exhibit five longitudinal *carinae* in the *escutellum*, with vertex of the head wider than long (Emeljanov 1971). Identification of *R. panzeri* and differentiation from closely related congeners primarily rely on the shape of the aedeagus, although molecular methods for species identification are also available (Bertin et al. 2010). This species is found in south-central and south-eastern Europe, Asia Minor, and Caucasus region (Holzinger et al. 2003); records of this cixiid occurrence in the Mediterranean basin are probably erroneous misidentification of a closely related species, *R. quinquecostatus* (Dufour), as it was shown for specimens recorded in England (Webb et al. 2013). Little is known about the biology of *R. panzeri*, except in the maize fields of South Banat region of Serbia (Jović et al. 2009; Cvrković et al. 2014) where its life cycle is about 1 year (univoltine species), with the eggs being laid in July/August, the nymph period develops between September of a year and June of the following year, and adults are observed in June/July. This species goes through winter in the nymph stage feeding on the roots of the plants (Jović et al. 2009). Regarding the mechanisms of transmission, it was observed that adults deposit their eggs in the soil surrounding roots of infected maize. The nymphs feed on these roots, acquire the phytoplasma, and spend the winter on the roots of wheat planted in fields previously cultivated with maize. Adults emerge infected with phytoplasma the next year (Jović et al. 2009).

Crucial for MR disease appearance, incidence, and epiphytotic is adaptation of *R. panzeri* life cycle to maize and winter wheat as feeding and developmental plants, enabling very successful life cycle completion and leading to high population numbers infected with "stolbur" phytoplasma. Due to the short life span of cixiid adults (usually 1 month), phytoplasma transmission relies on pathogen acquisition by the nymph stage of the vectors. In the case of MR disease epidemiology, *R. panzeri* adults and early instar nymphs preferably feed on maize leaves and roots, respectively, while overwintering nymphs feed on the roots of the winter wheat planted upon harvested maize field, as well as on Johnson grass, the most common weed in maize crops. All three preferable host plants of *R. panzeri* nymphs in MR-affected maize fields are proven "stolbur" phytoplasma sources for acquisition by the nymphs (Jović et al. 2009). *R. panzeri* adults emerge between middle June and early July from the winter wheat fields, and they are at that time already infected with "stolbur" phytoplasma and infective as vectors. The time of adult emergence coincides with the winter wheat harvest period; hence, they massively shift to the neighboring maize fields and aggregate on the maize plants in rows bordering the wheat field. Later, adults disperse within the field, but due to the ongoing adult's emergence period, higher numbers of *R. panzeri* specimens are always found on the edge rows of the fields during the flight period. As a consequence of this phenomenon, MR symptoms, usually appearing at the end of July, are always first recorded on the plants in the edge rows of the field where the most severely affected plants are usually located. At the same time of symptom appearance, *R. panzeri* females lay eggs in the soil surrounding the roots of infected maize plants. Therefore, when nymphs hatch (late August/early September), they start to feed on infected maize roots. Phytoplasma acquisition by *R. panzeri* nymphs continues even after the maize is harvested because vigorous remnant roots also serve as an appropriate food source and they are confirmed to be the source of "stolbur" phytoplasma inoculum (Jović et al. 2009). In mid-October, winter wheat is seeded over maize, which represents a good food source for early-stage nymphs (Jović et al. 2009). In addition to the importance of wheat as a feeding plant fostering a high population of emerging adults, it is also found infected by "stolbur" phytoplasmas by the feeding nymphs and represents an additional source of inoculum leading to higher vector infection rate. And last, but not least important, in spring time, *R. panzeri* nymphs are also found to aggregate on Johnson grass roots, which represents a reservoir plant for the "stolbur" phytoplasma.

There are several implications of the MR epidemiological cycle. One is that maize itself is an important source of primary infection of *R. panzeri* populations and high levels of MR infection in 1 year lead to a high percentage of "stolbur" infection in the vector population for the following year (Jović et al. 2009). Secondly, maize-winter wheat rotation intensifies MR disease problems by fostering high *R. panzeri* population levels, which could lead to epiphytotics in the following years. Environmental factors play a significant role both in the intensity and incidence of the disease (Bekavac et al. 2007; Jović et al. 2009). More severe disease and higher incidence of symptoms are associated with warm springs and summers, which facilitate earlier emergence of the adult vectors and consequently earlier infection of maize with phytoplasma. Thus, in the presence of the vector and the "stolbur" phytoplasma, favorable environmental conditions and maize-winter wheat rotation play a significant role in epiphytotic disease appearance.

Maize redness disease has major economic importance in years of its epiphytotics, such as in the early 1960s and 2000s in Serbia where up to 90% of plants were symptomatic in the most severely affected areas. However, probably more important is the cumulative annual yield reduction in maize due to the persistent occurrence of the disease in lower percentages. Due to the damage of ear development, grain filling, and overall seed numbers, MR seems to be important for seed production. MR disease is up to now recorded in Serbia, Hungary, Romania, Bulgaria, Bosnia and Herzegovina, north Italy, Russia, and India (Marić and Kosovac 1959; Penchich et al. 1972; Duduk and Bertaccini 2006; Jović et al. 2007; Calari et al. 2010; Acs et al. 2011; Genov et al. 2014; Kovačević et al. 2014; Grigoryevna 2015; Rao et al. 2017a). In some of these countries, its occurrence is known for more than 50 years (Serbia, Romania, Bulgaria, Russia), and it coincides with agricultural intensification, use of fertilizers, and maize-wheat rotation. Therefore, it seems that introduction of another crop in rotation, or avoiding winter crop, as well as seeding the larger fields to avoid edge effect and use of early maturing maize hybrids, may control the intensity of disease occurrence although its persistence will remain significant.

3.4 Millet [*Pennisetum glaucum* (L.) R. Br.]

Pearl millet, used as human and animal food, is a cereal crop adapted to arid and semiarid regions. A disease named "green ear" (GE) of pearl millet was recently reported in India (Kumar et al. 2010) associated with a phytoplasma member of group 16SrI-B ('*Ca.* P. asteris'). Diseased plants showed a broom-like appearance of panicles bearing leafy structures in place of florets, and panicles looked much greener in comparison to the healthy inflorescences. The vector of the GE agent was not identified.

3.5 Napier Grass (*Pennisetum purpureum* Schum.)

Napier grass is considered the most important forage crop in Africa, and a phytoplasma disease named napier grass stunt (NGS) has been reported causing severe losses. *Sorghum versicolor* was identified as an alternative host of the NGS phytoplasma (a representative of the 16SrXI group) (Asudi et al. 2016). Although harboring phytoplasmas, the sampled plants did not exhibit symptoms. Experimentally, it was also demonstrated that plants of the species *Sorghum bicolor* served as reservoirs of the phytoplasma associated with napier grass stunt (Asudi et al. 2015). Leafhoppers of the species *Maiestas banda* (Kramer) transmitted the NGS phytoplasma from infected napier plants to healthy plants of *S. bicolor*. Interestingly, inoculated plants displayed no symptoms.

3.6 Oat (*Avena sativa* L.)

Oat is an economically important grain crop. The common oat plant is a species of cereal grain grown for its seed. Oats are used for human consumption as oatmeal and rolled oats and also used as livestock feed. Oats make up a large part of the diet of horses and are regularly fed to cattle as well. Oats are susceptible to a large number of diseases; however, only a few reports are available about phytoplasma diseases (Fedotina 1977; Jomantiene et al. 2002; Urbanavičienė et al. 2007, 2008).

Oat Proliferation (OatP), Oat Stunt (OatSt), and Oat Yellows (OatY) Diseased plants of oat exhibiting abnormal proliferation of sterile spikelets (Fig. 3.6a) were observed in the field of Raseiniai region, in Lithuania. Other plants, in Vilnius, exhibited symptoms including stunting, development of numerous short tillers at the plant's base, and sterile deformed spikes (Fig. 3.6b), while both Kaunas and Vilnius oats exhibited symptoms of sterile, deformed, and yellow spikes (Fig. 3.6c). Phytoplasma ribosomal DNA was amplified in nested PCR, indicating that the plants contained phytoplasmas that were designated as oat proliferation (OatP), oat stunt (OatSt), and oat yellows (OatY) (Table 3.1). The RFLP patterns of amplified 16S rDNAs were analyzed, and phytoplasmas were classified according to Lee et al. (1998), Lee et al. (2004) and Marcone et al. (2000) in the group 16SrI ('*Ca.* P. asteris' related). In particular the RFLP patterns of the OatP 16S rDNA were indistinguishable from that from tomato big bud (BB) phytoplasma subgroup 16SrI-A; the OatSt phytoplasma was identified as subgroup 16SrI-B, while the OatY phytoplasma belongs to 16SrI-L subgroup.

In 1977, Fedotina reported electron microscopy of a mycoplasma-like organism in pseudo-rosette-diseased oat plants in Siberia, but the identity of that phytoplasma remains unknown. Subgroups 16SrI-A, 16SrI-B, and 16SrI-L are geographically widespread and have been found in numerous plant species. The discovery of phytoplasmas in diseased oats in Lithuania provokes questions concerning possible

Fig. 3.6 Naturally infected oat exhibiting symptoms of abnormal proliferation of sterile spikelets (oat proliferation), healthy plant on the left (**a**). Oat stunt disease symptoms: stunting, development of numerous short tillers at the plant's base, and sterile deformed spikes (**b**). Naturally infected oat exhibit symptoms of sterile, deformed, and yellow spikes (oat yellows): healthy plant is on the left (**c**)

Table 3.1 '*Ca*. P. asteris'-related phytoplasma strains infecting oats

Disease	Phytoplasma strain	Ribosomal subgroup	Distribution	Reference
Oat proliferation	OatP	16SrI-A	Raseiniai, Lithuania	Jomantiene et al. 2002
Oat stunt	OatSt	16SrI-B	Vilnius, Lithuania	Urbanavičienė et al. 2007, 2008
Oat yellows	OatY	16SrI-L	Vilnius, Kaunas, Lithuania	Urbanavičienė et al. 2007, 2008

impacts of this phytoplasma on oat cultivation in central Europe and other regions. Further study is needed to establish vectors, damage, and geographical distribution.

3.7 Rice (*Oryza sativa* L.)

Rice, together with wheat and maize, have been recognized as three major food crops in the world. However, rice represents the most important staple food to sustain inhabitants of underdeveloped or developing countries. Rice is a relevant source of carbohydrate and provides appropriate nutrition for more than half of the current human population. Diseases caused by diverse pathogens are serious threats to the rice yield. Phytoplasmas are involved with two diseases known as rice orange leaf and rice yellow dwarf.

Rice Orange Leaf (ROL) The disease has only been reported to date in some countries of eastern and southeastern Asia, where it occurs sporadically in rice fields (Valarmanthi et al. 2013; Li et al. 2015). Although of sporadic occurrence, in the 1980s, rice orange leaf (ROL) caused severe damage to rice production in South China and is currently considered a potential threat (Li et al. 2015). Symptoms observed in young plants are characterized by short and malformed leaves, commonly with twisted tips (Webster and Gunnell 1992). Later, the leaves become yellow-orange and, sometimes, totally orange, roll inward, and dry out. Poor growth is also a symptom observed in infected plants (Li et al. 2015). ROL disease is confused with tungro, caused by virus, since the symptoms are similar and both pathogens are transmitted by *Recilia dorsalis* (Motschulsky) (Hemiptera: Cicadellidae), a vector known as zigzag leafhopper (Webster and Gunnell 1992). Phytoplasmas affiliated with 16SrI group were molecularly identified in association with symptomatic rice plants in India (Valarmanthi et al. 2013) where recently 16SrI-B subgroup was identified in affected plants and zigzag leafhoppers present in rice fields (Valarmanthi et al. 2015).

ROL disease occurring in southern and southeastern Asia (Hibino et al. 1987) is associated with a phytoplasma transmitted by the leafhopper *R. dorsalis*, although recently *Nephotettix cincticeps* (Uhler) has been also suggested as a ROL vector in southern China (Shu et al. 2015). *R. dorsalis* are small insects measuring between 3.2 and 3.8 mm with a light gray coloration. They also present a light gray crown and pronotum with light brown spots and gray wings with conspicuous brown broad band, longitudinally, forming a zigzag pattern (Nielson 1968; Wilson and Claridge 1991). *R. dorsalis* is distributed throughout Asia and Australia (Nielson 1968; Pathak and Khan 1994). At temperatures close to 30°C, this leafhopper takes from 18 to 24 days to develop from egg to adult (Pongprasert 1974; Mathur and Chaturvedi 1980), overwinter in egg stage, and is able to produce up to four generations per year (Nasu 1967; Mathur and Chaturvedi 1980; Pathak and Khan 1994). Transmission studies showed that *R. dorsalis* can acquire ROL after 3 days of feeding and the incubation period in the vector ranged from 15 to 33 days with an average of 22 days (Arocha-Rosete and Jones 2010; Pandian et al. 2015).

Rice Yellow Dwarf (RYD) RYD is a disease known since the 1910s that continues to cause serious problems in rice-growing areas in various countries of Asia (Jung et al. 2003). Diseased plants exhibit stunting, proliferation of tillers, general chlorosis, pale green or yellow leaves, and gradual decay and can generate no panicles or malformed panicles (Webster and Gunnell 1992). Plants infected at the beginning of development frequently die.

The disease was initially believed to be a virus until 1967, when mycoplasma like organisms (now phytoplasmas) were associated with infected plants by electron microscopy. The RYD agent was characterized as a distinct phytoplasma based on 16S rDNA sequences, geographical distribution, and specificity of host and vector (Jung et al. 2003). This phytoplasma was identified to be closely related to phytoplasmas associated with diseases reported in various gramineous species such as sugarcane, bluegrass, Bermuda grass, and brachiaria grass and named '*Ca.* P. oryzae', which represents the 16SrXI-A group (Lee et al. 1998).

Leafhoppers belonging to three species of the genus *Nephotettix* (Matsumura) are involved with transmission of RYD disease (Jung et al. 2003). Since these species are present only in Asia, the disease is naturally limited to some Asian countries. Rice plants and grass *Alopecurus aequalis* (Sobol) seem to be the only host for the phytoplasma under natural conditions, since transmission from rice to other gramineous species is not yet confirmed (Arocha-Rosete and Jones 2010; Jung et al. 2003). Thus, alternative hosts are not important reservoirs or sources of inoculum due to the specificity of the pathogen, and, consequently, they have no role in the epidemiology of the disease. '*Ca.* P. oryzae' is transmitted by three species of leafhopper: *Nephotettix cincticeps* (Uhler), *N. nigropictus* (Stål), and *N. virescens* (Distant) (Nakashima et al. 1993). The specimens of *N. cincticeps* measure between 4.3 and 5.6 mm and present a general green to gray color, and in males a brown or black band is observed on the tip of wings. They have a light green crown with a black transverse line near the anterior margin, pronotum with anterior half-light green and the posterior half dark green, and light green-colored wings. *N. nigropictus* measures between 4.7 and 5.3 mm and presents in both sexes marginal and submarginal black or dark brown stripes well marked in the head. It also presents bright green wings with deep black markings along the commissure and a long irregular spot on the corium, near to the middle of the claval suture. Females present light brown band on the tip of wings. *N. virescens* measures between 4.2 and 5.1 mm, with the head, pronotum, and scutellum presenting usually green. In the male individuals, it is possible to observe in the forewings a distinct round spot that does not touch the claval suture and dark tip of wings. Females do not have spots. The precise identification of the species, however, is based on the morphology of the male genitalia (Nielson 1968; Ghauri 1971; Wilson and Claridge 1991). The three species are widely distributed in Asia, and *N. nigropictus* can be found in Australia as well (Ghauri 1971). Although variable with temperature, the biological cycle of these species is relatively short. *N. cincticeps* develops from egg to adult between 27 and 33 days; *N. nigropictus*, between 21 and 28 days; and *N. virescens* between 18 and 28 days, with *N. cincticeps*, for example, presenting three to four generations per year (Fukushi 1934; Abalos 1939; Kabir and Hossain 1986; Salim 2002). *Nephottetix* species overwinter as nymphs (Pathak and Khan 1994). These leafhoppers are able to transmit the phytoplasma persistently. They acquire the pathogen within 1 to 3 h of feeding on an infected plant. The latent period in the vector varies from 20 to 39 days, and the leafhoppers are able to inoculate the phytoplasma with less than 1 h of feeding on a healthy plant (Arocha-Rosete and Jones 2010).

3.8 Sorghum spp.

The genus *Sorghum* (Moench) belongs to the family Poaceae, encompassing numerous species and varieties used for pasture, grain production, and fodder, as well as for the extraction of alcohol and sugar. Sorghum originates from Africa and is broadly distributed in tropical and subtropical areas, since it is a plant tolerant to drought and high temperatures, adapted to acid and nutrient-poor soils, and able to respond to high intensity of radiation with high rate of growth. These features make sorghum one of the five top cereal crops in the world. The association between phytoplasmas and sorghum plants has been mainly investigated to determinate whether species of this genus may be alternative hosts for phytoplasmas having an important role in epidemiology as a reservoir for pathogens, allowing their survival during the intercrop period or even as source of inoculum, when the pathogen has efficient insect vector. Asymptomatic sorghum plants harboring phytoplasmas have been frequently found in field surveys. However, affected plants may exhibit various types of symptoms, which may occur singly, such as the production of small, white leaves, or set of symptoms, as in sorghum grassy shoot disease, which is characterized by numerous small and thin tillers and narrow leaves totally green or with a diverse degree of chlorophyll including albinism. In a field survey performed with various species sampled in different locations in Australia, phytoplasmas were detected in plants of *Sorghum stipoideum* (Ewart & Jean White) C. A. Gardner and C. E. Hubb. that exhibited symptoms of shoot proliferation (Schneider et al. 1999). The phytoplasma strain was closely related to the phytoplasma associated with the sugarcane white leaf (SCWL), a representative of the 16SrXI-B group. Subsequently, a further field survey was carried out in geographic areas of the northern Australia (Tran-Nguyen et al. 2000). Six species of sorghum were collected, which displayed predominant symptoms of grassy shoots and deformed side shoots. Phytoplasmas were detected only in association with grassy shoots present in plants of *S. stipoideum*. Molecular characterization discovered the presence of two strains, one of them identical to SGS phytoplasma (group 16SrXI-B) and the other, a distinct strain denominated sorghum bunchy shoot phytoplasma whose ribosomal sequence was reported as a reference phytoplasma for the 16SrXXIV group (Wei et al. 2007). Further investigations aiming to find grasses harboring phytoplasmas as potential reservoirs revealed that sorghum grassy shoot phytoplasma was the most frequently detected in five gramineous species grown in Australia, including the sorghum species *S. stipoideum* (Blanche et al. 2003).

Phytoplasmas were also found in asymptomatic plants of the species *S. halepense* growing spontaneously near a sugarcane crop in Cuba (Arocha-Rosete et al. 2005). The phytoplasma was identified as belonging to the 16SrXVI-A group, whose reference phytoplasma is associated with sugarcane yellow leaf syndrome. In fields located in Iran, a 16SrVI-A phytoplasma was molecularly characterized in association with naturally infected plants of *S. halepense*, showing symptoms of little leaf (Zibadoost and Rastgou 2016).

Fig. 3.7 Triticosecale plant infected by phytoplasma. Healthy spike on the left

3.9 Triticale [Triticosecale Wittm. Ex A. Camus (*Triticum* L. × *Secale* L.)]

Triticale is a polyploid species resulting from a cross between *Triticum* L. and rye, *Secale* L. Triticale is an economically important cereal grain crop, grown in many countries of Europe, North America, and Asia. The main triticale world producers are Poland, Germany, France, China, Belgium, and Russia where it is well suited to local soils. It is an economically important plant with valuable biochemical properties. Triticale is mainly a livestock feed. It is also used for making whole-grain breads and other food for human consumption. Triticale grains contain more proteins than wheat and rye (Lazauskas 1998).

Triticosecale Stunt (TrSt) The disease was first described in the Vilnius region of Lithuania in 2004 (Urbanavičienė et al. 2004, 2007). Diseased plants showing characteristic symptoms of phytoplasma infection (yellowing, stunting, sterile deformed spikes, twisted awns, shorten spikes) were collected from naturally infected plants. Based on symptoms in the plant host, the disease was designated as Triticosecale stunt (TrSt) (Fig. 3.7). The phytoplasma was classified on the basis of RFLP analysis of 16S rDNA amplified in PCR, according to the classification scheme established by Lee et al. (1998) and Marcone et al. (2000). Restriction fragment length polymorphism (RLFP) and nucleotide sequence analyses of a 1.2 kbp DNA fragment of the 16S rRNA gene of Triticosecale stunt (TrSt) phytoplasma revealed that it is a member of subgroup 16SrI-B (aster yellows). The vector of the TrSt disease has not been identified.

3.10 Wheat (*Triticum aestivum* L.)

Wheat, especially common wheat, is a cereal widely grown around the world and is used for human food as bread and for pasture. Recently, Rao et al. (2017b) identified '*Ca.* P. oryzae'-related phytoplasmas associated with streak, yellowing, and stunting disease (Fig. 3.8) in bread and durum wheat varieties in central India. No vector is known so far to transmit this disease in India.

Wheat Blue Dwarf (WBD) Wheat blue dwarf phytoplasma of winter wheat in northwestern China is transmitted by the leafhopper *Psammotettix striatus* (L.) (Hemiptera: Cicadellidae). This phytoplasma belongs to 16SrI-C phytoplasma subgroup. WBD-infected wheat plants exhibit severe stunting and dwarfing, with numerous short tillers developing at the base of the plants. Infected central leaves show chlorotic mottling with curling and fail to expand. Lower leaves and stems are dark blue-green and gradually became stiff and thickened. In severely infected wheat plants, spikes are both deformed and sterile, or they do not develop at all (Wu et al. 2010). Hollingsworth et al. (2008) described symptoms in wheat plants from chlorotic leaf blotches to complete plant collapse in North America.

The leafhopper *P. striatus* is one of the most economically important pests in the wheat crop in arid and semiarid regions of Western China due to the transmission of the WBD agent (Wu et al. 2010; Chen et al. 2014). They are small insects with length that can vary from 3.6 to 4.0 mm and present general brown coloration, with dark brown stains on the crown and pronotum, light wings, inner margins of dark brown cells, and yellow or ivory veins. This species is widely distributed in Europe and Asia (Nielson 1968). *P. striatus* can present up to four generations per year and overwinters in the egg stage that is laid in dead leaves or in the sheaths of weed or

Fig. 3.8 Symptoms of wheat streak, yellows, and stunting diseases (Courtesy of G.P. Rao)

wheat leaves. This insect vector transmits WBD persistently, and there is no evidence of transovarial transmission (Zhang et al. 1996; Peiwen et al. 2004). In transmission studies, it was observed that *P. striatus* presents a 7-day acquisition access period, a latent period of 15–17 days and an inoculation access period of 2–3 days (Peiwen et al. 2004).

3.11 Conclusions

Insect vectors of phytoplasmas are primarily phloem-feeding insects of the order Hemiptera. Nearly 100 species are recognized as phytoplasma vectors. These are distributed in eight families (Cicadellidae, Cixiidae, Delphacidae, Derbidae, Flatidae, Psyllidae, Pentatomidae, Tingidae), with Cicadellidae, mainly subfamily Deltocephalinae, being the taxa with the largest number of species registered as vectors (>70 species) (Lee et al. 2000; Weintraub and Beanland 2006; Wilson and Weintraub 2007; Weintraub and Wilson 2010). Insect vectors and alternative host plants represent natural phytoplasma reservoirs playing a key role in the epidemiology of the diseases associated with the presence of these pathogens. Weintraub and Beanland (2006) described the main features that make the Hemiptera an efficient group in the transmission of phytoplasma: (a) they are hemimetabolous insects, with nymphs and adults feeding in the same place and both capable of transmitting the phytoplasma; (b) they feed on specific parts of the plant in a nondestructive manner, efficiently inoculating phytoplasmas into vascular tissues without activating defensive plant responses; and (c) they have a persistent and propagative relationship with the phytoplasmas.

Although the importance of diseases associated with phytoplasma presence in several species of plants is recognized, currently, for many cereal crops such as oats, barley, sorghum, millet, and triticale, the insect vectors have not yet been identified. In particular 11 diseases associated with phytoplasmas (BaDef, MBS, MR, GE, OatP, OatSt, OatY, ROL, RYD, TrSt, WBD) have been reported infecting cereals cultivated in different regions of the world. All these diseases result in significant losses in grain production. On the other hand, some cereals are hosts of phytoplasmas associated with diseases of other *Poaceae* species. For example, sorghum is a host of phytoplasmas associated with diseases of sugarcane and napier grass. Also, weed species of the *Poaceae* are hosts of phytoplasmas associated with diseases of cultivated cereal species, such as Johnson grass which is a reservoir of the MR phytoplasma.

However, there is still a lack of knowledge about the insect vectors of some of these phytoplasmas, about their transmission, about the factors that affect their dissemination, as well about their variability and potential to adapt to different *Poaceae* species. Research about these and other aspects, like the diseases cycle, could provide important and useful results for their management.

References

Abalos R (1939) Rice leafhoppers. 1. *Nephotettix apicalis* Motch. 2. *Nephotettix bipunctatus Fabr*. Local names: hanip (Tagalog), sip-sip (Ilocano), po-or or po-ol (Pangasinan). *Philippine Journal of Agriculture* **10**, 205–209.

Acs Z, Jović J, Ember I, Cvrković T, Nagy Z, Talaber C, Gergely L, Toševski I, Kolber M(2011) First report of maize redness disease in Hungary. *Bulletin of Insectology* **64**(Supplement), S229–S230.

Alivizatos AS, Markham PG (1986) Acquisition and transmission of corn stunt spiroplasma by its leafhopper vector *Dalbulus maidis*. *Annals of Applied Biology* **108**, 535–544.

Arocha-Rosete Y, López M, Piñol B, Fernández M, Picornell B, Almeida R, Palenzuela I, Wilson MR, Jones P (2005) '*Candidatus* Phytoplasma graminis' and '*Candidatus* Phytoplasma caricae', two novel phytoplasmas associated with diseases of sugarcane, weeds and papaya in Cuba. *International Journal of Systematic and Evolutionary Microbiology* **55**, 2451–2463.

Arocha-Rosete Y, Jones P (2010) Phytoplasma diseases of gramineae. In: Phytoplasmas: genomes, plant hosts, and vectors. Eds Weintraub GP and Jones P. CABI Publishing, Wallingford, United Kingdom, 170-187 pp.

Asudi GO, van den Berg J, Midega CAO, Pickett JA, Khan ZR (2015) The significance of napier grass stunt phytoplasma and its transmission to cereals and sugarcane. *Journal of Phytopathology* **164**, 378–385.

Asudi GO, van den Berg J, Midega CAO, Schneider B, Seemüller E, Pickett JA, Khan ZR (2016) Detection, identification and significance of phytoplasmas in wild grasses in East Africa. *Plant Disease* **100**, 108–115.

Banttari EE, Moore MB (1960) Transmission of the aster yellows virus to barley. *Plant Disease Reporter* **44**, 154.

Bascopé-Quintanilla JB (1977) Agente causal de la llamada "raza mesa central" del achaparramiento del maiz. Dissertation. Escuela Nacional de Agricultura, Colégio de Postgraduados, Chapingo, México, 55 pp.

Bedendo IP, Davis RE, Dally EL (1997) Molecular evidence for the presence of maize bushy stunt phytoplasma in corn in Brazil. *Plant Disease* **81**, 957.

Bedendo IP, Davis RE, Dally EL (2000) Detection and identification of the maize bushy stunt phytoplasma in corn plants in Brazil using PCR and RFLP. *International Journal of Pest Management* **46**, 73–76.

Bekavac G, Purar B, Jocković D (2007) Corn reddening: the disease and breeding for resistance. *Journal of Plant Pathology* **89**, 397–404.

Bertin S, Picciau L, Acs Z, Alma A, Bosco D (2010) Molecular differentiation of four *Reptalus* species (Hemiptera: Cixiidae). *Bulletin of Entomological Research* **100**, 551–558.

Blanche KR, Tran-Nguyen L, Gibb KS (2003) Detection, identification and significance of phytoplasmas in grasses in northern Australia. *Plant Pathology* **52**, 505–512.

Calari A, Contaldo N, Ardizzi S, Bertaccini A (2010) Phytoplasma detection in corn with reddening in Italy. In: Current status and perspectives of phytoplasma disease research and management. Eds Bertaccini A, Laviña A, Torres E. Sitges, Spain, 5.

Carloni E, Carpane P, Paradell S, Laguna I, Pecci MG (2013) Presence of *Dalbulus maidis* (Hemiptera: Cicadellidae) and of *Spiroplasma kunkelii* in the temperate region of Argentina. *Journal of Economic Entomology* **106**, 1574–1581.

Chen W, Li Y, Wang Q, Wang N, Wu Y (2014) Comparative genome analysis of wheat blue dwarf phytoplasma, an obligate pathogen that causes wheat blue dwarf disease in China. *PloS One* **9**, e96436.

Chiykowski LN (1963) *Endria inimical* (SSay), a new leafhopper vector of a celery-infecting strain of aster yellows virus in barley and wheat. *Canadian Journal of Botany* **41**, 669–672.

Costa AS, Kitajima EW, Arruda SC (1971) Moléstias de vírus e de micoplasma no milho em São Paulo. *Revista da Sociedade Brasileira de Fitopatologia* **4**, 39–41.

Cvrković T, Jović J, Mitrović M, Krstić O, Toševski I (2014) Experimental and molecular evidence of *Reptalus panzeri* as a natural vector of "bois noir". *Plant Pathology* **63**, 42–53.
Duduk B, Bertaccini A (2006) Corn with symptoms of reddening: new host of "stolbur" phytoplasma. *Plant Disease* **90**, 1313–1319.
Emeljanov AF (1971) New genera of leafhoppers of the families Cixiidae and Issidae (Homoptera, Auchenorrhyncha) in the USSR. *Entomological Review* **50**, 350–354.
Fedotina VL (1977) Virus und mykoplasmaanlicheorganismen in zellen von hafer, der von der pseudorosettenkrankheit befallen ist. *Archiv fur Phytopathologie und Pflanzenshutz* **3**, 177–191.
Fukushi T (1934) Studies on the dwarf disease of rice plant. *Hokkaido University – Journal of the Faculty of Agriculture* **37**, 164.
Genov N, Mitrović J, Genov M, Duduk B (2014) First report of corn reddening caused by '*Candidatus* Phytoplasma solani' in Bulgaria. *Plant Disease* **98**, 991–991.
Ghauri MSK (1971) Revision of the genus *Nephotettix* Matsumura (Homoptera: Cicadelloidea: Euscelidae) based on the type material. *Bulletin of Entomological Research* **60**, 481–512.
Gonzales V, Gámez R (1974) Algunos factores que afectam la transmisión del virus del rayado fino del maíz por *Dalbulus maidis* DeLong & Wolcott. *Turrialba* **24**, 51–57.
Grigoryevna, VN (2015) Phytoplasmatic disease of maize under the conditions of Crimea. *Izvestiya Orenburgskogo Gosudarstvennogo Agrarnogo Universiteta* **5**, 14–17.
Haas ICR (2010) Potenciais hospedeiros alternativos para o fitoplasma, o espiroplasma, agentes do enfezamento do milho, e alterações bioquímicas em plantas infectadas por espiroplasma. Thesis. Escola Superior de Agricultura "Luiz de Queiroz", Universidade de São Paulo, Piracicaba.
Hibino H, Jonson GB, Santa Cruz FC (1987) Association of mycoplasma-like organisms with rice orange leaf in the Philippines. *Plant Disease* **71**, 792–794.
Hollingsworth CR, Atkinson LM, Samac DA, Larsen JE, Motteberg CD, Abrahamson MD, Glogoza P, MacRae IV (2008) Region and field level distributions of aster yellows phytoplasma in small grain crops. *Plant Disease* **92**, 623–630.
Holzinger WE, Kammerlander I, Nickel H (2003) Fulgoromorpha, Cicadomorpha, excluding Cicadellidae. The Auchenorrhyncha of Central Europe. Vol. 1. Brill, Boston, USA.
Hruska A, Gladstone SM, Obando R (1996) Epidemic roller coaster: maize stunt in Nicaragua. *American Entomologist* **42**, 248–252.
Jomantiene R, Davis RE, Alminaite A, Valiunas D, Jasinskaite R (2002) First report of oat as host of a phytoplasma belonging to group 16SrI, subgroup A. *Plant Disease* **86**, 443.
Jović J, Cvrković T, Mitrović M, Krnjajić S, Redinbaugh MG, Pratt RC, Gingery RE, Hogenhout SA,Toševski I (2007) Roles of "stolbur" phytoplasma and *Reptalus panzeri* (Cixiinae, Auchenorrhyncha) in the epidemiology of maize redness in Serbia. *European Journal of Plant Pathology* **118**, 85–89.
Jović J, Cvrković T, Mitrović M, Krnjajić S, Petrović A, Redinbaugh MG, Pratt RC, Hogenhout SA, Toševski I (2009) "Stolbur" phytoplasma transmission to maize by *Reptalus panzeri* and the disease cycle of maize redness in Serbia. *Phytopathology* **99**, 1053–1061.
Jović J, Cvrković T, Mitrović M, Krnjajić S, Krstić O, Redinbaugh MG, Pratt RC, Toševski I (2011) Hosts of "stolbur" phytoplasma in maize redness affected fields. *Bulletin of Insectology* **64**(Supplement), S155–S156.
Jung HY, Sawayanagi T, Wongkaew P, Kakizawa S, Nishigawa H, Wei W, Oshima K, Miyata S, Ugaki M, Hibi T, Namba S (2003) '*Candidatus* Phytoplasma oryzae', a novel phytoplasma taxon associated with rice yellow dwarf disease. *International Journal of Systematic and Evolutionary Microbiology* **53**, 1925–1929.
Kabir SMH, Hossain SKAKM (1986) Some aspects of biology of rice leafhopper, *Nephotettix virescens* (Distant). *Bangladesh Journal of Zoology* **14**, 25–32.
Kitajima EW, Costa AS (1972) Microscopia eletrônica de microrganismos tipo micoplasma nos tecidos de milho afetado pelo enfezamento e nos órgãos da cigarrinha vectora portadora. *Bragantia* **31**, 75–87.

Kosovac A, Radonjić S, Hrnčić S, Krstić O, Toševski I, Jović J (2016) Molecular tracing of the transmission routes of "bois noir" in Mediterranean vineyards of Montenegro and experimental evidence for the epidemiological role of *Vitex agnus-castus* (Lamiaceae) and associated *Hyalesthes obsoletus* (Cixiidae). *Plant Pathology* **65**, 285–298.

Kovačević M, Đurić Z, Jović J, Perković G, Lolić B, Hrnčić S, Toševski I, Delić D (2014) First report of "stolbur" phytoplasma associated with maize redness disease of maize in Bosnia and Herzegovina. *Plant Disease* **98**, 418–418.

Kumar S, Singh V, Lakhanpaul S (2010) First report of '*Candidatus* Phytoplasma asteris' associated with green ear disease of "bajra" (pearl millet) in India. *New Disease Reports* **22**, 27.

Lazauskas J (1998) AugalinkystėLietuvoje (Agriculture in Lithuania) 1895–1995, SPAB "Aušra", Kaunas, Lithuania, 388 pp.

Lee I-M, Gundersen-Rindal DE, Davis RE, Bartoszyk IM (1998) Revised classification scheme of phytoplasmas based on RFLP analyses of 16S rRNA and ribosomal protein gene sequences. *International Journal of Systematic Bacteriology* **48**, 1153–1169.

Lee I-M, Davis RE, Gundersen-Rindal DE (2000) Phytoplasmas: phytopathogenic mollicutes. *Annual Reviews in Microbiology* **54**, 221–255.

Lee I-M, Gundersen-Rindal DE, Davis RE, Bottner KD, Marcone C, Seemüller E (2004) '*Candidatus* Phytoplasma asteris', a novel phytoplasma taxon associated with aster yellows and related diseases. *International Journal of Systematic and Evolutionary Microbiology* **54**, 1037–1048.

Legrand AI, Power AG (1994) Inoculation and acquisition of maize bushy stunt mycoplasma by its leafhopper vector *Dalbulus maidis*. *Annals of Applied Biology* **125**, 115–122.

Li S, Hao W, Lu G, Huang J, Liu C, Zhou G (2015) Occurrence and identification of a new vector of rice orange leaf phytoplasma in South China. *Plant Disease* **99,** 1483–1487.

Maixner M (2010) Phytoplasma epidemiological systems with multiple plant hosts. In: Phytoplasmas: genomes, plant hosts, and vectors. Eds Weintraub GP, Jones P, CABI Publishing, Wallingford, United Kingdom, 213–232 pp.

Marcone C, Lee I-M, Davis RE, Ragozzino A, Seemüller E (2000) Classification of aster yellows-group phytoplasmas based on combined analyses of rRNA and *tuf* gene sequences. *International Journal of Systematic and Evolutionary Microbiology* **50**, 1703–1713.

Marić A, Kosovac Z (1959) Proučavanje uzroka i štetnosti crvenila kukuruza u Vojvodini. *Savremena Poljoprivreda* **7**, 1028–1043.

Marín R (1987) Biologia y comportamiento de *Dalbulus maidis* (Homoptera-Cicadellidae). *Revista Peruana de Entomologia* **30**, 113–117.

Massola Júnior NS, Bedendo I, Amorim L, Lopes JRS (1999) Quantificação de danos causados pelo enfezamento vermelho e enfezamento pálido do milho em condições de campo. *Fitopatologia Brasileira* **24**, 136–142.

Mathur KC, Chaturvedi DP (1980) Biology of leaf and planthoppers, the vectors of rice. *Proceedings of the Indian National Science Academy* **B46**, 797–812.

Mori N, Mitrović J, Smiljković M, Duduk N, Paltrinieri S, Bertaccini A, Duduk B (2013) *Hyalesthes obsoletus* in Serbia and its role in the epidemiology of corn reddening. *Bulletin of Insectology* **66**, 245–250.

Moya-Raygoza G, Nault LR (1998) Transmission biology of maize bushy stunt phytoplasma by the corn leafhopper (Homoptera: Cicadellidae). *Annals of the Entomological Society of America* **91**, 668–676.

Nakashima K, Kato S, Iwanami S, Murata N (1993) DNA probes reveal relatedness of rice yellow dwarf mycoplasma-like organisms (MLOs) and distinguish them from other MLOs. *Applied and Environmental Microbiology* **59**, 1206–1212.

Nasu S (1967) Rice leafhoppers. Taxonomy, distribution, host range, life cycle and control of rice leafhoppers. In: Major insect pests of the rice plant. Symposium at the International Rice Research Institute. Johns Hopkins Press, Baltimore, USA, 493–523 pp.

Nault LR (1980) Maize bushy stunt and corn stunt: a comparison of disease symptoms, pathogen host ranges, and vectors. *Phytopathology* **70**, 659–662.

Nault LR (1990) Evolution of insect pest: maize and leafhopper, a case study. *Maydica* **35**, 165–175.

Nielson MW (1968) The leafhopper vectors of phytopathogenic viruses (Homoptera: Cicadellidae) taxonomy, biology, and virus transmission. United States of Department of Agriculture, Washington (Technical Bulletin, 1382), 386 pp.

Oliveira CM, Querino RB (2017) Taxonomic, biological and ecological aspects of corn leafhopper *Dalbulus maidis* (Hemiptera: Cicadellidae). In: Diseases in maize: insect vectors, mollicutes and virus. Eds Oliveira CM, Sabato EO. Embrapa, Brasília, Brazil, 139-151 pp.

Oliveira CM, Lopes JR, Nault LR (2013) Survival strategies of *Dalbulus maidis* during maize off-season in Brazil. *Entomologia Experimentalis et Applicata* **147**, 141–153.

Oliveira CM, Querino RB, Frizzas MR (2017) Leafhoppers and planthoppers in maize crop in Brazil. In: Diseases in maize: insect vectors, mollicutes and viruses. Eds Oliveira CM, Sabato EO. Embrapa, Brasília, Brazil, 71–95 pp.

Oliveira E, Waquil JM, Fernandes FT, Paiva E, Resende RO, Kitajima WE (1998) Enfezamento pálido e enfezamento vermelho na cultura do milho no Brasil Central. *Fitopatologia Brasileira* **23**, 45–47.

Oliveira E, Carvalho RV, Duarte AP, Andrade RA, Resende RO, Oliveira CM, Reco PC (2002a) Molicutes e vírus em milho na safrinha e na safra de verão. *Revista Brasileira de Milho e Sorgo* **1**, 38–46.

Oliveira E, Magalhães PC, Gomide RL, Vasconcelos CA, Souza IRP, Cruz I, Shafert R (2002b) Growth and nutrition of mollicute infected maize. *Plant Disease* **86**, 945–949.

Oliveira E, Santos JC, Magalhães PC, Cruz I (2007) Maize bushy stunt phytoplasma transmission by *Dalbulus maidis* is affect by spiroplasma acquisition and environmental conditions. *Bulletin of Insectology* **60**, 229–230.

Oliveira E, Landau EC, Sousa SM de (2015) Simultaneous transmission of phytoplasma and spiroplasma by Dalbulus maidis leafhopper and symptoms of infected maize. *Phytopathogenic Mollicutes* **5**(1-Supplement), S99–S100.

Olivier CY, Seguin-Swartz G, Galka B, Olfert OO (2011) Aster yellows in leafhoppers and field crops in Saskatchewan, Canada, 2001–2008. *The Americas Journal of Plant Science and Biotechnology* **5**, 88–94.

Oman PW (1948) Distribution of *Dalbulus maidis* (Delong & Wolcott). *Proceedings of the Entomological Society of Washington* **50**, 34.

Pandian V, Velazhahan R, Suresh S, Robin S, Rabindran R, Rao GP (2015) Subgroup level identification of rice orange leaf phytoplasma and its natural transmission through zigzag leafhopper (*Reciliadorsalis*) in India. *Phytopathogenic Mollicutes* **5**, 107–112.

Pathak MD, Khan ZR (1994) Insect pests of rice. International Rice Research Institute, Los Banos, Philippines, 89 pp.

Peiwen G, Yunfeng W, Keke W, Xingan H (2004) A study on the characteristics of wheat blue dwarf phytoplasma transmitted by leafhopper Psammotettix striatus. *Plant Protection* **4**, 24–28.

Penchich M, Elenich D, Radenovich C, Zarich L, Kerechki B (1972) The study of some physiological aspects of the phenomenon of "reddening" in corn plants. *Sel'skokhozyaistvennaya Biologia* **3**, 461–464.

Pérez-López E, Olivier CY, Luna-Rodríguez M, Rodríguez Y, Iglesias LG, Castro-Luna A, Adame-García J, Dumonceaux TJ (2016) Maize bushy stunt phytoplasma affects native corn at high elevations in Southeast Mexico. *European Journal of Plant Pathology* **145**, 963–971.

Pongprasert S (1974) Resistance to the zigzag leafhopper, *Recilia dorsalis* (Motschulsky), in rice varieties. Thesis. University of the Philippines, Los Baños, Philippines, 76 pp.

Quaglino F, Zhao Y, Casati P, Bulgari D, Bianco PA, Wei W, Davis RE (2013) '*Candidatus* Phytoplasma solani', a novel taxon associated with "stolbur" and "bois noir"-related diseases of plants. *International Journal of Systematic and Evolutionary Microbiology* **63**, 2879–2894.

Rao GP, Kumar M, Madhupriya, Singh AK (2017a) First report of '*Candidatus* Phytoplasma asteris' (16SrI-B subgroup) associated with a maize leaf redness disease in India. *Phytopathogenic Mollicutes* **7**, 52–56.

Rao GP, Prakasha TL, Madhupriya, Thorat V, Kumar M, Baranwal VK, Mishra AN, Yadav A (2017b) First report of association of '*Candidatus* Phytoplasma cynodontis' (16SrXIV-B subgroup) with streak, yellowing, and stunting disease in durum and bread wheat genotypes from central India. *Plant Disease* **101**, 1314.

Sabato EO (2017) Corn stunting diseases. In: Diseases in maize: insect vectors, mollicutes and viruses. Eds Oliveira CM, Sabato EO. Embrapa, Brasília, Brazil, 11-23 pp.

Salim M (2002) Biology of rice green leafhopper, *Nephotettix virescens* (Distant) under laboratory conditions. *Pakistan Journal of Agricultural Research* **17**, 49–54.

Schneider B, Padovan A, De La Rue S, Eichner R, Davis R, Bernuetz A, Gibb K (1999) Detection and differentiation of phytoplasmas in Australia: an update. *Australian Journal of Agricultural Research* **50**, 333–342.

Shu L, Weijia H, Guanghua L, Jilei H, Chuanhe L, Guohui Z (2015) Occurrence and identification of a new vector of rice orange leaf phytoplasma in South China. *Plant Disease* **99**, 1483–1487.

Šutić D, Tošić M, Starović M, Stanković R, Tomić T (2002) Redness of maize. *Plant Protection* **53**, 57–73.

Tran-Nguyen L, Blanche KR, Egan B, Gibb KS (2000) Diversity of phytoplasmas in northern Australian sugarcane and other grasses. *Plant Pathology* **49**, 666–679.

Triplehorn BW, Nault LR (1985) Phylogenetic classification of the genus *Dalbulus* (Homoptera: Cicadellidae), and notes on the phylogeny of the Macrostelini. *Annals of the Entomological Society of America* **78**, 291–315.

Tsai JH (1988) Bionomics of *Dalbulusmaidis* (DeLong & Wolcott): a vector of mollicutes and virus (Homoptera: Cicadellidae). In: Mycoplasma diseases of crops: basic and applied aspects. Eds Maramorosch K, Raychaudhuri SP. Springer-Verlag, New York. USA, 209–221 pp.

Urbanavičienė L, Jomantienė R, Valiūnas D, Davis RE (2004) Molecular detection of phytoplasmas in oats, barley, and triticosecale and their classification based on 16S rRNA gene polymorphisms. *ŽemėsŪkioMokslai* **3**, 15–19.

Urbanavičienė L, Jomantienė R, Davis RE (2005) First report of barley as host of a phytoplasma belonging to group 16SrI, subgroup B, and ribosomal protein subgroup rpI-B in Lithuania. *Plant Disease* **89**, 339.

Urbanavičienė L, Jomantienė R, Valiūnas D, Davis RE (2007) Molecular identification of 16SrI-A, 16SrI-B, 16SrI-C, and 16SrI-L subgroups of phytoplasmas in gramineous plants in Lithuania. *Bulletin of Insectology* **60**, 127–128.

Urbanavičienė L, Valiūnas D, Jomantienė R (2008) Molecular identification of agents causing yellows diseases in oats (*Avena sativa* L.). *Zemdirbyste-Agriculture* **95**, 286–292.

Valarmanthi P, Rabindran R, Velazhahan R, Suresh S, Robin S (2013) First report of rice orange leaf disease phytoplasma (16SrI) in rice (*Oryza sativa*) in India. *Australasian Plant Disease Notes* **8**, 141–143.

Valarmanthi P, Rethinasamy V, Seetharaman S, Sabariappan R, Ramalingam R, Rao GP (2015) Subgroup level identification of rice orange leaf phytoplasma and its natural transmission through zigzag leafhopper (*Recilia dorsalis*) in India. *Phytopathogenic Mollicutes* **5**, 107–112.

van Nieuwenhove GA, Frías EA, Virla EG (2015) Effects of temperature on the development, performance and fitness of the corn leafhopper *Dalbulus maidis* (DeLong) (Hemiptera: Cicadellidae): implications on its distribution under climate change. *Agricultural and Forest Entomology* **18**, 1–10.

Waquil JM, Viana PA, Cruz I, Santos JP (1999) Aspectos da biologia da cigarrinha-do-milho, *Dalbulus maidis* (DeLong & Wolcott) (Hemiptera: Cicadellidae). *Anais da Sociedade Entomológica do Brasil* **28**, 413–420.

Webb MD, Ramsay AJ, Lemaître VA (2013) Revealing the identity of some early described European Cixiidae (Hemiptera: Auchenorrhyncha) – a case of forensic taxonomy; two new combinations and a name change for *Reptalus panzeri* in Britain. *Acta Musei Moraviae, Scientiae Biologicae (Brno)* **98**, 57–95.

Webster RK, Gunnell PS (1992) Compendium of rice diseases. APS Press, Saint Paul, USA, 44 pp.

Wei W, Davis RE, Lee I-M, Zhao Y (2007) Computer-simulated RFLP analysis of 16S rRNA genes: identification of ten new phytoplasma group. *International Journal of Systematic and Evolutionary Microbiology* **57**, 1855–1867.

Weintraub PG, Beanland L (2006) Insect vectors of phytoplasmas. *Annual Review of Entomology* **51**, 91–111.

Weintraub PG, Wilson MR (2010) Control of phytoplasma diseases and vectors. In: Phytoplasmas: genomes, plant hosts and vectors. Eds Weintraub PG, Jones P. CABI, Wallingford, United Kingdom, 233-249 pp.

Wilson MR, Claridge MF (1991) Handbook for the identification of leafhoppers and planthoppers of rice. CAB International, Wallingford, United Kingdom.

Wilson MR, Weintraub PG (2007) An introduction to Auchenorrhyncha phytoplasma vectors. *Bulletin of Insectology* **60**, 177.

Wu Y, Hao X, Li Z, Gu P, An F, Xiang J, Wang H, Luo Z, Liu J, Xiang Y (2010) Identification of the phytoplasma associated with wheat blue dwarf disease in China. *Plant Disease* **94**, 977–985.

Zhang QF, Xiang JY, Yang Y, Zhang Q (1996) The primary infection of wheat mycoplasma-like organism blue dwarf disease (WMBD). *Acta Phytopathologica Sinica* **11**, 107–110.

Zibadoost S, Rastgou M (2016) Molecular identification of phytoplasmas associated with some weeds in West Azarbaijan province, Iran. *Acta Agriculturae Slovenica* **107**, 129–136.

Chapter 4
Phytoplasma Diseases of Industrial Crops

Govind Pratap Rao, Elizabeth Alvarez, and Amit Yadav

Abstract Phytoplasmas are associated with diseases in several hundred plant species, including many economically important industrial crops like sugarcane, sugar beet, cassava, and cotton. A number of phytoplasma diseases are associated with sugarcane. Originally restricted to Asian countries, they are spreading rapidly to newer locations with the help of infected seed material and leafhopper vectors. In cassava, the two phytoplasma diseases causing serious yield losses are cassava frog skin in Latin America and cassava witches' broom in Asia. Because of unreliable and nonspecific symptoms, the identification and characterization of the phytoplasmas associated with these and other industrial crops at an early stage of plant growth are challenging. Here the progress made in understanding biology, economic importance, symptomatology, diagnosis, epidemiology, and control of phytoplasmas infecting sugarcane, sugar beet, cassava, and cotton crops are summarized.

Keywords Sugarcane · Sugarbeet · Cassava · Cotton · Identification · Epidemiology · Genetic diversity

4.1 Introduction

Phytoplasmas infect various industrials crops where they cause serious economic losses. This chapter presents the historical background, geographical distribution, economic loss, characterization, genetic diversity, transmission, and management aspects of phytoplasma disease of sugarcane, sugar beet, cassava, and cotton crops in the world.

G. P. Rao (✉)
Division of Plant Pathology, Advance Center for Plant Virology, Indian Agricultural Research Institute, New Delhi 110012, India
e-mail: gprao_gor@rediffmail.com

E. Alvarez
Fitopatóloga, Cassava Program, International Center for Tropical Agriculture (CIAT)-Headquarters and Regional Office for Latin America and the Caribbean, Cali, Colombia

A. Yadav
National Centre for Microbial Resource, (Formerly, Microbial Culture Collection), National Centre for Cell Science, Pune, India

G. P. Rao et al. (eds.), *Phytoplasmas: Plant Pathogenic Bacteria - I*,
https://doi.org/10.1007/978-981-13-0119-3_4

4.2 Cassava or Manioca

Cassava (*Manihot esculenta* Crantz) is the most important energy and food source in tropical regions of the world. Native to South America, cassava was domesticated 5000 years ago and has been cultivated extensively since. Today, it is a staple food in the diet of 17.8 million people worldwide (FAOSTAT 2015) as well as being an industrial crop with high potential in the socio-economic development of the regions where it is produced. Cassava is primarily a vegetatively propagated crop and prone to several biotic and abiotic stresses among which the phytoplasma diseases are a potential threat to food security for millions of people (Jackson 2014) since they cause important yield losses wherever cassava is grown. Phytoplasma diseases are economically important in Asia, Latin America, and the Caribbean. The most important cassava diseases associated with phytoplasma infection are cassava frog skin disease (CFSD) reported in 1971 in Colombia (Pineda et al. 1983) and cassava witches' broom (CBW) discovered in 2010 in Southeast Asia (Alvarez et al. 2013). These pathogens are spread through cuttings and also spread by insect vectors. Losses caused by cassava phytoplasma diseases vary greatly depending on the region, local environmental conditions, and cassava variety. Existing reports have documented losses in commercial crop yield of up to 90% in production areas (Alvarez et al. 2009).

Cassava Frog Skin Disease The origin of cassava frog skin disease (CFSD) is most probably the Amazon region of Colombia, Peru, or Brazil. It was described for the first time in Colombia in 1971, in the cassava-growing region of Quilcacé, Cauca Department, and is considered to be one of the most problematic diseases for cassava cultivation that affects the production of roots (Pineda et al. 1983). CFSD has been reported in Colombia (Alvarez et al. 2009), Costa Rica (Pardo et al. 2014), Paraguay (Cardozo et al. 2016), Brazil (Santos de Oliveira et al. 2014), Panama, Peru, and Venezuela (Pineda et al. 1983; Chaparro-Martínez and Trujillo-Pinto 2001). CFSD affects root production and can cause severe economic loss. In Colombia and Costa Rica, yield losses of more than 90% have been reported. Different groups and subgroups of phytoplasmas associated with CFSD were reported. The CFSD-associated phytoplasmas were identified as group 16SrIII-L by restriction fragment length polymorphism (RFLP) (Fig. 4.1a) and sequence analyses of amplified rDNA products in Colombia (Alvarez et al. 2009). A 16SrIII-A phytoplasma (de Souza et al. 2014), and a phytoplasma affiliated to subgroup 16SrIII-L were also described (Santos de Oliveira et al. 2014), the latter subgroup was also reported in Paraguay and Costa Rica (Pardo et al. 2014) associated with symptoms similar to those reported by Alvarez et al. (2009). A quantitative PCR was developed, and a TaqMan probe was designed for the phytoplasma detection in field material based on the *rp* gene (16SrIII-L phytoplasma) (Fig. 4.1b) enabling the increase of detection sensitivity from 100- to 1000-fold than that obtained from PCR (Alvarez et al. 2010).

The disease causes deep lesions, decreased diameter, and increased woodiness in the roots. Symptoms consist of small longitudinal fissures distributed throughout the root. As roots increase in diameter, the fissures tend to heal, giving the injuries a

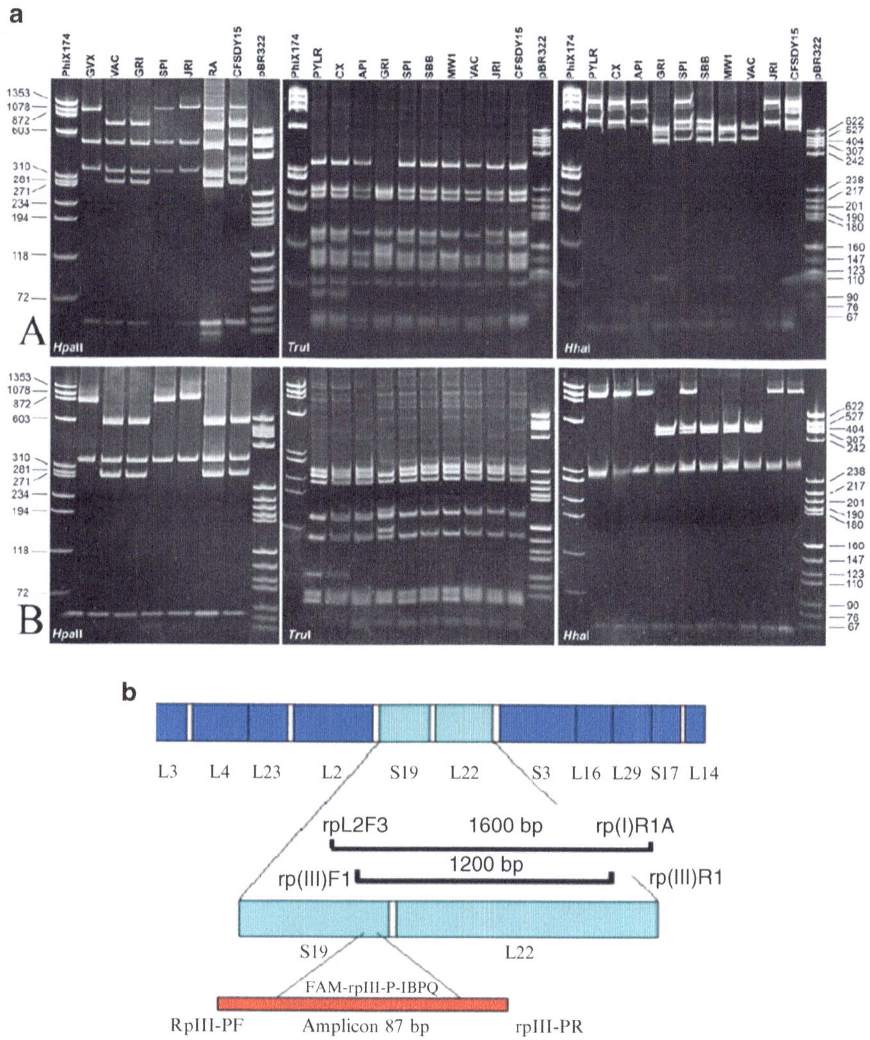

Fig. 4.1 (**a**) Polyacrylamide gels showing the restriction fragment length polymorphism profiles of 16S rDNA amplified by PCR with **A,** primers P1/P7, and **B,** in nested PCR with primers R16F2n/R16R2 from representative phytoplasma strains and cassava frog skin disease phytoplasma strain CFSDY15 (from Alvarez et al. 2009, with permission of APS). (**b**) *rp* gene region from which a TaqMan probe and primers for qPCR of CFSD phytoplasmas were designed

lip form. Symptoms in the roots are a woody aspect and a thickened peel that is cork-like, fragile, and opaque. The peel also presents liplike slits, creating a netlike or honeycomb pattern (Alvarez et al. 2009). The expression of CFSD symptoms is influenced by temperature and host genotype. Depending on the severity of symptoms, the depth and number of lesions increase until the root becomes deformed (Pineda et al. 1983; Alvarez et al. 2003). The symptoms on the aerial parts of the

Fig. 4.2 Characteristic symptoms of cassava frog skin disease in Colombia (**a**, **b**, **c**, and **d**); symptoms of CFSD in Costa Rica (**e**), in Paraguay (**f**), and in Brazil (**g**); healthy cassava roots (**h**)

plant are uncommon and mostly observed when the plants are harvested (Fig. 4.2). In Paraguay, symptoms including longitudinal liplike fissures on roots and a thick cork-like appearance on root peel and the diseased cassava roots were observed. The disease roots contained very low starch content (Cardozo et al. 2016). Disease symptoms observed in Paraguay were less severe than those observed in Costa Rica, where phytoplasma group 16SrIII-L was associated greater disease severity. The disease is exponentially propagated through stem cuttings. Since stems of diseased plants are thicker than those of healthy ones, diseased plants are selected for propagation, creating a demand for disease-free planting materials to prevent the dissemination of disease (Pardo et al. 2014). Most cassava varieties infected with CFSD express no leaf or stem symptoms. Molecular tests carried out on plants of cassava and periwinkle after dodder transmission trials confirmed the presence of phytoplasmas of group 16SrIII. Graft transmission could transfer phytoplasmas from infected to healthy cassava plants (CIAT 2005), and CFSD is transmitted by vectors of families of hemiptera, such as Cicadellidae and Delphacidae (Mejía et al. 2011). Insects in the *Scaphytopius* genus, in particular *S. fuliginosus* (Osborn), (Granada 1979), and *S. marginelineatus* (Stål) were shown to carry the phytoplasmas associated with CFSD (CIAT 2003, 2005). Recently, it has been demonstrated that *S. marginelineatus* is able to acquire the CFSD phytoplasma from infected plants and successfully transmit it to healthy plants (Alvarez and Betancourth 2016). Further

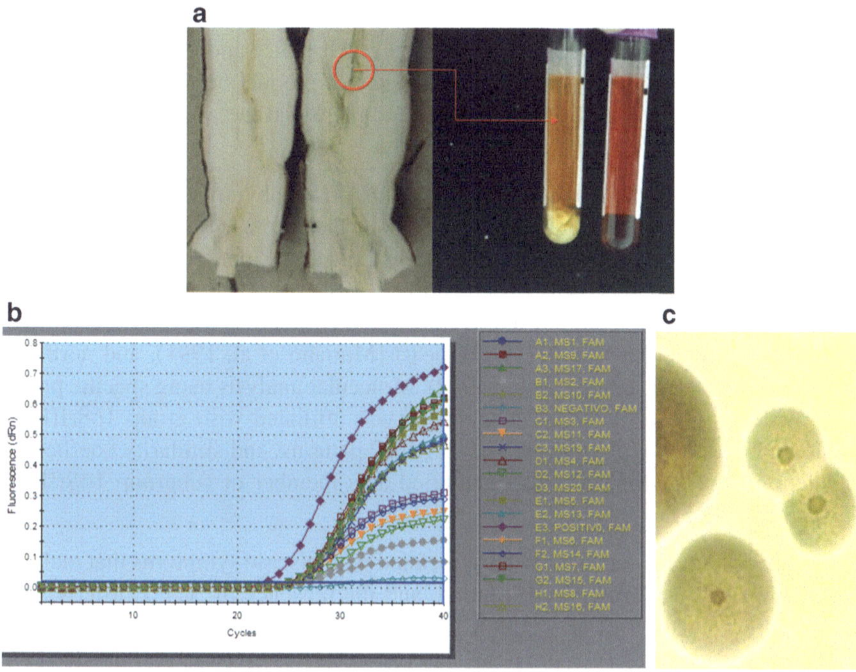

Fig. 4.3 On the left, symptoms in vascular tissue: the red circle indicates the vascular beams of phloem, and on the right the tube on the left shows symptomatic root fragments deposited in the liquid medium 8 days after sowing, and on the right, control tube (**a**). TaqMan quantitative PCR with 16SrIII-specific primers detecting phytoplasma DNA from colonies grown in solid medium (**b**). Phytoplasma colonies under the optical microscope grown in solid medium in an atmosphere of 95% N_2 and 5% CO_2 (magnification 40 X) (**c**)

studies are required, to determine the epidemiological significance of this insect as a vector of the disease in field transmission trials.

In Colombia, isolation of a phytoplasma associated with CFSD was accomplished using a method described by Contaldo et al. (2012). Fragments of roots, petioles, stems, leaves, and embryos from diseased cassava plants were used as a source for phytoplasma isolation in liquid medium, and colonies were observed in the solid medium; the presence of phytoplasmas was verified using qPCR (Fig. 4.3). Other molecular tests showed 450-bp bands with primers M1/M2, polymorphic patterns referable to those of group 16SrIII, and sequences with 99% homology with CFSD phytoplasma from both media (Betancourth et al. 2014). Pathogenicity was proven using stem injection of phytoplasma cells; five clones (CM 2952, Col 1, Col 896, Bra 184, SM 909-25) exhibited severe CFSD and typical root symptoms 6 months after inoculation (Alvarez et al. 2017).

The disease is managed mainly by using cuttings from healthy-looking plants that bear normal roots as propagation material. Improved quarantine procedures are required to prevent long-range disease spread. A heat treatment method has been developed and implemented in Colombia for the mass propagation of cassava planting material and subsequently applied in Brazil, Costa Rica, and Paraguay. Thermal chamber systems reportedly raised the efficiency of production to 90% in cassava compared with traditional farming systems and increased the availability of stalks of commercial genotypes for farmers.

Cassava Witches' Broom Disease Another important phytoplasma disease associated with cassava is the witches' broom (CWB) that is present in diverse tropical regions, mainly in Central and South America, including Cuba (Arocha et al. 2008) and Brazil (Lozano 1992). In Brazil, this disease was described in the 1940s in the southeast region (Silberschmidt and Campos 1944); later, it was observed in the central (Kitajima and Costa 1971), northeast (Mariano et al. 1991), and northern regions (Lozano et al. 1981; Lozano 1992). Molecular analysis using specific primers provided evidence that the phytoplasma was affiliated with group 16SrIII (X disease). Based on the phylogeny, virtual RFLP patterns, and similarity coefficient calculations, the phytoplasma was classified as a member of subgroup 16SrIII-B (Flôres et al. 2013).

In Cuba, cassava plants have been associated with disease symptoms that include leaf yellowing, small fruits, and stunting, and the presence of '*Ca.* P. asteris' has been reported (Arocha et al. 2009b). In Brazil, in areas of the northeast region, CWB was present in 85% of the fields and caused losses up to 70% (Fukuda et al. 1990). The disease also provokes yield losses of cuttings for planting, considering that affected plants display reduced size and excessive bud burst (EMBRAPA 2012). Recently, the disease was observed in fields located in the state of São Paulo (SP), southeastern Brazil. The symptoms included stunting, shoot proliferation, foliar chlorosis, and leaf malformation (Flôres et al. 2013). In Southeast Asia, CWB disease was observed in 2010 with incidence levels of 32% (Graziosi et al. 2016). Aster yellows phytoplasmas have been detected, and RFLP analyses of nested PCR-amplified fragments from Vietnamese and Cambodian CWB phytoplasmas indicated the presence of differentiable strains all related to the 16SrI group (Alvarez et al. 2013). The major symptoms are small yellow leaves, short internodes, sprout proliferation, stem vascular necrosis, and/or stunting (Fig. 4.4) in Vietnam. Stakes produce only a few dwarf and weak spindly sprouts that never reach a normal size, and, when the affected cassava is uprooted, the roots are thinner and smaller, with rough-textured skins and drastically reduced starch content (Alvarez et al. 2012). In 2010, the disease was observed in Quang Ngai, Dong Nai, and Yen Bai provinces of Vietnam, where more than 60,000 ha were affected, with crop losses as high as 80%, and reductions in yield and starch content reaching 30% (Alvarez et al. 2013; Hoat et al. 2015). In Thailand, cassava witches' broom disease was first reported in 2008. It spread rapidly and is now widely distributed across the region. Affected plants show bunches of shoots with short internodes, small yellowish leaves at the top of the plants, brown vascular tissues, and poor storage root development (Jackson 2014). Similarly, many cassava farms in Cambodia were affected in 2012, with losses of up to 50% (Alvarez et al. 2014). Currently, cassava witches' broom is spreading in Southeast Asia (Cambodia, China, Indonesia, Laos, the Philippines,

Fig. 4.4 Symptoms observed in cassava plants with witches' broom phytoplasma disease in Asia. (**a**, **b**, and **c**) Characteristic symptoms of CWB disease in Cambodia (**a**, **b**, **c**), phytoplasma disease in cassava in Thailand (**d**, **e**) and in Vietnam (**f**)

Thailand, and Vietnam) (Jackson 2014). In Wallis and Futuna Islands in the South Pacific, phytoplasmas found in cassava plants with symptoms associated with CWB disease have 100% identity with the '*Ca.* P. aurantifolia' (16SrII) (Davis et al. 2005). CWB has no known insect vector, but its spread is known to be mediated by the use of planting material obtained from infected stems. Key aspects of disease etiology, epidemiology, and control remain to be investigated (Graziosi et al. 2016).

During the crop growth, it is recommended to remove plants with phytoplasma symptoms as soon as they are seen and to destroy them by burning (Jackson 2014). Restricting the movement of cassava planting stakes, especially from infected areas, and restricting the movement of related species such as *Jatropha* spp. are also the best way to minimize disease incidence. Varietal resistance also exists but is not a significant management practice (Alvarez et al. 2012). Vector control could constitute an effective way of managing this disease and slowing its spread in fields. Also, heat therapy is successfully used to control phytoplasma diseases in cassava. Abiotic and biotic resistance inducers could carry ample potential for phytoplasma control in cassava. As a systemic pathogen, phytoplasmas can modulate plant hormones and down-regulate plant defenses, thus opening an avenue to external application of phytohormones to prime defense systems and induce resistance (Graziosi et al. 2016). Also, an accurate and rapid detection is a key component of disease management strategies, and a loop-mediated isothermal amplification (LAMP) assay has been developed to allow specific detection of CWB phytoplasmas from field-collected samples (Vu et al. 2016).

Cassava Phytoplasma Antholysis Jayasinghe et al. (1983) observed antholysis in some experimental clones of cassava in southwestern Colombia in 1981. This disease with no economic importance occurs in Brazil, the Caribbean, Central America, Colombia, and Venezuela (Frison and Feliu 1991). It was associated with the presence of phytoplasmas, and dissemination is reported mainly by vegetative propagation (Jayasinghe et al. 1984). The first observed symptoms were virescence followed by phyllody in the inflorescence.

Infected inflorescences commonly exhibit a very swollen gynophore and develop internodes in the floral receptacle, a phenomenon known as apostasis. Furthermore, elongation of the receptacle occurs above the insertion of the pistil, with the development of sprouts. Flower fertility is lost, resulting in nonfunctional flowers that abort prematurely. Transmission is 100% by stakes. Under greenhouse conditions, symptoms of antholysis appear within 1 month from planting. No vector transmission is known. The disease is reduced by selecting stakes from healthy plants. Varietal resistance also exists. Treatment with penicillin (500 to 1,000 ppm) did not reduce symptoms, whereas tetracycline reduced antholysis by 90%. This sensitivity and detection by Dienes' stain allowed to associate the disease to phytoplasma presence (Jayasinghe et al. 1983). Recently in Argentina disease symptoms associated with phytoplasmas were observed in a cassava field in the Misiones province. Typical symptoms of "superbrotamiento" are shown in Fig. 4.5. Stem cuttings from a diseased cassava plant generated new plants exhibiting the disease symptoms (H. Ceballos, 2017 unpublished data).

Cassava Phytoplasma Disease in Africa In Uganda, symptoms of leaf yellowing, chlorosis, shortening of internodes, and stunting were observed in cassava fields in Kawanda. The 16S rDNA sequence and restriction profile comparison obtained after RFLP assays of PCR amplicons identified the associated phytoplasma as a strain of '*Ca.* P. aurantifolia' (16SrII group) (Arocha et al. 2009a). In Côte d'Ivoire, cassava production is second to yams, and it is widely used as a typical food side dish called "attieké." Moreover, cassava-based flour or starch is widely used by private companies, and this has made cassava one of the most important Ivorian industrial crops. Cassava could become an alternative income source for women in the south littoral of the coconut-growing area of Grand-Lahou. Since the Côte d'Ivoire lethal yellowing phytoplasma (CILY, 16SrXXII-B, '*Ca.* P. palmicola'-related strains) devastated more than 400 hectares of coconut plantations in this area (Arocha Rosete et al. 2017). However, Kra et al. (2017) reported phytoplasmas of

Fig. 4.5 Cassava "superbrotamiento" disease with symptoms of multiple branching (**a**), healthy plants (left) and diseased plants (right) (**b**), healthy cassava stems (left) vs diseased cassava stems (right) (**c**)

16SrXXII-B subgroup affecting cassava in Côte d'Ivoire. Symptoms of leaf curling and yellowing were observed in cassava orchards located in the coconut-growing villages in Grand-Lahou, which are currently affected by CILY.

4.3 Cotton

A few reports are available on phytoplasma disease associated with cotton (*Gossypium hirsutum* L.) crops. Symptomatology is characterized by floral abnormalities, virescence, phyllody, and shoot proliferation. Disease symptoms also include little leaf, leaf yellowing, shortening of internodes, and stunting (Cousin et al. 1969; Kumar et al. 2010). The first report of the cotton phytoplasma disease (termed as "stenosis") was published by Uppal et al. (1944); in India and association with phytoplasmas (MLO at the time) was reported by Capoor et al. (1972). Recently, '*Ca.* P. asteris' was reported in cotton from Delhi, India (Kumar et al. 2010), while '*Ca.* P. aurantifolia'-related strains (16SrII-C, 16SrII-F) were detected in Mali and Burkina Faso (Martini et al. 2007; Marzachì et al. 2009).

4.4 Sugarbeet

Sugarbeet (*Beta vulgaris* L. subsp. *vulgaris* var. *altissima*) is a biennial, sugar-producing tuber crop grown in different parts of the world. It is an alternative crop to sugarcane and sweet sorghum for sugar production and contributes about 40% of total sugar production globally (Leilah et al. 2005). Sugarbeet cultivation in different parts of the world has been threatened from time to time by various pernicious plant pathogens. A phytoplasma disease of sugarbeets with symptoms of stunted growth with numerous small, narrow leaves (Fig. 4.6) and reduced tuber size was

Fig. 4.6 Little leaf symptoms in sugar beet associated with the presence of peanut witches' broom phytoplasma in India. (Courtesy of L. Rajendran)

observed during 2008 in Tamil Nadu, India (Rasu et al. 2011). Phytoplasma association of peanut witches' broom group (16SrII) was confirmed using PCR (Thilagavathi et al. 2011). In 1990 in Italy, a rosette disease of sugarbeet was associated with the presence of phytoplasmas by transmission electron microscopy (Canova et al. 1990). Later, Mumford et al. (2000) recorded unusual symptoms in sugar beet in 1999 in Hungary where symptoms included a pineapple-shaped crown, along with stunted, chlorotic, and necrotic leaves and petioles.

The disease incidence was recorded up to 60%, and the analyses indicate phylogenetic relatedness with aster yellows (16SrI) phytoplasmas. The observed symptoms were similar to the "low sugar disease," a condition recorded in France, associated with a phytoplasma of the "stolbur" group vectored by the leafhopper *Pentastiridius beieri* (Wagner) (Munchembled et al. 1999). Salehi and Izadpanah (2005) identified a strain of peanut witches' broom phytoplasma (16Srll) associated with witches' broom disease of sugarbeet from Chahgeer region in Abarque (Yazd Province of Iran) during 1998–2000. The associated agent was transmitted from sugar beet to sugarbeet, periwinkle, and eggplant and from periwinkle to sugarbeet via dodder and from periwinkle to periwinkle and from eggplant to eggplant, ornamental eggplant, and tomato by grafting. PCR using universal phytoplasma primer pairs consistently amplified segments of the expected size from the symptomatic sugar beet samples. Phylogenetic and putative restriction site analyses and similarity values showed that sugarbeet witches' broom phytoplasma was closest to members of peanut witches' broom phytoplasma group (16SrII).

The presence of a yellow wilt disease of sugar beet reported in Chile may cause 100% yield loss and seemed to have disappeared from 2001 to 2012 (IANSAGRO 2012). The evidence for the presence of phytoplasmas in sugarbeet was by Hepp and Sandoval (1999); later a phytoplasma belonging to ribosomal group 16SrIII was identified (Castro et al. 2000). Fiore et al. (2015) detected the sporadic occurrence of the disease and associated with it the presence of 16SrIII-J phytoplasmas. These phytoplasmas are present in different weed species and crops of agronomic interest in Chile and are transmitted by the leafhopper *Paratanus exitiosus* (Beamer) (Hepp and Vargas 2002; González et al. 2010, 2011; Longone et al. 2011; Fiore et al. 2012).

The sugarbeet disease called the "basses richesses" syndrome of sugar beet (SBR) was first described in 1991 in eastern France (Richard-Molard et al. 1995) and then repeatedly appeared in epidemic forms from 1991 to 1992 and from 1996 to 1998. It caused loss of root sugar content with dramatic economic consequences in 1992, with income loss nearly 50% over 1000 ha. SBR symptoms appear in late summer; affected plants showed new shoots with small, narrow, chlorotic leaves; and old leaves are yellow and necrotic. A brownish discoloration of vascular tissues, seen after cutting the tap root, is the most characteristic symptom of plants affected by SBR. Epidemiological studies in sugarbeet plots affected by SBR have shown that *P. beieri* can be infected by "stolbur" phytoplasma and may transmit it to sugar beet (Gatineau et al. 2001). The etiology of the disease remains unclear. Despite the association with a "stolbur" phytoplasma, several observations and studies indicated that this phytoplasma did not play a major etiological role in the disease. Preliminary microscopic observations of affected roots suggested that another phloem-limited organism (a bacterium-like organism: BLO) was involved. Further experiments confirmed that a BLO, related to '*Ca.* Phlomobacter fragariae' (agent of marginal chlorosis in strawberry), was present in symptomatic sugarbeet and could experimentally be associated with disease symptoms. Also, it was observed that *P. beieri* was an effective vector of this BLO. Further, Bressan et al. (2007) confirmed that the syndrome "basses richesses" of sugarbeet in France is associated with two phloem-restricted uncultured bacteria: a "stolbur" phytoplasma and γ- proteobacteria. The vector of proteobacteria is *P. leporinus* (Hemiptera, Cixiidae), formerly shown to transmit both the prokaryotes. The role of *P. leporinus* and two other planthopper species, *Cixius wagneri* China and *Hyalesthes obsoletus* Signoret, in spreading the two pathogens to sugarbeet were compared and quantified. Because of its abundance and high infection rates with proteobacterium, *P. leporinus* was confirmed to be the dominant vector of SBR disease. Symptoms associated with the two prokaryotes were similar, but "stolbur" was associated with a stronger reduction in taproot biomass and sugar content than proteobacteria. Other plant pathogenic phloem-restricted bacteria are proteobacteria. Marginal chlorosis of strawberry and syndrome "basses richesses" (SBR) of sugarbeet are associated with two related γ-3 proteobacteria in the Arsenophonus clade, i.e., '*C.* Phlomobacter fragariae' and SBR proteobacterium (SBRpr), transmitted, respectively, by *C. wagneri* and *P. leporinus* (Danet et al. 2002; Gatineau et al. 2002; Sémétey et al. 2007b). SBR can also be associated with a "stolbur" phytoplasma, which is also transmitted by *P. leporinus* and causes no differentiable symptoms in affected sugarbeets (Gatineau et al. 2001, 2002; Sémétey et al. 2007a). Experiments showed that *P. beieri* could transmit "stolbur" to periwinkle and sugarbeet.

4.5 Sugarcane

The phytoplasma diseases of sugarcane are more widespread than previously known and are of significant economic importance. Phytoplasmas infecting sugarcane (*Saccharum* spp. hybrids) are reported to be associated with several diseases including sugarcane grassy shoot (SCGS), sugarcane white leaf (SCWL), sugarcane green grassy shoot (SCGGS), sugarcane leaf yellows (SCLY), and Ramu stunt. These diseases cause more or less similar symptoms but differ in the identity of the associated phytoplasmas, vector relationship, and geographic distribution. SCWL, SCGS, and SCLY are the most important as causing significant economic losses to sugarcane yield and sugar recovery in Asian countries. These diseases have been spreading rapidly to newer locations by the use of infected propagation material and by leafhopper vectors. Both are associated with a specific phytoplasma that is a member of the rice yellow dwarf group (16SrXI) and appears to infect only sugarcane (Yadav et al. 2013); the SCWL and SCGS phytoplasmas could be differentiated by RFLP analysis of 16S ribosomal DNA using suitable restriction endonucleases. Sugarcane green grassy shoot is a recently discovered phytoplasma disease of sugarcane in Thailand. The SCLY disease has been reported in several Asian, African, American, and Australian countries and is associated with phytoplasmas in 16SrI, 16SrIII, 16SrXI, and 16SrXII groups. Ramu stunt disease of sugarcane (SCRS) is known to occur in Papua New Guinea. Moreover, mixed infections of SCGS+SCWL and SCGS+SCYL and/or SCYLV+SCLYP are reported from India and Thailand associated with serious yield decline in quality and quantity (Rao et al. 2012). Since the reported phytoplasma diseases are regularly emerging and re-emerging, hence it would be important to detect and manage them at an early stage of sugarcane growth to avoid further spread and significant losses.

Grassy Shoot SCGS was first reported in 1949 in India (Chona 1958), and then it has been recorded in most of the sugarcane-growing states of India and also in Thailand and Vietnam (Wongkaew et al. 1997; Sdoodee et al. 1999; Sdoodee 2001; Nasare et al. 2007; Viswanathan and Rao 2011; Rao et al. 2012; Hoat et al. 2012; Yadav et al. 2017). Symptoms similar to those of SCGS have been observed in Bangladesh, Malaysia, Nepal, Pakistan, Sri Lanka, and Sudan (Rishi and Chen 1989; Viswanathan 1997, 2001; Rao et al. 2003; Ariyarathna et al. 2007). SCGS disease is characterized by the production of a large number of thin, slender, adventitious tillers from the base of the affected stools. This profuse growth gives rise to a dense or crowded bunch of tillers bearing pale yellow or chlorotic leaves which remain thin and narrow, reduced in size, and have a soft texture (Fig. 4.7a). Each stalk that is produced from the affected stool shows shortened internodes and the development of side shoots from the bottom to the top (Fig. 4.7b). Affected plants do not produce millable canes. The disease is particularly pronounced in the ratoon crop where the clusters of slender tillers with reduced leaves, usually growing erect, give the appearance of perennial grass (Fig. 4.7c) (Rao et al. 2008).

Fig. 4.7 Symptoms of sugarcane grassy shoot disease: grassy shoot like chlorotic leaves of the affected clump of variety CoS 767 (**a**), grassy shoot symptoms with no tiller at 6 months of the affected clump (**b**), chlorotic leaves emerging in ratoon crops (**c**)

DAPI (4′-6-diamidino-2-phenylindole) stains are commonly used for quick and inexpensive phytoplasma detection using fluorescence and immunoflorescence microscopy (Viswanathan 2000). The phloem cells of the infected material showed a strong fluorescence, brighter than the typical of nuclei of parenchymal cells when stained using DAPI (Yadav et al. 2013). Universal phytoplasma-specific primer pairs were mostly used in nested PCR assays that successfully detect the SCGS phytoplasma in sugarcane and its reported leafhopper vectors (Viswanathan et al. 2005; Srivastava et al. 2006; Rao et al. 2014; Tiwari et al. 2017). Multilocus genes such as *secA*, *secY*, *poC*, *gyrA*, *gyrB*, and *dnaB* were also utilized for characterization of SCGS phytoplasmas (Nasare et al. 2007; Manimekalai et al. 2015, 2016; Rao et al. 2014; Kumar et al. 2017). Nucleotide sequence analysis of the 16S rRNA genes revealed that the SCGS phytoplasma affecting sugarcane crops in India is very closely related to RYD phytoplasma group (16SrXI group). Although there were significant variations in symptomatology and in the genetics of the detected phytoplasmas, no correlation could be established between symptoms and phytoplasma strain (Nasare et al. 2007; Viswanathan et al. 2011; Yadav et al. 2017). So far three phytoplasma subgroups 16SrXI-B, 16SrXI-D, and 16SrXI-F were found associated with SCGS disease (Rao et al. 2014, 2017; Zhang et al. 2016; Yadav et al. 2017) in India and China (Fig. 4.8). Nasare et al. (2007) concluded that the 16S rRNA gene and 16S–23S rRNA spacer region sequence identity among the SCGS-associated phytoplasma strains in India are more than 99%, and these results are confirmed by other studies (Viswanathan et al. 2011). Srivastava et al. (2006) demonstrated that the leafhopper *Deltocephalus vulgaris* was the vector of SCGS; later Rao et al. (2014) demonstrated *Exitianus indicus* as a putative vector for SCGS phytoplasma. Recently, Tiwari et al. (2017) reported two additional leafhopper vectors *Maiestas portica* (Melichar) and *Cofana unimaculata*.

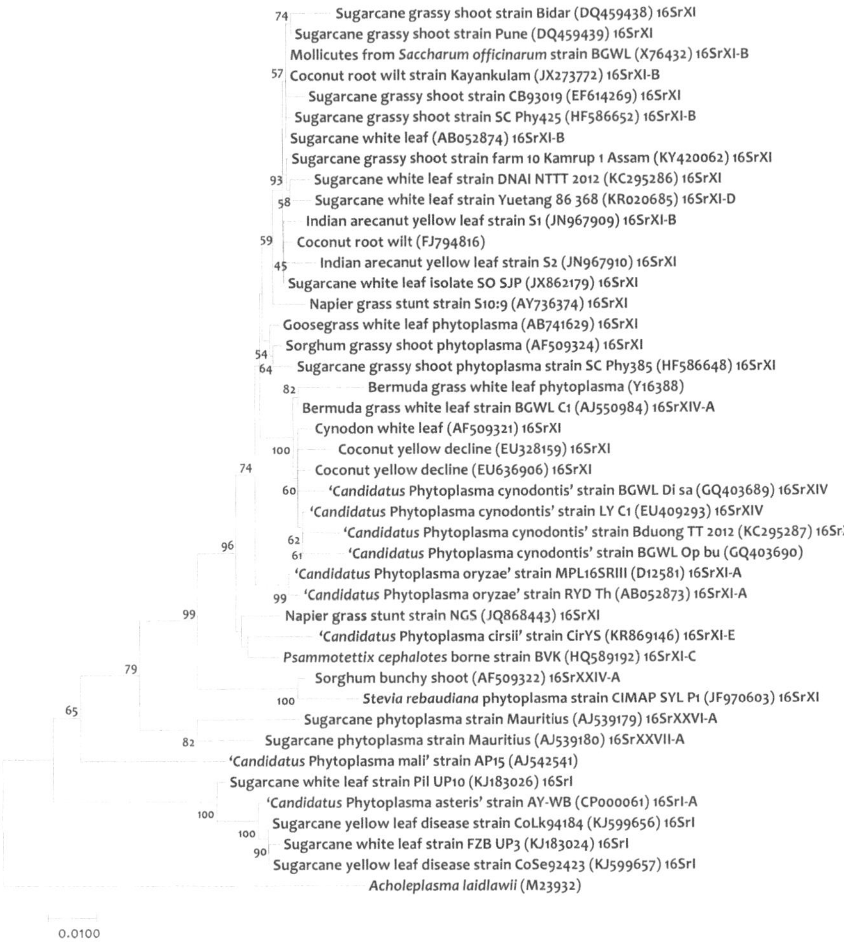

Fig. 4.8 Neighbor-joining tree showing the evolutionary relationship of representative phytoplasma strains of 16SrXI and 16SrXIV groups from different sugarcane cultivation regions

Deterioration in yields of many promising varieties of sugarcane by SCGS has been reported (Dhumal 2001; Viswanathan and Rao 2011; Tiwari et al. 2012; Gogoi et al. 2017). If diseased setts are used for planting, the germination percentage is reduced of 30–60% (Madan et al. 1981; Dhumal 1983); Bachchav et al. (1979) have reported 40–90% loss in sugarcane yield due to SCGS. The affected canes have very poor milling quality; the juice shows a reduction in brix, pol, CCS percentage, and purity but increases in the invert sugar (Dhumal and Nimbalkar 1983; Usmani and Rao 1991; Rao et al. 2000; Gogoi et al. 2017). Dhumal and Nimbalkar (1982) reported that the SCGS-infected canes had greater quantities of phosphorus, potassium, iron, copper, and zinc; on the contrary the contents of nitrogen, magnesium, manganese, and silica were lower. The reduced nitrogen content adversely affects growth parameters, photosynthesis efficiency, and carbohydrate and chlorophyll contents, leading to altered morphology and physiology of diseases cane plants. The high content of phosphorus may be responsible for stunted growth, chlorotic leaves, premature and profuse tillering, and stimulation of invertase activity, which adversely affects the juice quality. The higher content of potassium, copper, and zinc induces higher auxin concentration, resulting in premature sprouting of buds and profuse tillering, and influences the accumulation of reducing sugars by stimulating the invertase activity that deteriorates the juice quality. The reduction in silica content in diseased cane leaves may be responsible for soft and papery leaves and reduction in disease resistance, making the host more prone to other pathogens (Dhumal 2001). The infected leaves showed reduction in the contents of magnesium and manganese resulting into loss of chlorophylls (Usmani and Rao 1991), reduced photosynthetic efficiency (Rao et al. 1992), and inhibition of sucrose-synthesizing enzymes like sucrose synthetase and carboxylating enzymes like PEPCase and RuBPcase (Dhumal and Nimbalkar 1983; Dhumal and Hedge 1984). The lower activity of sucrose synthase and sucrose P synthase may be attributed to the lower contents of magnesium and manganese which are acting as cofactors in these reactions (Madan et al. 1981, Dhumal and Nimbalkar 1983). Studies on photosynthetic enzymes in diseased plants indicate that the phytoplasma has adverse effects on carboxylating enzymes like PEPCase, RuBPcase, and pyruvate Pi kinase. A similar trend was reported in the activity of NADP malic (decarboxylating) enzyme, while the activities of NADP-malate dehydrogenase, aspirate amino transferase, and alanine amino transferase were highly stimulated as compared to healthy plants. Leaf photorespiration was also reported higher in SCGS-affected leaves (Dhumal and Nimbalkar 1983a, b).

White Leaf Sugarcane white leaf (SCWL) is one of the most destructive sugarcane diseases in Sri Lanka and Thailand. It was first described in 1954 in the Lampang province in the northern part of Thailand (Mangelsdorf 1962), and 4 years later it was also discovered in Taiwan (Ling 1962). In Thailand, the disease subsequently spread to all important sugarcane-growing areas in the north, northeast, and east, resulting in one of the most lethal diseases of sugarcane. Currently, it seems present in all areas of Asian Countries (Thailand, Sri Lanka, Bangladesh, Vietnam, China, Nepal) where the crop is grown (Rishi and Chen 1989; Sarindu and Clark 1993; Nakashima et al. 1994, 1996; Wongkaew et al. 1997; Hoat et al. 2013). SCWL was also recorded in 1986 in Japan, in the Tanegashima island, but later it disappeared (Nakashima and Murata 1993; Nakashima et al. 2001). The most characteristic symptoms of SCWL are the presence of leaves with total chlorosis, proliferating tillers, and pronounced stunting. The leaves are narrower, chlorotic, and smaller than those of healthy plants, with a soft texture (Fig. 4.9). Severely diseased plants fail to produce millable canes. SCWL phytoplasma strains are very closely related to the SCGS agents and phylogenetically related to other phytoplasmas in grasses such as RYD and SGS (Seemüller et al. 1994) but could be assigned to a different subgroup by RFLP analyses of 16S rDNA (Lee et al. 1998). SCWL is transmitted by the leafhopper *Matsumuratettix hiroglyphicus* Matsumura (Matsumoto et al. 1968). The minimum acquisition and inoculation feeding periods are 3 h and

Fig. 4.9 Symptoms of sugarcane white leaf disease: field view of SCWL disease in ratoon crops of variety in Sri Lanka (**a**); sugarcane clump showing grassy shoot and white leaf symptoms (**b**); total chlorotic leaves of a sugarcane variety (**c**) (Courtesy of S. Thushari)

30 min, respectively (Chen 1978). The incubation period of SCWL phytoplasma in the insect vector is 25–35 days while on the host plant is 70–90 days (Matsumoto et al. 1968). Lee and Chen (1972) reported the optimum temperature for vector transmission is 25 °C. According to the studies of Chen (1978), female adults seem to be more efficient than the males in the transmission of this disease. *M. hiroglyphicus* is widely distributed in central and southern parts of Taiwan, Sri Lanka, and Thailand. The vector population is particularly abundant from July through October, and then it declines rapidly and remains low until April; six overlapping generations may occur in a year (Yang and Pan 1979). Disease incidence is correlated with the

population trend of the vector in the field. The females of *M. hiroglyphicus* usually lay their eggs in the soil to a depth of about 0.5 cm, but sometimes eggs are laid in the leaf sheath near the ground. Sandy soils are preferred for oviposition, and this may be one of the reasons why the disease is often more severe in sandy soils. Sugarcane and *Saccharum spontaneum* are the preferred known plant hosts of SCWL. Up to 100% incidence of SCWL has been recorded resulting in complete yield loss. Serious recent epidemics of SCWL have been recorded from Udon Thoni in 2000 and Burirum Districts in 2002 (Kusalwong et al. 2002). With PCR assays, phytoplasma DNAs were detected in SCWL diseased plants collected from Thailand (Nakashima et al. 1996; Wongkaew et al. 1997; Sdoodee et al. 1999; Hanboonsong et al. 2002, 2006) and in insect vectors *M. hiroglyphicus* and *Yamatotettix flavovittatus* (Nakashima et al. 1994; Hanboonsong et al. 2006). The transmission efficiency of *M. hiroglyphicus* (55%) was higher than that of *Y. flavovittatus* (45%). These two species peak at different times of the year and therefore complement each other in the transmission of SCWL disease; therefore their management requires the control of both insects (Hanboonsong et al. 2006; Kaewmanee and Hanboonsong 2011). Eight other leafhopper species *Balclutha rubrostriata* (Melichar), *Bhatia olivacea* (Melichar), *Exitianus indicus* Distant, *Macrosteles striifrons* Anufriew, *Recilia distincta* (Motschulsky), *Recilia dorsalis* (Motschulsky), *Thaia oryzivora* Ghauri, and *Xestocephalus* sp. were reported as putative vectors of SCWL in northeastern Thailand (Hanboonsong et al. 2006). The disease is spread through stem cuttings from healthy-looking and latently infected plants (Cronjé et al. 1998; Tran-Nguyen et al. 2000). Antibodies against SCWL phytoplasma had been generated (Sarindu and Clark 1993), but universal antigenic targets for the different phytoplasma strains need to be developed to avoid false negatives. Cultural practices, disease awareness, and farmer understanding remain as the major factors for successful planting of sugarcane with minimum losses from phytoplasma.

Weed grasses have been suggested to be a reservoir of SCWL but no molecular evidence is available to prove this hypothesis. The fact that SCWL occurs mainly in Asia and not in other sugarcane-growing countries in the world strongly suggests that quarantine barriers should be reinforced to prevent its spread and restrict its movement to other areas. Manimekalai et al. (2010) confirmed that the phytoplasma associated with arecanut palm (*Areca catechu*) has 99% identity with sugarcane white leaf and coconut root wilt disease phytoplasmas (16SrXI) and 98% identity with Bermudagrass white leaf phytoplasma (16SrXIV). The phylogenetic analysis confirmed the clustering of the yellow leaf disease phytoplasma of arecanut palms with 16SXI and 16SXIV groups. This indicates very close relationships of arecanut palm phytoplasma with SCWL and other related Bermudagrass phytoplasma and suggests that they could be a threat for the possibility of transfer and harbor this phytoplasma as alternative and collateral hosts. Use of resistant clones to control the SCWL disease is limited due to the lack of varieties combining high yield with disease resistance. Planting disease-free cuttings, rouging of diseased plants, and the prohibition of ratooning in infected fields are, therefore, recommended to control the disease. In Thailand, the disease is under control in infected areas by the routine use of healthy planting material, hot water treatment of cuttings for 2 h at 50°C, micro-propagation of disease-free plantlets, strict quarantine regulations, and various soil amendments (Chen and Kusalwong 2000; Kaewmanee and Hanboonsong 2011).

Yellow Leaf Sugarcane yellow leaf syndrome (SCYLS), characterized by a yellowing of the midrib and lamina, was first reported in the 1960s from East Africa (Rogers 1969) and later from Hawaii (Schenck 2001), South Africa (Cronjé et al. 1998), and Cuba (Peralta et al. 1999). It is now widely distributed in most sugarcane-growing countries from all continents. Losses from 30% to over 60% of susceptible varieties have been reported (Comstock et al. 1994, 1998; Arocha 2000). Symptoms of yellow leaf have been attributed to many causes, both biotic and abiotic, but the biotic agents are luteovirus or phytoplasmas in Hawaii, Brazil, Australia, South Africa, Cuba, the USA, and Mauritius (Vega et al. 1997; Cronjé et al. 1998; Matsuoka and Meneghin 2000; Arocha et al. 1999; Scagliusi and Lockhart 2000; Aljanabi et al. 2001; Rott et al. 2008). Phytoplasmas have been consistently associated with SCYLS, but latent infections also occur (Bailey et al. 1996; Cronjé et al. 1998; Arocha 2000; Aljanabi et al. 2001). Phytoplasma infection was reported associated with YLS of sugarcane in Africa, and a phytoplasma member of the X-disease group (16SrIII) was detected (Cronjé et al. 1998). It was also reported in Cuba, India, and Australia (Viswanathan 1997; Arocha et al. 1999; Tran-Nguyen et al. 2000). Parmessur et al. (2002) reported *Saccharosydne saccharivora* as the most abundant leafhopper species found in Cuban sugarcane plantations and responsible for SCYL transmission. In some cases, mixed infection of both viruses and phytoplasmas has been observed (Gaur et al. 2008; Rao et al. 2017). The major symptoms consist of yellowing of leaves with a bright yellow midrib, often when the rest of the lamina is still green (Fig. 4.10). Guerra and Cano (2005) detected YLS phytoplasma using DAPI staining. Later studies employing sequence analysis of 16S/23S rDNA spacer region and RFLP analysis of PCR-amplified 16S rDNA sequences revealed that two different phytoplasmas are associated with SCYL in nine African countries, although the plants were symptomatically similar (Cronjé et al. 1998, 1999; Aljanabi et al. 2001). The prevalent agent is a member of the X-disease group which showed a sequence identity of 98.8% with the X-disease phytoplasma. Detection and molecular characterization of AY phytoplasmas (subgroup 16SrI-A) from SCYL-diseased sugarcane plants from Cuba were also confirmed (Arocha et al. 1999). In Australia, a great genetic diversity among SCYL phytoplasmas was determined by RFLP and sequence analyses of PCR-amplified 16S rDNA (Tran-Nguyen et al. 2000). Yellow leaf syndrome (YLS) of sugarcane is a widely distributed disease syndrome in many sugarcane producing countries of the world and causes significant losses in yield and quality (Lockhart and Cronje 2000; Viswanathan et al. 2011). Losses of over 60% are reported in highly susceptible varieties (Arocha 2000). A strain of SCYL belonging to 16SrXII group ("stolbur") was shown to be associated with sugarcane leaf yellows in India (Gaur et al. 2008; Viswanathan et al. 2008). Recently, a 16SrI-B subgroup phytoplasma has been confirmed associated with sugarcane leaf yellows from two commercial sugarcane varieties from Lucknow, Uttar Pradesh, India (Kumar et al. 2015). The major symptoms associated

Fig. 4.10 Symptoms associated with sugarcane leaf yellows phytoplasmas

with the disease were midrib yellowing and irregular yellow patches on leaf lamina. These findings indicated a great phenotype and genetic diversity of phytoplasmas associated with leaf yellows disease. Peralta et al. (1999) observed histopathological alterations in SCYLS sugarcane leaves such as chloroplast disorganization, starch accumulation, and increasing number of mitochondria; biochemical alterations like a decrease in amylase activity, alterations of juice quality, and an increase in invertase activity were also reported. Higher levels of sucrose (Peralta et al. 1999; Arocha 2000) have also been found, which may be influenced by a proportional increase in some non-sugar carbohydrates. SCYLS incidence in different commercial cultivars in India was also reported as being responsible for the reduction in sugarcane production and sugar recovery in India (Rao et al. 2000; Viswanathan 2002, 2004; Gaur et al. 2008). Fontaniella et al. (2003) observed that SCYLS infection alters the contents and composition of polysaccharides, phenols, and polyamines in the juice of infected plants (cv. Cuba 120-78) in Cuba. The disease was associated with an increase in the concentration of reducing sugars, glucose index, and glycoproteins recovered in juice, whereas the amount of sucrose decreases. Sugarcane juice obtained from both healthy and SCYLS-affected Cuba 120-78 cultivars of sugarcane contained putrescine (PUT), cadaverine (CAD), spermidine, and spermine (SPM) as free and macromolecule conjugated compounds. Only CAD and SPM appeared as acid-soluble conjugates to small molecules, whereas PUT and CAD are the major polyamines (PAs) conjugated to macromolecules, mainly to high molecular mass glycoproteins. The disease was associated with an increase in total PA fraction. Arginase and ornithine decarboxylase activities, responsible for the synthesis of PUT, were higher in SCYLS juice than in those obtained from healthy plants. CAD and SPM presumably conjugated mostly to chlorogenic, syringic, and ferulic acids in juice from SCYLS plants. Many methods have been developed for the generic and specific detection of SCYL phytoplasmas, either based on nested PCR or NAH assays (Arocha et al. 2004a, b; Rott et al. 2008). Nucleic acid hybridization assay has been established for the generic detection of phytoplasmas (Harrison et al. 1994; Kirkpatrick and Smart 1995; Arocha et al. 2004a). Arocha et al. (2005) confirmed the vector status of the delphacid planthopper, *Saccharosydne saccharivora*, associated with SCYLS phytoplasma in Cuba. A new strain of SCYLS agent belonging to the X-disease group was shown to be present in Mauritius. This group previously described in sugarcane from South Africa was detected in both sugarcane *Sorghum verticilliflorum* and the planthopper *Perkinsiella saccharicida* in Mauritius. The presence of a closely related phytoplasma in the planthopper *P. saccharicida* indicates the possible involvement of the delphacid in the transmission of SCYLS phytoplasma (Joomun et al. 2007).

Green Grassy Shoot Disease Sugarcane green grassy shoot (SCGGS) is a recently discovered phytoplasma disease in Thailand (Pliansinchai and Prammanee 2000). The symptoms are very similar to those of SCGS disease; however, in SCGGS-affected sugarcane plants, the leaves do not become chlorotic. The result from PCR detection showed that SCGGS agent is genetically related to a phytoplasma infecting periwinkle and the SCWL phytoplasma (Pliansinchai and Prammanee 2000). Further study on the DNA sequence is required to characterize the phytoplasma associated with SCGGS disease. The disease could be transmitted through the canes with transmission percentages up to 15–100% (Pliansinchai and Suchato 1995). The highest percentage of infection was obtained when the basal stalk of the affected canes was planted (Pliansinchai et al. 1998). Sett transmission plays a major role in the disease spread, and insect vectors involved in transmission have not yet been identified.

Ramu Stunt Ramu stunt disease (RStD) of sugarcane was first observed in the late 1980s in the Ramu Valley of Papua New Guinea (PNG) causing severe crop losses in commercial sugarcane varieties (Eastwood 1990). The cultivar Ragnar proved to be highly susceptible. Since that time, the replacement of susceptible cultivars with resistant ones has kept the disease under control. The RStD disease is restricted to Papua New Guinea (Braithwaite 2010). The most common symptom of the disease is pronounced stunting. Leaves show a yellow mottled striping with short, erect, and stiff texture. In some cultivars, excessive tillering and grassy shoot appearance are also present. Affected plants die within 1 year after the appearance of the first symptoms. Kunita et al. (1994) and Cronjé et al. (1999) reported experimental transmission of RStD agent by the leafhopper *Eumetopina flavipes* Muir. A SCWL-related phytoplasma was identified by sequence analysis of 16S/23S rDNA spacer region and RFLP analysis of PCR-amplified 16S rDNA sequences that showed a sequence identity of 95.98% with the SCWL agent. It caused up to 40% loss in total sugarcane production (Suma and Jones 2000). This disease is a major quarantine threat particularly to the neighboring sugarcane industries in Australia and Indonesia.

Ramu stunt is a very severe, rapidly spreading, systemic disease with a range of symptoms. The rapid spread is due to the airborne insect vector, *E. flavipes* (Kunita et al. 1994). The most striking effect of Ramu stunt is its ability to suddenly and rapidly reduce growth, seen as a shortening of internodes and stunting. Diseased canes are thinner than healthy canes. Stools are severely stunted, and there is progressive death of stalks. Diseased stools ratoon poorly. In the field, infection in a susceptible variety can lead to total ratoon failure. Root systems are severely reduced and stunted. Older roots collapse and become necrotic (Braithwaite 2010). Ramu stunt is transmitted mainly through infected cuttings, and it has been suggested that its transmission also involve *E. flavipes*. The main commercial control used in PNG crops is the planting of resistant varieties and the destruction of infected crops. Control of the leafhopper may be another effective control strategy. In PNG, due to the widespread distribution and persistence of *E. flavipes* across multiple wild and cultivated hosts, management effort should focus on the planting of new, resistant varieties and vigilant surveillance for new outbreaks of the disease (Magarey et al. 2002; Magarey 2008).

For SCGS and SGWL diseases, treatment of cuttings with moist hot air at 54°C for 4 h and hot water treatment at 54°C for 2 h are recommended, respectively (Friso and Putter 1993). Leaf yellows can successfully be eliminated by tissue culture technique. Parmessur et al. (2002) reported the elimination of sugarcane yellows phytoplasma by regenerating plantlets from callus derived from young leaf rolls. Since the alternative/reservoir plants harboring the SCGS phytoplasma are not confirmed yet, understanding the host range of leafhoppers as well as other potential insect vector insects is also desirable for planning sustainable management strategies for SCGS disease. In Australia, phytoplasmas related to SCWL disease were observed in weeds growing near sugarcane fields (Blanche et al. 2003).

4.6 Conclusion and Perspectives

The phytoplasma diseases of industrial crops are widespread and of significant economic importance. SCWL and SCGS diseases seem to occur only in the Southeast Asian regions, and their agents have never been identified in plants other than sugarcane; they also seem to have strict insect vector specificity. In contrast, SCYL disease occurs in all the continents and is associated with a number of different phytoplasmas. Although some of these diseases have attracted significant research attention, many of their associated phytoplasmas are only partially characterized, and many research gaps still need to be addressed. The limited progress in research and management of diseases of these crops is at least partly due to the nature of cultivation of the crop. A number of sugarcane and cassava diseases including those associated with phytoplasmas have in the past been disseminated through the exchange of germplasm. Although movement of germplasm has been beneficial and is desirable, the potential risk of introducing new diseases should be considered. In order to prevent the spread of sugarcane and cassava phytoplasmas, it is necessary for countries to reinforce their inspection and quarantine facilities by acquiring molecular diagnostic tools. The use of resistant clones is of limited value; with the advancement of molecular detection methods, phytoplasmas can be detected in crops at an early stage resulting in timely disease management. No single approach can provide effective and long-lasting management of these phytoplasma diseases considering also the large extension of these crop cultivations. Judicious integration of phytoplasma-free planting material, appropriate cultural practices, and resistant clones can provide ideal management of phytoplasma associated diseases. International movement of phytoplasma-free germplasm as cryopreserved stocks could be a way to decrease inadvertent dispersal of phytoplasmas (Wang and Valkonen 2008; Wang et al. 2009).

References

Aljanabi SM, Parmessur Y, Moutia Y, Saumtally S, Dookun A (2001) Further evidence of the association of a phytoplasma and a virus with yellow leaf syndrome in sugarcane. *Plant Pathology* **50**, 628–636.

Alvarez E, Betancourth CA (2016) Transmission of a 16SrIII-L phytoplasma by leafhoppers (*Scaphytopius marginelineatus*) to cassava (*Manihot esculenta* Crantz) in Colombia. www.apsnet.org/meetings/Documents/2016_meeting_abstracts/aps2016_47.htm. Accessed 9 October 2016.

Alvarez E, Mejia JF, Loke JB, Hernández L, Llano GA (2003) Detecting the phytoplasma frogskin disease association in cassava (Manihot esculenta Crantz) in Colombia. *Phytopathology* **93**, S4.

Alvarez E, Mejía JF, Llano G, Loke J, Calari A, Duduk B, Bertaccini A (2009) Detection and molecular characterization of phytoplasmas associated with frogskin disease in cassava. *Plant Disease* **93**, 1139–1145.

Alvarez E, Mejía JF, Pardo JM (2010) Development of a real-time PCR assay, to detect and quantify a 16SrIII-L phytoplasma associated with cassava frogskin disease (CFSD). *Phytopathology* **100**, S5.

Alvarez E, Llano GA, Mejía JF (2012) Cassava diseases in Latin America, Africa and Asia. In: The cassava handbook: a reference manual based on the Asian regional cassava training course. Ed Howeler RH, Bangkok, Thailand. CIAT, Cali, Colombia, 258–304 pp.

Alvarez E, Pardo JM, Mejía JF, Bertaccini A, Thanh HD, Hoat TX (2013) Detection and identification of a 16SrI group phytoplasma associated with witches' broom disease of cassava in Vietnam. *Phytopathogenic Mollicutes* **3**, 77–81.

Alvarez E, Pardo JM, Truke MJ (2014) Detection and identification of '*Candidatus* Phytoplasma asteris'-related phytoplasma associated with a witches' broom disease of cassava in Cambodia. *Phytopathology* **104**, S3.7.

Alvarez E, Betancourth C, Muñoz J (2017) Pathogenicity of a 16SrIII-L phytoplasma associated with frogskin disease of cassava (*Manihot esculenta* Crantz) in Colombia. *Phytopathology* **107**, S2.5.

Ariyarathna HACK, Everard JMDT, Karunanayake EH (2007) Diseased sugarcane in Sri Lanka is infected with sugarcane grassy shoot and/or sugarcane white leaf phytoplasma. *Australasian Plant Disease Notes* **2**, 123–125.

Arocha Y (2000) Detection and characterization de los fitoplasmas associados al sindrome del amarilleamiento foliar (YLS) de la cana de azucar en Cuba. PhD thesis, University of Habana, Cuba, 120 pp.

Arocha Y, Gonzalez L, Peralta EL (1999) First report of virus and phytoplasma pathogens associated with yellow leaf syndrome of sugarcane in Cuba. *Plant Disease* **83**, 1177.

Arocha Y, Horta D, Peralta E, Jones P (2004a) Development of a non-radioactive methodology for the generic diagnostic of phytoplasmas in Cuba. Revista de Protection Vegetal **19**, 118–122.

Arocha Y, Peralta E, Jones P (2004b) Validation of a molecular diagnostic system for phytoplasmas associated with sugarcane yellow leaf syndrome (YLS) in Cuba and its comparison with a YLS field diagnostic method. *Revista de Protection Vegetal* **19**, 19–25.

Arocha Y, Horta D, Piñol B, Palenzuela I, Picornell S, Almeida R, Jones P (2005) First report of a phytoplasma associated with Bermuda grass white leaf disease in Cuba. *Plant Pathology* **54**, 233.

Arocha Y, Echodu R, Talengera D, Muhangi J, Rockefeller E, Asher O, Nakacwa R, Serugga R, Gumisiriza G, Tripathi J, Kabuye D, Otipa M, Vutseme Lukanda K, Boa E (2008) Occurrence of '*Candidatus* Phytoplasma aurantifolia' (16SrII group) in cassava and four other species in Uganda. *New Disease Reports* **17**, 28.

Arocha Y, Piñol B, Almeida R, Acosta K, Quiñones B, Zayas T, Varela M, Marrero E, Boa E, Lucas JA (2009a) First report of phytoplasmas affecting organoponic crops in central and eastern Cuba. *Plant Pathology* **58**, 793.

Arocha Y, Echodu R, Talengera D, Muhangi J, Rockefeller E, Asher O, Nakacwa R, Serugga R, Gumisiriza G, Tripathi J, Kabuye D, Otipa M, Vutseme K, Lukanda M, Boa E (2009b) Occurrence of '*Candidatus* Phytoplasma aurantifolia' (16SrII group) in cassava and four other species in Uganda. *Plant Pathology* **58**, 390.

Arocha Rosete Y, Diallo HA, Konan Konan JL, Yankey N, Saleh M, Pilet F, Contaldo N, Paltrinieri S, Bertaccini A, Scott J (2017) Detection and differentiation of the coconut lethal yellowing phytoplasma in coconut growing villages of Grand-Lahou, Côte d'Ivoire. *Annals of Applied Biology* **170**, 333–347.

Bachchav MB, Patil PG, Hapase DG, Ghure TK (1979) Quantitative and qualitative losses caused by smut and grassy shoot disease of sugarcane. *Proceeding of Annual Convention Deccan Sugar Technologists Association* **29**, 1–6.

Bailey R, Bechet G, Cronjé P (1996) Notes on the occurrence of yellow leaf syndrome of sugarcane in southern Africa. *Proceedings South Africa Sugar Technologists Association* **70**, 3–6.

Betancourth CA, Pardo JM, Truke MJ, Muñoz JE, Alvarez E (2014) In *vitro* isolation of a phytoplasma associated with frogskin disease of cassava (*Manihot esculenta* Crantz). *Phytopathology* **105**, S1.6.

Blanche KR, Tran-Nguyen TT, Gibb KS (2003) Detection, identification and significance of phytoplasmas in grasses in northern Australia. *Australasian Plant Pathology* **52**, 505–512.

Braithwaite KS (2010) Ramu stunt. In: Pest and Diseases Image Library. Available online: http://www.padil.gov.au (accessed August 20, 2017).

Bressan A, Sémétey O, Nusillard B, Boudon-Padieu E (2007) The syndrome "basses richesses" of sugarbeet in France is associated with different pathogen types and insect vectors. *Bulletin of Insectology* **60**, 395–396.

Canova A, Bellardi MG, Bertaccini A, Vicchi V (1990) Rosette disease and witches' broom in sugar beet and spinach in Italy. *Phytopathologia mediterranea* **29**, 39–43.

Capoor SP, Pande PK, Sinha RC (1972) Mycoplasma-like bodies found in cells of "small leaf" affected cotton plants. *Hindustan Antibiotics Bulletin* **15**, 40.

Cardozo L, Pardo JM, Zacher M, Torres A, Alvarez E (2016) First report of a 16SrIII phytoplasma associated with frogskin disease in cassava (*Manihot esculenta* Crantz) in Paraguay. *Plant Disease* **100**, 1492.

Castro S, Hepp R, Romero J (2000). La marchitez amarilla de la remolacha azucarera de Chile es producida por un fitoplasma del grupo 16SrIII. X Congress of SEF, Valencia, Spain, 50.
Chaparro-Martínez EI, Trujillo-Pinto G (2001) First report of frogskin disease in cassava (*Manihot esculenta*) in Venezuela. *Plant Disease* **85**, 1285.
Chen CT (1978) Vector-pathogen relationships of sugarcane white leaf disease. *Taiwan Sugar* **25**, 50–54.
Chen CT, Kusalwong A (2000) White leaf. In: A Guide to Sugarcane Diseases. Eds Rott P, Bailey RA, Comstock JC, Croft BJ, Saumtally AS. CIRAD, ISSCT, France, 231–236 pp.
Chona BL (1958) Some disease of sugarcane reported from India in recent years. *Indian Phytopathology* **11**, 1–9.
CIAT (2003) Crop and agroecosystem health management: Annual report. Project PE -1, CIAT, Cali, Colombia.
CIAT (2005) Annual Report 2005. Project PE-1. Crop and Agroecosystem Health Management. CIAT, Cali, Colombia, 51–53 pp.
Comstock J, Irvine J, Miller J (1994) Yellow leaf syndrome appears on the United States mainland. *Sugar Journal* **56**, 33–35.
Comstock, J., M. Irey, B. Lockhart, Z. Wang (1998) Incidence of yellow leaf syndrome in CP cultivars based on polymerase chain reaction and serological techniques. *Sugarcane* **4**, 21–24.
Contaldo N, Bertaccini A, Paltrinieri S, Windsor H, Windsor G (2012) Axenic culture of plant pathogenic phytoplasmas. *Phytopathologia Mediterranea* **51**, 607–617.
Cousin MT, Maillet PL, Gourret JP (1969) La virescence du cotonnier (*Gossypium hirsutum* L.) nouvelle maladie à mycoplasmes. *Compte Rendues Académie des Sciences Série D* **268**, 2382–2384.
Cronjé CPR, Tymon AM, Jones P, Bailey RA (1998) Association of a phytoplasma with a yellow leaf syndrome of sugarcane in Africa. *Annual of Applied Biology* **133**, 177–186.
Cronjé CPR, Bailey RA, Jones P, Suma S (1999) The phytoplasma associated with ramu stunt disease of sugarcane is closely related to the white leaf phytoplasma group. *Plant Disease* **83**, 588.
Danet J, Foissac X, Zreik, Salar P, Verdin E, Nourisseau JG, Garnier M (2002) '*Candidatus* Phlomobacter fragariae' is the prevalent agent of marginal chlorosis of strawberry in French production fields and is transmitted by the planthopper *Cixius wagneri* (China). *Phytopathology* **93**, 644–649.
Davis RI, Arocha Y, Jones P, Malau U (2005) First report of the association of phytoplasmas with-plant diseases in the territory of Wallis and Futuna. *Australasian Plant Pathology* **34**, 417–418.
de Souza AN, da Silva FN, Bedendo IP, Carvalho CM (2014) A phytoplasma belonging to a 16SrIII-A subgroup and dsRNA virus associated with cassava frogskin disease in Brazil. *Plant Disease* **98**, 771–779.
Dhumal KN (1983) Physiological studies in sugarcane. PhD Thesis, Shivaji University, Kolhapur, Maharashtra, India, 212 pp.
Dhumal KN (2001) Grassy shoot disease: host-pathogen interactions and management. In: Sugarcane Pathology Vol. II Virus and Phytoplasmas. Eds Rao GP, Ford RE, Tosic M, Teakle DS, Science Publishers Inc., Huston, USA, 319–332 pp.
Dhumal KN, Hedge BA (1984) Biochemical studies on grassy shoot disease affected sugarcane leaves of clones Co419 and Co 740. *Proceedings Annual Convention of Deccan Sugar Technologists Association* **34**, 26–32.
Dhumal KN, Nimbalkar JD (1982). Physiological studies on grassy shoot disease infected sugar cane cultivars. *Indian Phytopathology* **35**, 341–343.
Dhumal KN, Nimbalkar JD (1983a). Oxidative enzymes and some organic constituents in grassy shoot disease affected sugarcane leaves of clones Co419 and Co 740. *Proceedings Annual Convention of Deccan Sugar Technologists Association* **33**, 1–8.
Dhumal KN, Nimbalkar JD (1983b) Studies in grassy shoot disease affected sugarcane cultivar Co 419 and Co 740. *Indian Phytopathology* **36**, 448–452.

Eastwood D (1990) Ramu stunt disease. Development and consequences at Ramu Sugar Ltd. *Sugarcane* **2**, 15–19.

Embrapa (2012) Embrapa Mandioca e Fruticultura. Sistemas de Produção. Available at: http://sistemasdeproducao.cnptia.embrapa.br/FontesHTML/Mandioca/mandioca_cerrados/doencas.htm. Accessed on 18 August 2016.

FAOSTAT (2015) www.faostat.fao.org.

Flôres D, Haas IC, Canale MC, Bedendo IP (2013) Molecular identification of a 16SrIII-B phytoplasma associated with cassava witches' broom disease. *European Journal of Plant Pathology* **137**, 237–242.

Fontaniella B, Vicente C, Estrella Legaz M, de Armas R, Rodríguez CW, Martínez M, Piñón D, Acevedo R, Solas MT (2003) Yellow leaf syndrome modifies the composition of sugarcane juices in polysaccharides, phenols, and polyamines. *Plant Physiology and Biochemistry* **41**, 1027–1036.

Fiore N, Longone V, González F, Zamorano A, Pino AM, Araya J, Picciau L, Alma A, Paltrinieri S, Contaldo N, Bertaccini A (2012) Transmission of 16SrIII-J phytoplasma by *Paratanus exitiosus* leafhopper. 17th Congress of ICVG, Davis, CA, USA, 230–231.

Frison EA, Feliu E (1991) FAO/IBPGR Technical guidelines for the safe movement of cassava germplasm. FAO, Rome, 48 pp.

Fukuda C, Filho JT, Sa MFP, Carneiro JS, Paiva AB, Pereira CTC, Lima HA (1990) Levantamento da ocorrência do superbrotamento da mandioca na microrregião de Ibiapaba, CE. Congresso Brasileiro De Mandioca 6, Londrina, 49.

Frison EA, Putter CAJ (1993) FAO/IBPGR Technical guidelines for the safe movement of sugarcane germplasm. Food and Agriculture Organization of the United Nations/ International Board for Plant Genetic Resources, Rome, Italy, 44 pp.

González F, Zamorano A, Pino AM, Fiore N (2010) Caracterización molecular de nuevos fitoplasmas en la vid en Chile. XIX Congreso SOCHIFIT, Pucón, Chile, 103.

González F, Zamorano A, Pino AM, Paltrinieri S, Bertaccini A, Fiore N (2011) Identification of phytoplasma belonging to X disease group in cherry in Chile. *Bulletin of Insectology* **64**(Supplement), S235–S236.

Gatineau F, Larrue J, Clair D, Lorton F, Richard-Molard M, Boudon-Padieu E (2001) A new natural planthopper vector of stolbur phytoplasma in the genus *Pentastiridius* (Hemiptera: Cixiidae). *European Journal of Plant Pathology* **107**, 263–271.

Gatineau F, Jacob N, Vauterin S, Larrue J, Lherminier J, Richard-Molard M, Boudon-Padieu E (2002) Association with the syndrome "basses richesses" of sugar beet of a phytoplasma and a bacterium-likeorganism transmitted by a *Pentastiridius* sp. *Phytopathology* **92**, 384–392.

Gaur RK, Raizada R, Rao GP (2008) Sugarcane yellow leaf phytoplasma associated for the first time with sugarcane yellow leaf syndrome in India. *Plant Pathology* **57**, 772.

Gogoi AK, Madhupriya, Kalita MK, Rao GP, Nath PD (2017) Sugarcane grassy shoot disease: incidence, molecular characterization and effect on growth and juice quality of sugarcane in North east region of India. *Proceedings of Annual Convention of STAI* **75**, 263–278.

Granada GA (1979) Machismo disease of soybeans: I. Symptomatology and transmission. *Plant Disease Reporter* **63**, 47–50.

Graziosi I, Minato N, Alvarez E, Ngo DT, Hoat TX, Aye TM, Wyckhuys KA (2016) Emerging pests and diseases of South-east Asian cassava: a comprehensive evaluation of geographic priorities, management options and research needs. *Pest Management Science* **72**, 1071–1089.

Guerra MF, Cano JR (2005) DAPI staining, an efficient technique for the diagnosis of sugarcane yellow leaf syndrome. *Petria* **15**, 253–282.

Hanboonsong Y, Choosai C, Panyim S, Damak S (2002) Transovarial transmission of sugarcane white leaf phytoplasma in the insect vector *Matsumuratettix hiroglyphicus* (Matsumura). *Insect Molecular Biology* **11**, 97–103.

Hanboonsong Y, Ritthison W, Choosai C, Sirithorn P (2006) Transmission of sugarcane white leaf phytoplasma by *Yamatotettix flavovittatus*, a new leafhopper vector. *Journal of Economic Entomology* **99**, 1531–1537.

Harrison NA, Richardson PA, Jones P, Tyson AM, Edengreen SJ, Mpunami AA (1994) Comparative investigation of MLOs associated with Caribbean and African lethal decline diseases by DNA hybridization and PCR assays. *Plant Disease* **78**, 507–511.

Hepp R, Sandoval C (1999) La enfermedad marchitez amarilla de la remolacha azucarera en Chile está asociada a Fitoplasmas. *Fitopatologia* **34**, 183.

Hepp R, Vargas M (2002) Detection by PCR of the causal agent of yellow wilt of sugar beet in leafhoppers associated with sugar beet crop. *Fitopatologia* **37**, 73.

Hoat TH, Bon NG, Quan MV, Hien VD, & Thanh ND Dickinson M (2012) Detection and molecular characterization of sugarcane grassy shoot phytoplasma in Vietnam. *Phytoparasitica* **40**, 351–359.

Hoat TH, Nung LTT, Thanh DVT, Bon NG, Duong CA, Ha TN, Kumararinghe NC (2013) Molecular detection and identification of sugarcane white leaf in Vietnam. *International Sugar Journal* **115**, 505–511.

Hoat TX, Quan MV, Lan Anh DT, Cuong NN, Vuong PT, Alvarez E, Nguyen DT, Thuy TD, Wyckhuys K, Paltrinieri S, Pardo JM, Mejía JF, Thanh ND, Dickinson M, Duong CA, Kumasaringhe NC, Bertaccini A (2015) Phytoplasma diseases on major crops in Vietnam. *Phytopathogenic Mollicutes* **5**(1-Supplement), S69–S70.

IANSAGRO (2012) Amarillez virosa y marchitez amarilla antiguosy silenciosos enemigos de la remolacha. *Gestión y Tecnología* **22**, 21–25.

Jackson G (2014) Phytoplasma diseases of cassava: various phytoplasmas. http://africasoilhealth.cabi.org/wpcms/wp-content/uploads/2015/02/6-tubers-phytoplasma-cassava.pdf.

Jayasinghe U, Pineda B, Lozano JC (1983) Antólisis en yuca (*Manihot esculenta* Crantz), asociada con organismos similares a micoplasmas. *Fitopatologia Brasilera* **9**, 51–57.

Jayasinghe U, Pineda B, Lozano JC (1984) Antholysis in cassava (*Manihot esculenta* Crantz) possibly caused by mycoplasma-like organisms. *Phytopathology Zeitschrift* **109**, 295–300.

Joomun N, Suamtally Dookun A, Saumtally S, Genshan S (2007) Sugarcane leaf yellows phytoplasmas in Mauritius: molecular characterization, transmission and alternative hosts. *Proceedings of International Sugarcane Technologists* **26**, 1005–1013.

Kaewmanee C, Hanboonsong Y (2011) Evaluation of the efficiency of various treatments used for sugarcane white leaf phytoplasma control. *Bulletin of Insectology* **64**(Supplement), S197–S198.

Kirkpatrick BC, Smart CD (1995) Phytoplasmas: can phylogeny provide the means to understand pathogenicity. *Advances in Botanical Research* **21**, 187–212.

Kitajima EW, Costa AS (1971) Corpúsculos do tipo micoplasmaassociados a moléstias de plantas, do grupo amarelo, no ESP. *Ciência e Cultura* **23**, 285–291.

Kra KD, Toualy YMN, Kouamé AC, Diallo HA, Arocha Rosete Y (2017) First report of a phytoplasma affecting cassava orchards in Côte d'Ivoire. *New Disease Reports* **35**, 21.

Kumar S, Singh V, Lakhanpaul S (2010) First report of '*Candidatus* Phytoplasma asteris' (16SrI) associated with little leaf of cotton and luffa in India. *Australasian Plant Disease Notes* **5**, 117–119.

Kumar S, Tiwari AK, Holkar S, Dattamajumder SK, Rao GP (2015) Characterization of a 16SrI-B group phytoplasma associated with sugarcane leaf yellows disease in India. *Sugar Tech* **17**, 156–161.

Kumar S, Jadon J, Rao GP (2017) Use of *secA* gene for characterization of phytoplasmas associated with sugarcane grassy shoot disease in India. *Sugar Tech* **19**, 632–637.

Kunita LS, Young GR, Pais E, Jones P, Nagaraja H (1994) Preliminary observation on *Eumetopina* sp. (Hemiptera: Delpahcidae) as a vector for ramu stunt, a new sugarcane disease in Papua New Guinea. *Journal of Australian Entomological Society* **33**, 185–186.

Kusalwong A, Singh M, Rao GP, Lehrer AT (2002) Current status of sugarcane virus and phytoplasma diseases of sugarcane in Thailand. The First International Conference on Tropical and Sub-tropical plant Diseases, Chiang Mai, Thailand, 78.

Lee CS, Chen CT (1972) Preliminary studies on transmission characteristics of sugarcane white leaf disease by *Matsumuratettix hiroglyphicus Mats*. *Report of the Taiwan Sugar Experiment Station* **56**, 57–62.

Lee I-M, Gundersen-Rindal DE, Davis RE, Bartoszyk IM (1998) Revised classification scheme of phytoplasmas based on RFLP analyses of 16S rRNA and ribosomal protein gene sequences. *International Journal of Systematic and Evolutionary Microbiology* **48**, 1153–1169.

Leilah AA, Badawi MA, Said EM, Ghonema MH, Abdou MAE (2005) Effect of planting dates, plant population and nitrogen fertilization on sugarbeet productivity under the newly reclaimed sandy soils in Egypt. *Basic Applied Science* **3**, 95–110.

Ling KC (1962) White leaf disease of sugarcane. *Taiwan Sugar* **9**, 1–5.

Lockhart BEL, Cronje CPR (2000) Yellow leaf syndrome. In: A guide to sugarcane disease. Eds Rott P, Bailey RA, Comstock JC, Croft BJ, Saumtally SA. La Librairie du CIRAD, Montpellier, France, 291–295 pp.

Longone V, González F, Zamorano A, Pino AM, Araya J, Díaz V, Paltrinieri S, Calari A, Bertaccini A, Picciau L, Alma A, Fiore N (2011). Epidemiological aspects of phytoplasmas in Chilean grapevines. *Bulletin of Insectology* **64**(Supplement), S91–S92.

Lozano JC (1992) Overview of integrated cassava diseases. *Fitopatologia Brasilera* **17**, 18–22.

Lozano JC, Bellotti A, Reyes JA, Howeler R, Leihner D, Doll J (1981) Problemas en el cultivo de la yuca. Cali, Colombia, Centro Internacional de Agricultura Tropical (CIAT), 208 pp.

Magarey RC (2008) Management of *Eumetopina flavipes*: the vector of ramu stunt disease of sugarcane in Papua New Guinea, Project Report ID: PC/2006/017, Tully Experimental Station, Australia.

Magarey RC, Suma S, Irawan Kuniata LS, Allsopp PG (2002) Sik na binatang bilong suka – diseases and pests encountered during a survey of *Saccharum germplasm* "in the wild" in Papua New Guinea. *Proceedings of the Australian Society of Sugar Cane Technologists* **20**, 219–227.

Mangelsdorf AJ (1962) A research program for the Thailand sugar industry. Department of Agriculture, Bangkok, 16 pp.

Manimekalai R, Sathish Kumar R, Soumya VP, Thomas GV (2010) Molecular detection of phytoplasma associated with yellow leaf disease in areca palms (*Areca catechu*) in India. *Plant Disease* 94, 1376.

Manimekalai R, Roshna M, GangaRaj KP, Viswanathan R, Rao GP (2016) ABC Transporter from sugarcane grassy shoot phytoplasma: gene sequencing and sequence characterization. *Sugar Tech* **18**, 407–413.

Mariano RLR, Laranjeiras D, Santos EO, Padovan IP, Peixoto AR (1991) Superbrotamento da mandioca em Pernambuco. *Fitopatologia Brasilera* **16**, 42.

Martini M, Lee I-M, Bottner KD, Zhao Y, Botti S, Bertaccini A, Harrison N, Carraro L, Marcone C, Khan AJ, Osler R (2007) Ribosomal protein gene-based phylogeny for finer differentiation and classification of phytoplasmas. *International Journal of Systematic and Evolutionary Microbiology* **57**, 2037–2051.

Marzachì C, Coulibaly A, Coulibaly N, Sangare A, Diarra M, De Gregorio T, Bosco D (2009) Cotton virescence phytoplasma and its weed reservoir in Mali. *Journal of Plant Pathology* **91**, 717–721.

Matsuoka S, Meneghin S (2000) Yellow leaf syndrome and alleged pathogens: a casual but not a causal relationship. *Proceedings of the International Society of Sugarcane Technologists* **24**, 382–388.

Matsumoto T, Lee CS, Teng WS (1968) Studies on sugarcane white leaf disease of Taiwan with special references to transmission by leafhopper, *Epitettix higroglyphicus* Mats. *Proceedings of the International Society of Sugarcane Technologists* **13**, 1090–1098.

Mejía JF, Alvarez E, Ocampo RO, Llano GA, Luna JM, Pardo JM,de Ávila D, Restrepo E (2011) Determinación de vectores aéreos y/o del suelo asociados con la transmisión de la enfermedad del cuero de sapo en yuca (*Manihot esculenta Crantz*) en los departamentos de Cauca y Sucre, en Colombia. *Fitopatologia Colombiana* **35**, 15.

Mumford RA, Potyondi L, Harju VA, Henry CM (2000) The identification of a phytoplasma from the aster yellows group infecting sugar beet in Hungary. *Plant Pathology* **49**, 806.

Munchembled C, Garressus S, Ecalle F, Boudon-Padieu E, Gatineau F (1999) Le syndrome des basses richesses. 5th Conference Internationale Sur les Ravageurs En Agriculture. Montpellier, France. Paris: ANPP 529, 36.

Madan VK, Agnihotri VP, Singh K, Pande HP, Saxena YR (1981) Biochemical studies on sugarcane diseases. *Sugar Journal* **44**, 19–20.

Nakashima K, Murata N (1993) Destructive plant diseases caused by mycoplasma-like organisms in Asia. *Outlook on Agriculture* **22**, 53–58.

Nakashima K, Challepron W, Wongkaew P, Sirithorn P (1994) Detection of mycoplasma-like organism associated with white leaf disease of sugarcane in Thailand using DIVA probes. *JIRCAS Journal* **1**, 1–57.

Nakashima K, Hayashi T, Chaleeprom W, Wongkaew P, Sirithorn P (1996) Complex phytoplasma flora in Northeast Thailand as revealed by 16S rDNA analysis. *Annals of Phytopathological Society of Japan* **62**, 57–60.

Nakashima K, Wungkaew P, Sirithorn P (2001) Molecular detection and characterization of sugarcane white leaf phytoplasmas. In: Sugarcane Pathology II, Virus and Phytoplasma Diseases. Eds Rao GP, Ford RE, Tosic M, Teakle DS, Science Publishers Inc., Enfield, USA, 157–175 pp.

Nasare K, Yadav A, Singh AK, Shivasharanappa KB, Nerkar YS, Reddy VS (2007) Molecular and symptom analysis reveal the presence of new phytoplasmas associated with sugarcane grassy shoot disease in India. *Plant Disease* **91**, 1413–1418.

Fiore N, González X, Zamorano A, Quiroga N, Paillalef R, Pino AM (2015) Phytoplasmas associated with yellow wilt disease of sugar beet in Chile. *Phytopathogenic Mollicutes* **5**(1-Supplement), S63–S64.

Pardo JM, Truke MJ, Cardozo L, Varela I, Alvarez E (2014) A real-time PCR assay to detect and quantify 16SrIII-L and 16SrI phytoplasmas associated with cassava frogskin disease in Costa Rica and Paraguay. *Phytopathology* **105**, S1.3.

Parmessur Y, Aljanabi S, Saumtally S, Dookun-Saumtally A (2002) *Sugarcane yellow leaf virus* and sugarcane yellows phytoplasma: elimination by tissue culture. *Plant Pathology* **51**, 561–566.

Peralta EL, Chinea A, Ortega E, Arocha Y, Piñón D (1999) Estudio, diagnóstico y control del síndrome del amarilleamiento foliar de la caña de azúcar (YLS) en Cuba. Informe Final de Proyecto. Academia Nacional de Ciencias de Cuba, 50 pp.

Pineda B, Jayasinghe UV, Lozano JC (1983) La enfermedad cuero de sapo en yuca (*Manihot esculenta Crantz*). *ASIAVA (Colombia)* **4**, 10–12.

Pliansinchai U, Prammanee S (2000) Green grassy shoot. In: A Guide to Sugarcane Diseases. Eds Rott P, Bailey RA, Comstock JC, Croft BJ, Saumtally AS, CIRAD ISSCT, Montpellier, France, 221–225 pp.

Pliansinchai U, Laebwan U, Suchato W (1998) Sugarcane green grassy shoot disease: epidemiology, transmission and control. 3rd National Cane and Sugar Conference, Thailand, 24.

Pliansinchai U, Suchato W (1995) Sett transmission of sugarcane green grassy shoot disease. 2nd National Plant Protection Congress, Chiang Mai, Thailand, 243–248.

Rao GP, Sinha SK, Singh RP (1992) Bio-chemical changes in grassy shoot disease affected plants of sugarcane. *Proceeding Sugar Technologists Association of India* **54**, 79–82.

Rao GP, Ford RE, Tosic M, Teakle DS (2000) Sugarcane Pathology II: Virus and Phytoplasma Disease, Science Publisher's Inc., New Hampshire, USA, 377 pp.

Rao GP, Gaur RK, Singh M, Sharma SR, Berry S (2003) Molecular characterization of phytoplasma associated with grassy shoot disease and yellow leaf syndrome in India. *Sugar Tech* **5**, 301–304.

Rao GP, Srivastava S, Gupta PS, Singh A, Singh M, Marcone C (2008). Detection of sugarcane grassy shoot phytoplasma infecting sugarcane in India and its phylogenetic relationships to closely related phytoplasmas. *Sugar Tech* **10**, 74–80.

Rao GP, Mall S, Marcone C (2012) Recent biotechnological approaches in diagnosis and management of sugarcane phytoplasma diseases. *Recent Trends in Biotechnology and Microbiology* **2**, 19–29.

Rao GP, Madhupriya, Tiwari AK, Kumar S, Baranwal VK (2014) Identification of sugarcane grassy shoot-associated phytoplasma and one of its putative vectors in India. *Phytoparasitica* **42**, 349–354.

Rao GP, Madhupriya, Thorat V, Manimekalai R, Tiwari AK, Yadav A (2017) A century progress of research on phytoplasma diseases in India. *Phytopathogenic Mollicutes* **7**, 1–38.

Rasu T, Rajendran R, Lingan R, Sundarasu S, Gandhi K, Sevugapperumal N, Ramalingam R, Ramasamy S (2011) First report of little leaf disease associated with phytoplasma on sugar beet (*Beta vulgaris* L. subsp. *vulgaris* var. altissima Doll) in India. *Journal of General Plant Pathology* **77**, 139–141.

Richard-Molard M, Garressus S, Malatesta G, Orny G, Valentin P, Reinbold C, Gerst M, Blech F, Fonne G, Putz C, Grousson C, Boudon-Padie, E (1995) Le syndrome des basses richesses investigations au champ et tentatives d'identification de l'agent pathogène et du vecteur. 58ème Congrès de l'Institut International de Recherches Betteravières, Dijon-Beaune, France, 299–309.

Rishi N, Chen CT (1989) Grassy shoot and white leaf disease. In: Diseases of Sugarcane: Major Diseases. Eds Ricaud C, Egan BT, Elsevier, Amsterdam, The Netherlands, 289–300 pp.

Rogers PF (1969) Proceedings of a Meeting on the Yellow Wilt Condition of Sugarcane. June 25th–26th, Nairobi, Kenya: East African Specialist Committee on Sugarcane Research.

Rott P, Mirkov TE, Schenck S, Girard JC (2008) Recent advances in research on Sugarcane yellow leaf virus, the causal agent of sugarcane yellow leaf. *Sugarcane International* **26**, 18–22.

Sémétey O, Bressan A, Gatineau F, Boudon-Padieu E (2007a) Development with RISA of a specific assay for detection of the bacterial agent of the syndrome "basses richesses" of sugarbeet. *Plant Pathology* **56**, 797–804.

Sémétey O, Gatineau F, Bressan A, Boudon-Padieu E (2007b) Characterization of a γ-3 Proteobacteria responsible for the syndrome "basses richesses" of sugarbeet transmitted by *Pentastiridius sp.* (Hemiptera: Cixiidae). *Phytopathology* **97**, 72–78.

Santos de Oliveira SA, Medeiros Abreu EF, Silva Araujo MT, Oliveira EJ, Chumbinho de Andrade E, Pardo Garcia JM, Alvarez E (2014) First report of a 16SrIII-L phytoplasma associated with frogskin disease in cassava (*Manihot esculenta Crantz*) in Brazil. Plant Disease **98**, 153.

Sarindu N, Clark MF (1993) Antibody production and identification of MLOs associated with sugarcane white leaf disease and Bermuda grass white leaf disease from Thailand. *Plant Pathology* **42**, 396–402.

Scagliusi S, Lockhart B (2000) Transmission, characterization and serology of a luteovirus associated with yellow leaf syndrome of sugarcane. *Phytopathology* **90**, 120–124.

Schenck S (2001) Sugarcane yellow leaf syndrome: history and current concepts. In: Sugarcane Pathology II: Virus and Phytoplasma Diseases. Eds Rao GP, Ford RE, Tosi¢ M, Teakle DS. Science Publishers Inc., Enfield, USA, 25–35 pp.

Sdoodee R (2001) Recent studies on white leaf and grassy shoot phytoplasma of sugarcane. In: Sugarcane Pathology II. Virus and Phytoplasma Diseases. Eds Rao GP, Ford RE, Tosi¢ M, Teakle DS. Science Publishers Inc., Enfield, USA, 235–244 pp.

Sdoodee R, Schneider B, Padovan A, Gibbs K (1999) Detection and genetic relatedness of phytoplasmas associated with plant diseases in Thailand. *Journal of Biochemistry Molecular Biology and Biophysics* **3**, 133–142.

Seemüller E, Schneider B, Mauree R, Ahrens U, Daire X, Kison H, Lorenz KH, Firrao G, Avinent L, Sears BB, Stackebrant E (1994) Phylogenetic classification of phytopathogenic mollicutes by sequence analysis of 16SrDNA. *International Journal of Systematic Bacteriology* **44**, 440–446.

Silberschmidt K, Campos AR (1944) Estudos relativos à doença superbrotamento ou en vassouramento da mandioca. *Arquivos do Instituto Biológico* **15**, 1–26.

Srivastava S, Singh V, Gupta PS, Sinha OK, Baitha A (2006) Nested PCR assay for detection of sugarcane grassy shoot phytoplasma in the leafhopper vector *Deltocephalus vulgaris*: A first report. *Plant Pathology* **22**, 25–32.

Suma S, Jones P (2000) Ramu stunt. In: A guide to sugarcane diseases. Eds Rott P, Bailey RA, Comstock JC, Croft BJ, Saumtally AS, CIRAD ISSCT, Montpellier, France, 226–230 pp.
Salehi M, Izadpanah KA (2005) Partial characterization of a phytoplasma associated with witches' broom disease of sugar beet in Iran. *Iranian Journal of Plant Pathology* **41**, 121–123.
Tiwari AK, Vishwakarma SK, Rao GP (2012) Increasing incidence of sugarcane grassy shoot disease in Uttar Pradesh and its impact on yield and quality of sugarcane. *Phytopathogenic Mollicutes* **2**, 63–67.
Tiwari AK, Kumar S, Mall S, Jadon V, Rao GP (2017) New efficient natural leafhopper vectors of sugarcane grassy shoot phytoplasma in India. *Sugar Tech* **19**, 191–197.
Tran-Nguyen L, Blanche KR, Egan B, Gibb KS (2000) Diversity of phytoplasma in northern Australian sugarcane and other grasses. *Plant Pathology* **49**, 666–679.
Uppal BN, Capoor SP, Raychaudhuri SP (1944) Small leaf disease of cotton. *Current Science* **11**, 284–285.
Usmani TA, Rao GP (1991) Pigments, primary productivity and yieold loss in grassy shoot affected sugarcane. *International Journal Tropical Plant Disease* **9**, 269–276.
Vega J, Scagliusi SM, Ulian EC (1997) Sugarcane yellow leaf disease in Brazil: evidence of association with a luteovirus. *Plant Disease* **81**, 21–26.
Viswanathan R (1997) Detection of phytoplasmas associated with grassy shoot disease of sugarcane by ELISA techniques. *Journal of Plant Disease and Protection* **104**, 9–16.
Viswanathan R (2000) Detection of phytoplasmas causing grassy shoot disease in sugarcane by immunofluorescence technique. *Indian Phytopathology* **53**, 475–477.
Viswanathan R (2001) Different aerated steam therapy regimes on the development of grassy shoot disease symptoms in sugarcane. *Sugar Tech* **3**, 83–91.
Viswanathan R (2002) Sugarcane yellow leaf syndrome in India: incidence and effect on yield parameters. *Sugarcane International* **20**, 17–23.
Viswanathan R (2004) Ratoon stunting disease infection favors severity of yellow leaf syndrome caused by Sugarcane yellow leaf virus in sugarcane. *Sugarcane International* **22**, 3–7.
Viswanathan R, Rao GP (2011) Disease scenario and management of major sugarcane diseases in India. *Sugar Tech* **13,** 336–353.
Viswanathan R, Balamuralikrishnan M, Poongothai M (2005) Detection of phytoplasmas causing grassy shoot disease in sugarcane by PCR technique. *Sugar Tech* **7**, 71–73.
Viswanathan R, Balamuralikrishan M, Karuppaiah R (2008) Identification of three genotypes of sugarcane yellow leaf virus causing yellow leaf disease from India and their molecular characterization. *Virus Genes* **37**, 368–379.
Viswanathan R, Chinnaraja C, Karuppaiah R, Ganesh Kumar V, Jenshi Rooba JJ, Malathi P (2011) Genetic diversity of sugarcane grassy shoot (SCGS)-phytoplasmas causing grassy shoot disease in India. *Sugar Tech* **13**, 220–228.
Vu NT, Pardo JM, Alvarez E, Le HH, Wyckhuys K, Nguyen KL, Le DT (2016) Establishment of a loop-mediated isothermal amplification (LAMP) assay for the detection of phytoplasma-associated cassava witches' broom disease. *Applied Biological Chemistry* **59**, 151–156.
Wang QC, Valkonen JPT (2008) Efficient elimination of sweet potato little leaf phytoplasma from sweet potato by cryotherapy of shoot tips. *Plant Pathology* **57**, 338–347.
Wang QC, Bart P, Engelmann F, Lambardi M, Valkonen JPT (2009) Cryotherapy of shoot tips: a technique for pathogen eradication to produce healthy planting material and for cryopreservation of healthy plant genetic resources. *Annals of Applied Biology* **154**, 351–363.
Wongkaew P, Hanboonsong Y, Sirithon P, Choosai C, Boonkrongs, Tinnangwattana T, Kitchareon Panya R, Damak S (1997) Differentiation of phytoplasmas associated with sugarcane and gramineous weed white leaf disease and sugarcane grassy shoot by RFLP analysis and sequencing. *Theoretical and Applied Genetics* **95**, 660–663.
Yadav A, Thorat V, Deokule S, Shouche Y, Prasad DT (2017) New subgroup 16SrXI-F phytoplasma strain associated with sugarcane grassy shoot (SCGS) disease in India. *International Journal of Systematic and Evolutionary Microbiology* **67**, 374–378.

Yadav A, Deokule S, Prasad DT (2013) Detection, diversity and differential expression studies of Sugarcane Grassy Shoot (SCGS) phytoplasma. Ph. D. Thesis. University of Pune, India. http://shodhganga.inflibnet.ac.in/handle/10603/106870

Yang SL, Pan YS (1979) Ecology of *Matsumuratettix hiroglyphicus* (Matsumura), an insect vector of sugarcane white leaf disease. In: Proceedings of Republic of China-US Cooperation Science. Seminar on Mycoplasma Disease of Plants, 11–19.

Zhang RY, Li WF, Huang YK, Wang XY, Shan HL, Luo ZM, Yin J (2016) Group 16SrXI phytoplasma strains, including subgroup 16SrXI-B and a new subgroup, 16SrXI-D, are associated with sugarcane white leaf. *International Journal of Systematic and Evolutionary Microbiology* **66**, 487–491.

Chapter 5
Grapevine Phytoplasmas

Elisa Angelini, Fiona Constable, Bojan Duduk, Nicola Fiore, Fabio Quaglino, and Assunta Bertaccini

Abstract The diseases associated with phytoplasmas in grapevine are collectively called yellows and occur in the majority of grapevine-growing regions over the world. At first, a short overview of symptoms and damage associated with the presence of grapevine phytoplasmas is reported. Then, vectors, alternative host plants, and epidemiological cycles, where known, are discussed for the main grapevine yellows in the different continents. Moreover, potential insect vectors and host plants, together with molecular characterization of the associated phytoplasmas, are reported.

Keywords Characterization · Epidemiology · Grapevine yellows · Symptoms · *Vitis vinifera*

E. Angelini (✉)
CREA Viticulture and Enology, Conegliano, TV, Italy
e-mail: elisa.angelini@crea.gov.it

F. Constable
Agriculture Victoria, AgriBio, Bundoora, VIC, Australia

B. Duduk
Institute of Pesticides and Environmental Protection, Belgrade, Serbia

N. Fiore
Departamento de Sanidad Vegetal, Universidad de Chile, La Pintana, RM – Santiago, Chile

F. Quaglino
Department of Agricultural and Environmental Sciences – Production, Landscape, Agroenergy, University of Milan, Milan, Italy

A. Bertaccini
Department of Agricultural and Food Sciences, *Alma Mater Studiorum* – University of Bologna, Bologna, Italy

G. P. Rao et al. (eds.), *Phytoplasmas: Plant Pathogenic Bacteria - I*,
https://doi.org/10.1007/978-981-13-0119-3_5

5.1 Introduction

Phytoplasmas in grapevine are associated with the diseases called "grapevine yellows" (GY), which occur in the majority of grapevine-growing regions. All GY diseases show similar symptoms in *Vitis vinifera*, independent from the phytoplasma present, though the agents of GY are often different phytoplasmas, mainly according to the geographic areas (Table 5.1). GY diseases associated with diverse phytoplasmas can also occur in the same area or even in the same vineyard, as is the case of the main GY diseases in Central-Southern Europe. For this reason, the identification of the phytoplasmas in a symptomatic grapevine is not possible by visual observation but only by laboratory diagnostic methods. In this chapter, the symptomatology are described for GY diseases, whereas epidemiology and characterization are treated separately per continent: Europe, Australasia, the Americas, and the rest of the world.

5.2 Symptoms of Grapevine Yellows

The main symptoms associated with GY diseases are present on the leaves, bunches, and canes (Fig. 5.1). The symptoms on the leaves and canes are generally the first ones to appear in the most seriously infected plants as soon as the sprouting has started: internodes are short, and canes appear rubbery and not lignified or can show dark pustules, depending on the variety; necrosis of terminal buds may occur.

The flowers of infected plants can wither and fall; severely diseased plants can show a strong decline, having only one or a few canes alive. In the spring, it is easy to confuse GY symptoms with other diseases or physiological problems. As the vegetative season advances, symptoms become more typical and easier to recognize: part or all of the leaves become red in the red varieties and yellow in the white varieties including the main veins. Leaves are often, but not always, crispy, brittle, and downward rolling, depending on the variety; in the late-season infections, the bunches wither and shrivel. However, partially infected plants, or those belonging to low susceptible varieties, can maintain good production at harvest time. A partial or total lack of lignification in the canes of the infected grapevines occurs at the end of the summer; thus, canes are more susceptible to death from the winter cold.

Not all the grapevine species and varieties show the same susceptibility toward phytoplasma infection, when growing in the same environment and exposed to the same disease pressure. In general, rootstock varieties are poorly susceptible and rarely show GY symptoms (Moutous 1977; Caudwell et al. 1994; Borgo et al. 2009; Eveillard et al. 2016). Moreover symptom expression, phytoplasma concentration, and occurrence of infected plants under field and laboratory conditions vary according to the *V. vinifera* variety (Borgo et al. 2006; Bressan et al. 2005; Constable 2010; Jagoueix-Eveillard et al. 2012; Eveillard et al. 2016) (Fig. 5.2). The differential susceptibility could depend on genetic or epigenetic features of the grapevine variety/species or by the feeding behavior of the insect vectors. Moreover, the symptom severity could also depend on the phytoplasma.

Table 5.1 Summary of phytoplasmas reported to be associated with GY

Disease (acronym)	Continent	Countries	*Candidatus* species (16Sr group/subgroup)	Relevant literature
"Bois noir" (BN)	Europe	France, Germany, Italy, Slovenia, Croatia, Greece, Hungary, Serbia, Ukraine, Bosnia and Herzegovina, Austria, Spain, Bulgaria, FYR Macedonia, Czech Republic, Romania, Portugal, Georgia, Montenegro	'*Ca*. P. solani' (16SrXII-A, 16SrXII-F, 16SrXII-G, 16SrXII-J, and 16SrXII-K)	Caudwell (1957); Maixner et al. (1995); Bertaccini et al. (1995); Saric et al. (1997); Davis et al. (1997); Kolber et al. (2003); Duduk et al. (2004); Milkus et al. (2005); Delić et al. (2006); Riedle-Bauer et al. (2006); Sabaté et al. (2007); Sakalieva et al. 2007; Mitrev et al. (2007); Starý et al. (2013); Chireceanu et al. (2013); De Sousa et al. (2013); Quaglino et al. (2014); and Kosovac et al. (2016)
"Flavescence dorée" (FD)		France, Italy, Serbia, Spain, Portugal, Hungary, Slovenia, Croatia, Austria, Switzerland	(16SrV-C, 16SrV-D)	Caudwell (1957); Martini et al. (1999); Duduk et al. (2004); Torres et al. (2005); De Sousa et al. (2010); Ember et al. (2011); Mehle et al. (2011); Seruga-Music et al. (2011); Reisenstein and Steffek (2011); Radonjić et al. (2013); and Casati et al. (2017)
Others		Italy, Hungary, Serbia	'*Ca*. P. asteris' (16SrI-B); '*Ca*. P. prunorum' (16SrX-B); '*Ca*. P. ulmi' (16SrV-A)	Alma et al. (1996); Saric et al. (1997); Varga et al. (2000); Duduk et al. 2004; and Dermastia et al. (2017)
Australian GY (AGY)	Australia		'*Ca*. P. australiense' (16SrXII-B)	Padovan et al. (1995)
			'*Ca*. P. australasia' (16SrII-D)	Gibb et al. (1999)
			(16SrXXIV)	Constable et al. (2002)

(continued)

Table 5.1 (continued)

Disease (acronym)	Continent	Countries	*Candidatus* species (16Sr group/subgroup)	Relevant literature
North America GY (NAGY)	America	USA	'*Ca*. P. asteris' (16SrI-A)	Davis et al. (1998, 2015)
			'*Ca*. P. pruni' (16SrIII-Aα, 16SrIII-Aβ)	
"Bois noir" BN		Canada, Chile	'*Ca*. P. solani' (16SrXII-A)'	Rott et al. (2007) and Gajardo et al. (2009)
Others		Canada, Chile, Peru	'*Ca*. P. asteris' (16SrI-B and 16SrI-C), (16SrIII-J); '*Ca*. P. ulmi' (16SrV-A); '*Ca*. P. fraxini' (16SrVII-A); '*Ca*. P. brasiliense' 16SrXV-A	Gajardo et al. (2009); Olivier et al. (2009a, b); Fiore et al. (2015b); and Wei et al. (2017)
"Bois noir" BN	Africa	South Africa	'*Ca*. P. solani' (16SrXII-A)	Botti and Bertaccini 2006b
Others		Tunisia, South Africa	'*Ca*. P. asteris' (16SrI-B); '*Ca*. P. aurantifolia' (16SrII-B)	M'hirsi et al. (2004) Engelbrecht et al. (2010) and Botti and Bertaccini (2006b)
"Bois noir" BN	Asia	Israel, Lebanon, China, Turkey, Jordan, Iran	'*Ca*. P. solani' (16SrXII-A)	Orenstein et al. (2001); Choueiri et al. (2002); Duduk et al. (2010); Canik et al. (2011); Salem et al. (2013); Ertunc et al. (2015); and Mirchenari et al. (2015)
Others		Syria, Turkey, Lebanon, Iran	'*Ca*. P. trifolii' (16SrVI); '*Ca*. P. asteris' (16SrI-B); (16SrIX-C); '*Ca*. P. aurantifolia' (16SrII-B); '*Ca*. P. fraxini' (16SrVII-A)	Contaldo et al. (2011); Canik et al. (2011); Ertunc et al. (2015); Salehi et al. (2016); Casati et al. (2016); and Zamharir et al. (2017)

Fig. 5.1 Symptoms of GY on red (left) and white (right) varieties

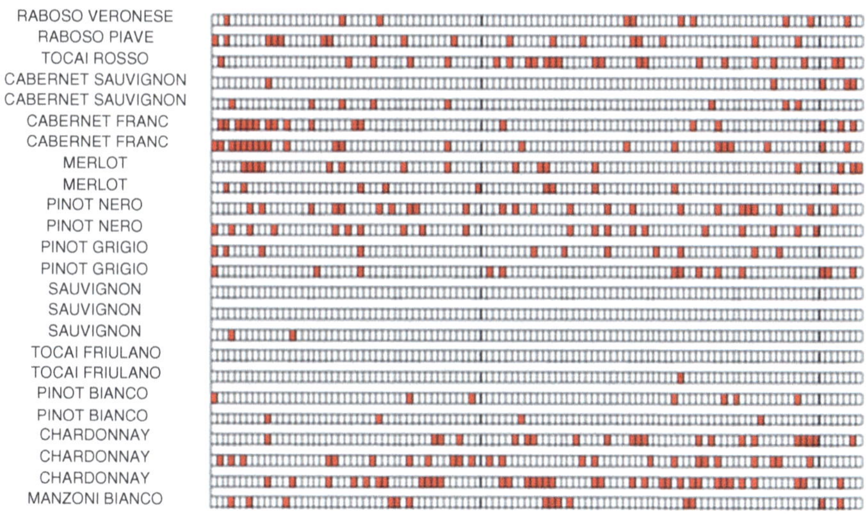

Fig. 5.2 Schematic map of a vineyard (Tezze, Italy) heavily infected with "flavescence dorée", where several different varieties were planted. The variety planted in each row is listed. Red squares point out infected grapevine plants; white squares point out healthy plants. The different susceptibility to the phytoplasma infection is evident

Generally, grapevine can also recover from GY disease resulting in disappearance of symptoms. It was reported that in the recovered grapevines, the phytoplasmas could not be detected and the insect vector(s) could not acquire phytoplasmas from asymptomatic and recovered grapevines (Morone et al. 2007; Galetto et al. 2014). However, the recovery rate depends on reinoculation by infective vectors and on grapevine variety (Osler et al. 2003; Bellomo et al. 2007; Morone et al. 2007; Maggi et al. 2017), and it seems not to be dependent on the phytoplasma strain

(Constable 2010). Although the underlying physiological and molecular mechanisms are not yet fully understood, the recovery phenomenon has been linked to accumulation of H_2O_2, resulting in lower phytoplasma titer (Musetti et al. 2007; Gambino et al. 2013).

5.3 Grapevine Yellows in Europe

"Flavescence dorée" (FD) This is the most important phytoplasma disease of grapevine in the European countries that induces losses of up to 100% in the case of severe epidemics (Fig. 5.3). The FD phytoplasma belongs to the 16SrV group (Bertaccini et al. 1995, 1997; Martini et al. 1999); although other phytoplasmas in this group are genetically closely related, they occupy quite different ecological niches (Bertaccini et al. 2014; Lee et al. 2004b). FD and related phytoplasmas have been classified into two distinct subgroups, 16SrV-C and 16SrV-D, (FD-C and FD-D, respectively) on the basis of sequence differences in their 16S rRNA gene. FD-D is considered widespread, and it was reported in France, Spain, Portugal, Switzerland, and northern Italy regions, while FD-C is present in some Italian areas and in Slovenia, Serbia, Croatia, Austria, Hungary, and Switzerland (Martini et al. 1999; Angelini et al. 2001; Duduk et al. 2003, 2004; Seruga-Music et al. 2011; Reisenstein and Steffek 2011; Radonjić et al. 2013; Casati et al. 2017). Besides grapevine, 16SrV-C phytoplasmas were also reported in *Clematis vitalba*, *Ailanthus altissima*, and *Corylus avellana* plants near vineyards (Angelini et al. 2003; Filippin et al. 2009, 2011; Casati et al. 2017). Due to the importance of the FD disease, significant efforts have been devoted to develop multilocus sequence typing (MLST)

Fig. 5.3 Grapevine severely infected by FD: red coloration of leaves and downward curling, in the middle typical lack of lignification due to phytoplasma presence

tools for finer differentiation of closely related strains of epidemiologic relevance. Variability analysis of the *secY* and *rpsC* genes showed that FD phytoplasma variants belong to differentiable strain clusters and seemed closely related to phytoplasmas infecting European alder or Palatinate grapevine yellows (Angelini et al. 2001, 2003; Martini et al. 2002). The MLST further divide FD strains classified in 16SrV-C and 16SrV-D groups into genetically distinct types designated as mapFD1 (16SrV-C) with high genetic variability, mapFD2 (16SrV-D) with lower genetic variability, and mapFD3 (16SrV-C) (Arnaud et al. 2007; Quaglino et al. 2010). Further FD strain differentiation has been achieved with the *rp* and *secY* genes (Bertaccini et al. 2003, 2009; Botti and Bertaccini 2006a, 2007). Bioinformatic analysis of the concatenated sequences of the *tuf*, *rpsC-rplV*, *rplF-rplR*, and *map* genes of FD and other 16SrV phytoplasmas infecting grapevine in Europe indicate that these form a single cluster of strains that is distinct from other 16SrV phytoplasmas, and they are thought to have a common origin in Europe (Arnaud et al. 2007; Malembic-Maher et al. 2011). Although genetic diversity in FD strains has been useful for the identification of their potential origins and dispersal within the environment, the biological differences among them are not so far clarified.

FD is transmitted by the leafhopper *Scaphoideus titanus* Ball introduced in Europe from North America that only feeds on grapevine (Schvester et al. 1962, 1969; Mori et al. 2002; Bressan et al. 2005). The pathogen can be introduced to new areas (long-distance transmission) by infected propagation material, and to a vineyard, from outside sources such as planting material or insect vectors. The incubation period in the plant usually lasts one year, and the development of disease symptoms normally takes place in the following year(s) (Boudon-Padieu 2000, 2003; Bressan et al. 2005). The disease can be quite devastating, and sometimes incidences of over 80% have been reported (Bressan et al. 2005). The biology of *S. titanus* is synchronized with the phenology of grapevine, and it has one generation per year (Schvester et al. 1962; Caudwell 1990; Bressan et al. 2005). *S. titanus* overwinters as eggs laid in the bark of 2-year-old grapevine canes. Since *S. titanus* lives and feeds only on grapevine, it can be a very efficient FD vector, especially when no measures to control its presence are applied. This vector transmits the phytoplasma in a persistent manner, with a latency period of 4–5 weeks (Schvester et al. 1963; Caudwell et al. 1970). The leafhopper can acquire the phytoplasma at all growth stages, but early in the season acquisition efficacy seems to be low, because the first and second instars do not acquire the phytoplasma with the same efficacy as the older stages. Once infected, it remains infectious for the rest of its life (Boudon-Padieu 2000, 2003; Galetto et al. 2016). *S. titanus* is native to North America and found in woodland or hedge row vegetation near vineyards, while it is restricted to grapevine in Europe and Virginia (Maixner et al. 1993; Beanland et al. 2006). Studies showed that FD-infectious *S. titanus* could be found on symptomless wild *V. riparia* in European woodlands: it is therefore possible that they may act as a source of inoculum in FD epidemiology (Lessio et al. 2007). Adult *S. titanus* have been found sporadically on *C. vitalba*, but can survive only for a short time and cannot acquire FD from, and transmit to, *C. vitalba* or from *C. vitalba* to grapevine (Angelini et al. 2004; Filippin et al. 2007). *Orientus ishidae* is another potential insect vector, which can occur in FD-infected vineyards and can harbor 16SrV-C and 16SrV-D strains (Mehle et al. 2010, 2011; Gaffuri et al. 2011; Trivellone et al. 2016). It was also shown to be capable of transmitting FD phytoplasmas from broad bean to grapevine (Lessio et al. 2016). Some other insect vectors were shown to experimentally transmit FD, such as *Euscelidius variegatus* Kirschbaum to herbaceous hosts, including *Vicia faba* (Caudwell et al. 1972), and *Dictyophara europaea* from wild clematis to grapevine (Filippin et al. 2009); however the *C. vitalba* role in FD epidemiology must be further investigated.

"Bois noir" (BN) disease was first reported in 1961 in vineyards of northeastern France. Its symptoms were undistinguishable to those of FD, but, because BN was spreading more slowly, it was considered a non-epidemic form of FD. However on the basis of its non-transmissibility by *S. titanus*, it was established that BN was a disease distinct from FD. A few years later, similar symptoms had been observed in vineyards of the Mosel and the Rhein Valley in Germany, where *S. titanus* did not occur. Experimental evidence proved that this disease, originally named "Vergilbungskrankheit" (VK), is transmitted by the plant hopper *Hyalesthes obsoletus* Signoret; now BN and VK are considered to be the same disease (Belli et al. 2010). BN spreads in all grapevine-growing countries where it is sometimes responsible for serious crop losses (Belli et al. 2010; Foissac et al. 2013). On the basis of unique biological properties and exclusive molecular markers within multiple genes (*tufB*, *rplV-rpsC*, *secY*), the phytoplasma associated with BN disease has been identified as '*Ca.* P. solani' (Quaglino et al. 2013). The '*Ca.* P. solani' strains associated with BN in Europe were assigned to the subgroup 16SrXII-A (Bertaccini et al. 1995), but later more diversity was found, and subgroups 16SrXII-F, 16SrXII-G, 16SrXII-J, and 16SrXII-K were also differentiated (Quaglino et al. 2009, 2017). Moreover, sequence analysis of the *tufB* gene revealed that three BN *tuf*-types are present in grapevines and alternative plant hosts, according to ecological diverse pathosystems: (i) bindweed, *H. obsoletus*, grapevine tuf-type b, (ii) nettle – *H. obsoletus* – grapevine tuf-type a, and (iii) *Calystegia sepium* – *H. obsoletus* – grapevine *tuf*-type c tuf-type c (Langer and Maixner 2004). *C. arvensis* and *U. dioica* have been reported as being the main host plants of *H. obsoletus* in Germany, northern Italy, Spain, and Austria (Mori et al. 2013). Recently, in Austria, Aryan et al. (2014) detected a high incidence of a tuf-type b with a distinguished *Hpa*II-restriction profile designed as tuf-type b2 that appears to have different ecological features. The biological complexity of BN disease has stimulated research on molecular markers for verification of possible genetic diversity related to pathogenicity of the strains. MLST on variable genes, such as *secY*, *vmp1*, and *stamp*, provided evidence of high variability among BN strains within the tuf-types (Foissac et al. 2013; Murolo and Romanazzi 2015). For example, based on *Rsa*I-RFLP analyses of *vmp1* gene amplicons, '*Ca.* P. solani' populations show 23 profiles (Pacifico et al. 2009; Murolo et al. 2014) (Fig. 5.4). Based on phylogenetic analysis of concatenated nucleotide sequences of the genes *vmp1* and *stamp* for 76 '*Ca.* P. solani' strains, 49 *vmp1*/*stamp* sequence variants were grouped into five *vmp1*/*stamp* clusters. The cluster *vmp1*/

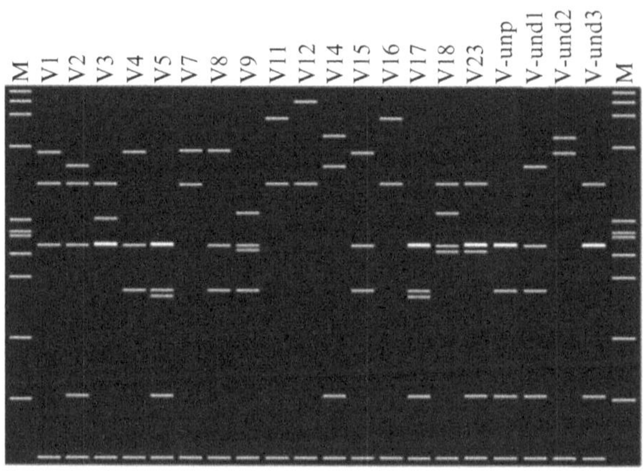

Fig. 5.4 Virtual *Rsa*I-RFLP profiles of *vmp1* gene fragments obtained from BN phytoplasmas. Designation of *vmp1 Rsa*I-RFLP profiles is according to the SEE-ERANET nomenclature (X. Foissac, INRA, Bordeaux, France) and Quaglino et al. (2016). M, marker phiX174 digested with the enzyme *Hae*III

stamp-4 included BN strains (tuf-type a) associated with nettle-related biological cycle, while the other four clusters included BN strains (tuf-type b) associated with bindweed-related biological cycle (Quaglino et al. 2016) (Fig. 5.5). Several weeds, such as *Chenopodium album* and *Malva sylvestris*, host the '*Ca.* P. solani' in or around infected vineyards and can therefore play a role in BN diffusion (Marchi et al. 2015; Mori et al. 2015; Oliveri et al. 2015). In the Euro-Mediterranean regions, the main BN insect vector is *H. obsoletus*, a polyphagous cixiid living preferentially on nettle (*Urtica dioica*), bindweed (*Convolvulus arvensis*), mugwort (*Artemisia vulgaris*), and chaste tree (*Vitex agnus-castus*) inside and/or around vineyards (Alma et al. 1988; Sforza et al. 1998; Langer and Maixner 2004; Sharon et al. 2005). Recently, *Reptalus panzeri* and *R. quinquecostatus* have been reported as vectors of BN in Serbian and France vineyards, respectively (Cvrković et al. 2014; Chuche et al. 2016), and *Anaceratagallia ribauti* was reported as vector of "stolbur" phytoplasmas even if not to grapevine (Riedle-Bauer et al. 2008).

Molecular epidemiology approaches, using *vmp1*- and *stamp*-based markers, increased the knowledge of the population variability of BN throughout vineyards and their surroundings in the Mediterranean area (Fialová et al. 2009; Fabre et al. 2011; Foissac et al. 2013; Murolo et al. 2014; Landi et al. 2015; Murolo and Romanazzi 2015). Moreover recent studies reported the direct epidemiological role of *V. agnus-castus* in the *H. obsoletus*-mediated BN transmission to grapevine (Kosovac et al. 2016) and the ability of *R. panzeri* to transmit '*Ca.* P. solani' from corn with reddening disease to grapevine (Cvrković et al. 2014). The complexity of BN disease epidemiology renders it difficult to design efficient control strategies. Insecticides applied to the grapevine canopy influence neither the disease nor the presence of *H. obsoletus* (Maixner 2007; Mori et al. 2008).

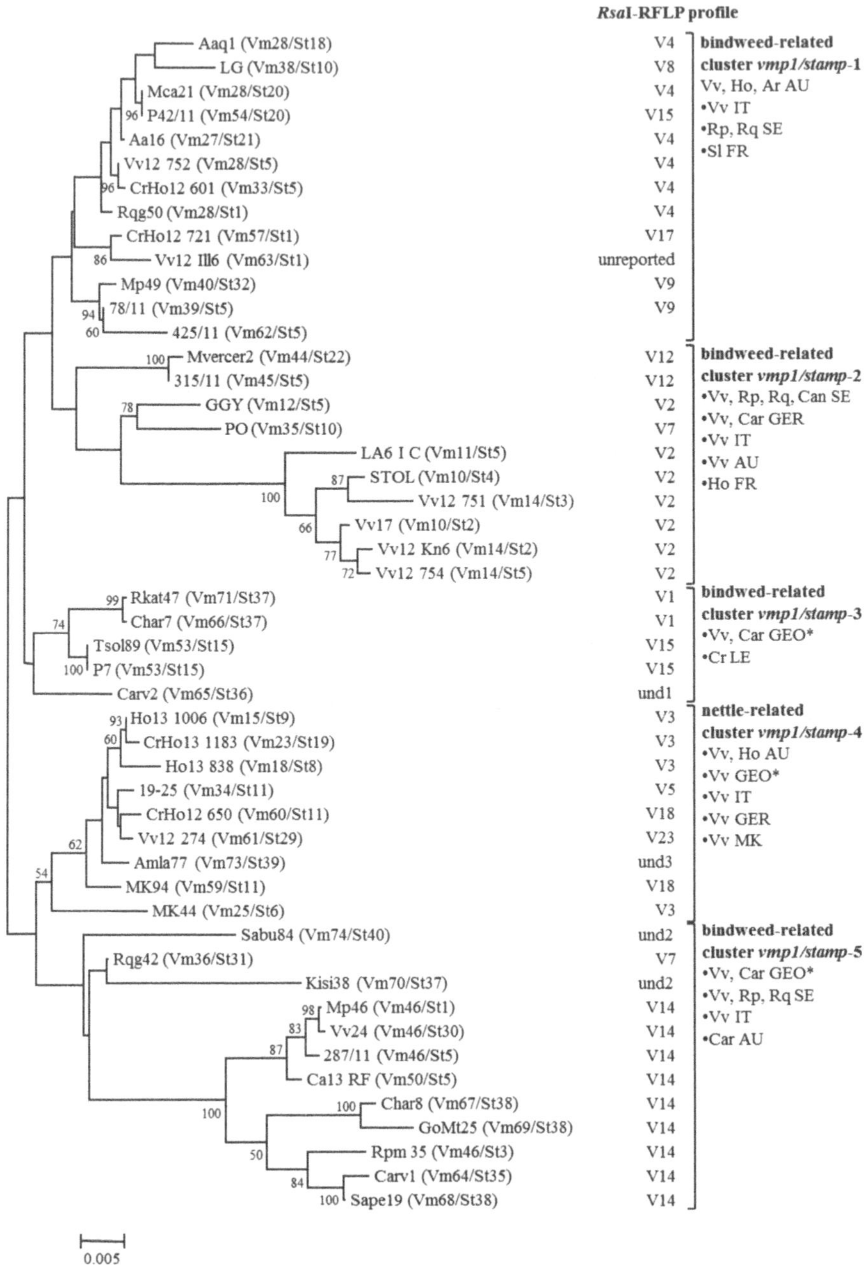

Fig. 5.5 Unrooted phylogenetic tree inferred from '*Ca*. P. solani' strains based on concatenated nucleotide sequences of the genes *vmp1* and *stamp*. Minimum evolution method was employed using the Jukes-Cantor model and bootstrap replicated 1,000 times. One phytoplasma strain representative of each *vmp1*/*stamp* type was included. Clusters are delimitated by parentheses. Acronyms within clusters indicated phytoplasma hosts and origin. Hosts: Ar, *Anaceratagallia ribauti*; Can, *Capsicum annuum*; Car, *Convolvulus arvensis*; Cr, *Catharanthus roseus*; Ho, *Hyalesthes obsoletus*; Rp, *Reptalus panzeri*; Rq, *Reptalus quinquecostatus*; Sl, *Solanum lycopersicum*; Vv, *Vitis vinifera*. Origin: AU, Austria; FR, France; GEO, Georgia; GER, Germany; IT, Italy; LE, Lebanon; MK, Makedonia; SE, Serbia; SL, Slovenia

The management of *H. obsoletus* host plants in the vineyards and surrounding areas is therefore considered crucial for BN control (Mori et al. 2012). Thus, preventive measures such as checking the sanitary status of propagation material and treating diseased mother plants through thermotherapy are applied to limit long-distance dissemination and infield spread of the disease. Other strategies for reducing BN spread or incidence are based on (i) preventive removal of the grapevines' suckers on which *H. obsoletus* could feed after grass mowing, (ii) trunk cutting above the grafting point on the symptomatic grapevines, and (iii) treatments by resistance inducers (Mori et al. 2015).

Other Phytoplasmas In Germany, a disease named Palatinate GY associated with 16SrV-C group phytoplasmas was detected in the grapevine-growing area bordering France. Unlike FD, in this case the grapevine is a secondary and incidental host since the disease, transmitted by *Oncopsis alni*, is not known to be epidemic (Maixner et al. 1995). The 16SrI group (aster yellows) phytoplasmas are reported among the viticultural regions around the world. In Europe they occur sporadically; in particular 16SrI-B phytoplasmas have been found in grapevine in Italy, Slovenia, and Croatia, while 16SrI-C phytoplasmas have only been reported in Italy (Alma et al. 1996; Saric et al. 1997; Mikec et al. 2006; Landi et al. 2013; Jezič et al. 2013). 16SrI phytoplasmas were detected in declining Syrah grapevines in France, but their subgroup was not determined (Renault-Spilmont et al. 2006). Four leafhopper species can transmit 16SrI-B phytoplasmas to grapevine including *Euscelis incisus* and *Macrosteles quadripunctulatus*, which are the most efficient vectors, and *Euscelidius variegatus* and *S. titanus*, which are less efficient (Alma et al. 2001). It was reported that *M. quadripunctulatus* does not survive on grapevine (Batlle et al. 2008). Other possible vectors infected by 16SrI-B and 16SrI-C phytoplasmas were found in Italian vineyards, although the phytoplasmas were not detected in the grapevines (Landi et al. 2015). However, these insects would appear to be completing their life cycle on other plant species near to or within vineyards, and probably grapevine is just their incidental host. Phytoplasmas of the 16SrX group, mainly 16SrX-B ('*Ca*. P. prunorum', European stone fruit yellows), were also sporadically detected in Italy, Hungary, and Serbia (Bertaccini et al. 1996; Varga et al. 2000; Duduk et al. 2004), and in Italy also 16SrV-A phytoplasmas ('*Ca*. P. ulmi') were detected in some plants (Martini et al. 2002; Dermastia et al. 2017); recently in northern Italy, 16SrVII-A ('*Ca*. P. fraxini'-related) and 16SrVI ('*Ca*. P. trifolii'-related) subgroups were also identified in both symptomatic and asymptomatic grapevine samples and in some of the reported insect vectors (Zambon et al. 2018); however their epidemiologic significance remains to be determined.

5.4 Grapevine Yellows in Australia

Australian grapevine yellows (AGY) disease exhibits symptoms that are indistinguishable from those associated with GY. However AGY disease in Australia is associated with three distinct phytoplasmas that are not reported to occur in grapevines in any other country. They include '*Ca.* P. australiense' (16SrXII-B), '*Ca.* P. australasia', or Australian tomato big bud (TBB) phytoplasma (16SrII-D) and Buckland Valley grapevine yellows (BVGY) phytoplasma (16SrXXIII) (Padovan et al. 1995; Davis et al. 1997; Gibb et al. 1999; Constable et al. 2002; Wei et al. 2007). AGY is observed in many Australian grape-growing regions, although it occurs with greatest incidence in the warmer inland areas including the Murray Valley in New South Wales (NSW) and Victoria, the Riverina in NSW, and the Riverland in South Australia (Magarey and Wachtel 1986b; Bonfiglioli et al. 1995; Constable et al. 2004). The disease most frequently affects Chardonnay and Riesling, but it has been observed also in other white and red varieties (Magarey and Wachtel 1986a; Bonfiglioli et al. 1995). Symptoms begin to develop in late spring (October–November) and are visible until late summer (February). The disease can be epidemic occurring with high incidence in one year and then decreasing in subsequent years (Magarey and Wachtel 1986b; Constable et al. 2004). An incidence of 73.4% has been recorded in a single year, and over a period of 6 years, 95% of grapevines were affected (Constable et al. 2004). Clusters of AGY disease occurring within vineyards are thought to be caused by migration of infectious vectors through a vineyard and feeding in clusters rather than spread from grapevine to grapevine, since alternative hosts are not observed and potential insect vectors have rarely been found in Australian vineyards (Constable et al. 2004; Magarey et al. 2006). High AGY disease incidence occurring in grapevine growing beside wasteland areas or riparian vegetation compared to grapevines growing in other parts of the vineyard also supports an external source of phytoplasmas (Magarey et al. 2006). Significant economic impact was observed in some Chardonnay and Riesling vineyards that in the 1970s and 1980s experienced 13% yield losses and in some grapevine plants, which exhibited severe symptoms over a large portion of the canopy, producing 54% less compared to the healthy plants (Emmett et al. 1983; Magarey and Wachtel 1983). Later studies, using Chardonnay vineyards in which minimal pruning techniques were used, revealed no significant difference in yield between unaffected and AGY-diseased grapevines in some vineyards and a moderately significant reduction in yield (43–44%) in others. Yield loss worsened (54% yield) when grapevines exhibited AGY symptoms coupled with restricted growth (Glenn 2000). One hypothesis for the difference in the economic impact of AGY disease between early and later yield studies is a change in canopy management during the 1980s in warmer inland areas from harder pruning styles to minimal pruning, the latter producing larger canopies and greater fruit production (Clingeleffer 1988; Sommer et al. 1993). Larger canopies and fruit production may offset the losses incurred from AGY disease, as typically only a portion of a grapevine and perhaps only a few shoots are affected by AGY.

It has been suggested that restricted growth (RG) and late-season leaf curl (LSLC) symptoms in Australia might also be associated with phytoplasmas, but there is no consistent evidence to support this hypothesis (Bonfiglioli et al. 1997). PCR testing could not correlate RG and LSLC symptoms with phytoplasma presence (Bonfiglioli et al. 1995; Gibb et al. 1999; Padovan et al. 1995; Constable et al. 2003a). A statistical analysis of 6 years of AGY, LSLC, and RG disease incidence data from three affected vineyards showed that the diseases were not always associated with each other (Constable et al. 2004).

The tissue type, titer, and time of year affect phytoplasma detection in Australian grapevines: they are most reliably detected by PCR methods in symptomatic shoots during mid to late summer (January–February), before harvest (Gibb et al. 1999; Constable et al. 2003a). Phytoplasmas can be detected also in the roots, trunks, and cordons of AGY-affected grapevines throughout the year indicating that they may spread within and persistently infect grapevines although they can be unevenly distributed or in uneven titer (Gibb et al. 1999; Constable et al. 2003a, b). Expression of AGY symptoms in grapevines held under netting to prevent reinfection events further supports persistent infection by Australian grapevine phytoplasmas (Magarey et al. 2006). However, phytoplasmas associated with AGY disease are rarely detected in symptomless shoots, and cases where the plants may not express symptoms in the subsequent years are observed even if they are persistently infected (Magarey and Wachtel 1986b; Constable et al. 2003a, b). An analysis of disease survey data from 6 years showed that grapevines affected by AGY disease were at statistically greater risk of displaying the symptoms again in following years (Constable et al. 2016). Recurrence of disease is likely to be linked to persistent infection as well as reinfection events. It is postulated that symptom expression in Australian grapevines is strongly linked to the presence and possibly titer of phytoplasmas in an affected shoot, but when phytoplasmas occur in other grapevine tissues, they may be symptomless (Gibb et al. 1999; Constable et al. 2003b).

'*Ca*. P. australiense' is found in most grape-growing regions where AGY disease occurs and is most frequently detected in diseased grapevines (Gibb et al. 1999). TBB phytoplasma occurs less frequently, and interestingly it has been found in mixed infections with '*Ca*. P. australiense' without changes to typical AGY symptom expression (Gibb et al. 1999; Constable et al. 2003a).

The BVGY phytoplasma is restricted to the cool-climate Alpine Valleys wine region of northeastern Victoria, which includes the Buckland Valley, and within this region, it has only been detected in two Chardonnay vineyards that were within 4 km of one another (Constable et al. 2003b). The occurrence of the BVGY phytoplasma in both vineyards could not be traced to planting material originating within the region or to other regions, and there was no link between the two vineyards, except their proximity. Therefore, it seems likely that the phytoplasma has spread from a local alternative plant host by an insect vector, although both of them have not been found. The BVGY phytoplasma is the most frequently detected in this region, while the TBB phytoplasma has never been detected, and '*Ca*. P. australiense' has only been detected once (Constable et al. 2003b; Dermastia et al. 2017). Both '*Ca*. P. australiense' and TBB phytoplasma have a broad geographic range within Australia, which includes tropical regions in the north to cool temperate regions in the south (Schneider et al. 1999; Constable 2010) infecting native plant species, weeds, domestic ornamental and crop plants, and economically important commercial crops.

In Australia '*Ca*. P. australiense' has been detected in *Acaena novae-zelandiae* (bidgee widgee), *Carica papaya* (papaya), *Catharanthus roseus* (periwinkle),

Cucurbita maxima (pumpkin), *C. moschata* (pumpkin), *Einardia nutans* (climbing saltbush), *Enchylaena tomentosa* (ruby saltbush), *Euphorbia terracina* (false caper), *Exocarpus cupressiformis* (cherry ballart), *Fragaria x ananassa* (strawberry), *Gomphocarpus fruticosus* (cottonbush, swan plant), *Melilotus indicus* (Hexam sp., Hexham scent), *Jacksonia scoparia* (winged broom pea), *Liquidambar styraciflua* (sweetgum), *Paulownia fortunei* (paulownia), *Phaseolus vulgaris* (bean), *Maireana brevifolia* (Yanga bush), *Medicago sativa* (alfalfa), *M. polymorpha*, and *Vigna radiata* (mung bean) (Schneider et al. 1999; Davis et al. 2003; Pilkington et al. 2003; Bayliss et al. 2005; Streten and Gibb 2005, 2006; Streten et al. 2005; Magarey et al. 2006; Getachew et al. 2007; Habili et al. 2007; Constable et al. 2016). *M. brevifolia*, *E. tomentosa*, *E. terracina*, and *E. nutans* occur near AGY-affected vineyards, but it is not yet known if they have a role in the epidemiology of AGY disease (Magarey et al. 2006). The importance of other hosts in the AGY disease cycle is not known, although commercial crops that also host '*Ca.* P. australiense' are unlikely to be important to AGY disease epidemiology as they are not frequently grown near vineyards (Dermastia et al. 2017). TBB phytoplasma and related strains have a similar geographic distribution within Australia to '*Ca.* P. australiense.' They also have a very broad host range, infecting more than 50 plant species, representing 20 plant families (Schneider et al. 1999; Davis et al. 2003; Streten and Gibb 2006; Saqib et al. 2007; Lee et al. 2010; Aryamanesh et al. 2011; Yang et al. 2013). The role of these plant hosts in AGY disease epidemiology is also unknown.

Insect vectors of Australian grapevine infecting phytoplasmas remains unknown, despite intensive surveys and some transmission studies. No insect vector has been identified for '*Ca.* P. australiense' in any Australian crop, although it was detected using PCR in specimens of the common brown leafhopper, *Orosius argentatus* (Evans), collected from an Australian vineyard. This insect was able to acquire the phytoplasma from grapevine under experimental conditions, but transmission was not demonstrated (Glenn 2000; Beanland 2002). '*Ca.* P. australiense' was also detected in several leafhopper and plant hopper specimens collected from Australian strawberry farms, including *Orosius* sp., *Xestocephalus* sp., *Thanatodictya* sp., and an undetermined species of the tribe Gaetuliini, although the latter three species have not been recorded in vineyards (Constable et al. 2016). The leafhoppers *Arawa variegata* and *Recilia hospes* captured on New Zealand strawberry plants were infected with '*Ca.* P. australiense', but their vector status is unknown (Charles et al. 2002). *Arawa* sp. and *R. hopses* have been found in Australian vineyards affected by AGY disease, but '*Ca.* P. australiense' was not detected in these specimens (Beanland 2002). The TBB phytoplasma is transmitted by *O. argentatus*, *Batrachomorphus punctatus*, and *Austroagallia torrida* in other Australian crops (Hill 1941, 1943; Helson 1951; Hutton and Grylls 1956; Grylls 1979; Osmelak 1986; Pilkington et al. 2004). All three insect species have been found in Australian vineyards (Glenn 2000; Beanland 2002). *Orosius orientalis* can acquire TBB phytoplasma from grapevine and then transmit it to faba bean, but back-transmission to grapevine was not demonstrated (Beanland 2002). Acquisition and transmission of TBB phytoplasma in Australian grapevines by *B. punctatus* and *A. torrida* were not demonstrated.

5.5 Grapevine Yellows in the Americas

The first information about the presence of phytoplasmas in grapevine in the American continent dates back to 1971, specifically in the Elqui Valley in Chile, where GY symptoms were observed (Caudwell et al. 1971; Gärtel 1972; Caudwell 1980). Similar symptoms were then reported in Northern New York State (NNYS) in the USA (Uyemoto 1976; Uyemoto et al. 1977; Pearson et al. 1985) and in Argentina in 1980 (Caudwell 1990). Since 1987 GY symptoms have been registered again in the USA in Virginia (Wolf et al. 1994). The first detection of phytoplasmas in grapevine in Canada occurred in 2006 (Rott et al. 2007); finally very recently GY was reported in 2015 in vineyards from Perú (Wei et al. 2017).

North America In NNYS, grapevine leaf curl and berry shrivel symptoms have been reported in two De Chaunac vineyards since 1974, and it was observed that by eliminating the affected plants, the symptoms did not reappear: it was not possible to determine the disease etiology, although the symptoms resembled those of FD (van Heerden 1978; Magarey 1986). In Virginia GY was observed in Chardonnay and Riesling; it is a lethal disease; however, it spreads slowly (Wolf et al. 1994; Davis et al. 1998). Phytoplasmas belonging to groups 16SrI (aster yellows) and 16SrIII (X-disease) were detected (Prince et al. 1993), later classified as 16SrI-A and 16SrIII-I subgroups (Davis et al. 1998). Subsequently, the identification of the 16SrIII phytoplasmas in grapevine was revised indicating the presence of two 16SrIII-A sequevars, named 16SrIIIα and 16SrIIIβ, distinct from '*Ca.* P. pruni' (Davis et al. 2015), but no epidemiological information is provided about these two variants in infected vineyards. This disease named North American GY (NAGY) was only found in the USA, in particular in Maryland, Missouri, southeast Pennsylvania, Ohio, Virginia, and the Finger Lakes region of New York State infecting Chardonnay, Pinot Gris, Viognier, Petit Manseng, Cabernet Sauvignon, Malbec, and Black Malvasia (Davis et al. 2012, 2015). The NAGY phytoplasma alternative hosts are *Vitis* spp., *Ulmus americana*, *Platanus occidentalis*, *Prunus serotina*, *Taraxacum* sp., *Trifolium pratense*, *T. repens*, and *Lespedeza* spp. (Stoepler and Wolf 2014). The role of these host plants in the NAGY epidemiology is not known; however affected plants have been found in cluster inside the vineyards or on the edge of them. This situation, together with the identification of several species of leafhopper potential vectors of phytoplasmas and the detection of 16SrI and 16SrIII phytoplasmas in asymptomatic *V. riparia* plants, leads to the conclusion of possible horizontal dissemination in the vineyards (Prince et al. 1993; Davis et al. 1998; Beanland et al. 2006; Stoepler and Wolf 2014). It was proven that *S. titanus* is able to transmit NAGY phytoplasmas to broad bean (*Vicia faba*) plants, although at the time of the trial the phytoplasmas involved in NAGY disease had not yet been identified (Maixner et al. 1993). Undoubtedly it would be necessary to carry out further transmission trials to confirm *S. titanus* in the role of vector of the phytoplasmas involved in the NAGY disease. Indeed, the leafhopper has high probability to be vector of NAGY phytoplasmas, because it is known that it is able to experimentally transmit to grapevine phytoplasmas belonging to the ribosomal group 16SrI (Alma et al. 2001). In addition, *S. titanus*, native of the USA, prefers to feed on *V. labrusca* L. and *V. riparia*, which have been found near the vineyards infected by NAGY phytoplasmas (Maixner et al. 1993; Beanland et al. 2006; Chuche and Thiéry 2014; Davis et al. 2015).

The first phytoplasma detected in grapevine in Canada, in vineyards from Ontario and British Columbia, was '*Ca.* P. solani' (16SrXII-A) (Rott et al. 2007). From 2006 to 2008, phytoplasmas belonging to the 16SrI-A, 16SrI-B, and 16SrI-C subgroups and 5 new strains or subgroups were reported in 22 symptomatic and symptomless grapevine cultivars in British Columbia, Ontario, and Quebec, 16SrI-A being the most frequently phytoplasma detected (Olivier et al. 2009b, 2014). Sporadically, the presence of phytoplasmas belonging to the 16SrIII ribosomal group has also been reported (Saguez et al. 2015). The highest percentages of infection was observed in Sauvignon Blanc, Cabernet Franc, Syrah, and Cabernet Sauvignon cultivars (Olivier et al. 2014; Vincent et al. 2015). About 90% of the positive samples were collected from asymptomatic plants: this high percentage of symptomless infection is not reflected in any of the other grapevine-growing countries in the world (Olivier et al. 2009b, 2014). The most likely explanation for this situation is associated with the low seasonal temperatures of the wine-growing regions of Canada. It has been demonstrated that in tricolor chrysanthemum, *Ismelia carinata* (Schousb.) Sch. Bip. 1844, infected with phytoplasmas belonging to the 16SrI ribosomal group, the rapid onset of symptoms is favored only with high temperatures (Maggi et al. 2014). In Canada 11 phytoplasma insect vectors have been identified, but a total of 37 leafhopper species have been found positive for phytoplasmas. Nine out of these 11 insect species are known as 16SrI ribosomal group phytoplasma vectors: *Amplicephalus inimicus* (Say, 1830), *Aphrodes bicinctus* (Schrank), *Colladonus geminatus* (Van Duzee), *Euscelis maculipennis* (DeLong and Davidson 1934), *Gyponana hasta* (DeLong 1942), *Macrosteles quadrilineatus* (Forbes 1885), *Paraphlesius irroratus* (Say), *S. titanus*, and *Scaphytopius acutus* (Say) (Lee et al. 2004a; Olivier et al. 2009a, 2014; Saguez et al. 2014). All the vector species found in Canada, except for *S. titanus* and *M. quadrilineatus*, feed preferentially on weeds; however the latter one is abundantly present in Canadian vineyards and, together with the other insect vector species, its ability to transmit 16SrI phytoplasmas to and from grapevine is unknown (Nielson 1968; Alma et al. 1996; Lee et al. 2004a; Olivier et al. 2014; Saguez et al. 2014). 16SrI phytoplasmas were detected in several alternative hosts throughout Canada, but in the same plants collected near to or within vineyards, 16SrI phytoplasmas were never detected. This situation suggests that phytoplasmas belonging to the 16SrI ribosomal group are endemic in Canada and may be transmitted to grapevine from these reservoir plants (Olivier et al. 2009a).

South America In Argentina typical symptoms of GY were observed in vineyards by Hewitt in 1980 (Caudwell 1990), but phytoplasmas involved have not been identified. The presence of GY diseases in *V. vinifera* L. in Chile was reported in 1971, based on the observation of symptoms; however the first laboratory evidence for the presence of phytoplasmas in grapevine occurred in 2003 by electron microscopy and molecular tools (Gajardo et al. 2003; Herrera and Madariaga 2003; Bertaccini et al. 2004). To date six phytoplasma subgroups have been detected in grapevine in Chile: 16SrI-B, 16SrI-C, 16SrIII-J, 16SrV-A, 16SrVII-A, and 16SrXII-A (Gajardo et al. 2009; Fiore et al. 2015b). There is no information about the prevalence of GY in Chilean vineyards. The 16SrIII-J is the most widespread phytoplasma, infecting not only grapevine but also several other woody and herbaceous species (González et al. 2011; Fiore et al. 2015b; Quiroga et al. 2015), and its draft genome sequence is now available (Zamorano and Fiore 2016). Transmission trials have shown that *Paratanus exitiosus* (Beamer) and *Bergallia valdiviana* Berg 1881 can transmit 16SrIII-J phytoplasma to periwinkle and the first insect also to grapevine (Fiore et al. 2012, 2015a; Quiroga et al. 2015). *S. titanus* has never been found in Chile, while *P. exitiosus* and *B. valdiviana* are widely distributed throughout the country. They are commonly captured on weeds in Chilean vineyards and occasionally feed on grapevine plants (Fiore et al. 2015a; Quiroga et al. 2015). The leafhopper *Amplicephalus curtulus* Linnavuori & De Long, in which phytoplasmas belonging to 16SrI-B and 16SrXII-A subgroups were detected, also has been frequently captured in weeds occurring in Chilean vineyards, but it was not possible to determine whether it transmits these phytoplasmas to grapevine (Longone et al. 2011). Some weeds present in or around the vineyards were found positive for phytoplasmas detected also in grapevine: *Galega officinalis* was found infected with 16SrVII-A phytoplasma; *C. arvensis* with 16SrI-B, 16SrVII-A, and 16SrXII-A; and *Polygonum aviculare* with 16SrI-B and 16SrVII-A (Longone et al. 2011). In the fall of 2015, grapevine plants exhibiting symptoms of leaf discoloration (yellowing), vein necrosis, and in some cases berry shriveling were observed in Peru, and '*Ca.* P. brasiliense'-related phytoplasmas (16SrXV-A) were identified (Wei et al. 2017).

5.6 Grapevine Yellows in Africa and Asia

In Asian countries grapevine has been domesticated and grown for a long time; however there are only scattered reports about the presence of GY. So far there are only a few reports of 16SrXII-A phytoplasmas (BN) in grapevine in Lebanon, in China, and in Georgia (Choueiri et al. 2002; Duduk et al. 2010; Quaglino et al. 2014); the same phytoplasma was also detected in Jordan, Iran, and Israel (Salem et al. 2013; Mirchenari et al. 2015; Sharon et al. 2005), while in Lebanon and in Iran, 16SrIX-C phytoplasmas occur in grapevine and at least 12 wild plant species (Casati et al. 2016; Salehi et al. 2016). A strain related to "stolbur" but differentiable from it at the RFLP level on 16S rDNA was also reported in Iran (Karimi et al. 2009). Phytoplasmas in the 16SrIX group were also reported in a few GY-affected grapevines in Turkey where the main phytoplasmas detected were 16SrXII-A and 16SrI (Canik et al. 2011; Ertunc et al. 2015) and more recently in Iran (Zamharir et al. 2017). In Iran other phytoplasmas such as strains '*Ca.* P. fraxini'-related

(16SVII-A) and '*Ca.* P. aurantifolia'-related (16SrII-B) were recently reported in grapevine showing symptoms of decline and leaf scorch (Zamharir et al. 2017).

GY diseases were first reported in Africa in Tunisia in 2004 and were associated with 16SrI-B phytoplasmas (M'hirsi et al. 2004). In 2006 a GY disease in South Africa was associated with a mixed infection of 16SrXII-A and 16SrII-B phytoplasmas (Botti and Bertaccini 2006b). Subsequent studies have shown that the epidemic of GY disease is associated with the presence of 16SrI-B phytoplasmas in South African grapevines. The disease has been observed in several cultivars, although Chardonnay was the most frequently affected, showing rapid decline and eventual death of the grapevines with yield losses of up to 30% (Carstens 2014). The yearly incidence of GY disease is often lowest in the first year and increases in subsequent years. A cumulative incidence of up to 37.7% over 4 years has been reported for some vineyards, which indicates that new infections are likely to occur each season (Carstens et al. 2011; Carstens 2014). The 16SrI-B phytoplasmas have also been detected in 11 other plant species within and around vineyards, including *Catharanthus roseus*, *Bidens bipinnata*, *Erigeron bonariensis*, *Sonchus oleraceus*, *Raphanus raphanistrum*, *Cucurbita* sp., *Setaria verticillata*, *Triticosecale* sp., *Zea mays*, and *Urtica urens* (Krüger et al. 2015a, b); it is possible that some of these species act as reservoirs from which the phytoplasmas spread to grapevines. The vector of 16SrI-B phytoplasmas to grapevine in South Africa is *Mgenia fuscovaria*, leafhopper native to South Africa (Krüger et al. 2011). It is the most abundant leafhopper in vineyards, and adults are present all year round (Krüger et al. 2015b). Grapevine is not yet a proven acquisition host, but if *M. fuscovaria* can acquire 16SrI-B phytoplasmas from grapevine and transmit them to other grapevines, then GY disease incidence might have a significant impact on production of susceptible varieties.

5.7 Conclusions

GY are spread in almost all of the grapevine-growing areas worldwide, though the typical symptoms or the associated phytoplasmas have not yet been identified in some countries. Associated phytoplasmas and epidemiological cycles can be very diverse. The main information gathered so far is summarized in Table 5.1. Host plants and vectors can be many; however GY can also be transmitted by grafting, with variable rate due to the uneven distribution of phytoplasmas among and along the canes, difference in strains infecting the plants, and poor survival rate of infected scions or rootstocks (Borgo et al. 2007). Latency in grapevine is particularly concerning, as it allows collection of scions from not (yet) symptomatic, but infected grapevines. Moreover, *Vitis* sp., which is used as rootstock, is generally a symptomless host of the phytoplasma or can show very little symptoms (Boudon-Padieu 2000; Caudwell et al. 1994; Constable 2010; Eveillard et al. 2016). While in some areas, like Europe, many efforts have been carried out to identify and characterize the phytoplasmas associated with GY, their host and plant vectors, further studies are required to identify the epidemiological cycles and the insect vectors in countries in which they are not yet studied.

References

Alma A., Arnò C., Arzone A., Vidano C (1988) New biological reports on Auchenorrhyncha in vineyards. Proceedings of 6th Auchenorrhyncha Meeting. Turin, Italy, 509–516.

Alma A, Davis RE, Vibio M, Danielli A, Bosco D, Arzone A, Bertaccini A (1996) Mixed infection of grapevines in Northern Italy by phytoplasmas including 16S rRNA RFLP subgroup 16SrI-B strains previously unreported in this host. *Plant Disease* **80**, 418–421.

Alma A, Palermo S, Boccardo G, Conti M (2001) Transmission of chrysanthemum yellows, a subgroup 16SrI-B phytoplasma, to grapevine by four leafhopper species. *Journal of Plant Pathology* **83**, 181–187.

Angelini E, Clair M, Borgo A, Bertaccini A, Boudon-Padieu E (2001) "Flavescence dorée" in France and Italy: occurrence of closely related phytoplasma isolates and their near relationships to palatinate grapevine yellows and an alder yellows phytoplasma. *Vitis* **40**, 79–86.

Angelini E, Negrisolo E, Clair D, Borgo M, Boudon-Padieu E (2003) Phylogenetic relationships among "flavescence dorée" strains and related phytoplasmas determined by heteroduplex mobility assay and sequence of ribosomal and nonribosomal DNA. *Plant Pathology* **52**, 663–672.

Angelini E, Squizzato F, Luchetta G, Borgo M (2004) Detection of a phytoplasma associated with grapevine "flavescence dorée" in *Clematis vitalba*. *European Journal of Plant Pathology* **110**, 193–201.

Arnaud G, Malembic-Maher S, Salar P, Bonnet P, Maixner M, Marcone C, Boudon-Padieu E, Foissac X (2007) Multilocus sequence typing confirms the close genetic interrelatedness of three distinct "flavescence dorée" phytoplasma strain clusters and group 16SrV phytoplasmas infecting grapevine and alder in Europe. *Applied and Environmental Microbiology* **73**, 4001–4010.

Aryamanesh N, Al-Subhi AM, Snowball R, Yan G, Siddique KHM (2011). First report of *Bituminaria* witches' broom in Australia caused by a 16SrII phytoplasma. Plant Disease **95**, 226.

Aryan A, Brader G, Mörtel J, Pastar M, Riedle-Bauer M (2014) An abundant "stolbur" tuf b phytoplasma strain is associated with grapevine, stinging nettle and *Hyalesthes obsoletus*. *European Journal of Plant Pathology* **140**, 213–227.

Batlle A, Altabella N, Sabate J, Lavina A (2008) Study of the transmission of "stolbur" phytoplasma to different crop species, by *Macrosteles quadripunctulatus*. *Annals of Applied Biology* **152**, 235–242.

Bayliss KL, Saqib M, Dell B, Jones MGK, Hardy GEStJ (2005) First record of '*Candidatus* Phytoplasma australiense' in paulownia trees. *Australasian Plant Pathology* **34**, 23–24.

Beanland L (2002) Developing AGY management strategies. GWRDC Final Report DNR00/02.

Beanland L, Noble R, Wolf, TK (2006) Spatial and temporal distribution of North American grapevine yellows disease and of potential vectors of the causal phytoplasmas in Virginia. *Environmental Entomology* **35**, 332–344.

Belli G, Bianco PA, Conti M (2010) Grapevine yellows: past, present and future. *Journal of Plant Pathology* **92**, 303–326.

Bellomo C, Carraro L, Ermacora P, Pavan F, Osler R, Frausin C, Governatori G (2007) Recovery phenomena in grapevines affected by grapevine yellows in Friuli Venezia Giulia. *Bulletin of Insectology* **60**, 235–236.

Bertaccini A, Vibio M, Stefani E (1995) Detection and molecular characterization of phytoplasmas infecting grapevine in Liguria (Italy). *Phytopathologia Mediterranea* **34**, 137–141.

Bertaccini A, Murari E, Vibio M, Danielli A, Davis RE, Borgo M, Consolaro R, Sancassani P (1996) Identificazione molecolare dei fitoplasmi presenti in viti affette da giallumi nel Veneto. *L'Informatore Agrario* **20**, 55–59.

Bertaccini A, Vibio M, Schaff DA, Murari E, Martini M, Danielli A (1997) Geographical distribution of elm yellows-related phytoplasmas in grapevine "flavescence dorée" outbreaks in Veneto (Italy). 12th ICVG Meeting, Lisbon, Portugal, 57–58.

Bertaccini A, Mori N, Botti S, Castiglioni A, Cavallini G, Malossi A (2003) Survey on "bois noir" phytoplasmas spreading in vineyards of Modena province (Italy). 14th ICVG Meeting, Locorotondo, Bari, Italy, 104–105.

Bertaccini A, Botti S, Fiore N, Gajardo A, Montealegre J (2004) Identification of a new phytoplasma(s) infecting grapevine with yellows in Chile. 15 Congress International Organization for Mycoplasmology, Athens, Georgia, USA, 63–64.

Bertaccini A, Paltrinieri S, Dal Molin F, Mitrović J, Duduk B (2009) Molecular identification and geographic distribution of "flavescence dorée" phytoplasma strains. *Le Progrès agricole et viticole* **HS**, 135–136.

Bertaccini A, Duduk B, Paltrinieri S, Contaldo N (2014) Phytoplasmas and phytoplasma diseases: a severe threat to agriculture. *American Journal of Plant Sciences* **5**, 1763–1788.

Bonfiglioli RG, Magarey PA, Symons RH (1995) PCR confirms an expanded symptomatology for Australian grapevine yellows. *Australian Journal of Grape Wine Research* **1**, 71–75.

Bonfiglioli RG, Carey CT, Schleifert LF, Kinnear AJ, Symons RH (1997) Description and progression of symptoms associated with grapevine yellows disease in young Chardonnay vines in the Sunraysia district. *The Australian Grapegrower Winemaker* **400**, 11–15.

Borgo M, Angelini E, Bellotto D (2006) *Vitis vinifera* sensitivity to grapevine yellows phytoplasmas in Italy. XXIX Congrés Mondial de l'OIV, 34–35, CD 1–11.

Borgo M, Angelini E, Filippin L, Bellotto D, Forte V, Stringher L, Bazzo I (2007) Critical points in prevention and control strategies against grapevine yellows. XXX Congrés Mondial de l'OIV, CD: 1–6.

Borgo M, Filippin L, Bertazzon N, Angelini E (2009) Detection of phytoplasmas associated with grapevine yellows in rootstocks. *Le Progrès Agricole et Viticole* **HS**, 162–163.

Botti S, Bertaccini A (2006a) FD-related phytoplasmas and their association with epidemic and non epidemic situations in Tuscany (Italy). 15 ICVG Meeting, Stellenbosch, South Africa, 163–164.

Botti S, Bertaccini A (2006b) First report of phytoplasmas in grapevine in South Africa. *Plant Disease* **90**, 1360.

Botti S, Bertaccini A (2007) Grapevine yellows in Northern Italy: molecular identification of "flavescence dorée" phytoplasma strains and of "bois noir" phytoplasmas. *Journal of Applied Microbiology* **103**, 2325–2330.

Boudon-Padieu E. (2000) Recent advances on grapevine yellows: detection, etiology, epidemiology and control strategies. 13 ICVG Meeting, Adelaide, Australia, 87–88.

Boudon-Padieu E. (2003) The situation of grapevine yellows and current research directions: distribution, diversity, vectors, diffusion and control. 14th ICVG Meeting, Locorotondo, Bari, Italy, 12–17 September, 47–53.

Bressan A, Girolami V, Boudon Padieu E (2005) Reduced fitness of the leafhopper vector *Scaphoideus titanus* exposed to "flavescence dorée" phytoplasma. *Entomologia experimentalis et applicata* **115**, 283–290.

Canik D, Ertunc F, Paltrinieri S, Contaldo N, Bertaccini A (2011) Identification of different phytoplasmas infecting grapevine in Turkey. *Bulletin of Insectology* **64**(Supplement), S225-S226.

Carstens R (2014) The incidence and distribution of grapevine yellows disease in South African vineyards. Dissertation (MSc), Stellenbosch University. http://hdLhandlEnet/10019.1/86683

Carstens R, Petersen Y, Stephan D, Burger J (2011) Current status of aster yellows disease in infected vineyards in the Vredendal grape producing area of South Africa. *Phytopathogenic Mollicutes* **1**, 83–85.

Casati P, Quaglino F, Abou-Jawdah Y, Picciau L, Cominetti A, Tedeschi R, Jawhari M, Choueiri E, Sobh H, Molino Lova M, Beyrouthy M, Alma A, Bianco PA (2016) Wild plants could play a role in the spread of diseases associated with phytoplasmas of pigeon pea witches' broom group (16SrIX). *Journal of Plant Pathology* **98**, 71–81.

Casati P, Jermini M, Quaglino F, Corbani G, Schaerer S, Passera A, Bianco PA, Rigamonti IE (2017) New insights on "flavescence dorée" phytoplasma ecology in the vineyard agroecosystem in southern Switzerland. *Annals of Applied Biology* **171**, 37–51.

Caudwell A (1957) Deux années d'étude sur la flavescence dorée, nouvelle maladie grave de la vigne. *Annales de l'Amélioration de Plant* **4**, 359–393.

Caudwell A (1980) L'amarilliamento de Elqui, nouvelle jaunisse de la vigne au Chili. ICVG Meeting, Niagara Falls, Ontario, Canada, 9–13.

Caudwell A (1990) Epidemiology and characterization of "flavescence dorée" (FD) and other grapevine yellows. *Agronomie* **10**, 655–663.

Caudwell A, Kuszala C, Bachelier JC, Larrue J (1970) Transmission de la flavescence dorée de la vigne aux herbacées par l'allongement du temps d'utilisation de la cicadelle *Scaphoideus littoralis* Ball et l'étude de sa survie sur un grand nombre d'espèces végétales. *Annales de Phytopathologie* **2**, 415–428.

Caudwell A, Larrue J, Kuzsala C, Bachelier JC (1971) Pluralité des jaunisses de la vigne. *Annales de Phytopathologie* **3**, 95–105.

Caudwell A, Kuszala C, Larrue J Bachelier JC (1972) Transmission de la flavescence dorée de la fève à la fève par des cicadelles des genres *Euscelis* et *Euscelidius*. *Annales de Phytopathologie* **HS**, 181–189.

Caudwell A, Larrue J, Tassart V (1994) Caractère "porteur de la flavescence dorée" chez les vignes porte-greffes, en particulier le 3309 Couderc et le Fercal. *Agronomie* **14**, 83–94.

Charles JG, Allan DJ, Andersen MT, Langford G, Mossop D (2002) The search for a vector of strawberry lethal yellows (SLY) in New Zealand. *New Zealand Plant Protection* **55**, 385–389.

Choueiri E, Jreijiri F, El Zammar S, Verdin E, Salar P, Danet JL, Bové J, Garnier M (2002) First report of grapevine "bois noir" disease and a new phytoplasma infecting solanaceous plants in Lebanon. *Plant Disease* **86**, 697.

Chuche J, Thiéry D (2014) Biology and ecology of the "flavescence dorée" vector *Scaphoideus titanus*: a review. *Agronononomic Sustainable Development* **34**, 381–403.

Chuche J, Danet J-L, Salar P, Thiery D (2016) Transmission of '*Candidatus* Phytoplasma solani' by *Reptalus quinquecostatus* (Hemiptera: Cixiidae). *Annals of Applied Biology* **169**, 214–223.

Chireceanu C, Ploaie PG, Petrescu A, Cazacu S (2013) Molecular detection of "stolbur" phytoplasma from natural *foci* associated with grapevine yellows in Romania. 3rd European "Bois Noir" Workshop, Barcelona, Spain, 40.

Clingeleffer PR (1988) Response of Riesling clones to mechanical hedging and minimal pruning of cordon trained vines (MPCT) – implications for clonal selection. *Vitis* **27**, 87–93.

Constable FE (2010) Phytoplasma epidemiology: Grapevines as a model In: Phytoplasmas: Genomes, Plant Hosts and Vectors. Eds Weintraub P, Jones P, Wallingford. CAB International, United Kingdom, 188–212 pp.

Constable FE, Whiting JR, Jones J, Gibb KS, Symons RH (2002) A new grapevine yellows phytoplasma from the Buckland Valley of Victoria, Australia. *Vitis* **41**, 147–154.

Constable FE, Gibb KS, Symons, RH (2003a) The seasonal distribution of phytoplasmas in Australian grapevines. *Plant Pathology* **52**, 267–276.

Constable, FE; Whiting, JR; Jones, J, Gibb, KS and Symons, RH (2003b). The distribution of grapevine yellows disease associated with the Buckland Valley grapevine yellows phytoplasma. *Journal of Phytopathology* **151**, 65–73.

Constable FE, Jones J, Gibb KS, Chalmers YM, Symons RH (2004) The incidence distribution and expression of Australian grapevine yellows, restricted growth and late season curl diseases in selected Australian vineyards. *Annals of Applied Biology* **144**, 205–218.

Constable FE, Nancarrow N, Molero Obregon SJ, Kelly G, Rodoni B (2016) Developing molecular diagnostics for the detection of strawberry viruses. Final report for Horticulture Innovation Australia Limited Project Number: BS11008.

Contaldo N, Soufi Z, Bertaccini A (2011) Preliminary identification of phytoplasmas associated with grapevine yellows in Syria. *Bulletin of Insectology* **64**(Supplement), S217-S218.

Cvrković T, Jović J, Mitrović M, Krstic O, Tosevski I (2014) Experimental and molecular evidence of *Reptalus panzeri* as a natural vector of "bois noir". *Plant Pathology* **63**, 42–53.

Davis RE, Dally EL, Gundersen DE, Lee I-M, Habili N (1997) '*Candidatus* Phytoplasma australiense', a new phytoplasma taxon associated with Australian grapevines yellows. *International Journal of Systematic Bacteriology* **47**, 26–29.

Davis RE, Jomantiene R, Dally EL, Wolf TK (1998) Phytoplasmas associated with grapevine yellows in Virginia belong to group 16SrI, subgroup A (tomato big bud phytoplasma subgroup) and group 16SrIII, new subgroup I. *Vitis* **37**, 131–137.

Davis RE, Dally EL, Zhao Y, Lee I, Wolf TK, Beanland L, Fiola JA, Chien M, Walter-Peterson, HC (2012) Grapevine yellows from New York to Virginia: multilocus molecular genetic characterization of phytoplasmas affecting commercial vineyards. *Phytopathology* **102**, S3.2.

Davis RE, Dally EL, Zhao Y, Ming Lee I-M, Wei W, Wolf TK, Beanland L, Johnson DA, Fiola JA, Walter-Peterson H, Dami I, Chien M (2015) Unraveling the etiology of North American grapevine yellows (NAGY): novel NAGY phytoplasma sequevars related to '*Candidatus* Phytoplasma pruni'. *Plant Disease* **99**, 1087–1097.

Davis RI, Jacobson SC, De La Rue SJ, Tran-Nguyen L, Gunua TG, Rahamma S (2003). Phytoplasma disease surveys in the extreme north of Queensland, Australia, and the island of New Guinea. *Australian Plant Pathology* **32**, 269–277.

Delić D, Martini M, Ermacora P, Carraro L, Myrta A (2006) First report of grapevine "bois noir" in Bosnia and Herzegovina. *Journal of Plant Pathology* **88**, 225–229.

Dermastia M, Bertaccini A, Constable F, Mehle N (2017) Grapevine yellows diseases and their phytoplasma agents: biology and detection. Springer Briefs in Agriculture, 99 pp.

De Sousa E, Casati P, Cardoso F, Baltazar C, Durante G, Quaglino F, Bianco PA (2010) "Flavescence dorée" phytoplasma affecting grapevine (*Vitis vinifera*) newly reported in Portugal. *Plant Pathology* **59**, 398.

De Sousa E, Mimoso C, Cardoso F (2013) "Bois noir" identification on grapevine in Portugal. 3rd European "Bois Noir" Workshop, Barcelona, Spain, 48.

Duduk B, Ivanovic M, Dukic N, Botti S, Bertaccini A (2003) First report of an elm yellows subgroup 16SrV-C phytoplasma infecting grapevine in Serbia. *Plant Disease* **87**, 599.

Duduk B, Botti S, Ivanovic M, Krstic B, Dukic N, Bertaccini A (2004) Identification of phytoplasmas associated with grapevine yellows in Serbia. *Journal of Phytopathology* **152**, 575–579.

Duduk B, Tian JB, Contaldo N, Fan XP, Paltrinieri S, Chen QF, Zhao QF, Bertaccini A (2010) Occurrence of phytoplasmas related to "stolbur" and to '*Candidatus* Phytoplasma japonicum' in woody host plants in China. *Journal of Phytopathology* **158**, 100–104.

Ember I, Acs Z, Salar P, Danet J-L, Foissac X, Kolber M, Malembic-Maher S (2011) Survey and genetic diversity of phytoplasmas from the 16SrV-C and -D subgroups in Hungary. *Bulletin of Insectology* **64**(Supplement), 33–34.

Emmett RW, Pywell M, Holland SA, Magarey PA, Wachtel MF (1983) Australian vine yellows III. Effect on the yield of two grapevine cultivars – preliminary results. 4th International Congress of Plant Pathology, University of Melbourne, Melbourne, Victoria, Australia, 453.

Engelbrecht M, Joubert J, Burger TJ (2010) First report of aster yellows phytoplasma in grapevines in South Africa. *Plant Disease* **94**, 373.

Ertunc F, Canik Orel D, Bayram S, Paltrinieri S, Bertaccini A, Topkaya Ş, Soylemezoglu G (2015) Occurrence and identification of grapevine phytoplasmas in main viticultural regions of Turkey. *Phytoparasitica* **43**, 303–310.

Eveillard S, Jollard C, Labroussaa F, Khalil D, Perrin M, Desque D, Salar P, Razan F, Hevin C, Bordenave L, Foissac X, Masson JE, Malembic-Maher S (2016) Contrasting susceptibilities to "flavescence dorée" in *Vitis vinifera*, rootstocks and wild *Vitis* species. *Frontiers in Plant Science* **7**, 1762.

Fabre A, Danet JL, Foissac X (2011) The "stolbur" phytoplasma antigenic membrane protein gene stamp is submitted to diversifying positive selection. *Gene* **472**, 37–41.

Fialová R, Válová P, Balakishiyeva G, Danet JL, Šafářová D, Foissac X, Navrátil M (2009) Genetic variability of "stolbur" phytoplasma in annual crop and wild plant species in south Moravia. *Journal of Plant Pathology* **91**, 411–416.

Filippin L, Jović J, Forte V, Cvrković T, Toševski I, Borgo M, Angelini E (2007) Occurrence and diversity of phytoplasmas detected in clematis and their relationships with grapevine "flavescence dorée" phytoplasmas. *Bulletin of Insectology* **60**, 327–328.

Filippin L, Jović J, Cvrković T, Forte V, Clair D, Toševski I, Boudon-Padieu E, Borgo M, Angelini E (2009) Molecular characteristics of phytoplasmas associated with "flavescence dorée" in

clematis and grapevine and preliminary results on the role of *Dictyophara europaea* as a vector. *Plant Pathology* **58**, 826–837.
Filippin L, De Prà V, Zottini M, Borgo M, Angelini E (2011) Nucleotide sequencing of imp gene in phytoplasmas associated to "flavescence dorée" from *Ailanthus altissima*. *Bulletin of Insectology* **64**(Supplement), S49-S50.
Fiore N, Longone V, González F, Zamorano A, Pino AM, Araya J, Picciau L, Alma A, Paltrinieri S, Contaldo N, Bertaccini A (2012) Transmission of 16SrIII-J phytoplasma by *Paratanus exitiosus* leafhopper. 17th ICVG Meeting, Davis, California, USA, 230–231.
Fiore N, Longone V, Gonzalez X, Zamorano A, Pino AM, Quiroga N, Picciau L, Alma A, Paltrinieri S, Contaldo N, Bertaccini A (2015a) Transmission of 16SrIII-J phytoplasma by *Paratanus exitiosus* (Beamer) leafhopper in grapevine. *Phytopathogenic Mollicutes* **5**(1-Supplement), S43–S44.
Fiore N, Zamorano A, Pino AM (2015b) Identification of phytoplasmas belonging to the ribosomal groups 16SrIII and 16SrV in Chilean grapevines. *Phytopathogenic Mollicutes* **5**, 32–36.
Foissac X, Carle P, Fabre A, Salar P, Danet J-L, the Stolbur-Euromedconsortium (2013) '*Candidatus* Phytoplasma solani' genome project and genetic diversity in the Euro-Mediterranean basin. 3rd European "Bois Noir" Workshop. Barcelona, Spain, 11–13.
Gaffuri F, Sacchi S, Cavagna B (2011) First detection of the mosaic leafhopper, *Orientus ishidae*, in northern Italian vineyards infected by the "flavescence dorée" phytoplasma. *New Disease Reports* **24**, 22.
Gajardo A, Botti S, Montealegre J, Fiore N, Bertaccini A (2003) Survey on phytoplasmas identified in Chilean grapevines. 14th ICVG Meeting, Locorotondo, Bari, Italy, 85–86.
Gajardo A, Fiore N, Prodan S, Paltrinieri S, Botti S, Pino AM, Zamorano A, Montealegre J, Bertaccini A (2009) Phytoplasmas associated with grapevine yellows disease in Chile. *Plant Disease* **93**, 789–796.
Galetto L, Miliordos D, Roggia C, Rashidi M, Sacco D, Marzachì C, Bosco D (2014) Acquisition capability of the grapevine "flavescence dorée" by the leafhopper vector *Scaphoideus titanus* Ball correlates with phytoplasma titre in the source plant. *Journal of Pest Science* **87**, 671–679.
Galetto L, Miliordos DE, Pegoraro M, Sacco D, Veratti F, Marzachì C, Bosco D (2016) Acquisition of "flavescence dorée" phytoplasma by Scaphoideus titanus Ball from different grapevine varieties. *International Journal of Molecular Sciences* **17**, 1563.
Gambino G, Boccacci P, Margaria P, Palmano S, Gribaudo I (2013) Hydrogen peroxide accumulation and transcriptional changes in grapevines recovered from "flavescence dorée" disease. *Phytopathology* **103**, 776–784.
Gärtel W (1972) Enfermedades y plagas de la viticultura chilena II parte. Universidad Catolica Chile Facultad Agronomica Monografia, Santiago, Chile, 33 pp.
Getachew MA, Mitchell A, Gurr GM, Fletcher MJ, Pilkington LJ, Nikandrow A (2007) First report of a '*Candidatus* Phytoplasma australiense' related strain in lucerne (*Medicago sativa*). Australia. *Plant Disease* **91**, 111.
Gibb KS, Constable FE, Moran JR, Padovan AC (1999) Phytoplasmas in Australian grapevines – Detection, differentiation and associated diseases. *Vitis* **38**, 107–114.
Glenn D (2000) National program for the management of phytoplasmas in Australian grapevines. Final report CRCV 95/2 for Grape and Wine Research and Development Corporation.
González F, Zamorano A, Pino AM, Paltrinieri S, Bertaccini A, Fiore N (2011) Identification of phytoplasma belonging to X-disease group in cherry in Chile. *Bulletin of Insectology* **64**(Supplement), S235–S236.
Grylls NE (1979) Leafhopper vectors and the plant disease agents they transmit in Australia. In: Leafhopper Vectors and Plant Disease Agents. Eds Maramorosch K, Harris FK, Academic Press, New York, USA, 179–214 pp.
Habili N, Farrokhi N, Randles JW (2007) First detection of '*Candidatus* Phytoplasma australiense' in *Liquidambar styraciflua* in Australia. *Plant Pathology* **56**, 346.
Helson GAH (1951) The transmission of witches' broom virus disease of lucerne by the common brown leafhopper, *Orosius argentatus* (Evans). *Australian Journal of Science Research Ser B* **4**, 115–124.

Herrera G, Madariaga M (2003) Evidencias inmunológicas, microscópicas y moleculares de la presencia de fitoplasmas en vides. *Agricultura Técnica* **63**, 15–22.

Hill AV (1941) Yellow dwarf of tobacco in Australia. II. Transmission by the jassid *Thamnotettix argentata* (Evans). *Journal of Council Sciences Industrial Research* **14**, 181–186.

Hill AV (1943) Insect transmission and host plants of virescence (big bud of tomato). *Journal of Council Sciences Industrial Research* **16**, 85–90.

Hutton EM, Grylls NE (1956) Legume "little leaf", a virus disease of subtropical pasture species. *Australian Journal of Agricultural Research* **7**, 85–97.

Jagoueix-Eveillard S, Labroussaa F, Salar P, Danet JL, Hevin C, Perrin M, Masson J, Foissac X, Malembic-Maher S (2012) Looking for resistance to the "flavescence dorée" disease among *Vitis vinifera cultivars* and other *Vitis* species. 17th ICVG Meeting, Davis, California, USA, 234–235.

Jezič M, Karoglan Kontić J, Preiner D, Maletić E, Ćurković-Perica M (2013) Grapevine yellows affecting the Croatian indigenous grapevine cultivar Grk. *Acta Botanica Croata* **72**, 287–294.

Karimi M, Contaldo N, Mahmoudi B, Duduk B, Bertaccini A (2009) Identification of "stolbur"-related phytoplasmas in grapevine showing decline symptoms in Iran. *Le Progrès Agricole et Viticole* **HS**, 208–209.

Kolber M, Ember I, Varga K, Botti S, Martini M, Lázár J, Bertaccini A (2003) Six-year survey of grapevine yellows distribution in Hungary. 14th ICVG Meeting, Locorotondo, Bari, Italy, 99–100.

Kosovac A, Radonjić S, Hrnčić S, Krstic O, Tosevski I, Jovic J (2016) Molecular tracing of the transmission routes of "bois noir" in Mediterranean vineyards of Montenegro and experimental evidence for the epidemiological role of *Vitex agnus-castus* (Lamiaceae) and associated *Hyalesthes obsoletus* (Cixiidae). *Plant Pathology* **65**, 285–298.

Krüger K, De Klerk A, Douglas-Smit N, Joubert J, Pietersen G, Stiller M (2011) Aster yellows phytoplasma in grapevines: identification of vectors in South Africa. *Bulletin Insectology* **64**(Supplement), S137-S138.

Krüger K, Pietersen G, Smit N, Carstens R (2015a) Epidemiology of aster yellows phytoplasma: alternate host plants and the vector *Mgenia fuscovaria* (Hemiptera: Cicadellidae) in South Africa. 18th ICVG Meeting, Ankara, Turkey, 126–127.

Krüger K, Venter F, Schröder ML (2015b) First insights into the influence of aster yellows phytoplasmas on the behaviour of the leafhopper *Mgenia fuscovaria. Phytopathogenic Mollicutes* **5**(1-Supplement), S41-S42.

Landi L, Isidoro N, Riolo P (2013) Natural phytoplasma infection of four phloem-feeding Auchenorrhyncha across vineyard agroecosystems in Central-Eastern Italy. *Journal of Economic Entomology* **106**, 604–613.

Landi L, Riolo P, Murolo S, Romanazzi G, Nardi S, Isidoro N (2015) Genetic variability of stolbur phytoplasma in *Hyalesthes obsoletus* (Hemiptera: Cixiidae) and its main host plants in vineyard agroecosystems. *Journal of Economic Entomology* **108**, 1506–1515.

Langer M, Maixner M (2004) Molecular characterisation of grapevine yellows associated phytoplasmas of the "stolbur"-group based on RFLP-analysis of non-ribosomal DNA. *Vitis* **43**, 191–200

Lee E, Wylie SJ, Jones MGK (2010) First report of '*Candidatus* Phytoplasma aurantifolia' associated with severe stunting and necrosis on the invasive weed *Pelargonium capitatum* in Western Australia. *Plant Disease* **94**, 1264.

Lee I-M, Gundersen-Rindal DE, Davis RE, Bottner KD, Marcone C, Seemuller E (2004a) '*Candidatus* Phytoplasma asteris', a novel phytoplasma taxon associated with aster yellows and related diseases. *International Journal of Systematic and Evolutionary Microbiology* **54**, 1037–1048.

Lee I-M, Martini M, Marcone C, Zhu SF (2004b) Classification of phytoplasma strains in the elm yellows group (16SrV) and proposal of '*Candidatus* Phytoplasma ulmi' for the phytoplasma associated with elm yellows. *International Journal of Systematic and Evolutionary Microbiology* **54**, 337–347.

Lessio F, Tedeschi R, Alma A (2007) Presence of *Scaphoideus titanus* on American grapevine in woodlands, and infection with "flavescence dorée" phytoplasmas. *Bulletin of Insectology* **60**, 373–374.
Lessio F, Picciau L, Gonella E, Mandrioli M, Tota F, Alma A (2016) The mosaic leafhopper *Orientus ishidae*: host plants, spatial distribution, infectivity, and transmission of 16SrV phytoplasmas to vines. *Bulletin of Insectology* **69**, 277–289.
Longone V, Gonzáles F, Zamorano A, Pino AM, Araya J, Díaz V, Paltrinieri S, Calari A, Bertaccini A, Picciau L, Alma A, Fiore N (2011) Epidemiological aspects of phytoplasmas in Chilean grapevines. *Bulletin of Insectology* **64**(Supplement), S91-S92.
M'hirsi S, Acheche H, Fattouch S, Boccardo G, Marrakchi M, Marzouki N (2004) First report of phytoplasmas in the aster yellows group infecting grapevine in Tunisia. *Plant Pathology* **53**, 521.
Magarey PA (1986) Grapevine yellows aetiology, epidemiology and diagnosis. *South African Journal of Enology and Viticulture* **7**.
Magarey PA, Wachtel, MF (1983) Australian vine yellows, I Inference of a mycoplasma-like aetiology for a new disease of Australian grapevine. 4th International Congress of Plant Pathology, Melbourne, Australia, 118.
Magarey PA, Wachtel MF (1986a) Grapevine yellows – a widespread, apparently new disease in Australia. *Plant Disease* **70**, 694.
Magarey PA, Wachtel, MF (1986b) Australian grapevine yellows. *International Journal of Tropical Plant Disease* **4**, 1–14.
Magarey PA, Gillett SR, Dixon JR, Magarey CC, Thiele SA, Cass J, Fletcher MJ, Habili N, Maixner MW, Gurr GM, Pilkington L (2006) Australian grapevine yellows: source, spread and control. Final Report to Grape and Wine Research and Development Corporation. Project Number: SAR 02/03.
Maggi F, Galetto L, Marzachi C, Bosco D (2014) Temperature-dependent transmission of '*Candidatus* Phytoplasma asteris' by the vector leafhopper *Macrosteles quadripunctulatus* Kirschbaum. *Journal of General Applied Entomology* **2**, 87–94.
Maggi F, Bosco D, Galetto L, Palmano S, Marzachì C (2017) Space-time point pattern analysis of "flavescence dorée" epidemic in a grapevine field: disease progression and recovery. *Frontiers in Plant Science* **7,** 1987.
Maixner M (2007) Biology of *Hyalesthes obsoletus* and approaches to control this soilborne vector of Bois noir disease. *Bulletin OILB/SROP* **30**, 3–9.
Maixner M, Pearson RC, Boudon-Padieu E, Caudwell A (1993) *Scaphoideus titanus*, a possible vector of grapevine yellows in New York. *Plant Disease* **77**, 408–413.
Maixner M, Ahrens U, Seemüller E (1995) Detection of the German grapevine yellows (Vergilbungskrankheit) MLO in grapevine, alternative hosts and a vector by a specific PCR procedure. *European Journal of Plant Pathology* **101**, 241–250.
Malembic-Maher S, Salar P, Filippin L, Carle P, Angelini E, Foissac X (2011) Genetic diversity of European phytoplasmas of the 16SrV taxonomic group and proposal of '*Candidatus* Phytoplasma rubi' as new taxa. *International Journal of Systematic and Evolutionary Microbiology* **61**, 2129–2134.
Marchi G, Cinelli T, Rizzo D (2015) Occurrence of different phytoplasma infections in wild herbaceous dicots growing in vineyards affected by "bois noir" in Tuscany (Italy). *Phytopathologia Mediterranea* **54**, 504–515.
Martini M, Murari E, Mori N, Bertaccini A (1999) Identification and epidemic distributions of two "flavescence dorée"-related phytoplasmas in Veneto (Italy). *Plant Disease* **83**, 925–930.
Martini M, Botti S, Marcone C, Marzachì C, Casati P, Bianco PA, Benedetti R, Bertaccini A (2002) Genetic variability among "flavescence dorée" phytoplasmas from different origins in Italy and France. *Molecular and Cellular Probes* **16**, 197–208.
Mehle N, Seljak G, Rupar M, Ravnikar M, Dermastia M (2010) The first detection of a phytoplasma from the 16SrV (elm yellows) group in the mosaic leafhopper *Orientus ishidae*. *New Disease Reporter* **22**, 11.

Mehle N, Rupar M, Seljak G, Ravnikar M, Dermastia M (2011) Molecular diversity of "flavescence dorée" phytoplasma strains in Slovenia. *Bulletin of Insectology* **64**(Supplement), S29–S30.

Mikec I, Križanac I, Budinščak Ž, Šeruga Musić M, Krajačić M, Škorić D (2006) Phytoplasmas and their potential vectors in vineyards of indigenous Croatian varieties 15th ICVG Meeting, Stellenbosch, South Africa, 255–257.

Milkus B, Clair D, Idir S, Habili N, Boudon-Padieu E (2005) First detection of "stolbur" phytoplasma in grapevines (*Vitis vinifera* cv. Chardonnay) affected with grapevine yellows in the Ukraine. *Plant Pathology* **54**, 236.

Mirchenari SM, Massah A, Zirak L (2015) "Bois noir": new phytoplasma disease of grapevine in Iran. *Journal of Plant Protection Research* **55**,88–93.

Mitrev S, Nakova E, Pejčinovski F, Angelini E (2007) Geographical distribution of "bois noir" phytoplasmas infecting grapevines in the Republic of Macedonia. *Bulletin of Insectology* **60**, 155–156.

Mori N, Bressan A, Martini M, Guadagnini M, Girolami V, Bertaccini A (2002) Experimental transmission by *Scaphoideus titanus* Ball of two "flavescence dorée"-type phytoplasmas. *Vitis* **41**, 99–102.

Mori N, Pavan F, Bondavalli R, Reggiani N, Paltrinieri S, Bertaccini A (2008) Factors affecting the spread of "bois noir" disease in north Italy vineyards. *Vitis* **47**, 65–72.

Mori N, Pavan F, Reggiani N, Bacchiavini M, Mazzon L, Paltrinieri S, Bertaccini A (2012) Correlation of "bois noir" disease with nettle and vector abundance in northern Italy vineyards. *Journal of Pest Science* **85**, 23–28.

Mori N, Mitrović J, Smiljković M, Duduk N, Paltrinieri S, Bertaccini A, Duduk B (2013) *Hyalesthes obsoletus* in Serbia and its role in the epidemiology of corn reddening. *Bulletin of Insectology* **66**, 245–250.

Mori N, Quaglino F, Tessari F, Pozzebon A, Bulgari D, Casati P, Bianco PA (2015) Investigation on "bois noir" epidemiology in north-eastern Italian vineyards through a multidisciplinary approach. *Annals of Applied Biology* **166**, 75–89.

Morone C, Boveri M, Giosue S, Gotta P, Rossi V, Scapin I, Marzachì (2007) Epidemiology of "flavescence dorée" in vineyards in northwestern Italy. *Phytopathology* **97**, 1422–1427.

Moutous G (1977) Définitions des symptômes de la flavescence dorée sur les variétés porte-greffes. *Revue Zoologique Agricole Pathologie Végétale* **76**, 90–98.

Murolo S, Romanazzi G (2015) In-vineyard population structure of '*Candidatus* Phytoplasma solani' using multilocus sequence typing analysis. *Infectious Genetic Evolution* **31**, 221–230.

Murolo S, Mancini V, Romanazzi G (2014) Spatial and temporal "stolbur" population structure in a cv. Chardonnay vineyard according to *vmp1* gene characterization. *Plant Pathology* **63**, 700–707.

Musetti R, Marabottini R, Badiani M, Martini M, Sanità di Toppi L, Borselli S, Borgo M, Osler R (2007) On the role of H_2O_2 in the recovery of grapevine (*Vitis vinifera*, cv Prosecco) from "flavescence dorée" disease. *Functional Plant Biology* **34**, 750–758.

Nielson MW (1968) The leafhopper vectors of phytopathogenic viruses (Homoptera, Cicadellidae): taxonomy, biology, and virus transmission. *USDA Technical Bulletin* **1382**, 1–386.

Oliveri C, Pacifico D, D'Urso V, La Rosa R, Marzachì C, Tessitori M (2015) "Bois noir" phytoplasma variability in a Mediterranean vineyard system: new plant host and putative vectors. *Australasian Plant Pathology* **44**, 235–244.

Olivier CY, Lowery DT, Stobbs LW (2009a) Phytoplasma diseases and their relationships with insect and plant hosts in Canadian horticultural and field crops. *Canadian Entomology* **141**, 425–462.

Olivier CY, Lowery DT, Stobbs LW, Vincent C, Galka B, Saguez J, Bittner L, Johnson R, Rott M, Masters C, Green M (2009b). First report of aster yellows phytoplasmas ('*Candidatus* Phytoplasma asteris') in Canadian grapevines. *Plant Disease* **93**, 669.

Olivier CY, Saguezb J., Stobbsc L., Lowery T, Galka B, Whybourne K, Bittner L, Chen X, Vincent C (2014) Occurrence of phytoplasmas in leafhoppers and cultivated grapevines in Canada. *Agriculture, Ecosystems and Environment* **195**, 91–97.

Orenstein S, Zahavi T, Weintraub PG (2001) Distribution of phytoplasmas in wine grapes in the Golan Heights, Israel and development of a new universal primer. *Vitis* **40**, 219–223.
Osler R, Carraro L, Ermacora P, Ferrini F, Loi N, Loschi A, Martini M, Mutton PB, Refatti E (2003) Roguing: a controversial practice to eradicate grape yellows caused by phytoplasmas. 14th ICVG Meeting, Locorotondo, Bari, 68.
Osmelak JA (1986) Predicting vector occurrences and disease incidence in tomato crops: a control strategy. 2nd International Workshop on Leafhoppers and Planthoppers of Economic Importance, Brigham Young University, Provo, Utah, USA, 161–174.
Pacifico D, Alma A, Bagnoli B, Foissac X, Pasquini G, Tessitori M, Marzachì C (2009) Characterization of "bois noir" isolates by restriction fragment length polymorphism of a "stolbur"-specific putative membrane protein gene. *Phytopathology* **99**, 711–715.
Padovan AC, Gibb KS, Bertaccini A, Vibio M, Bonfiglioli RE, Magarey PA, Sears BB (1995) Molecular detection of the Australian grapevine yellows phytoplasma and comparison with grapevine yellows phytoplasmas from Italy. *Australian Journal of Grape and Wine Research* **1**, 25–31.
Pearson RC, Pool RM, Gonsalves D, Goffinet MC (1985) Occurrence of "flavescence dorée"-like symptoms on White Riesling grapevines in New York, USA. *Phytopathologia Mediterranea* **24**, 82–87.
Pilkington LJ, Gibb KS, Gurr GM, Fletcher MJ, Nikandrow A, Elliot E, van de Ven R, Read DMY (2003) Detection and identification of a phytoplasma from lucerne with Australian lucerne yellows disease. *Plant Pathology* **52**, 754–762.
Pilkington LJ, Gurr GM, Fletcher MJ, Nikandrow A, Elliott E (2004) Vector status of three leafhopper species for Australian lucerne yellows phytoplasma. *Australian Journal of Entomology* **43**, 366–373.
Prince JP, Davis RE, Wolf TK, Lee IM, Mogen BD, Dally EL, Bertaccini A, Credi R, Barba M (1993) Molecular detection of diverse mycoplasma-like organisms (MLOs) associated with grapevine yellows and their classification with aster yellows, X-disease, and elm yellows MLOs. *Phytopathology* **83**, 1130–1137.
Quaglino F, Zhao Y, Bianco PA, Wei W, Casati P, Durante G, Davis RE (2009) New 16Sr subgroups and distinct single nucleotide polymorphism lineages among grapevine "bois noir" phytoplasma populations. *Annals of Applied Biology* **154**, 279–289.
Quaglino F, Contaldo N, Duduk B, Pacifico D, Marzachì C, Foissac X, Zhao Y, Bianco PA, Wei W, Casati P, Davis RE, Bertaccini A (2010) Insight into the genetic diversity among phytoplasmas in the "stolbur" group. COST ACTION FA0807, Workshop WG1-WG4 Chianciano Terme (Siena), Italy, 246.
Quaglino F, Zhao Y, Casati P, Bulgari D, Bianco PA, Wei W (2013) '*Candidatus* Phytoplasma solani', a novel taxon associated with "stolbur" and "bois noir" related diseases of plants. *International Journal of Systematic and Evolutionary Microbiology* **63**, 2879–2894.
Quaglino F, Maghradze D, Chkhaidze N, Casati P, Failla O, Bianco PA (2014) First report of '*Candidatus* Phytoplasma solani' and '*Candidatus* Phytoplasma convolvuli' associated respectively with grapevine bois noir and bindweed yellows in Georgia. *Plant Disease* **98**, 1151
Quaglino F, Maghradze D, Casati P, Chkhaidze N, Lobjanidze M, Ravasio A, Passera, Venturini AG, Failla O, Bianco PA (2016) Identification and characterization of new '*Candidatus* Phytoplasma solani' strains associated with bois noir disease in *Vitis vinifera* L. cultivars showing a range of symptoms severity in Georgia, the Caucasus region. *Plant Disease* **100**, 904–915.
Quaglino F, Murolo S, Zhao Y, Casati P, Durante G, Wei W, Bianco PA, Romanazzi G, Davis RE (2017) Identification of new -J and -K 16SrXII subgroups and distinct single nucleotide polymorphism genetic lineages among '*Candidatus* Phytoplasma solani' strains associated with bois noir in Central Italy. *Australasian Plant Pathology* **46**, 31–34.
Quiroga N, González X, Zamorano A, Pino AM, Picciau L, Alma A, Paltrinieri S, Contaldo N, Bertaccini A, Fiore N (2015) Transmission of 16SrIII-J phytoplasma by *Bergallia valdiviana* Berg 1881 leafhopper. *Phytopathogenic Mollicutes* **5**(1-Supplement), S47–S48.

Radonjić S, Hrnčić S, Krstić O, Cvrković T, Mitrović M, Jović J, Toševski I (2013) First report of alder yellows phytoplasma infecting common and grey alder (*Alnus glutinosa* and *A. incana*) in Montenegro. *Plant Disease* **97**, 686–686.

Reisenstein H, Steffek R (2011) First outbreaks of grapevine "flavescence dorée" in Austrian viticulture. *Bulletin of Insectology* **64**(Supplement), S223–S224.

Renault-Spilmont A-S, Beccavin I, Grenan S (2006) Detection of a phytoplasma belonging to group I in declining Syrah. 15 ICVG Meeting, Stellenbosch, South Africa, 195–196.

Riedle-Bauer M, Tiefenbrunner W, Otreba J, Hanak K, Schildberger B, Regner F (2006) Epidemiological observations on "bois noir" in Austrian vineyards. *Mitteilungen Klosterneuburg* **56**, 177–181.

Riedle-Bauer M, Sára A, Regner F (2008) Transmission of a "stolbur" phytoplasma by the Agalliinae leafhopper *Anaceratagallia ribauti* (Hemiptera, Auchenorrhyncha, Cicadellidae). *Journal of Phytopathology* **156**, 687–690.

Rott M, Johnson R, Masters C, Green M (2007) First report of "bois noir" phytoplasma in grapevine in Canada. *Plant Disease* **91**, 1682.

Sabaté J, Laviña A, Legorburu J, Fortanete J, Perez De Obanos J, Perez Marín JL, Reyes J, Batlle A (2007) Incidence of "bois noir" phytoplasma in different wine-growing regions of Spain and its relation to *Hyalesthes obsoletus*. *Bulletin of Insectology* **60**, 367–368.

Saguez J, Olivier C, Hamilton A, Lowery T, Stobbs L, Lasnier J, Galka B, Chen X, Mauffette Y, Vincent C (2014) Diversity and abundance of leafhoppers in Canadian vineyards. Journal of Insect Science **14**. DOI: https://doi.org/10.1093/jis/14.1.73

Saguez J, Olivier C, Lasnier J, Hamilton A, Stobbs L, Vincent C (2015) Biologie et lutte intégrée des cicadelles et des maladies à phytoplasmes des vignobles de l'est du Canada. *Bulletin Technique, Agriculture et agro-alimentaire Canada*, 68 pp.

Salehi E, Salehi M, Taghavi SM, Izadpanah K (2016) First report of a 16SrIX group (pigeon pea witches' broom) phytoplasma associated with grapevine yellows in Iran. *Journal of Plant Pathology* **98**, 376.

Salem NM, Quaglino F, Abdeen A, Casati P, Bulgari D, Alma A, Bianco PA (2013) First report of '*Candidatus* Phytoplasma solani' strains associated with grapevine "bois noir" in Jordan. *Plant Disease* **97**, 1505.

Sakalieva D, Paltrinieri S, Calari A, Bertaccini A (2007) Molecular identification of "bois noir" phytoplasmas in grapevine in Bulgaria. *Bulletin of Insectology* **60**, 153–154.

Saqib M, Smith BD, Parrish JL, Ramsdale RA, Jones MGK (2007) Detection of phytoplasma in *Allocasuarina fraseriana* and *Acacia saligna* in Kings Park. *Journal of Research Society West Austalia* **90**, 175–178.

Saric A, Skoric D, Bertaccini A, Vibio M, Murari E (1997) Molecular detection of phytoplasmas infecting grapevines in Slovenia and Croatia. 12 ICVG Meeting, Lisbon, Portugal, 77–78.

Schneider B, Padovan A, De La Rue S, Eichner R, Davis R, Bernuetz A, Gibb, K (1999) Detection and differentiation of phytoplasmas in Australia: an update. *Australian Journal of Agricultural Research* **50**, 333–342.

Schvester D, Carle P, Moutous G (1962) Sur la trasmission de la flavescence dorée des vigne par une cicadelle. *Compte Rendu de l'Académie d'Agriculture de France* **47**, 1021–1024.

Schvester D, Carle P, Moutous G (1963) Transmission de la flavescence dorée de la vigne par *Scaphoideus littoralis* Ball. *Annales de l'Epiphytie* **14**, 175–198.

Schvester D, Carle P, Moutous G (1969) Nouvelles données sur la transmission de la flavescence dorée des vignes par *Scaphoideus littoralis* Ball.. *Annales Zoologique Ecologique Animal* **1**, 445–465.

Seruga-Music M, Škorić D, Haluška I, Križanac I, Plavec J, Mikec I (2011) First report of "flavescence dorée"-related phytoplasma affecting grapevines in Croatia. *Plant Disease* **95**, 353.

Sforza R, Clair D, Daire X, Larrue J, Boudon-Padieu E (1998) The role of *Hyalesthes obsoletus* (Hemiptera: Cixiidae) in the occurrence of "bois noir" of grapevine in France. *Phytopathology* **146**, 549–556.

Sharon R, Soroker V, Wesley SD, Zahavi T, Harari A, Weintraub PG (2005) *Vitex agnus-castus* is a preferred host plant for *Hyalesthes obsoletus*. *Journal of Chemical Ecology* **31**, 1051–1063.

Sommer KJ, Clingeleffer PR, Ollat N. (1993) Effects of minimal pruning on grapevine canopy development, physiology and cropping level in both cool and warm climates. *Viticultural and Enological Sciences* **48**, 135–139.

Starý M, Válová P, Šafářová D, Lauterer P, Ackermann P., Navrátil M (2013) Survey and molecular detection of "bois noir" in vineyards of the Czech Republic *Horticultural Science (Prague)* **40**, 83–87.

Stoepler TM, Wolf TK (2014) Identification of candidate vectors of North American grapevine yellows, a lethal wine grape disease. 2nd Hemipteran-Plant Interactions Symposium, UC Riverside, Riverside, California, USA.

Streten C, Gibb KS (2005) Genetic variation in '*Candidatus* Phytoplasma australiense'. *Plant Pathology* **54**, 8–14.

Streten C, Gibb KS (2006) Phytoplasma diseases in subtropical and tropical Australia. *Australasian Plant Pathology* **35**, 129.

Streten C, Herrington ME, Hutton DG, Persley DM, Waite GK, Gibb KS (2005) Plant hosts of the phytoplasmas and rickettsia-like-organisms associated with strawberry lethal yellows and green petal diseases. *Australasian Plant Pathology* **34**, 165–173.

Torres E, Botti S, Rahola J, Martin MP, A Bertaccini (2005) Grapevine yellows diseases in Spain: eight years survey of disease spread and molecular characterization of phytoplasmas involved. *Anales Jardin Botanico de Madrid* **62**, 127–133.

Trivellone V, Filippin L, Narduzzi-Wicht B, Angelini E (2016) A regional-scale survey to define the known and potential vectors of grapevine yellow phytoplasmas in vineyards South of Swiss Alps. *European Journal of Plant Pathology* **145**, 915–927.

Uyemoto JK (1976) A new disease affecting the grapevine variety De Chaunac. *Proceedings American Phytopathological Society* **1**, 146.

Uyemoto JK, Cummins JR, Abawi GS (1977) Virus and virus-like diseases affecting grapevines in New York vineyards. *American Journal of Enology and Viticulture* **28**, 131–136.

van Heerden HP (1978) Plant protection and phytosanitary control in Southern Africa. *The Deciduous Fruit Grower* **28**, 396–409.

Varga K, Kölber M, Martini M, Pondrelli M, Ember I, Tökés G, Lázár J, Mikulás J, Papp E, Szendrey G, Schweigert Á, Bertaccini A (2000) Phytoplasma identification in Hungarian grapevines by two nested-PCR systems. 13th ICVG Meeting, Adelaide, Australia, 113–115.

Vincent C, Olivier CY, Lasnier J, Vincent C (2015) Dix questions sur le système cicadelles/phytoplasmes/vignes. *Antennae* **22**, 3–6.

Wei W, Davis RE, Lee I-M, Zhao Y (2007) Computer-simulated RFLP analysis of 16S rRNA genes: identification of ten new phytoplasma groups. *International Journal of Systematic and Evolutionary Microbiology* **57**, 1855–1867.

Wei W, Pérez López E, Bermúdez Díaz L, Davis RE, Granda-Wong C, Zhao Y (2017) First report of a new grapevine yellows disease in Peru and its association with infection by a '*Candidatus* Phytoplasma brasiliense' related phytoplasma strain. *Plant Disease* **101**, 502.

Wolf TK, Prince JP, Davis RE (1994) Occurrence of grapevine yellows in Virginia vineyards. *Plant Disease* **78**, 208.

Yang S, Habili N, Aoda A, Dundas I, Paull JG, Randles JW (2013) Three group 16SrII phytoplasma variants detected in co-located pigeon pea, lucerne and tree medic in South Australia. *Australasian Plant Disease Notes* **8**, 125–129.

Zambon Y, Canel A, Bertaccini A, Contaldo N (2018) Molecular diversity of phytoplasmas associated with grapevine yellows disease in North-Eastern Italy. *Phytopathology* **108**, 206-214.

Zamharir MG, Paltrinieri S, Hajivand S, Taheri M, Bertaccini A (2017) Molecular identification of diverse '*Candidatus* Phytoplasma' species associated with grapevine decline in Iran. *Journal of Phytopathology* **165**, 407–413.

Zamorano A, Fiore N (2016) Draft genome sequence of 16SrIII-J phytoplasma, a plant pathogenic bacterium with a broad spectrum of hosts. *Genome Announcements* **4**, e00602–16.

Chapter 6
Fruit Crop Phytoplasmas

Nicola Fiore, Assunta Bertaccini, Piero A. Bianco, Mirosława Cieślińska, Luca Ferretti, Trinh Xuan Hoat, and Fabio Quaglino

Abstract The present chapter focuses on phytoplasmas and phytoplasma diseases affecting stone, pome, and small fruit worldwide. An outlook is also provided on other fruit tree species growing mainly in tropical and subtropical areas that are oftern infected by phytoplasma diseases usually associated with crop losses or loss of fruit quality and marketability.

Keywords Stone fruit · Pome fruit · Small fruit · Detection · Phytoplasma identification · Symptoms

6.1 Introduction

Phytoplasma detection in fruit crops and in insect vectors is now mainly based on molecular tools as for all phytoplasma diseases. For a number of these species grown in temperate areas, phytoplasma specific protocols that allow a sensitive and

N. Fiore (✉)
Departamento de Sanidad Vegetal, Universidad de Chile, La Pintana, RM – Santiago, Chile
e-mail: nfiore@uchile.cl

A. Bertaccini
Department of Agricultural and Food Sciences, *Alma Mater Studiorum* – University of Bologna, Bologna, Italy

P. A. Bianco · F. Quaglino
Department of Agricultural and Environmental Sciences – Production, Landscape, Agroenergy, University of Milan, Milan, Italy

M. Cieślińska
Research Institute of Horticulture, Skierniewice, Poland

L. Ferretti
CREA – Research Centre for Plant Protection and Certification, Rome, Italy

T. X. Hoat
Plant Protection Research Institute, Hanoi, Vietnam

G. P. Rao et al. (eds.), *Phytoplasmas: Plant Pathogenic Bacteria - I*,
https://doi.org/10.1007/978-981-13-0119-3_6

reliable detection were developed and validated. For phytoplasmas in the 16SrX group, specific primers (Lorenz et al. 1995; Lee et al. 1995a) that usually also require RFLP analysis of the amplified fragments can be employed. Specific primer pairs are also available that target non-ribosomal genes (Jarausch et al. 1998; Martini et al. 2007a). Several qPCR assays based either on SybrGreen® (Martini et al. 2007b; Yvon et al. 2009; Jarausch et al. 2010) or TaqMan® (Pignatta et al. 2008; Thébaud et al. 2009; Nikolić et al. 2010) technologies have been developed; group specific qPCR systems (Torres et al. 2005; Lee and Lin 2011) can also be used. Finally, a LAMP-based protocol has been recently developed (de Jonghe et al. 2017). Several of the above mentioned PCR and qPCR protocols have been validated through interlaboratory ring tests (The EUPHRESCO FruitPhytoInterlab group 2011; Pasquini et al. 2013). In the pome and stone fruits, several other non species-specific phytoplasmas were detected worldwide as summarized in Table 6.1.

6.2 Pome Fruit Phytoplasmas

The main phytoplasma historically associated with diseases in apple and pear are '*Candidatus* Phytoplasma mali' and '*Ca.* P. pyri', respectively (Seemüller and Schneider 2004). They both belong to 16SrX-A and 16SrX-C subgroups (Lee et al. 1998a, b; Pastore et al. 1996). Due to the ecological complexity of '*Ca.* P. mali' biological cycle, including plant hosts and insect vectors, several studies focused on the identification of molecular markers allowing the distinction of strains related to specific biological features. Sequence analyses of a non-ribosomal genomic fragment (nitroreductase-like gene) and of the *rplV-rpsC* genes (coding the ribosomal proteins L22 and S3, respectively) distinguished three AP phytoplasma subtypes (AT-1, AT-2, and AP-15) and four rpX subgroups (rpX-A, -B, -C, and -D), respectively (Jarausch et al. 2000b; Martini et al. 2007b, 2008). Moreover, nucleotide sequence analyses of multiple genomic regions revealed genetic diversity among '*Ca.* P. mali' populations (Paltrinieri et al. 2010; Casati et al. 2010, 2011; Franova et al. 2013; Cieślińska et al. 2015). The full genome of '*Ca.* P. mali' has been sequenced and as for '*Ca.* P. pyri' was determined to be linear and not circular, as reported for the other available phytoplasma genomes. Genome analysis revealed the presence of SAP11 (effector)-like protein and other phage-related elements putatively involved in the interaction with hosts (Kube et al. 2008). Full length genome sequencing also identified genes coding AAA+ ATPases and HflB/FstH proteases, suitable for distinguishing closely related strains showing a range of virulence during the infection of apple trees (Seemüller and Schneider 2007; Schneider and Seemüller 2009; Seemüller et al. 2013).

Apple '*Ca.* P. mali' is associated with apple proliferation (AP) disease, one of the most important diseases of apple trees. Under intense cultivation the disease is the major constraint for the sustainability of this crop. '*Ca.* P. mali' is a quarantine pathogen for EPPO in Europe and NAPPO in North America, depleting vigor of the

Table 6.1 Non species-specific phytoplasmas detected in stone and pome fruits

Host species	Ribosomal group/ subgroup	Geographic origin	Reference
Almond (*Prunus dulcis*, syn. *Prunus amygdalus*)	16SrIX-C; 16SrII-B; 16SrXII-A; 16SrVI-A	Iran	Salehi et al. (2006); Zirak et al. (2009b)
	16SrIII-A	USA	Uyemoto and Kirkpatrick (2011)
	16SrX-B	Turkey	Sertkaya et al. (2005)
Apple (*Malus* Mill spp.)	16SrI-B, 16SrI-C, 16SrXII-A	United Kingdom, Czech Republic	Bertaccini et al. (2001)
Apricot (*Prunus armeniaca* L.)	16SrI-B	Italy	Lee et al. (1995a)
	16SrX-A	Slovenia	Mehle et al. (2007)
	16SrIII-A	USA	Uyemoto and Kirkpatrick (2011)
European plum (*Prunus domestica* L.)	16SrI-B	Czech Republic; Lithuania	Navrátil et al. (2001a); Valiūnas et al. (2007)
	16SrX-A	Slovenia	Mehle et al. (2007)
Japanese plum (*Prunus salicina* Lindl.)	16SrI-B; 16SrIII-B; 16SrV; 16SrX-A; 16SrXII-A	Italy	Lee et al. (1995a); Paltrinieri et al. (2004)
	16SrII; 16SrXII-A	Iran	Zirak et al. (2009a)
Nectarine [*Prunus persica* (L.) Batsch]	16SrX-A; 16SrX-C	Poland	Cieślińska and Morgaś (2011)
	16SrXII-A; 16SrX-A	Italy	Paltrinieri et al. (2001)
Peach [*Prunus persica* (L.) Batsch]	16SrIII-A	USA, Canada	Uyemoto and Kirkpatrick (2011)
	16SrI-A; 16SrV-B	India	Singh et al. (2014); Thakur et al. (1998); Khan et al. (2013)
	16SrX-C	Czech Republic	Navrátil et al. (2001a)
	16SrXII-B	Bolivia	Jones et al. (2005)
	16SrV-B; 16SrVII-A, 16SrXII-A	Italy	Paltrinieri et al. (2003, 2006)
	16SrI-B	Croatia	Krizanac et al. (2010)
	16SrXV-A	Azerbaijan	Balakishiyeva et al. (2011)
	16SrVII-A; 16SrI-W	Canada	Zunnoon-Khan et al. (2010); Arocha-Rosete et al. (2011a, b)
	16SrI-C; 16SrV-B	China	Zhang et al. (2013); Li et al. (2014b)
	16SrIII-B; 16SrX-C	Argentina	Fernández et al. (2013, 2017)
	16SrX-C	Spain	Sabaté et al. (2014)
	16SrIX-B	Iran	Zamharir (2014)
	16SrXII-A	Chile	Paltrinieri et al. (2006)
Peach and nectarine	16SrIX-B	Lebanon	Abou-Jawdah et al. (2009)
Pear (*Pyrus* spp.)	16SrI-B;16SrXII-A	Croatia	Krizanac et al. (2010)
	16SrX-A	Hungary	Del Serrone et al. (1998)
Sour cherry (*Prunus cerasus* L.)	16SrI-R; 16SrI-Q	Lithuania	Jomantiene et al. (2011); Valiūnas et al. (2009a)
	16SrI-B	Poland	Cieślińska and Smolarek (2015)

(continued)

Table 6.1 (continued)

Host species	Ribosomal group/ subgroup	Geographic origin	Reference
Sweet and Chinese cherry	16SrV-B	China	Lee et al. (1995b); Zhu et al. (1997); Jung et al. (2003)
Sweet and sour cherry	16SrI-B; 16SrIII-T	Lithuania	Valiūnas et al. (2007, 2009b)
	16SrIII-J	Chile	González et al. (2011)
Sweet cherry (*Prunus avium* L.)	16SrIII-A	USA	Uyemoto and Kirkpatrick (2011)
	16SrX-C; 16SrXII-A; 16SrI-B; 16SrV-B; 16SrIII-B	Italy	Paltrinieri et al. (2001, 2008)
	16SrX-A	Czech Republic	Navrátil et al. (2001a)
	16SrX-A	Slovenia	Mehle et al. (2007)
	16SrX-C	Poland	Cieślińska and Smolarek (2015)
	16SrXII-A	Bulgaria	Avramov et al. (2011)
	16SrI-B	Turkey	Çaglayan et al. (2013)
	16SrI-S	China	Gao et al. (2011)

trees whose fruit cannot be commercialized because of their poor organoleptic qualities. AP was described for the first time in Italy in 1950 and occurs in many countries in the European areas: Albania, Austria, Bulgaria, Croatia, Czech Republic, France, Germany, Greece, Hungary, Moldova, the Netherlands, Norway, Poland, Romania, Southern Russia, Serbia and Montenegro, Slovakia, Slovenia, Spain, Switzerland, and Ukraine. In Asia AP has only been reported to occur in Turkey. Epidemics of the disease were reported in several Italian northern regions where *Malus domestica* cultivars are the main hosts of '*Ca*. P. mali' and most cultivars are susceptible. Apple cultivars known to be affected by AP phytoplasma include Belle de Boskoop, Gravestein, Golden Delicious, Winter Banana, Florina, Prima, Priscilla, Idared, McIntosh, Starking, Starkrimson, Roja de Benejama, Antonokova, Cortland, Spartan, Yellow Transparent, and Wealthy. In northern Italy severe epidemics have been reported in cultivars Golden Delicious, Florina, Canadian Renette, and Granny Smith grafted on different rootstocks. The recorded host range includes apple, periwinkle, bindweed, hazelnut, Bermuda grass, oriental lily, magnolia, cherry, apricot, plum, Japanese plum, pear, and rose (Kaminska and Sliwa 2004; Mehle et al. 2007). AP is associated in the most severe cases with productivity losses from 50% to 80%, depending on the variety, rootstock, environmental conditions, and agronomic management adopted (Zawadzka 1976; Osler and Loi 1986; Kunze 1989; Musetti et al. 2008). The main symptom is proliferation of the buds occurring mostly in the upper part of the plant due to development of axillary buds. The first symptoms appear the year after infection when the presence of the phytoplasma stimulates the growth of the secondary shoots, usually more sensitive to powdery mildew and then the witches' brooms. The disease then causes lack of vigor; in spring the flowers may show phyllody. The leaves are smaller, roll downward, and become brittle; also, they are finely and irregularly serrated, sometimes showing reddening or yellowing and often early fall. Petioles are shorter, but, at the

Fig. 6.1 Apple proliferation disease. Upper row from left to right. Symptoms of proliferation on the upper part of the plant with premature development of axillary buds on the trunk, and proliferation on secondary shoot with powdery mildew. Lower row left, leaves with abnormal stipules at right healthy leaf; right, fruit of smaller size and deformed elongated petioles

basis of the leaf, abnormally long stipules appear, and the fruits are usually smaller and the flavor is poor, since both sugar and acidity are very low (Fig. 6.1). '*Ca.* P. mali' is transmitted from infected plants to healthy ones by psyllids. In Italy, *Cacopsylla melanoneura* (Förster) is the main vector of '*Ca.* P. mali' in the northwestern regions (Tedeschi and Alma 2007), while *C. picta* (Förster) (main vector of '*Ca.* P. mali' in Germany) (Jarausch et al. 2007; Mayer et al. 2009) in the northeastern regions (Carraro et al. 2001). Additional putative vectors and natural plant hosts have been reported (Tedeschi and Alma 2006; Tedeschi et al. 2009). The long-distance spread of '*Ca.* P. mali' is however mainly due to the movement of infected saplings, cuttings, and rootstock materials. The disease is graft-transmissible, with different rates of efficiency; in addition, since the colonization of aerial parts of apple trees shows seasonal variations, transmission rates vary up to 30%. Spring was reported to be the season when scion wood was least likely to be infected (up to 0.08%) and the vegetatively propagated rootstocks are especially hazardous as they are generally symptomless (Pedrazzoli et al. 2008). Available tools are very poor for reliable and effective disease containment for AP. In the focus areas, eradication of infected plants and pesticides application against the insect vectors are mandatory.

Fig. 6.2 On the left pear decline infected pear tree (with red leaves) near to healthy pear; on the right a diseased tree producing just one fruit

Pear Pear decline (PD) associated with the presence of '*Ca*. P. pyri' is a serious disease that, after its first reports in North America, occurs in many of the countries where pear is cultivated. It had long been known as "moria" in northern Italy (Refatti 1964) where the phytoplasma was detected as PD and studied after 1970 (Giunchedi et al. 1982; Giunchedi and Poggi Pollini 1984; Bertaccini et al. 1994; Pastore et al. 1996; Camele et al. 1998; Guerrini et al. 2000) as well as in many other countries in the European and North American areas (Blattný and Váňa 1974b; Westwood and Cameron 1978; Behnke et al. 1980; Davies et al. 1985; Eleftheriou 1985; Del Serrone et al. 1998; Ember et al. 2004; Duduk et al. 2005a, b; Süle et al. 2007; Krizanac et al. 2008; Çaglayan et al. 2008; Hunter et al. 2010) and recently also in Iran, Turkey, and in Chile (Salehi et al. 2008; Sertkaya et al. 2008; Facundo et al. 2017). In the 1960s in California (USA), about two million plants died, and those surviving after the PD epidemic showed reduced yield of marketable fruits (Hibino and Schneider 1970). There are two symptom types, quick and slow decline. In the quick decline, the plants die quickly in summer time without showing previous symptoms. The slow decline reduces vigor and induces leaf reddening and rolling (Fig. 6.2).

The disease can spread through propagation with infected buds and/or rootstocks or by insects of the genus *Cacopsylla*: *C. pyri*, *C. pyricola*, and *C. pyritsuga* are the reported vectors (Davies et al. 1992; Carraro et al. 1998c; Sanchez and Ortín-Angulo 2011). Phytoplasmas are consistently detected in the PD diseased trees:

they are usually unevenly distributed (Garcia Chapa et al. 2003) and in low concentration. The clear correlation of phytoplasma presence with symptom expression is still unresolved for PD, and it has been shown that phytoplasmas could be present also in asymptomatic trees (Lee et al. 1995a). In an experimental trial, 105 pear plants showing symptoms possibly linked to phytoplasma presence were tested by molecular assays, and no correlation was found between phytoplasma presence and symptomatology. However pear cultivars grafted on Quince A rootstock were more susceptible to phytoplasma infection (58–85%), while cvs William and Decana del Comizio appear to be less susceptible, especially when grafted on "franco comune" rootstock (29–43% of infection). Plants of cv William grafted on Farold 40 and Farold 87 showed a similar degree of phytoplasma infection (57%), but all the phytoplasma-positive trees were asymptomatic (Pastore et al. 1998). It was shown that grafting PD infected materials in winter time allow PD transmission (Errea et al. 2002). However, a recent 3-year-long experiment on PD infected pear plants under insect-proof nets showed a reduced presence of symptoms over the years; moreover in the last year, all the plants were PD negative at both crown and root levels. These results suggest no seasonal translocation of '*Ca.* P. pyri' in the absence of the insect vector and confirmed phytoplasma elimination in the new phloem (Paltrinieri et al. 2017).

6.3 Stone Fruit Phytoplasmas

Almond Almond witches' broom (AlmWB) disease is associated with the presence of '*Candidatus* Phytoplasma phoenicium', 16SrIX-B (Verdin et al. 2003). The first epidemic of a lethal devastating almond [*Prunus dulcis* (Mill.) D.A.Webb] disease occurred in the south of Lebanon in the early 1990s; it was later reported in north Lebanon and in Iran starting in 1995 (Salehi and Izadpanah 1995; Abou-Jawdah et al. 2002; Molino Lova et al. 2011). During the last two decades, the outbreak of AlmWB has led to a rapid decline of almond trees in northern regions and in the Bekaa Valley in Lebanon (Choueiri et al. 2001; Abou-Jawdah et al. 2002) and in Fars province and in other southern provinces in Iran (Salehi et al. 2006). In Lebanon, the disease, rapidly spread and killed almost 100,000 trees over a period of 10 years from coastal areas to high mountainous areas. In 2009, '*Ca.* P. phoenicium' was also identified in association with a severe disease of peach (*P. persica*) and nectarine (*P. persica* var. *nucipersica*) in southern Lebanon (Abou-Jawdah et al. 2009), and more than 40,000 newly diseased almond trees were observed in 2010 throughout the country, in 16 out of 24 Lebanese districts (Molino Lova et al. 2011). In the case of peach and nectarine trees, the first symptom observed is early flowering, followed by development of buds in infected branches. In addition, phyllody at the flowering period and serrate, slim, light green leaves and witches' brooms developing from the trunk and the crown of the trees several months after are observed (Fig. 6.3). Even if the presence of witches' broom is more common in almond trees

Fig. 6.3 Symptoms due to '*Ca*. P. phoenicium' in peach and nectarine trees: early flowering and development of all the buds of the infected branches (**a**); phyllody at the flowering (**b**); abnormal fruits (**c**); serrate, slim, light green leaves (**d**); witches' broom (**e-f**)

than in peach/nectarine, in the latter phyllody, never recorded on almond is the differential symptomatology (Molino Lova et al. 2011). In Iran, '*Ca*. P. phoenicium' was not identified in peach and nectarine but in other plant species, such as GF-677 (*P. amygdalus* × *P. persica*) and wild almond (*P. scoparia*) (Salehi et al. 2015). A total loss of production happens 1–2 years after the initial appearance of the symptoms (Abou-Jawdah et al. 2002). AlmWB was found to infect properly managed orchards, abandoned orchards, and isolated wild trees. The most characteristic symptoms in almond trees are (i) shoot proliferation on the main trunk with the appearance of a witches' broom, (ii) perpendicular development of many axillary buds with small and yellowish leaves, and (iii) general tree decline with final dieback (Fig. 6.4). In Lebanon field surveys conducted in AlmWB almond and peach orchards and surroundings detected '*Ca*. P. phoenicium' in the leafhopper *Asymmetrasca decedens* (prevalent in almond) and in the cixiids *Cixius* sp., *Tachycixius* spp., and *Eumecurus* spp. (prevalent in *Smilax aspera* L. and *Anthemis* sp.) and in crops and wild plants where the insects were collected. Transmission trials demonstrated that *A. decedens*, *T. viperinus*, and *T. cf. cypricus* are able to transmit '*Ca*. P. phoenicium'. AlmWB epidemiological cycle in Lebanon involves *A. decedens*, possibly responsible for the transmission from almond to almond, and cixiids of the genus *Tachycixius*, possibly responsible for the transmission from weeds to almond (Abou-Jawdah et al. 2014; Tedeschi et al. 2015). Multiple gene typing analyses of '*Ca*. P. phoenicium' strains infecting almond, peach, and nectar-

Fig. 6.4 Main symptoms of AlmWB on almond trees: shoot proliferation on the main trunk (**a**); perpendicular development of axillary buds on the branches, with small and yellowish leaves (**b**); general decline of the tree with final dieback (**c**)

ine in Lebanon revealed a substantial genetic homogeneity within the analyzed phytoplasma populations based on housekeeping gene sequence analyses and allowed the identification of distinct AlmWB-associated phytoplasma strains from diverse host plants based on *inmp* (integral membrane protein) gene sequence analysis. This evidence, along with reports of multiple insect vectors suggests that AlmWB could be associated with phytoplasma strains derived from the adaptation of an original strain to diverse hosts (Quaglino et al. 2015). '*Ca*. P. phoenicium' strains, classified in 16SrIX-F and 16SrIX-G subgroups (Molino Lova et al. 2011), are considered as genetic variants of subgroup 16SrIX-B due to common biological traits (Casati et al. 2016). Healthy plant material and vector control are the main measures applied for AlmWB containment. Other phytoplasma diseases reported in almond worldwide are AlmWB-like diseases, inducing broomings in Iran (Verdin et al. 2003) in association with phytoplasmas classified in subgroup 16SrIX-C (Salehi et al. 2006).

Apricot The phytoplasma associated with the disease named apricot chlorotic leaf roll belongs to the 16SrX-B subgroup, also known as European stone fruit yellows (ESFY) or '*Ca*. P. prunorum' (Seemüller and Schneider 2004), and is considered the most destructive pathogen of apricot (Lederer and Seemüller 1992; Navrátil et al. 1998, 2001a; Jarausch et al. 2001; Torres et al. 2004; Sertkaya et al. 2005; Ambrožič Turk et al. 2008; Cieślińska and Morgaś 2011; Ludvíková et al. 2011; Mehle et al. 2011; Tarcali 2013; Žežlina et al. 2016). ESFY is associated with high mortality of apricot trees (Morvan 1977); a study carried out in France has shown that the average annual mortality of infected plants is 5% (Gentit et al. 1998). The prevalence of infected plants in each orchard depends on the variety, rootstock, age of orchard, environmental conditions, infecting strain, and insect vector populations. In some orchards in Germany, the prevalence of infected plants has been up to 80% (Jarausch et al. 2004). Prevalence values range from 5% to 40% (Desvignes et al. 1999; Kison and Seemüller 2001; Laimer da Camara Machado et al. 2001; Torres et al. 2004); in Italy and Switzerland, they reach 93% and 4%, respectively (Pastore et al. 1999; Genini and Ramel 2004). During the spring, infected plants initially produce leaves instead of flowers, while from the summer end until the middle of autumn, the main symptoms consist of yellows and upward rolling of the leaves, followed by leaf reddening, branch and phloem necrosis, and decline.

Peach and Nectarine The ESFY phytoplasma in Mediterranean is associated with yellows and decline diseases (Lorenz et al. 1994; Delic et al. 2005; Sertkaya et al. 2005; Ambrožič Turk et al. 2008; Cieślińska and Morgaś 2011; Ludvíková et al. 2011; Mehle et al. 2011; Tarcali 2013; Etropolska et al. 2015; Žežlina et al. 2016; Etropolska and Lefort 2017). Peach species, along with those of apricot and plum, are the most susceptible to ESFY (Kison and Seemüller 2001; Ermacora et al. 2010). Infected plants are less productive and vigorous and may die within a few years. In peach orchards in Spain, the prevalence of infected plants varied from 5% to 25% (Battle et al. 2012). In Northern-Central Italy, the prevalence of the disease varied between 1% and 4%, with the highest infection rates concerning the cultivars Venus and Super Crimson Gold, grafted on GF 677 (Poggi Pollini et al. 2001). In peach and nectarine during winter and early spring, the leaf buds of infected plants break early. From the end of the summer and during the autumn, in some cultivars early fall and moderate rolling of leaves occurs. In others, these symptoms appear later in the season and are accompanied by reddening, upward rolling of the leaves, and vein enlargement (Poggi Pollini et al. 2001).

Plum ESFY phytoplasma is responsible for several decline diseases described over the years in *Prunus* spp. in Europe (Ahrens et al. 1993; Lorenz et al. 1994). On plum, it is associated with decline diseases referred to also as plum leptonecrosis in Italy on *P. salicina* (Goidànich 1933; Giunchedi et al. 1978) and as "Molières decline" in France (Bernhard et al. 1977). In Europe, '*Ca*. P. prunorum' is the major and most economically important phytoplasma affecting plum. On Japanese plum it induces economically significant damages (Dosba et al. 1991), and it represents one of the major limiting factors in the production of this crop in countries bordering the Mediterranean Sea (Spain, France, Italy, Balkans), where the cultivation of Japanese plum and other susceptible *Prunus* species is widespread. A high percentage of naturally infected trees often associated with a high mortality rate of plants make orchards unproductive 8–10 years after planting. Infection percentages up to 100% have been recorded on susceptible Japanese plums cvs Ozark Premier and Shiro after 7 years of growth (Carraro et al. 1998a). Cultivar also affects the mortality rate of the infected plants that is generally higher on susceptible varieties than on plants grafted on susceptible rootstocks. A different response to '*Ca*. P. prunorum' infection is reported for some Japanese plum varieties under natural pressure of ESFY disease and cvs. Bragialla, Brarossa, Fortune, and Ruby Crunch are reported as not affected after 5 years of observations (Landi et al. 2010). The impact of the ESFY disease on the European plum is variable: severe symptoms resembling those

Fig. 6.5 Off-season growth in plum due to an early break of leaf bud in late winter–early spring: ESFY infected tree is already at the leaf stage, while the healthy is still covered by flowers (**b**); upward rolled leaves with a brownish-red coloration (**a**)

occurring on Japanese plum were first observed in Italy on "Susina di Dro" (Poggi Pollini et al. 1995). A 5-year monitoring period carried out in orchards located in a natural and severely ESFY-affected area showed that only 6 out of 39 cultivars/selections observed were found infected and 4 of these showed only slight symptoms (Landi et al. 2010). A different response to '*Ca.* P. prunorum' infection, ranging from little/not affected to moderately/highly susceptible, were also observed in experimental trials on French *P. domestica* cultivars (Jarausch et al. 2000a). Moreover, European plum can be latently infected and act as reservoirs of the pathogen. As for other *Prunus* species, symptoms induced on plum are different depending on the vegetative growth stage and season. The most typical symptom is off-season growth due to an early break of leaf bud before flowering that can be observed in late winter–early spring (Jarausch et al. 2008) (Fig. 6.5a). In summer, typical symptoms consist of smaller and narrowed leaves, upward and longitudinally rolled, and with a brownish-red coloration that became thick and brittle (Fig. 6.5b). Phloem necrosis can also appear, especially after winter frost. A poor fruit production is observed; fruits are smaller, ripen later than the healthy trees and may fall prematurely. Finally, an earlier defoliation generally occurs in summer or early fall. Initially, the disease can involve one branch, but in 2–3 years, the whole plant can be affected and die. An abnormal production of suckers from the rootstock usually occurs that can survive even after the aerial part of the tree has died. *P. salicina* is reported as highly susceptible to '*Ca.* P. prunorum' (Ferrini et al. 2002; Carraro et al. 2004), and the majority of Japanese plum varieties exhibit severe symptoms when infected. *P. domestica* is generally more tolerant (Carraro et al. 1998a) and can also be latently infected. Under experimental conditions considerable differences were observed among *Prunus* rootstocks in response to '*Ca.* P. prunorum' infection (Kison and Seemüller 2001). With regard to rootstocks commonly used in Europe for plum varieties, myrobolan (*P. cerasifera*) and Ishtara [(*P. cerasifera* × *P. persica*) × *P. salicina*] were found to be moderately susceptible, whereas GF677 (*P. dulcis* × *P. persica*) and *P.*'Marianna' GF 8/1 were only slightly affected.

The psyllid *Cacopsylla pruni* (Scopoli) is the only known vector of '*Ca.* P. prunorum' (Carraro et al. 1998b, 2001) to *Prunus* species, although the two known genetic groups of *C. pruni* (Sauvion et al. 2007) appear to constitute divergent biological species (Peccoud et al. 2013). A strong host preference is reported, and Japanese plum trees are considered to be the more suitable hosts (Thébaud et al. 2006). '*Ca.* P. prunorum' can be efficiently transmitted also by grafting throughout the year as it persists in the stem during winter (Seemüller et al. 1998). Infected rootstocks and/or scions can contribute to disease establishment and spread. Monitoring of ESFY in a plum growing area in Italy showed that myrobolan rootstocks infected by '*Ca.* P. prunorum' resulted in 100% infection of newly grafted scions within one winter season (Paltrinieri et al. 2004).

Sweet and Sour Cherry A ESFY-related disease, named "Molière decline", was for the first time observed in France in sweet cherry (Bernhard et al. 1977), but a clear relationship between '*Ca.* P. prunorum' presence and cherry "Molière decline" was not established. The symptoms observed were yellowing of leaves, decline of shoots, and death of the plant. The first indications about the destructive nature of '*Ca.* P. prunorum' are from Kison and Seemüller (2001). Later there was a report on five cherry rootstocks (Gisela 1, Gisela 3, Gisela 5, F 12/1, Weihroot 158) graft inoculated with a ESFY strain from flowering cherry (*P. serrulata*) showing decline symptoms (Lederer and Seemüller 1992).

Flowering cherry was used as scion on the rootstocks, and all the plants showed symptoms on leaves were less vigorous, and finally the entire plant or the scion died. '*Ca.* P. prunorum' was detected then in sweet and sour cherry in Poland and Hungary and only in sour cherry in the Czech Republic where out of ten sour cherry plants infected by '*Ca.* P. prunorum' only one was asymptomatic (Ludvíková et al. 2011). In the same country, a ESFY-infected sour cherry plant showed stunting, leaf rolling, and yellowing (Navrátil et al. 2001a). In Poland, three out of six sweet cherry plants infected by ESFY were asymptomatic, while the others showed chlorotic leaf roll, short internodes, and wilting, but in the same country, only one out of three sour cherry plants infected with '*Ca.* P. prunorum' exhibited short internodes and decline (Cieślińska and Morgaś 2011) and leaf roll and yellowing of the leaves in sweet cherry varieties Kordia II.14 and Trzebnica (Cieślińska and Smolarek 2015). In Hungary the situation is more dramatic, with a high death rate of sweet and sour cherry plants (Tarcali 2013).

Widespread epidemics of X-disease in cherry were reported in the USA associated with the presence of '*Ca.* P. pruni' which belongs to the ribosomal subgroup 16SrIII-A (Davis et al. 2013). The disease was first reported in California in sweet cherry plants and then in Connecticut in peach trees; X-disease was found only in peach in Canada besides the USA (Uyemoto and Kirkpatrick 2011). In addition to sweet and sour cherry, the phytoplasma infects peach, nectarine, almond, Japanese

Fig. 6.6 Cherry trees infected by phytoplasmas in Chile: from left to right, plant on the left showing decline infected by strain 16SrIII-J; plant with shortened internodes; dead plant; phloem necrosis in the trunk of infected plant

and European plum, apricot, sour cherry (*Prunus emarginata*), and chokecherry (*P. virginiana*). The latter species represents an important reservoir of the phytoplasma, along with some spontaneous herbaceous species such as *Erigeron canadensis*, *Solidago rugosa*, *Medicago hispida*, and *Asclepias syriaca* (Douglas 1986; Uyemoto and Kirkpatrick 2011). The severity of X-disease depends on climatic conditions, being more destructive in California than in Washington State. On the other hand, sweet cherry scions grafted on *P. mahaleb* show a rapid decline, because of the hypersensitivity reaction of this rootstock that is reported as resistant to the pathogen. In rootstocks susceptible to the disease, such as Mazzard, Colt, and Stockton Morello, the scion declines slowly, and the fruits have a reduced size and are conical and scarcely colored, with the peduncle thick and short. The canopy is less developed, with smaller and reddish leaves, whose margins folded upwards (Kirkpatrick et al. 1995; Uyemoto and Kirkpatrick 2011). The phytoplasma can be transmitted by several species of leafhopper, as *Colladonas clitellarius*, *C. montanus*, *C. geminatus*, *Euscelidius variegatus*, *Fieberiella florii*, *Graphocephala confluens*, *Gyponana lamina*, *Keonella confluens*, *Norvellina seminuda*, *Osbornellus borealis*, *Paraphlepsius irroratus*, *Scaphytopius delonghi*, and *S. acutus* (Rice and Jones 1972; McClure 1980; Larsen and Whalen 1988; Kirkpatrick et al. 1990). The disease has not been reported in the last decade after extensive uprooting of the symptomatic orchards. In Chile 16SrIII-J phytoplasmas were identified in plants showing dieback, shortened internodes, and decline (Fig. 6.6) (González et al. 2011).

6.4 Small Fruit Phytoplasmas

***Vaccinium* species** Blueberry stunt (BBS) was first described in New Jersey, USA as a virus disease (Wilcox 1942). In the mid-1900s, the disease became widespread in major blueberry production regions in the USA. It has been reported in eastern Canada and in several states of eastern and southeastern USA (Gocio and Dale

1982; Ramsdell and Stretch 1987; Bagadia et al. 2013; Arocha-Rosete et al. 2015). All cultivars of highbush blueberry and several other species of *Vaccinium* (*V. angustifolium*, *V. vacillans*, *V. atrococcum*, *V. stramineum*, *V. myrtilloides*) are susceptible to stunt disease (Ramsdell and Stretch 1987). Infected bushes of the most susceptible cultivars are usually less than half the size of healthy plants. Fruit production is reduced causing economical losses. Symptoms on highbush blueberry include overall dwarfing, witches' broom, and shortened internodes. Leaves of infected plants are small, spoon-shaped, and cupped slightly downward and turn red in summer or early autumn (Fig. 6.6). The sharp-nosed leafhopper [*Scaphytopius magdalensis* (Provancher)], and perhaps related species feeding on *Ericaceae* plants, transmits the BBS phytoplasma; it was also experimentally transmitted by *S. acutatus* and *S. frontalis* (Tomlinson et al. 1950; Maramorosch 1955; Tozzi et al. 1993), by dodder to periwinkle (Tomlinson et al. 1950), and by grafting of infected scions onto healthy blueberry plants (Ramsdell and Stretch 1987). Diseased samples collected from USA harbored phytoplasmas classified in the aster yellows phytoplasma subgroup 16SrI-E (Lee et al. 1998b); further study on BBS strains from Michigan and Quebec confirmed the presence of subgroup 16SrI-E phytoplasmas (Arocha-Rosete et al. 2015). Moreover BBS-diseased plants exhibiting typical stunt syndrome in New Jersey were infected with both 16SrI-E and 16SrIX-E subgroup phytoplasmas; the latter were identified in 4.4% of the bushes (Bagadia et al. 2013). Blueberry witches' broom is a disease reported in wild European blueberry (bilberry) plants (*Vaccinium myrtillus*) and described in The Netherlands, Germany, the then Czechoslovakia, Yugoslavia, Scotland, France, Lithuania, and Austria (Bos 1960; Blattńy and Vána 1974; de Leeuw 1975; Valiūnas et al. 2004; Borroto-Fernández et al. 2007).

Besides *V. myrtillus*, the disease affects *V. vitis-idaea* (lingonberry), *V. uliginosum* (bog bilberry), and *V. oxycoccus* (small cranberry); in the 1960s it had been a serious problem in *V. myrtillus* production in the former Czechoslovakia with 15% fruit losses (Blattńy and Blattńy 1970). Infected European blueberry plants exhibit symptoms of shoot proliferation and bushy growth (Fig. 6.7). The size of branches and leaves is reduced. The shoots of plants may have an erect position, while the normal, healthy plants have plagiotropic shoots. Sometimes the leaves may prematurely turn red, but they usually drop later in the autumn than leaves of healthy plants which may result in higher sensitivity to frost damage. Tomenius and Ahman (1983) reported little leaf disease of *V. vitis-idaea* and wild *V. myrtillus* in Sweden. The leafhopper *Idiodonus cruentatus* Panz. was reported as a vector of the witches' broom agent in *V. myrtillus* (Blattńy 1963); however the disease also occurred in *I. cruentatus*-free areas indicating that other leafhoppers can be involved in disease spread. On the basis of RFLP, nucleotide sequence and phylogenetic analyses of 16S rDNA, the phytoplasma strain detected in Lithuania was classified in the 16SrIII-F group (Valiūnas et al. 2004) as that detected in Germany (Paltrinieri et al. 1999). Phytoplasmas belonging to ribosomal group 16SrVI were also identified in *V. myrtillus* exhibiting symptoms of shoot proliferation in Austrian forests (Borroto-Fernández et al. 2007).

Fig. 6.7 Left: marginal yellowing and cupping of leaves and branch stunting of blueberry infected with 16SrI-E phytoplasma in Canada (Courtesy of Y. Arocha-Rosete). Right: shoot proliferation and bushy growth of blueberry infected by phytoplasma in Austria (Courtesy of M. Laimer)

Blueberry Reddening Disease Yellowing and reddening on the upper leaves, shoot proliferation, and uneven ripening of the fruits were observed on the blueberry plants cvs. Bluecrop, Duke, and Spartan grown in central Serbia (Starović et al. 2013). More than 20% of plants in the single field were infected, and the incidence of the disease increased up to 50% in the following year. "Stolbur" phytoplasmas (16SrXII-A subgroup) were detected; however its association with blueberry reddening disease was not confirmed.

Cranberry False Blossom The disease reported in several northern USA states (Chen 1970; Stretch 1987; Xu and Chen 1996) was described only in American cranberry (*V. macrocarpon*) and European cranberry (*V. oxycoccus*). During the 1920s and 1930s, increasing occurrence of CBFB and its vector *Limotettix vaccinii* (Dobroscky 1929) was a serious threat to cranberry production; however the control of the vector and the use of resistant cultivars have almost eliminated the disease. However, it reappeared in the late 1990s: infected plants exhibit abnormal floral structure including straight pedicels of the flowers, enlarged calyx, and shortened, discolored petals streaked with deep pink, red, or green, usually sterile flowers with abnormal pistils and stamens, and sometimes forming phyllody. Moreover, symptoms of upright growth above the level of normal vines and witches' broom of shoots, leaves closely fitting to the stem, premature reddening of the leaves, and enlarged terminal flower buds can be also observed (Stretch 1987). *L. vaccinii* and *S. magdalensis* are potential CBFB vectors (Lee et al. 2014). The CBFB phytoplasma was identified as belonging to 16SrIII-B subgroup (Xu and Chen 1996), but recent research assigns it to subgroup 16SrIII-Y (Lee et al. 2014).

Fig. 6.8 From left: numerous weak and erect shoots in rubus stunt-affected blackberry (healthy plant at left); phyllody of flowers of loganberry infected with 16SrIII phytoplasmas; premature reddening of the leaves in rubus stunt-affected blackberry cv Loch Tay

***Rubus* species** Several distinct phytoplasmas were associated with rubus stunt disease widely affecting *Rubus* species. The most common is '*Ca.* P. rubi' (Malembic-Maher et al. 2011) classified in subgroup 16SrV-E and reported in many European countries (Marani et al. 1977; Mäurer and Seemüller 1994; Bertaccini et al. 1995; Marcone et al. 1997; Davies 2000; Ermacora et al. 2003; Sertkaya et al. 2004; Vindimian et al. 2004; Valiūnas et al. 2007; Cieślińska 2001, 2011; Ramkat et al. 2014). Recently, a blackberry phytoplasma strain was reported in Portugal and has been characterized and assigned to a new subgroup of the same group (16SrV-I) (Franova et al. 2016). '*Ca.* P. asteris'-related phytoplasmas were identified in wild raspberry and blackberry growing in Austrian forests (Borroto-Fernández et al. 2007), in blackberry in Pakistan (Fahmeed et al. 2009), in Poland (Cieślińska 2011), and in the United Kingdom (Reeder et al. 2010). '*Ca.* P. solani', 16SrXII-A was reported only in Europe (Borroto-Fernández et al. 2007; Kuzmanović et al. 2011; Bobev et al. 2013). Phytoplasmas in *Rubus* are usually associated with stunting, small leaves, abnormal internodes (short or elongated), enlarged sepals, phyllody, virescence, proliferation of axillary buds, shoots and flowers, leaf reddening in early autumn, and fruit malformations (Fig. 6.8). The phytoplasma is transmitted by the leafhopper *Macropsis fuscula* (de Fluiter and van der Meer 1953) mostly to species of the genus *Rubus* but also to the dog rose *Rosa canina* (Mäurer and Seemüller 1994; Davies 2000; Jarausch et al. 2001). Phytoplasmas belonging to the X-disease group (16SrIII) were described in the United Kingdom (Davies 2000), Oregon, USA (Davis et al. 2001), and in Poland (Cieślińska 2011). PCR/RFLP analysis of the 16S rDNA fragment revealed that two distinct phytoplasmas belonging to 16SrIII and 16SrV groups were associated with rubus stunt in the United Kingdom. Results of a similar analysis of a phytoplasma found in *Rubus occidentalis* (black raspberry) in Oregon were the basis for its classification in the newly established subgroup 16SrIII-Q (Converse et al. 1982; Davis et al. 2001). "Stolbur" phytoplasma was detected in *R. idaeus* in Austria and in stunted *R. fruticosus* (cv. Evergreen Thornless) from central southern Bulgaria (Borroto-Fernández et al. 2007; Bobev et al. 2013) using PCR/RFLP or sequence analysis. Similar analyses showed that *R. fruticosus* from an Austrian forest was infected by phytoplasmas

Fig. 6.9 From left: abnormal flower buds (upper) of full blossom disease-affected white currant "Blanka" in comparison with healthy flower buds (lower); flower malformation in a FBD white currant cv Blanka (courtesy by T. Malinowski). Sterile malformed flowers of black currant showing symptoms of full blossom disease

belonging to ribosomal group 16SrI-B (Borroto-Fernández et al. 2007). The PCR/RFLP and comparative genomic analysis of the 16S rRNA and *secY* genes revealed that phytoplasmas found in Poland were phylogenetically closely related to '*Ca.* P. rubi', '*Ca.* P. asteris' and X-disease phytoplasma strains (Cieślińska et al. 2014). Rapid and simultaneous detection of different groups of phytoplasmas infecting *Rubus* species was achieved by multiplex qPCR assay using TaqMan probes (Linck et al. 2017). This assay broadens the availability of tools useful for screening of nursery plant material and provides a new technique for the study of phytoplasmas infecting *Rubus* species.

***Ribes* species** Full blossom disease (FBD) was described on red currant "Houghton Castle" in Bojnice in former Czechoslovakia. Pleomorphic bodies were observed by transmission electron microscopy in the phloem tissue (Rakús et al. 1974; Rakús 1978). Later it has also been recorded in white currant cvs. Blanka, Meridián, and Primus (Špak et al. 2001). Since 1998, multiple cultivars of red and white currants with FBD symptoms have been observed in the Czech Republic in germplasm collections, propagation material, and commercial plantations (Špak et al. 2006). FBD was also recorded in Poland in red currant cv. Jonkheer van Tets and white currant cv. Blanka mostly infected with Blackcurrant reversion virus (BRV) but not tested for phytoplasma presence. The presence of phytoplasmas of aster yellows group (16SrI) in black currant cv. Karlštejnský dlouhohrozen with symptoms of the severe form of blackcurrant reversion disease was firstly detected in the Czech Republic (Špak et al. 2004). The incidence of the disease in the Czech Republic varied from 1% to 70% of infected bushes in different cultivars and locations. Similarly, the intensity of symptoms in the field observed between 1999 and 2004 varied from a few malformed flowers up to a massive occurrence of such flowers on whole branches. Yield losses of up to 70% were recorded in severely infected bushes of red currant cv. Vitan, and fruit quality was poor. The presence of both phytoplasma and BRV was confirmed in FBD infected bushes (Špak et al. 2006). The typical symptoms are malformations of flowers (Fig. 6.9) including the absence of stamens, presence of more than one style, enlarged petals and sepals, sterile flowers, and malformed fruit (van der Meer 1987). Diseased bushes are reduced in size and produce sparse crops of small berries (Rakús and Maliarčiková 1975). To determine the

Fig. 6.10 From left: shoot proliferation and small berries from plant infected by phytoplasma (normal fruits from a healthy plant, at the top of the picture) in murtilla witches' broom disease. Strawberry fruit with leaves due to the phytoplasma presence (right)

role of phytoplasmas and BRV in FBD etiology and their potential to induce symptoms in *Ribes* sp., graft and dodder transmission experiments were conducted. A 16SrI-C phytoplasma was transmitted from FBD-symptomatic red currant by dodder to periwinkle, however at very low rate, and back transmission from periwinkle to red currant seedlings was by dodder as well as by grafting of cuttings from infected shoots to healthy red currant rootstocks (Špak et al. 2009; Přibylová et al. 2011). The problem with detection in symptomatic currants was associated with low concentration and erratic distribution of both BRV and phytoplasma (Špak et al. 2008). Phytoplasmas detected in red and white currant with full blossom symptoms in the Czech Republic were classified as subgroups 16SrI-B, 16SrX-A, 16SrX-B, and 16SrX-C (Navrátil et al. 2001b, 2004, 2007).

***Fragaria* species** Phytoplasma diseases of strawberry are mainly represented by virescence and phyllody symptomatology in both flowers and fruit (Fig. 6.10) that has been known for a long time (Posnette 1953) and determined as infectious disease transmissible by insect vectors or by grafting. At the end of the 1960s in New Zealand, symptoms on a lethal yellows disease were associated with phytoplasma presence (Stubbs 1968). Phytoplasmas were detected by transmission electron microscopy and molecular techniques in several countries worldwide. In Italy, Canada, and Czech Republic, 16SrI-C phytoplasmas were identified with losses reaching 30% in some severe cases (Franova Honetslegrova et al. 1996; Pastore et al. 2002; Contaldo et al. 2012). Phytoplasmas were detected in plants showing witches' broom symptoms (multiplier disease) and stunting in Florida and in other parts of the USA (Jomantiene et al. 1996, 1998). Phytoplasma strains distinguishable at the molecular level were also identified in strawberry with similar symptoms: 16SrI-J and 16SrVI groups in the USA (Harrison et al. 1997), 16SrI-B in Southern Italy (Bertaccini et al. 1997), and 16SrXII-B in New Zealand (Andersen et al. 1998). The strawberry lethal yellows (SLY) disease in Australia is associated with '*Ca.* P. australiense' and tomato big bud (Straten et al. 2005). '*Ca.* P. australiense' is also associated with strawberry green petal (SGP) disease.

In strawberry with a yellows disease in Lithuania, a new taxon named '*Ca.* P. fragariae' was identified and assigned to ribosomal group 16SrXII-E (Valiūnas et al. 2006). Recently phytoplasmas of group 16SrIII were reported from Bolivia in strawberry plants displaying rosette formation and small fruits (Arocha et al. 2010). Subgroup 16SrI-F was detected in Cuba in nurseries with plants showing symptoms of leaf yellowing and reddening and fruit deformation and stunting (Ferriol-Marchena et al. 2013). In Spain "stolbur" and aster yellows were detected in plants with yellow leaves, stunting, reduced leaf size, and virescence (Oviedo Delgado and Ibeas Corcelles 2017), while in Mexico a similar symptomatology was associated with the presence of '*Ca.* P. hispanicum' (16SrXIII-A subgroup) (Pérez-López and Dumonceaux 2016; Pérez-López et al. 2017). In strawberry red leaf in Argentina, 16SrXIII-F subgroup phytoplasmas were identified (Fernández et al. 2015).

Myrtaceae Family Murtilla (*Ugni molinae* Turcz) witches' broom disease is only reported from Chile. In spring and summer, the infected plants showed small and yellow leaves at the end of summer which turned red during the autumn season, and the twigs become necrotic and dry. The berries, if present, are smaller and poor in sugar and flavorings (Fig. 6.10). The first report of this disease, based on symptoms, has been made in the early 1980s (Novoa 1982); however the first laboratory evidence for phytoplasma presence was obtained recently (Andrade et al. 2009). '*Ca.* P. fraxini' and '*Ca.* P. ulmi' (ribosomal subgroups 16SrVII-A and 16SrV-A, respectively) were detected in symptomatic plants (Arismendi et al. 2011).

6.5 Tropical/Subtropical Fruit

Banana *Musa* spp. is native to tropical Indomalaya and Australia and is likely to have been first domesticated in Papua New Guinea. They are grown in 135 countries, primarily for their fruit, and to a lesser extent to make fiber, banana wine, banana beer, and as ornamental plants. Phytoplasmas were detected from wilted cooking banana (plantain) plants in the Solomon Islands and Papua New Guinea and are most closely related to phytoplasmas in the 16SrXXII-A ('*Ca.* P. palmicola') and 16SrIV-A and 16SrIV-C subgroups. The plants showed yellowing and/or leaf death and unfilled fruit bunches, discontinuous necrotic vascular streaks, and pockets of rot and discoloration were also observed (Davis et al. 2012, 2015).

Citrus All citrus varieties are quite susceptible to phytoplasma infection although very often their presence is in association with '*Candidatus* Liberibacter' species in a disease known as "huanglongbing" (HLB). Particularly in the Brazil epidemics, 16SrIX group phytoplasmas were detected, while in the Chinese epidemic, 16SrI-B group phytoplasmas were detected associated with HLB (Teixeira et al. 2008; Chen et al. 2009). Recently in Mexico, HLB was associated with the presence of 16SrI-B and 16SrI-S phytoplasmas in orange and lime (Arratia-Castro et al. 2014). In Puerto Rico 16SrIX group phytoplasmas are reported in *C. sinensis* and *C. limon* (Caicedo

Fig. 6.11 Jujube young tree with severe JWB symptoms in China (left and center). Right, lime witches' broom ('*Ca.* P. aurantifolia') in Oman

et al. 2015). In China, grapefruit (*C. paradisi*) with HLB showed 16SrII-A phytoplasma presence in some of the infected plants (Lou et al. 2014). Historically the first described '*Candidatus* Phytoplasma' species (Zreick et al. 1995) is that detected in lime affected by the witches' broom disease in Oman (Fig. 6.11), Iran, and in the United Arab Emirates (Garnier et al. 1991; Bové et al. 2000; Al-Yahyai et al. 2015; Al-Abadi et al. 2016), but the disease in citrus was described in China, and phytoplasma presence was reported a long time ago (Chen et al. 1979). Acid lime is the other citrus species reported to be affected by these phytoplasmas in India (Ghosh et al. 1999), Iran (Salehi et al. 2000; Djavaheri and Rahimian 2004). This phytoplasma was also detected in limequat tree in Iran (Faghihi et al. 2017). Sweet orange (*C. sinensis*) and mandarin (*C. reticulate*) showing typical symptoms of witches' broom disease were detected in Egypt, and phytoplasma presence was demonstrated without molecular identification (EL Banna et al. 2015). Aster yellows were reported in citrus in Pakistan (Fahmeed et al. 2009).

Jujube For several decades jujube witches' broom (JWB), associated with '*Ca.* P. ziziphi' has caused a devastating disease of the jujube (*Ziziphus jujuba* Mill) industry in China (Tsai et al. 1988). Transmission electron microscopy showed that a phytoplasma was associated with the disease (Yi and La 1973). Diseased trees show the precocious development of proliferating secondary shoots which have an overabundance of abnormally small and sometimes chlorotic leaves (Fig. 6.11). Phyllody is another characteristic of diseased trees and symptomatic limbs do not bear fruit. Symptoms are at first limited to one or more branches, but then the disease spreads progressively throughout the entire crown. Trees of all ages are susceptible and die within a few years after symptoms first appear (La and Chang 1979). Fan et al. (2008) verified the presence of this phytoplasma in jujube trees belonging to diverse cultivars in both symptomatic and asymptomatic samples. The phytoplasma associated with JWB is '*Ca.* P. ziziphi' (Jung et al. 2003) belonging to subgroup 16SrV-B, and it was reported in China and India infecting also stone fruit (Zhu et al. 1997; Tian et al. 2000; Khan et al. 2008). The known insect vector of the phytoplasma is *Hishimonoides chinensis* Anufriev and *Hishimonus sallatus* Uhler (La and Woo 1980).

Longan (*Dimocarpus longan* Lour.) (Sapindaceae) is found commonly in most of Asia, primarily in China, Taiwan, Vietnam, and Thailand, and in the latter two countries, it is one of the major fruit plants promoted for domestic consumption and export. The witches' broom syndrome is a devastating epidemic in the majority of the longan-growing areas of Vietnam. The main symptoms include small and upward rolled leaves, shortened internodes, and witches' broom; small, stunted shoots with curved, rolled up margins, deformed leaves with blisters, and hairy patches on the underside; and abnormal development of flower structures, flower abortion, and failure to produce fruit or develop small fruit. Phytoplasmas detected belonged to 16SrI, 16SrII, 16SrV, and 16SrXII-A groups/subgroups (Nguyen et al. 2012; Hoat et al. 2016). Phytoplasmas are also reported in plants showing similar symptoms in Thailand (Chantrasri et al. 1999). During a recent study, a partial relationship between the symptoms and *Eriophyes dimocarpi* mites presence was verified (Hoat et al. 2017).

Papaya In *Carica papaya* (Caricaceae), phytoplasma diseases are reported from Australia, Cuba, and other parts of the Caribbean, Ethiopia, Hawaii, India, Israel, Oman, and South Africa. Mosaic, yellow crinkle, and dieback diseases of papaya are reported in Australia (Gibb et al. 1996, 1998; De La Rue et al. 1999). While in yellow crinkle and mosaic (PpYC and PpM) diseases, '*Ca*. P. australasia' (16SrII-D) phytoplasma was identified; in the dieback disease, '*Ca*. P australiense' was detected (16SrXII-B) (White et al. 1998). PpYC disease consists of yellowing of leaves and bending down of the petioles. The crown leaves develop clear patches at the margins and between the veins, and these areas become brown and die; the older leaves die and fall leaving a bare stem with a few stunted leaves on top. The flowers may show abnormal distortion. In the PpM disease, stunted yellow leaves with clear margins and short stems with many side shoots are observed. The petioles and upper stems have "water-soaked" streaks, and the fruit may show light green areas. It could be problematic in some cases to distinguish among these two diseases. Papaya dieback (PpDB) disease shows inner crown leaves with bunched appearance, yellowing, followed by a slight bending of the stem tip. The affected leaves shrivel after a quick yellowing, the crown dies within 1–4 weeks, and the fruit falls off while still green or rot. PpDB dieback is an economically important disease limiting papaya production in southern and central Queensland, with outbreaks frequently causing losses of 10–100%. The phytoplasma associated with PpDB has been transmitted by dodder to tomato and other test plants. In the Caribbean a bunchy top disease is described beginning with yellow mottling of the upper leaves similar to the one present in the yellow crinkle disease, and then the leaves begin to die. Growth of leaves is slow, and the distance between them along the stem becomes shorter giving a "bunchy top" appearance. The oldest leaves fall down, leaving a few stunted leaves at the top of the stem. The plants may die, or new shoots appear from lower down the stem. In Cuba, '*Ca*. P. caricae' (16SrXVII group) (Arocha et al. 2005) and group 16SrI phytoplasmas (Acosta et al. 2011) were identified in diseased plants. Occurrence of aster yellows phytoplasmas is also reported in Sri Lanka in papaya dieback (Abeysinghe et al. 2016) and in a papaya yellows disease in Taiwan (Bau et al.

2011). Phytoplasmas belonging to different 16Sr groups have been identified in the Asian continent including groups 16SrI and 16SrII in India associated with similar dieback symptoms (Rao et al. 2011) and axillary shoot proliferation (Verma et al. 2012); and group 16SrXII in Israel (Gera et al. 2005).

Pomegranate *Punica granatum* (Lythraceae) is grown in many subtropical countries especially in the Mediterranean region for its nutritional and medical properties. Phytoplasma presence in pomegranate is associated with yellows symptoms in Turkey where two phytoplasma strains belonging to subgroups 16SrI-B and 16SrXII-A were identified (Gazel et al. 2016) and in Iran where a group 16SrIII was associated with a slow decline, reduction in yield and fruit size and yellowing of the leaves, and foliar reddening with thickening of veins (Karimi et al. 2015), and a 16SrII-D subgroup was associated with little leaf symptoms (Salehi et al. 2016).

6.6 Nuts

Chestnut In Japan a chestnut yellows disease (Shimada and Kouda 1954) was shown by transmission electron microscopy to be associated with phytoplasma presence (Okuda et al. 1974). In Korea in 1993, chestnut trees with little leaf disease (CLL) showing small, yellow leaves were also associated with phytoplasma presence (Han et al. 1997). More recently Japanese chestnut trees (*Castanea crenata* Sieb. and Zucc.) showing a witches' broom disease, including abnormally small leaves and yellowing of young leaves, was found associated with the presence of '*Ca.* P. castaneae' (Jung et al. 2002). The disease named chestnut witches' broom (CnWB) differed from Korean CLL and the chestnut yellows of Japan in symptom severity and led to severe crop losses especially in some provinces.

Hazelnut Decline and yellows are diseases reported for a long time in several European hazelnut (*Corylus avellana* L., Betulaceae) growing areas and were associated with 16SrX group phytoplasma presence (Marcone et al. 1996). In stunted but also in asymptomatic plants a 16SrIII-B phytoplasma was identified in Oregon (USA) (Jomantiene et al. 2000); '*Ca.* P. asteris' was reported in a few stunted plants of European hazelnut in Poland (Cieślińska and Kowalik 2011). In the United Kingdom, '*Ca.* P. fragariae' (16SrXII-E) was detected in 10–15-year-old plants showing yellowing of leaves, leaf scorch, a lack of density of the tree canopy, and proliferation of small thin branches (Hodgetts et al. 2015). In Chile, hazelnut plants cv. Barcelona 12-year-old showing leaf and catkin malformation and yellowing in some branches were positive for the presence of phytoplasmas in 16SrI, 16SrIII, and 16SrV groups (Pérez Fuentealba et al. 2017). Nothing is reported about disease epidemiology and/or possible insect vectors.

Peanut Peanut witches' broom is the only disease reported in *Arachis hypogaea* L. (Fabaceae); the phytoplasmas detected in China and Taiwan resulted closely phylogenetically related and all belonging to 16SrII-A subgroup (Chen et al. 1994; Li et al. 2014a). The disease agent is transmitted by the insect vector *Orosius orientalis* (Yang 1975).

Pecan Pecan bunch is a widespread disease reported in commercially grown southern pecan, *Carya illinoensis* (Wangenh) K. Koch (Juglandaceae), in the USA. The tree is grown mainly in the America continent where it is the tree symbol for the state of Texas. The symptoms on diseased plants are bushy growths of slender willowy shoots, resulting from an abnormal forcing of lateral buds. Symptoms may appear on only one branch or on many branches. Bunch disease is very conspicuous in the spring and early summer because the diseased shoots come out of dormancy earlier than non-infected shoots (Seliskar et al. 1974). The phytoplasmas detected in this species in Georgia belong to group 16SrIII-C (Lee et al. 1998b).

6.7 Conclusions

Management of phytoplasma diseases in fruit crops requires relevant efforts worldwide considering the huge number of diverse species that are affected or susceptible to these diseases. Early phytoplasma detection in mother plants together with the implementation of phytoplasma-free certified plants could greatly improve the sanitary status of these species and reduce the large diffusion of different phytoplasmas by propagation materials. It is very complicated to propose a scheme based on insect vector control since these are quite often not effective, and also their identity is quite often not known.

References

Abeysinghe S, Kumari WGSM, Arachchi IMM, Dickinson M (2016) Occurrence of phytoplasma diseases of papaya in Sri Lanka. *Acta Horticulturae* **1111**, 25–30.

Abou-Jawdah Y, Abdel Sater A, Jawhari M, Sobh H, Abdul-Nour H, Bianco PA, Molino Lova M, Alma A (2014) *Asymmetrasca decedens* (Cicadellidae, Typhlocybinae), a natural vector of '*Candidatus* Phytoplasma phoenicium'. *Annals of Applied Biology* **165**, 395–403.

Abou-Jawdah Y, Karakashian A, Sobh H, Martini M, Lee I-M (2002) An epidemic of almond witches' broom in Lebanon: classification and phylogenetic relationship of the associated phytoplasma. *Plant Disease* **86**, 477–484.

Abou-Jawdah Y, Sobh H, Akkary M (2009) First report of almond witches' broom phytoplasma ('*Candidatus* Phytoplasma phoenicium') causing a severe disease on nectarine and peach trees in Lebanon. *EPPO Bulletin* **39**, 94–98.

Acosta K, Zamora L, Fernández A, Arocha Y, Martinez Y, Santos ME, Méndez J, Chávez A, Leyva-López NE (2011) First report of '*Candidatus* Phytoplasmas asteris' (group 16SrI) affecting papaya in Cuba. *New Disease Reports* **24**, 29.

Ahrens U, Lorenz KH, Seemüller E (1993) Genetic diversity among mycoplasmalike organisms associated with stone fruit disease. *Molecular Plant-Microbe Interactions* **6**, 686–691.

Al-Abadi SY, Al-Sadi AM, Dickinson M, Al-Hammadi MS, Al-Shariqi R, Al-Yahyai RA, Kazerooni EA, Bertaccini A (2016) Population genetic analysis reveals a low level of genetic diversity of '*Candidatus* Phytoplasma aurantifolia' causing witches' broom disease in lime. *SpringerPlus* **5**, 1701.

Al-Yahyai RA, Al-Sadi AM, Al-Said FA, Al-Kalbani Z, Carvalho CM, Elliot SL, Bertaccini A (2015) Development and morphological changes in leaves and branches of acid lime (*Citrus aurantifolia*) affected by witches' broom disease. *Phytopathologia Meditteranea* **54**, 133–139.

Ambrožič Turk B, Mehle N, Brzin J, Nikolič P, Dermastia M, Boben J, Ravnikar M (2008) High infection pressure of ESFY phytoplasma threatens the cultivation of stone fruit species. *Journal of Central European Agriculture* **9**, 795–802.

Andersen MT, Longmore J, Wood GA, Sutherland PW, Beck DL, Forster RLS (1998) Phormium yellow leaf phytoplasmas is associated with strawberry lethal yellows disease in New Zealand. *Plant Disease* **82**, 606–609.

Andrade N, Villagra M, Arismendi N (2009) Evidencias microscópicas y moleculares de la presencia de fitoplasmas en plantas de murta (*Ugni molinae* Turcz.) afectadas por la enfermedad "escoba de bruja". *Tropical Plant Pathology* **34**, 245–249.

Arismendi N, González F, Zamorano A, Andrade N, Pino AM, Fiore N (2011) Molecular identification of '*Candidatus* Phytoplasma fraxini' in murta and peony in Chile. *Bulletin of Insectology* **64**(Supplement), S95–S96.

Arocha Y, Lopez M, Pinol B, Fernandez M, Picornell B, Almeida R, Palenzuela I, Wilson MR, Jones P (2005) '*Candidatus* Phytoplasma graminis' and '*Candidatus* Phytoplasma caricae', two novel phytoplasmas associated with diseases of sugarcane, weeds and papaya in Cuba. *International Journal Systematic and Evolutionary Microbiology* **55**, 2451–2463.

Arocha Y, Plata G, Franco J, Maín G, Veramendi S, Lazcano F, Crespo JL, Lino V, Calderón C, Llerena R, Andrew R, Antezana O, Gutiérrez A, Coca M, Boa E (2010) First report of a 16SrIII phytoplasma (X-disease group) affecting bell pepper, strawberry (frutilla), *Schinus molle* and *Arracacia xanthorrhiza* in Cochabamba, Bolivia. *Plant Pathology* **59**, 395–395.

Arocha-Rosete Y, Kent P, Agrawal V, Hunt D, Hamilton A, Bertaccini A, Scott J, Crosby W, Michelutti R (2011a) Identification of *Graminella nigrifrons* as a potential vector for phytoplasmas affecting *Prunus* and *Pyrus* species in Canada. *Canadian Journal of Plant Pathology* **33**, 465–474.

Arocha-Rosete Y, Zunnoon-Khan S, Krukovets I, Crosby W, Scott J, Bertaccini A, Michelutti R (2011b) Identification and molecular characterization of the phytoplasma associated with peach rosette-like disease at the Canadian Clonal Genebank based on the 16S rRNA gene analysis. *Canadian Journal of Plant Pathology* **33**, 127–134.

Arocha-Rosete YA, Schilder A, Lambert L, Scott J (2015) Identification and molecular characterization of the blueberry stunt phytoplasma in Canada. *Phytopathogenic Mollicutes* **5**(1-Supplement), S17–S18.

Arratia-Castro A, Santos-Cervantes ME, Fernández-Herrera E, Chávez-Medina JA, Flores-Zamora JL, Camacho-Beltrán E, Méndez-Lozano J, Leyva-López NE (2014) Occurrence of '*Candidatus* Phytoplasma asteris' in citrus showing "huanglongbing" symptoms in Mexico. *Crop Protection* **62**, 144–151.

Avramov Z, Contaldo N, Bertaccini B, Sakalieva D (2011) First report of "stolbur" phytoplasmas in *Prunus avium* in Bulgaria. *Bulletin of Insectology* **64**(Supplement), S71–S72.

Bagadia PG, Polashock J, Bottner-Parker KD, Zhao Y, Davis RE, Lee I-M (2013) Characterization and molecular differentiation of 16SrI-E and 16SrIX-E phytoplasmas associated with blueberry stunt disease in New Jersey. *Molecular and Cellular Probes* **27**, 90–97.

Balakishiyeva G, Qurbanov M, Mammadov A, Bayramov S, Aliyev J, Foissac X (2011) Detection of '*Candidatus* Phytoplasma brasiliense' in a new geographic region and existence of two genetically distinct populations. *European Journal of Plant Pathology* **130**, 457–462.

Battle A, Sabaté J, Iglesias I, Laviña A (2012) Identification and epidemiology of '*Candidatus* Phytoplasma prunorum' in Spanish peach orchards. Susceptibility of different rootstocks to the disease. *Acta Horticulturae* **962**, 443–448.

Bau HJ, Hung SC, Chang WC, Chen YK (2011) First report of group 16SrXII pytoplasma associated with papaya yellows in Taiwan. *Plant Disease* **95**, 1581.

Behnke HD, Schaper U, Seemüller E (1980) Association of mycoplasma-like organism with pear decline symptoms in the Federal Republic of Germany. *Phytophatologische Zeitschrift* **97**, 89–93.

Bernhard R, Marenaud C, Sechet J, Fos A, Moutous G (1977) A complex disease of certain *Prunus*: "Molieres decline". *Comptes Rendus des Seances de l'Academie d'Agriculture de France* **3**, 178–188.

Bertaccini A, Vibio M, de Luca G, Villani A (1994) Individuazione e caratterizzazione molecolare di micoplasmi presenti in coltivazioni di pero dell'Emilia-Romagna. *Frutticoltura* **9**, 83–86.

Bertaccini A, Vibio M, Gennari F, Guerrini A, Benni A, Lee I-M (1995) Detection of mycoplasmalike organisms (phytoplasmas) in *Rubus* by nested polymerase chain reaction (PCR). *Acta Horticulturae* **385**, 126–131.

Bertaccini A, Vibio M, Pastore M, Recupero S, Guerrini S, Grimaldi D (1997) Nested PCR assays for detection of phytoplasma in strawberry. *Acta Horticulturae* **493**, 787–790.

Bertaccini A, Davies D, Fialova R, Franova J, Karesova R, Martini M, Navratil M, Paltrinieri S, Brighetti M (2001) A molecular survey to identify phytoplasmas associated with apple trees showing different diseases symptoms. *Acta Horticulturae* **550**, 371–376.

Blattńy C Jr (1963) Die Übertragung der Hesenbesenkrankheit der Heidelbeeren (*Vaccinium myrtillus* L.) durch die Zikade *Idionus cruentatus* Panz. *Phytopathology Zeittschrift* **49**, 203–205.

Blattńy C Sr, Blattńy C Jr (1970) Bluebrry witches' broom. In: Virus diseases of small fruits and grapevines. Ed Frazier NW, University of California, Division of Agriculture Science, Berkeley, USA, 177–179 pp.

Blattńy C, Vana V (1974) Mycoplasmalike organisms in *Vaccinium myrtillus* L. infected with blueberry witches broom. *Biologia Plantarum* **16**, 476–478.

Blattný C, Váňa V (1974) Pear decline accompanied with mycoplasmalike organisms in Czechoslovakia. *Biologia Plantarum* **16**, 474–475.

Bobev SG, de Jonghe K, Maes (2013) First report of '*Candidatus* Phytoplasma solani' on blackberry (*Rubus fruticosus*) in Bulgaria. *Plant Disease* **97**, 282.

Borroto-Fernández EG, Calari A, Hanzer V, Katinger H, Bertaccini A, Laimer M (2007) Phytoplasma infected plants in Austrian forests: role as a reservoir? *Bulletin of Insectology* **60**, 391–392.

Bos L (1960) A witches' broom virus disease of *Vaccinium myrtillus* in The Netherlands, *Tijdsch Plantenzi* **66**, 259–263.

Bové JM, Danet JL, Bananej K, Hassanzadeh N, Taghizadeh M, Salehi M, Garnier M (2000) Witches' broom disease of lime (WBDL) in Iran. Fourteenth Conference IOCV Riverside, California, USA, 207–212.

Çaglayan K, Ulubaş Serçe Ç, Öztürk H, Gençer NS, Gazel M (2008) A preliminary account of the presence of pear decline in Marmara Region of Turkey. *Acta Horticulturae* **781**, 449–452.

Çaglayan K, Gazel M, Küçükgöl C, Paltrineri S, Contaldo N, Bertaccini A (2013) First report of '*Candidatus* phytoplasma asteris' (group 16SrI-B) infecting sweet cherries in Turkey. *Journal of Plant Pathology* **95**, S4.77.

Caicedo JD, Rivera-Vargas LI, Segarra AE, Davis RE (2015) Detection and molecular characterisation of a group 16SrIX phytoplasma infecting citrus (*Citrus sinensis* and *C. limon*), coffee (*Coffea arabica*), periwinkle (*Catharanthus roseus*), and tabebuia (*Tabebuia heterophylla*) in Puerto Rico. *Australasian Plant Disease Notes* **10**, 28.

Camele I, Vibio M, Rana GL, Bertaccini A (1998) La moria del pero minaccia la frutticoltura della Basilicata. *Informatore Fitopatologico* **9**, 35–38.

Carraro L, Loi N, Ermacora P, Osler R (1998a) High tolerance of European plum varieties to plum leptonecrosis. *European Journal of Plant Pathology* **104**, 141–145.

Carraro L, Osler R, Loi N, Ermacora P, Refatti E (1998b) Transmission of European stone fruit yellows phytoplasma by *Cacopsylla pruni*. *Journal of Plant Pathology* **80**, 233–239.

Carraro L, Loi N, Ermacora P, Osler R (1998c) Transmission of pear decline by using naturally infected *Cacopsylla pyri* L. *Acta Horticulturae* **472**,665–668.

Carraro L, Loi N, Ermacora P (2001) Transmission characteristics of the European stone fruit yellows phytoplasma and its vector *Cacopsylla pruni*. *European Journal of Plant Pathology* **107**, 695–700.

Carraro L, Ferrini F, Ermacora P, Loi N (2004) Transmission of European Stone Fruit Yellows phytoplasma to *Prunus* species by using vector and graft transmission. *Acta Horticulturae* **657**, 449–453.

Casati P, Quaglino F, Tedeschi R, Spiga FM, Alma A, Spadone P, Bianco PA (2010) Identification and molecular characterization of '*Candidatus* Phytoplasma mali' isolates in north-western Italy. *Journal of Phytopathology* **158**, 81–87.

Casati P, Quaglino F, Stern AR, Tedeschi R, Alma A, Bianco PA (2011) Multiple gene analyses reveal extensive genetic diversity among '*Candidatus* Phytoplasma mali' populations. *Annals of Applied Biology* **158**, 257–266.

Casati P, Quaglino F, Abou-Jawdah Y, Picciau L, Cominetti A, Tedeschi R, Jawhari M, Choueiri E, Sobh H, Molino Lova M, Beyrouthy M, Alma A, Bianco PA (2016) Wild plants could play a role in the diffusion of diseases associated with phytoplasmas of pigeon pea witches' broom group (16SrIX). *Journal of Plant Pathology* **98**, 71–81.

Chantrasri P, Sardsud V, Srichart W (1999) Transmission studies of phytoplasma, the causal agent of witches' broom disease of longan. 25th Congress on Science and Technology of Thailand; Pitsanulok, Thailand.

Chen TA (1970) Mycoplasmalike organisms in sieve tube elements of plants infected with blueberry stunt and cranberry false blossom. *Phytopathology* **61**, 233–236.

Chen J, Pu X, Deng X, Liu S, Li H, Civerolo E (2009) A phytoplasma related to '*Candidatus* phytoplasma asteris' detected in citrus showing "huanglongbing" (yellow shoot disease) symptoms in Guangdong, P. R. China. *Phytopathology* **99**, 236–242.

Chen MR, Zhang SG, Zheng GB (1994) Occurrence and control methods of peanut witches' broom disease happening in South China. *Guangdong Agricultural Science* **2**, 33–35.

Chen T-Y, Shen C-Y, Tao S-C, Kung T-H, Chen N-W, Tai Y-M (1979) Studies on the pathogens of "huanglongbing" (citrus yellow shoot disease) III. Mycoplasma-like organisms associated with "huanglongbing" in Kwangtung. *Acta Biochemica Biophisica Sinica* **11**, 191–193.

Choueiri E, Jreijiri F, Issa S, Verdin E, Bové J, Garnier M (2001) First report of a phytoplasma disease of almond (*Prunus amygdalus*) in Lebanon. *Plant Disease* **85**, 802.

Cieślińska M (2001) The preliminary results on detection of phytoplasmas associated with small fruit diseases in Poland. *Acta Horticulturae* **551**, 87–92.

Cieślińska M (2011) Detection and characterization of phytoplasmas associated with diseases of *Rubus* spp. in Ploand. *Journal of Plant Pathology* **93**, 51–56.

Cieślińska M, Kowalik B (2011) Detection and molecular characterization of '*Candidatus* Phytoplasma asteris' in European Hazel (*Corylus avellana*) in Poland. *Journal of Phytopathology* **159**, 585–588.

Cieślińska M, Morgaś H (2011) Detection and Identification of '*Candidatus* Phytoplasma prunorum', '*Candidatus* Phytoplasma mali' and '*Candidatus* Phytoplasma pyri' in stone fruit trees in Poland. *Journal of Phytopathology* **159**, 217–222.

Cieślińska M, Smolarek T (2015) Molecular diversity of phytoplasmas infecting cherry trees in Poland. *Phytopathogenic Mollicutes* **5**(1-Supplement), S31–S32.

Cieślińska M, Wójcik-Seliga J, Kowalik B (2014) Molecular diversity of phytoplasmas infecting *Rubus* sp. plants in Poland. In: Phytoplasmas and phytoplasma disease management: how to reduce their economic impact. Ed Bertaccini A, COST Action FA0807, 151–158 pp.

Cieślińska M, Hennig E, Kruczyńska D, Bertaccini A (2015) Genetic diversity of '*Candidatus* Phytoplasma mali' strains in Poland. *Phytopathologia Mediterranea* **54**, 477–487.

Contaldo N, Mejia JF, Paltrinieri S, Calari A, Bertaccini A (2012) Identification and *groEl* gene characterization of green petal phytoplasma infecting strawberry in Italy. *Phytopathogenic Mollicutes* **2**, 59–62

Converse RH, Clarke RG, Oman PW, Milbrath GM (1982) Witches' broom disease of black raspberry in Oregon. *Plant Disease* **66**, 949–951.

Davies D (2000) The occurrence of two phytoplasmas associated with stunted *Rubus* species in the UK. *Plant Pathology* **49**, 86–88.

Davies DL, Adams AM, Glenister M (1985) Mycoplasma-like organism (MLOs) in pear and apples. Report East Molling Research Station for 1984, 151.

Davies DL, Guise CM, Clark MF, Adams AN (1992) Parry's disease of pears is similar to pear decline and is associated with mycoplasma-like organisms transmitted by *Cacopsylla pyricola*. *Plant Pathology* **41**, 195–203.

Davis RE, Dally EL, Converse RH (2001) Molecular Identification of a phytoplasma associated with witches' broom disease of black raspberry in Oregon and its classification in group 16SrIII, new subgroup Q. *Disease Notes* **85**, 1121.

Davis R, Zhao Y, Dally EL, Lee IM, Jomantiene R, Douglas SM (2013) '*Candidatus* Phytoplasma pruni', a novel taxon associated with X-disease of stone fruits, *Prunus* spp.: multilocus characterization based on 16S rRNA, secY, and ribosomal protein genes. *International Journal of Systematic and Evolutionary Microbiology* **63**, 766–776.

Davis RI, Kokoa P, Jones LM, Mackie J, Constable FE, Rodoni BC, Gunua TG, Rossel JB (2012) A new banana wilt disease associated with phytoplasmas in Papua New Guinea. *Australasian Plant Disease Notes* **7**, 91–97.

Davis RI, Henderson J, Jones LM, McTaggart AR, O'Dwyer C, Tsatsia F, Fanai C, Rossel JB (2015) First record of a wilt disease of banana plants associated with phytoplasmas in Solomon Islands. *Australasian Plant Disease Notes* **10**, 1–6.

de Fluiter HJ, van der Meer FA (1953) Rubus stunt, a leafhopper borne virus disease. *Tijdschrift Over Plantenziekten* **59**, 195.

de Jonghe K, De Roo I, Maes M (2017) Fast and sensitive on-site isothermal assay (LAMP) for diagnosis and detection of three fruit tree phytoplasmas. *European Journal of Plant Pathology* **147**, 749–759.

De La Rue S, Schneider B, Gibb K (1999).Genetic variability in phytoplasmas associated with papaya yellow crinkle and papaya mosaic diseases in Queensland and the northern territory. *Australian Plant Pathology* **28**, 108–114.

de Leeuw GTN (1975) Presence of mycoplasma-like organisms in the phloem of witches' broom diseased *Vacinium myrtillus* plants in the Netherlands. *Phytopathologische Zeithscrift* **83**, 91–94.

Del Serrone P, La Starza S, Krystai L, Kolber M, Barba M (1998) Occurrence of apple proliferation and pear decline phytoplasmas in diseased pear trees in Hungary. *Journal of Plant Pathology* **80**, 53–58.

Delic D, Martini M, Ermacora P, Myrta A, Carraro L (2005) First report of fruit tree phytoplasmas and their psyllid vectors in Bosnia and Herzegovina. *Bulletin OEPP/EPPO Bulletin* **37**, 444–448.

Desvignes JC, Boyé R, Cornaggia D, Grasseau N (1999) Maladies dues à des phytoplasmes. In: Maladies à virus des arbes fruitiers. Ed Desvignes JC, CTIFL, Paris, France, 113–143 pp.

Djavaheri M, Rahimian H (2004) Witches' broom of bakraee (*Citrus reticulata* hybrid) in Iran. *Journal of Plant Disease* **88**, 683.

Dobroscky D (1929) Cranberry false-blossom disease spread by a leafhopper. *Science* **12**, 635.

Dosba F, Lansac M, Mazy K, Garnier M, Eyquard JP (1991) Incidence of different diseases associated with mycoplasmalike organisms in different species of *Prunus*. *Acta Horticulturae* **283**, 311–320.

Douglas SM (1986) Detection of mycoplasma-like organisms in peach and chokecherry with X-disease by fluorescence microscopy. *Phytopathology* **76**, 784–787.

Duduk B, Botti S, Trkulja V, Ivanović M, Stojčić J, Bertaccini A (2005a) Occurence of pear decline phytoplasma in Bosnia and Herzegovina. *Journal of Plant Pathology* **87**, 75.

Duduk B, Ivanović M, Obradović A, Paltrinieri S, Bertaccini A (2005b) First report of pear decline phytoplasma on pear in Serbia. *Plant Disease* **89**, 774.

EL Banna OM, Neven IT; Sahar AY, Shalaby AA (2015) Molecular and electron microscope evidence for an association of phytoplasma with citrus witches' broom disease. *International Journal of Scientific & Engineering Research* **6**, 127–133.
Eleftheriou EP (1985) A decline-like disorder of pear in Greece: pear decline or scion-rootstock incompatibility? *Scientia Horticulturae* **27**, 241–250.
Ember I, Nemeth M, Krizbai L, Kolber M, Szakál M, Botti S, Bertaccini A, Bohár GY, Zsovák-Hangyál R (2004) Identification of phytoplasmas on poamceous fruit tree species in Hungary. *Acta Horticulturae* **657**, 443–448.
Ermacora P, Carraro L, Martini M, Ferrini F, Loi N (2003) Presence of *Rubus* stunt in blackberry in Northeastern Italy. *Journal of Plant Pathology* **85**, 306.
Ermacora P, Loi N, Ferrini F, Loschi A, Martini M, Osler R, Carraro L (2010) Hypo- and hyper-virulence in apricot trees infected by ESFY. *Julius-Kühn-Archiv* **427**, 197–200.
Errea P, Aguelo V, Hormaza JI (2002) Seasonal variations in detection and transmission of pear decline phytoplasma. *Journal of Phytopathology* **150**, 439–443.
Etropolska A, Jarausch W, Jarausch B, Trenchev G (2015) Detection of European fruit tree phytoplasmas and their insect vectors in important fruit-growing regions in Bulgaria. *Bulgarian Journal of Agricultural Science* **21**, 1248–1253.
Etropolska A, Lefort F (2017) First detection of European stone fruit yellows phytoplasma in peach trees on the territory of Canton of Geneva, Switzerland. 24th ICVF International Conference. Thessaloniki, Greece, 95.
Facundo R, Quiroga N, Méndez P, Zamorano A, Fiore N (2017) First report of '*Candidatus* Phytoplasma pyri' on pear in Chile. *Plant Disease* **101**, 830.
Faghihi MM, Bagheri A, Askari Seyahooei M, Pezhman A, Faraji G (2017) First report of a '*Candidatus* Phytoplasma aurantifolia'-related strain associated with witches' broom disease of limequat in Iran. *New Disease Reports* **35**, 24.
Fahmeed F, Arocha Rosete Y, Acosta Pérez K, Boa E, Lucas J (2009) First report of '*Candidatus* Phytoplasma asteris' (group 16SrI) infecting fruits and vegetables in Islamabad, Pakistan. *Journal of Phytopathology* **157**, 639–641.
Fan XP, Paltrinieri S, Pastore M, Petriccione M, Wang X, Tian JB, Bertaccini A (2008) Molecular detection of '*Candidatus* Phytoplasma ziziphae' in different jujube varieties. *Acta Horticulturae* **772**, 207–210.
Fernández FD, Guzmán FA, Curzel V, Bejarano N, Conci LR (2013) Detection and molecular characterization of a phytoplasma affecting *Prunus persica* L. in Jujuy, Argentina. *European Journal of Plant Pathology* **135**, 627–631.
Fernández FD, Meneguzzi NG, Guzmán FA, Kirschbaum DS, Conci VC, Nome CF, Conci LR (2015) Detection and identification of a novel 16SrXIII subgroup phytoplasma associated with strawberry red leaf disease in Argentina. *International Journal of Systematic and Evolutionary Microbiology* **65**, 2741–2747.
Fernández FD, Marini D, Farrando R, Conci LR (2017) First report of a '*Candidatus* Phytoplasma pyri' strain in Argentina. *Australasian Plant Disease Notes* **12**, 8.
Ferrini F, Carraro L, Ermacora P, Loi N (2002) Vector and graft transmission of European stone fruit yellows phytoplasma to *Prunus* species. *Petria* **12**, 367–368.
Ferriol-Marchena X, Hernández-Rodríguez L, Luis-Pantoja M, García-García G, Banguela-Castillo A, Pérez JM (2013) First report of a '*Candidatus* Phytoplasma asteris' isolate affecting a strawberry nursery in Cuba. *New Disease Reports* **28**, 19.
Franova Honetslegrova J, Vibio M, Bertaccini A (1996) Electron microscopy and molecular identification of phytoplasmas associated with strawberry green petals in the Czech Republic. *European Journal of Plant Pathology* **102**, 831–836.
Franova J, Ludvíková H, Paprštein F, Bertaccini A (2013) Genetic diversity of Czech '*Candidatus* Phytoplasma mali' strains based on multilocus gene analyses. *European Journal of Plant Pathology* **136**, 675–688.
Franova J, de Sousa E, Koloniuk I, Cardoso F, Contaldo N, Paltrinieri S, Bertaccini A (2016) Multigene characterization of a new '*Candidatus* Phytoplasma rubi'-related strain associated

with blackberry witches' broom. *International Journal of Systematic and Evolutionary Microbiology* **66**, 1438–1446.

Gao R, Wang J, Zhao W, Li XD, Zhu SF, Hao YJ (2011) Identification of a phytoplasma associated with cherry virescence in China. *Journal of Plant Pathology* **93**, 465–469.

Garcia-Chapa M, Laviña A, Sanchez I, Medina V, Batlle A (2003) Occurrence, symptom expression and characterization of phytoplasma associated with pear decline disease in Catalonia (Spain). *Journal of Phytopathology* **151**, 584–590.

Garnier M, Zreik L, Bové JM (1991) Witches' broom, a lethal mycoplasmal disease of lime in the Sultanate of Oman and the United Arab Emirates. *Plant Disease* **75**, 546–551.

Gazel M, Çağlayan M, Pınar HB, Mejia JF, Paltrinieri S, Bertaccini A, Contaldo N (2016) Detection and identification of phytoplasmas in pomegranate trees with yellows symptoms. *Journal of Phytopathology* **164**, 136–140.

Genini M, Ramel ME (2004) Distribution of European stone fruit yellows phytoplasma in apricot trees in Western Switzerland. *Acta Horticulturae* **657**, 455–458.

Gentit P, Cornaggia D, Desvignes JC (1998) Identification and comparison of different *Prunus* phytoplasma diseases by indexing on GF305 peach seedlings in the greenhouse. *Acta Horticulturae* **472**, 723–729.

Gera A, Mawassi M, Zeidan A, Spiegel S, Bar-Joseph M (2005) A isolate of '*Candidatus* Phytoplasma australiense' group associated with Nivun Haamir dieback disease of papaya in Isreal. *Plant Pathology* **54**, 560.

Ghosh DK, Das AK, Singh S, Singh SJ, Ahlawat YA (1999) Occurrence of witches' broom, a new phytoplasma disease of acid lime (*Citrus aurantifolia*) in India. *Plant Disease* **83**, 302.

Gibb K, Persley D, Schneider B, Thomas J (1996) Phytoplasmas associated with papaya diseases in Australia. *Plant Disease* **80**, 174–178.

Gibb K, Schneider B, Padovan A (1998) Differential detectionand generic relatedness of phytoplasmas in papaya. *Plant Pathology* **47**, 323–332.

Giunchedi L, Poggi Pollini C (1984) La moria del pero: sintomi cause e possibili rimedi. *Informatore Fitopatologico* **2**, 24–28.

Giunchedi L, Marani F, Credi R (1978) Mycoplasma-like bodies associated with plum decline (leptonecrosis). *Phytopathologia Mediterranea* **17**, 205–209.

Giunchedi L, Credi R, Babini AR, Bertaccini A (1982) Mycoplasmalike organism associated with pear moria in Italy. *Phytopathologia Mediterranea* **21**, 20–22.

Gocio MJ, Dale JL (1982) Investigation of blueberry stunt disease in Arkansas. *Plant Disease* **66**, 244–245.

Goidànich G (1933) Un deperimento dei susini. *Bollettino Stazione di Patologia Vegetale di Roma* **13**, 160–173.

González F, Zamorano A, Pino AM, Paltrinieri S, Bertaccini A, Fiore N (2011) Identification of phytoplasma belonging to X-disease group in cherry in Chile. *Bulletin of Insectology* **64**(Supplement), S235–S236.

Guerrini S, Bissani R, Poggi Pollini C, Giunchedi L (2000) Studi epidemiologici sulla moria del pero. *Frutticoltura* **9**, 30–34.

Han SS, Kim YH, So IY, Chai JK (1997) Association of phytoplasma with chestnut (*Castanea crenata* Sieb. et Zucc.) little leaf disease in Korea. *Journal of Mycoplasmology Korea* **8**, 48–54.

Harrison NA, Legard DE, Dibonito R, Richardson PA (1997) Detection and differentiation of phytoplasmas associated with disease of strawberry in Florida. *Plant Disease* **81**, 230.

Hibino H, Schneider H (1970) Mycoplasmalike bodies in sieve tubes of pear trees affected with pear decline. *Phytopathology* **60**, 499–501.

Hoat TH, Contaldo N, Quan MV, Paltrinieri S, Satta E, Chi TT, Hao LV, Cuong NN, Bertaccini A (2016) Molecular detection and identification of 16SrII and aster yellows phytoplasmas associated with longan witches' broom syndrome in Vietnam. *Phytopathogenic Mollicutes* **6**, 10–15.

Hoat TX, Quan MV, Hao LV, Chi TT, Anh LDT, Son PV, Cuong NN, Bertaccini A (2017) Association of *Eriophyes dimocarpi* (Acari: Eriophyidae) with longan witches' broom disease in Vietnam. *Archives of Phytopathology and Plant Protection* **50**, 70–83.

Hodgetts J, Flint LJ, Davey C, Forde S, Jackson L, Harju V, Skelton A, Fox A (2015) Identification of '*Candidatus* Phytoplasma fragariae' (16SrXII-E) infecting *Corylus avellana* (hazel) in the United Kingdom. *New Disease Reports* **32**, 3.

Hunter DM, Svircev AM, Kaviani M, Michelutti R, Wang L, Thompson D (2010) First report of pear decline caused by '*Candidatus* Phytoplasma pyri' in Ontario, Canada. *Plant Disease* **94**, 634.

Jarausch B, Fuchs A, Schwind N, Krczal G, Jarausch W (2007) *Cacopsylla picta* as most important vector for '*Candidatus* Phytoplasma mali' in Germany and neighbouring regions. *Bulletin of Insectology* **60**, 189–190.

Jarausch B, Mühlenz I, Beck A, Lampe I, Harzer U, and Jarausch W (2008) Epidemiology of European stone fruit yellows in Germany. *Acta Horticulturae* **781**, 417–422.

Jarausch W, Lansac M, Saillard C, Broquaire JM, Dosba F (1998) PCR assay for specific detection of European stone fruit yellows phytoplasma and its use for epidemiological studies in France. *European Journal of Plant Pathology* **104**, 17–27.

Jarausch W, Eyquard JP, Lansac M, Mohns M, Dosba F (2000a). Susceptibility and tolerance of new French *Prunus domestica* cultivars to European stone fruit yellows phytoplasma. *Journal of Phytopathology* **148**, 489–493.

Jarausch W, Saillard C, Helliot B, Garnier M, Dosba F (2000b) Genetic variability of apple proliferation phytoplasmas as determined by PCR-RFLP and sequencing of a non-ribosomal fragment. *Molecular and Cellular Probes* **14**, 17–24.

Jarausch W, Jarausch-Wehrheim B, Danet JL, Broquaire JM, Dosba F, Saillard C, Garnier M (2001) Detection and identification of European stone fruit yellows and other phytoplasmas in wild plants in the surroundings of apricot chlorotic leaf roll-affected orchards in southern France. *European Journal of Plant Pathology* **107**, 209–217.

Jarausch W, Pecerella T, Schwind N, Jarausch B, Krczal G (2004) Establishment of the quantitative real-time PCR for the quantification of apple proliferation phytoplasmas in plants and insects. *Acta Horticulturae* **657**, 415–420.

Jarausch W, Fuchs A, Jarausch B (2010) Establishment of a quantitative real-time PCR assay for the specific quantification of '*Ca.* Phytoplasma prunorum' in plants and insects. *Julius-Kühn-Archiv* **427**, 392–394.

Jomantiene R, Davis RE, Maas J, Dally EL (1996) Phytoplasmas associated with disease of strawberry in Florida. *Phytopathology* **86**, S123.

Jomantiene R., Davis RE, Maas J, Dally EL (1998) Classification of new phytoplasmas associated with diseases of strawberry in Florida, based on analysis of 16S rRNA and ribosomal protein gene operon sequences. *International Journal of Systematic Bacteriology* **48**, 269–277.

Jomantiene R, Postman JD, Montano HG, Maas JL, Johnson KB (2000) First report of clover yellow edge phytoplasma in *Corylus* (hazelnut). *Plant Disease* **84**, 102.

Jomantiene R, Yhao Z, Lee IM, Davis RE (2011) Phytoplasmas infecting sour cherry and lilac represent two distinct lineages having close evolutionary affinities with clover phyllody phytoplasma. *European Journal of Plant Pathology* **130**, 97–107.

Jones P, Arocha Y, Antesana O, Montilliano E, Franco P (2005) First report of an isolate of '*Candidatus* Phytoplasma australiense' associated with a yellow leaf roll disease of peach (*Prunus persicae*) in Bolivia. *Plant Pathology* **54**, 558.

Jung H-Y, Sawayanagi T, Kakizawa S, Nishigawa H, Miyata S, Oshima K, Ugaki M, Lee J-T, Hibi T, Namba S (2002) '*Candidatus* Phytoplasma castaneae', a novel phytoplasma taxon associated with chestnut witches' broom disease. *International Journal of Systematic and Evolutionary Microbiology* **52**, 1543–1549.

Jung HY, Sawayanagi T, Kakizawa S, Nishigawa H, Wei W, Oshima K, Miyata S, Ugaki M, Hibi T, Namba S (2003) '*Candidatus* Phytoplasma ziziphi', a novel phytoplasma taxon associated with jujube witches' broom disease. *International Journal of Systematic and Evolutionary Microbiology* **53**, 1037–1041.

Kaminska M, Sliwa H (2004) First report of phytoplasma belonging to apple proliferation group in roses in Poland. *Plant Disease* **88**, 1283.

Karimi MR, Paltrinieri S, Contaldo N, Kamali H, Sajadinejad M, Ajami MR, Bertaccini A (2015) Phytoplasma detection and identification in declining pomegranate in Iran. *Phytopathogenic Mollicutes* **5**, 95–99.

Khan MS, Raj SK, Snehi SK (2008) Natural occurrence of '*Candidatus* Phytoplasma ziziphi' isolates in two species of jujube trees (*Ziziphus* spp.) in India. *New Disease Reports* **17**, 14.

Khan JA, Kumar J, Thakur PD, Handa A, Jarial K (2013) First report of a '*Candidatus* Phytoplasma ziziphi'-related strain associated with peach decline disease in India. *Journal of Plant Pathology* **95**, S4.76.

Kirkpatrick BC, Fisher GA, Fraser JD, Purcell AH (1990) Epidemiology and phylogenetic studies on western X-disease mycoplasma-like organisms. 7th Congress Interantional Organization for Mycoplasmology. Baden near Vienna, Austria 1988, 288–297.

Kirkpatrick BC, Uyemoto JK, Purcell AH (1995) X-disease. In: Compendium of stone fruit diseases. Eds Ogawa JM, Zehr EI, Bird GW, Ritchie DF, Uriu K, Uyemoto JK. APS, St. Paul, Minnesota, USA, 57–59 pp.

Kison H, Seemüller E (2001) Differences in strain virulence of European stone fruit yellows phytoplasma and susceptibility of stone fruit trees on various rootstocks to the pathogen. *Journal of Phytopathology* **149**, 533–541.

Krizanac I, Mikec I, Budinscak Z, Seruga Music M, Krajacic M, Skoric D (2008) Pomaceous fruit tre phytoplasmas and their potential vectors in Croatia. *Acta Horticulturae* **781**, 477–482.

Krizanac I, Mikec I, Budinscak Z, Seruga Music M, Skoric D (2010) Diversity of phytoplasmas infecting fruit trees and their vectors in Croatia. *Journal of Plant Disease and Protection* **117**, 206–213.

Kube M, Schneider B, Kuhl H, Dandekar T, Heitmann K, Migdoll AM, Reinhardt R, Seemüller E (2008) The linear chromosome of the plant-pathogenic mycoplasma '*Candidatus* Phytoplasma mali'. *BMC Genomics* **26**, 306.

Kunze L (1989) Apple proliferation. In: Virus and virus-like diseases of pome fruits and simulating noninfectious disorders. Ed Fridlund PR, Washington State University, Pullmann, Washington State, USA, 99–113 pp.

Kuzmanović S, Stavretovic M, Pavlovic S, Gavrilovic V, Aleksic G, Stojanovic S, Jošic D (2011) Detection of "stolbur" phytoplasma on blackberry – a new natural host in Serbia. *Genetika* **43**, 559–568.

La YJ, Chang MU (1979) Association of mycoplasma-like organisms with witches' broom disease of jujube. *R.O.C.-U.S. Seminar Mycoplasma Disease Plants* **1**, 21–24.

La Y, Woo K (1980) Transmission of jujube witches' broom mycoplasma by the leafhopper *Hishimonus sellatus* Uhler. *Journal of Korean Forestry Society* **48**, 29–39.

Laimer da Cámara Machado M, Heindrich M, Hanzer V, Arthofer W, Strommer S, Paltrinieri S, Martini M, Bertaccini A, Kummert J, Davies D (2001) Improved detection of viruses and phytoplasmas in fruit tree tissue culture. *Acta Horticulturae* **550**, 463–469.

Landi F, Prandini A, Paltrinieri S, Missere D, Bertaccini A (2010) Assessment of susceptibility to European stone fruit yellows phytoplasma of new plum variety and five rootstock/plum variety combinations. *Julius-Kühn-Archiv* **427**, 378–382.

Larsen KJ, Whalen ME (1988) Dispersal of *Paraphlepsius irroratus* (Say) (Homoptera: Cicadellidae) in peach and cherry orchards. *Environmental Entomology* **17**, 842–851.

Lederer W, Seemüller E (1992) Demonstration of mycoplasmas in *Prunus* species in Germany. *Journal of Phytopathology* **134**, 89–96.

Lee CY, Lin CP (2011) Multiplex detection and quantification of two pear decline phytoplasmas in Taiwan by real-time PCR. *Plant Pathology Bulletin* **20**, 11–20.

Lee I-M, Bertaccini A, Vibio M, Gundersen DE (1995a) Detection of multiple phytoplasmas in perennial fruit trees with decline symptoms in Italy. *Phytopathology* **85**, 728–735.

Lee I-M, Zhu S, Gundersen DE, Zhang C, Hadidi A (1995b) Detection and identification of a new phytoplasma associated with cherry lethal yellows in China. *Phytopathology* **85**, 1179.

Lee I-M, Gundersen-Rindal DE, Bertaccini A (1998a) Phytoplasma: Ecology and genomic diversity. *Phytopathology* **88**, 1359–1366.

Lee I-M, Gundersen-Rindal DE, Davis RE, Bartoszyk IM (1998b) Revised classification scheme of phytoplasmas based on RFLP analyses of 16S rRNA and ribosomal protein gene sequences. *International Journal of Systematic Bacteriology* **48**, 1153–1169.

Lee I-M, Polashock J, Bottner-Parker KD, Bagadia PG, Rodriguez-Saona C, Zhao Y, Davis RE (2014) New subgroup 16SrIII-Y phytoplasmas associated with false-blossom diseased cranberry (*Vaccinium macrocarpon*) plants and with known and potential insect vectors in New Jersey. *European Journal of Plant Pathology* **139**, 399–406.

Li Y, Piao C-G, Tian G-Z, Liu Z-X, Guo M-W, Lin C-L, Wang X-Z (2014a) Multilocus sequences confirm the close genetic relationship of four phytoplasmas of peanut witches' broom group 16SrII-A. *Journal of Basic Microbiology* **54**, 818–827.

Li Z, Song S, Zhang L, Gao L, Wu Y (2014b) Identification of the phytoplasma associated with peach yellows disease in northwest China. *Canadian Journal of Plant Pathology* **36**, 151–160.

Linck H, Krüger E, Reineke A (2017) A multiplex TaqMan qPCR assay for sensitive and rapid detection of phytoplasmas infecting *Rubus* species. *PLoS One* **12**, e0177808.

Lorenz KH, Dosba F, Poggi Pollini C, Llacer G, and Seemüller E (1994) Phytoplasma diseases of *Prunus* species in Europe are caused by genetically similar organisms. *Zeitschrift fur Pflanzenkrankheiten and Pflanzenschutz* **101**, 567–575.

Lorenz KH, Schneider B, Ahrens U, Seemüller E (1995). Detection of the apple proliferation and pear decline phytoplasmas by PCR amplification of ribosomal and nonribosomal DNA. *Phytopathology* **85**, 771–776.

Lou B, Bai X, Bai Y, Deng C, Roy Chowdhury M, Chen C, Song Y (2014) Detection and molecular characterization of a 16SrIIA phytoplasma in grapefruit (*Citrus paradisi*) with huanglongbing like symptoms in China. *Journal of Phytopathology* **162**, 387–395.

Ludvíková H, Fránová J, Suchá J (2011) Phytoplasmas in apricot, peach and sour cherry orchards in East Bohemia, Czech Republic. *Bulletin of Insectology* **64**(Supplement), S67–S68.

Malembic-Maher S, Salar P, Filippin L, Carle P, Angelini E, Foissac X (2011) Genetic diversity of European phytoplasmas of the 16SrV taxonomic group and proposal of 'Candidatus Phytoplasma rubi'. *International Journal of Systematic and Evolutionary Microbiology* **61**, 2129–2134.

Maramorosch K (1955) Transmission of blueberry stunt virus by *Scaphytopius magdalensis*. *Journal of Economic Entomology* **48**, 106.

Marani F, Pisi A, Bertaccini A (1977) The association of mycoplasma-like bodies with *Rubus* stunt disease. *Phytopathologia Mediterranea* **16**, 150–153.

Marcone C, Ragozzino A, Seemüller E (1996) European stone fruit yellows phytoplasma as the cause of peach vein enlargement and other yellows and decline diseases of stone fruits in southern Italy. *Phytopathology* **144**, 559–564.

Marcone C, Ragozzino A, Seemüller E (1997) Identification and characterization of the phytoplasma associated with elm yellows in southern Italy and its relatedness to other phytoplasmas of the elm yellows group. *European Journal of Forest Pathology* **27**, 45–54.

Martini M, Lee I-M, Bottner KD, Zhao Y, Botti S, Bertaccini A, Harrison N, Carraro L, Marcone C, Khan AJ, Osler R (2007a) Ribosomal protein gene-based phylogeny for finer differentiation and classification of phytoplasmas. *International Journal of Systematic and Evolutionary Microbiology* **57**, 2037–2051.

Martini M, Loi N, Ermacora P, Carraro L, Pastore M (2007b) A real-time PCR method for detection and quantification of '*Candidatus* Phytoplasma prunorum' in its natural hosts. *Bulletin of Insectology* **60**, 251–252.

Martini M, Ermacora P, Falginella L, Loi N, Carraro L (2008) Molecular differentiation of '*Candidatus* Phytoplasma mali' and its spreading in Friuli Venezia Giulia region (North-East Italy). *Acta Horticulturae* **781**, 395–402.

Mäurer R, Seemüller E (1994) Nature and genetic relatedness of mycoplasma-like organisms causing rubus stunt in Europe. *Plant Pathology* **44**, 244–249.

Mayer CJ, Jarausch B, Jarausch W, Jelkmann W, Vilcinskas A, Gross J (2009) *Cacopsylla melanoneura* has no relevance as vector of apple proliferation in Germany. *Phytopathology* **99**, 729–738.

McClure MS (1980) Spatial and seasonal distribution of leafhopper vectors of peach X-disease in Connecticut. *Environmental Entomology* **9**, 668– 672.
Mehle N, Brzin J, Boben J, Hren M, Frank J, Petrovič N, Gruden K, Dreo T, Žežlina I, Seljak G, Ravnikar M (2007) First report of '*Candidatus* Phytoplasma mali' in *Prunus avium, P. armeniaca and P. domestica. Plant Pathology* **56**, 721.
Mehle N, Ravnikar M, Seljak G, Knapiè V, Dermastia M (2011) The most widespread phytoplasmas, vectors and measures for disease control in Slovenia. *Phytopathogenic Mollicutes* **1**, 1–12.
Molino Lova M, Quaglino F, Abou-Jawdah Y, Choueiri E, Sobh H, Alma A, Tedeschi R, Casati P, Bianco PA (2011) Identification of new 16SrIX subgroups, -F and -G, among '*Candidatus* Phytoplasma phoenicium' strains infecting almond, peach and nectarine in Lebanon. *Phytopathologia Mediterranea* **50**, 273–282.
Morvan G (1977) Apricot chlorotic leaf roll. *EPPO Bulletin* **7**, 37–55.
Musetti R, Martini M, Ermacora P, Ferrini F, Loi N, Osler R (2008) '*Candidatus* Phytoplasma mali' causing apple proliferation. In: Characterization, diagnosis and management of phytoplasmas. Eds Harrison NA, Rao GP, Marcone C, Studium Press, Houston, Texas, USA, 35–58 pp.
Navrátil M, Válová P, Fialová R, Fránová J, Voráčková Z, Karešová R (1998) Occurrence of fruit tree phytoplasmas in the Czech Republic. *Acta Horticulturae* **472**, 649–653.
Navrátil M, Válová P, Fialová R, Petrova K, Franova J, Nebesarova J, Poncarova-Vorackova Z, Karesova R (2001a) Survey for stone fruit phytoplasmas in Czech Republic. *Acta Horticulturae* **550**, 377–382.
Navrátil P, Válová P, Fialová R, Pribylová J, Špak J,. Kubelková D. Karešová R (2001b) First attempt on identification of phytoplasma associated with full blossom of red and white currant in the Czech Republic. *Acta Horticuturae* **551**, 51–54.
Navrátil M, Válová P, Fialová R, Pribylová J, Špak J, Kubelková D, Karešová R (2004) Occurrence of phytoplasmas in red and white currant in the Czech Republic. *Acta Horticulturae* **656**, 119–124.
Navrátil M, Přibylová J, Válová P, Fialová R, Šafářová D, Špak J, Kubelková D, Petrzik K, Karešová R, Špaková V (2007) Detection and identification of phytoplasmas in *Ribes rubrum. Bulletin of Insectology* **60**, 123–124.
Nikolić P, Mehle N, Gruden K, Gruden K, Dermastia M (2010) A panel of real-time PCR assaya for specific detection of three phytoplasmas from the apple proliferation group. *Molecular and Cellular Probes* **24**, 303–309.
Nguyen TTD, Paltrinieri S, Mejia JF, Hoat TX, Bertaccini A (2012) Detection and identification of phytoplasmas associated with longan witches' broom in Vietnam. *Phytopathogenic Mollicutes* **2**, 23–27.
Novoa R (1982) Antecedentes sobre establecimiento y producción de frutos de la especie *Ugni molinae* Turcz. Tesis, Universidad Austral de Chile, Valdivia, Chile, 101 pp.
Okuda S, Doi Y, Yora K (1974) Mycoplasmalike bodies associated with chestnut yellows. *Annals Phytopathological Society Japan* **40**, 464–468.
Osler R, Loi N (1986) Apple proliferation disease. In: Fastidiosus Plant Prokaryotes: Cultivation, detection and associated economic problems. Eds Cappellini RA, Wells JM, Rutgers University, New Brunswick, New Jersey, USA, 59–73 pp.
Oviedo Delgado NJ, Ibeas Corcelles JI (2017) A new method for the identification of phytoplasmas in strawberries (*Fragaria fragaria*). *Biosaia* **6**.
Paltrinieri S, Caprara, Pondrelli M, Carraro L, Bertaccini A (1999) Differenziazione molecolare di fitoplasmi appartenenti al gruppo 16SrIII. *Petria* **10**, 103–104.
Paltrinieri S, Martini M, Stefani E, Pondrelli M, Fideghelli C, Bertaccini A (2001) Phytoplasma infection in peach and cherry in Italy. *Acta Horticulturae* **550**, 365–370.
Paltrinieri S, Davino S, Botti S, Davino M, Bertaccini A (2003) Identificazione molecolare di fitoplasmi associati ad un deperimento del pesco in Sicilia. Atti IV Convegno Nazionale sulla peschicoltura meridionale. Campobello di Licata, Agrigento, Italia, 293–296 pp.

Paltrinieri S, Bertaccini A, Lugli A, Monari W (2004) Three years of molecular monitoring of phytoplasma spreading in a plum growing area in Italy. *Acta Horticulturae* **657**, 501–506.
Paltrinieri S, Botti S, Bertaccini A, Dal Molin F, Mori N, Fiore N (2006) Are phytoplasmas involved in a severe peach decline? *Acta Horticulturae* **713**, 421–426.
Paltrinieri S, Bertaccini A, Lugaresi C (2008) Phytoplasmas in declining cherry plants. *Acta Horticulturae* **781**, 409–416.
Paltrinieri S, Duduk B, Dal Molin F, Mori N, Comerlati G, Bertaccini A (2010) Molecular characterization of '*Candidatus* Phytoplasma mali' strains in outbreaks of apple proliferation in north eastern Italy, Hungary, and Serbia. *Julius-Kühn-Archiv* **427**, 178–182.
Paltrinieri S, Troni K, Bortolotti PP, Nannini R, Bertaccini A (2017) Pear decline: the fate of phytoplasmas in an Italian orchard. 24th International Conference ICVF, Thessaloniki, Greece, 69.
Pasquini G, Bertaccini A, Bianco PA, Casati P, Costantini E, Ferretti L, Martini M, Marzachì C, Palmano S, Paltrinieri S (2013) Protocollo diagnostico per '*Candidatus* Phytoplasma prunorum' (ESFY). *Petria* **23**, 491–516.
Pastore M, Vibio M, Insero O, Danielli A, La Cara F, Bertaccini A (1996) Identification of a specific type of apple proliferation related phytoplasmas in pear in southern Italy. FEBS Meeting Barcellona, Spain, 198 n. 9.
Pastore M, Lee I-M, Vibio M, Santonastaso M, La Cara F, Bertaccini A (1998) Susceptibility to phytoplasma infection of three pear varieties grafted on different rootstocks. *Acta Horticulturae* **472**, 673–680.
Pastore M, Bertaccini A, Vibio M, Izzo P, Genovese MR, Santonastaso M, Murari E (1999) Identification by molecular techniques of phytoplasmas associated with apricot chlorotic leaf roll in Italy. *Acta Horticulturae* **488**, 779–782.
Pastore M, Bertaccini A, Graziano V, Cubicciotti V (2002) Phytoplasma infection in strawberry cultivation in Southern Italy. *Acta Horticulturae* **567**, 639–642.
Peccoud J, Labonne G, Sauvion N (2013) Molecular test to assign individuals within the *Cacopsylla pruni* complex. *PLOS ONE* **8**, e72454.
Pedrazzoli F, Filippi M, Deromedi M, Bragagna P, Battocletti I, Bianchedi PL, Ciccotti AM (2008) Apple proliferation transmission by grafting in different periods of the year. *Acta Horticulturae* **781**, 489–493.
Pérez Fuentealba S, Paltrinieri S, Guerrero J, Fiore N, Bertaccini A (2017) Detection and preliminary identification of phytoplasmas in hazelnut in Chile. 24th International Conference ICVF, Thessaloniki, Greece, 74.
Pérez-López E, Dumonceaux TJ (2016) Detection and identification of the heterogeneous novel subgroup 16SrXIII-(A/I)Iphytoplasma associated with strawberry green petal disease and Mexican periwinkle virescence. *International Journal of Systematic Evolutionary Microbiology* **66**, 4406–4415.
Pérez-López E, Rodríguez-Martínez D, Olivier CY, Luna-Rodríguez M, Dumonceaux TJ (2017) Molecular diagnostic assays based on cpn60 UT sequences reveal the geographic distribution of subgroup 16SrXIII-(A/I)I phytoplasma in Mexico. *Scientific Reports* **7**, 950.
Pignatta D, Poggi Pollini C, Giunchedi L, Ratti C, Reggiani N, Gobber M, Miorelli P, Forno F, Mattedi L, Ropelato E (2008) A real-time PCR assay for the detection of European Stone Fruit Yellows phytoplasma (ESFYP) in plant propagation material. *Acta Horticulturae* **781**, 499–503.
Poggi Pollini C, Bissani R, Giunchedi L, Vindimian E (1995) Occurrence of phytoplasma infection in European plums (*Prunus domestica*). *Journal of Phytopathology* **143**, 701–703.
Poggi Pollini C, Bissani R, Giunchedi L (2001) Occurrence of European stone fruit phytoplasma (ESFYP) infection in peach orchards in Northern-Central Italy. *Journal of Phytopathology* **149**, 725–730.
Posnette AF (1953) Green petal-a new virus disease of strawberries. *Plant Pathology* **2**, 17–18.
Přibylová J, Petrzik K, Špak J (2011) Association of aster yellows subgroup 16SrI-C phytoplasmas with a disease of *Ribes rubrum*. *Bulletin of Insectology* **64**(Supplement), S65–S66.
Quaglino F, Kube M, Jawhari M, Abou-Jawdah J, Siewert C, Choueiri E, Sobh H, Casati P, Tedeschi R, Molino Lova M, Alma A, Bianco PA (2015) '*Candidatus* Phytoplasma phoeni-

cium' associated with almond witches' broom disease: from draft genome to genetic diversity among strain populations. *BMC Microbiology* **15**, 148.

Rakús D (1978) The incidence of currant full blossom in the ČSSR. Seventh Czechoslovak Plant Protection Conference. Nitra, Slovakia, 132.

Rakús D, Maliarčiková V (1975) Harmfulness of the currant full blossom. *Sbor ÚVTI-Zahradnictví* **2**, 163–168

Rakús D, Králík O, Brčák J (1974) Mycoplazma v *Ribes houghtonianum* Jancz. Nakazené, plnokvetosti. *Sbor ÚVTI-Ochrana Rostlin* **10**, 307–309.

Ramkat R, Ruinelli M, Maghuly F, Schoder L, Laimer M (2014) Identification of phytoplasmas associated with *Rubus* spp. as prerequisite for their successful elimination. In: Phytoplasmas and phytoplasma disease management: how to reduce their economic impact. Ed Bertaccini A, COST Action FA0807, 159–166 pp.

Ramsdell DC, Stretch AW (1987) Blueberry stunt. In: Virus diseases of small fruits. Ed Converse HR. United States Department of Agriculture Handbook 631, 106–108 pp.

Rao GP, Chaturvedi Y, Priya M, Mall S (2011) Association of a 16SrII group phytoplasma with dieback disease of papaya in India. *Bulletin of Insectology* **64**(Supplement), S105–S106.

Reeder R, Kelly PL, Aroch, Z (2010) '*Candidatus* Phytoplasma asteris' identified in blackberry (*Rubus fruticosus*) in the United Kingdom. *Plant Pathology* **59**, 394.

Refatti E (1964) La moria del pero in Italia. *Notiziario delle Malattie delle Piante* **68**, 85–122.

Rice RE, Jones RA (1972) Leafhopper vectors of the western X-disease pathogen: collections in central California. *Environmental Entomology* **1**, 726–730.

Sabaté J, Laviña A, Battle A (2014). First report of '*Candidatus* Phytoplasma pyri' causing peach yellow leaf roll (PYLR) in Spain. *Plant Disease* **98**, 989.

Salehi M, Izadpanah K (1995) Almond brooming. *Iran Journal of Plant Pathology* **32**, 111–112.

Salehi M, Izadpanah K, Taghizadeh M (2000) Herbaceous host range of lime witches' broom phytoplasma in Iran. *Journal of Plant Pathology* **36**, 343–353.

Salehi M, Izadpanah K, Heydarnejad J (2006) Characterization of a new almond witches' broom phytoplasma in Iran. *Journal of Phytopathology* **154**, 386–391.

Salehi M, Izadpanah K, Taghavi SM, Rahimian H (2008) Characterization of a phytoplasma associated with pear decline in Iran. *Journal of Phytopathology* **156**, 493–495.

Salehi M, Salehi E, Abbasian M, Izadpanah K (2015) Wild almond (*Prunus scoparia*), a potential source of almond witches' broom phytoplasma in Iran. *Journal of Plant Pathology* **97**, 377–381.

Salehi M, Esmailzadeh Hosseini SA, Rasoolpour R, Salehi E, Bertaccini A (2016) Identification of a phytoplasma associated with pomegranate little leaf disease in Iran. *Crop Protection* **87**, 50–54.

Sanchez JA, Ortín-Angulo MC (2011) Sampling of *Cacopsylla pyri* (Hemiptera: Psyllidae) and *Pilophorus gallicus* (Hemiptera: Miridae) in pear orchards. *Journal of Economic Entomology* **104**, 1742–1751.

Sauvion N, Lachenaud O, Genson G, Rasplus J-Y, Labonne G (2007) Are there several biotypes of *Cacopsylla pruni*? *Bulletin of Insectology* **60**, 185–186.

Schneider B, Seemüller E (2009) Strain differentiation of '*Candidatus* phytoplasma mali' by SSCP and sequence analyses of the *hflb* gene. *Journal of Plant Pathology* **91**, 103–112.

Seemüller E, Schneider B (2004) '*Candidatus* Phytoplasma mali', '*Candidatus* Phytoplasma pyri' and '*Candidatus* Phytoplasma prunorum', the causal agents of apple proliferation, pear decline and European stone fruit yellows, respectively. *International Journal of Systematic and Evolutionary Microbiology* **54**, 1217–1226.

Seemüller E, Schneider B (2007) Differences in virulence and genomic features of strains of '*Candidatus* Phytoplasma mali', the apple proliferation agent. *Phytopathology* **97**, 964–970.

Seemüller E, Stolz H, Kison H (1998) Persistence of the European stone fruit yellows phytoplasma in aerial parts of *Prunus* taxa during the dormant season. *Journal of Phytopathology* **146**, 407–410.

Seemüller E, Sule S, Kube M, Jelkmann W, Schneider B (2013) The AAA plus ATPases and HflB/FtsH proteases of '*Candidatus* Phytoplasma mali': phylogenetic diversity, membrane topology, and relationship to strain virulence. *Molecular Plant-Microbe Interactions* **26**, 367–376.

Seliskar CE, Glenn E, Knight K, Bourne CE (1974) Mycoplasmalike organism associated with pecan bunch disease. *Phytopathology* **64**, 1269–1272.

Sertkaya G, Osler R, Musetti R Ermacora P, Martini M (2004) Detection of phytoplasmas in *Rubus* spp. by microscopy and molecular techniques in Turkey. *Acta Horticulturae* **656**, 181–186.

Sertkaya G, Martini M, Ermacora P, Musetti R, Osler R (2005) Detection and characterization of phytoplasmas in diseased stone fruits and pear by PCR-RFLP analysis in Turkey. *Phytoparasitica* **33**, 380–390.

Sertkaya G, Sertkaya E, Kaya K, Çağlayan K, ErtunÇ F (2008) Detection of pear decline disease in pear and quince in the eastern Mediterranean region of Turkey. *Acta Horticulturae* **781**, 511–515.

Shimada C, Kouda S (1954) Yellowing disease of chestnut. *Annals Phytopathological Society Japan* **19**, 95.

Singh J, Rani A, Kumar P, Baranwal VK, Saroj PL, Sirohi A, Pandey AN, Schenk PM (2014) New host record of a '*Candidatus* Phytoplasma asteris'-related strain infecting peach in India. *Australasian Plant Disease Notes* **9**, 125.

Špak J, Pribylová J, Kubelková D (2001) Outbreak of full blossom of red currant in the Czech Republic. *Acta Horticulturae* **551**, 55–58.

Špak J, Pribylová J, Kubelková D, Špaková V (2004) The presence of phytoplasma in black currant infected with the blackcurrant reversion disease. *Journal of Phytopathology* **152**, 600–605.

Špak J, Navrátil M, Karešová R, Pribylova J, Valova P. Kucerova J, Kubelkova D, Fialova R, Spakova V (2006) Occurrence, symptom variation and yield loss caused by full blossom disease in red and white currants in the Czech Republic. *Crop Protection* **25**, 446–453.

Špak J, Přibylová J, Kubelková D, Špaková V (2008) Transmission of phytoplasmas associated with full blossom disease in *Ribes rubrum*. *Acta Horticulturae* **780**, 67–70.

Špak J, Kubelková D, Přibylová J, Špaková V (2009) Elucidation of the roles of blackcurrant reversion virus and phytoplasma in the etiology of full blossom disease in currants. *Plant Disease* **93**, 832–838.

Starović M, Kojic S, Kuzmanovic ST, Stojanovic SD, Pavlovic S, Josic D (2013) First report of blueberry reddening disease in Serbia associated with 16SrXII-A ("stolbur") phytoplasma. *Plant Disease* **97**, 1653.

Straten C, Waited GK, Herrington ME, Hutton DG, Persley DM, Gibb KS (2005) Rickettsia-like-organisms and phytoplasmas associated with diseases in Australian strawberries. *Australasian Plant Pathology* **34**, 157–164.

Stretch AW (1987) Cranberry false blossom. In: Virus diseases of small fruits. Ed Converse HR. United States Department of Agriculture Handbook **631**, 110–111.

Stubbs LL (1968) Lethal yellows disease of strawberry. *Australian Plant Disease Records* **20**, 33–34.

Süle S, Jenser G, Szita È (2007) Management of pear decline caused by '*Candidatus* Phytoplasma pyri' in Hungary. *Bulletin of Insectology* **60**, 319–320.

Tarcali G (2013) Occurence of '*Ca.* Phytoplasma prunorum' in the Carpathian-basin. *Analele Universităţii din Oradea, Fascicula Protecţia Mediului* **XXI**, 281–289.

Tedeschi R, Alma A (2006) *Fieberiella florii* (Stäl) (*Homoptera*: *Auchenorrhyncha*) as a vector of '*Candidatus* Phytoplasma mali'. *Plant Disease* **90**, 284–290.

Tedeschi R, Alma A (2007) '*Candidatus* Phytoplasma mali': the current situation of insect vectors in northwestern Italy. *Bulletin of Insectology* **60**, 187–188.

Tedeschi R, Lauterer P, Brusetti L, Tota F, Alma A (2009) Composition, abundance and phytoplasma infection in the hawthorn psyllid fauna of northwestern Italy. *European Journal of Plant Pathology* **123**, 301–310.

Tedeschi R, Picciau L, Quaglino F, Abou-Jawdah Y, Molino Lova M, Jawhari M, Casati P, Cominetti A, Choueiri E, Abdul-Nour H, Bianco PA, Alma A (2015) A cixiid survey for natural potential vectors of '*Candidatus* Phytoplasma phoenicium' in Lebanon and preliminary transmission trials. *Annals of Applied Biology* **166**, 372–388.

Teixeira DC, Wulff NA, Martins EC, Kitajima EW, Bassanezi R, Ayres AJ, Eveillard S, Saillard C, Bové JM. (2008) A phytoplasma closely related to the pigeon pea witches' broom phytoplasma (16SrIX) is associated with citrus huanglongbing symptoms in the state of Sao Paulo, Brazil. *Phytopathology* **98**, 977–984.

Thakur PD, Handa A, Chowfla SC, Krczal G (1998) Outbreak of a phytoplasma disease of peach in the Northwestern Himalayas of India. *Acta Horticulturae* **472**, 737–742.

The EUPHRESCO FruitPhytoInterlab group (2011) European interlaboratory comparison and validation of detection methods for '*Candidatus* Phytoplasma mali', '*Candidatus* Phytoplasma prunorum' and '*Candidatus* Phytoplasma pyri': preliminary results. *Bulletin of Insectology* **64**(Supplement), S281–S284.

Thébaud G, Nicolas S, Chadoeuf J, Dufils A, Labonne G (2006) Identifying risk factors for European stone fruit yellows from a survey. *Phytopathology* **96**, 890–899.

Thébaud G, Yvon M, Alary R, Sauvion N, Labonne G (2009) Efficient transmission of '*Candidatus* Phytoplasma prunorum' is delayed by eight months due to a long latency in its host-alternating vector. *Phytopathology* **99**, 265–273.

Tian JB, Bertaccini A, Martini M, Paltrinieri S, Guo HP, Pastore M (2000) Molecular detection of jujube witches' broom phytoplasmas in micropropagated jujube shoots. *HortScience* **35**, 1274–1275.

Tomenius K, Ahman G (1983) Mycoplasmalike organisms in the phloem of little leaf diseased plants of *Vaccinium vitusidaea* and *V. myrtillus* in Sweden. *Sweden Journal of Agricultural Research* **13**, 205–209.

Tomlinson WE Jr, Maruci PE, Doehlert CA (1950) Leafhopper transmission of blueberry stunt disease. *Journal of Economic Entomology* **43**, 658–662.

Torres E, Martin MP, Paltrinieri S, Vila A, Masalles R, Bertaccini A (2004) Spreading of EFSY phytoplasmas in stone fruit in Catalonia (Spain). *Journal of Phytopathology* **152**, 432–437.

Torres E, Bertolini E, Cambra M, Montón C, Martín MP (2005). Real-time PCR for simultaneous and quantitative detection of quarantine phytoplasmas from apple proliferation (16SrX) group. *Molecular and Cellular Probes* **19**, 334–340.

Tozzi DCM, Ramsdell DC, Taboada O, Lee I-M, Davis RE (1993) Epidemiological studies on the stunt disease of highbush blueberry. *Annals of Applied Biology* **123**, 579–599.

Tsai JH, Chen XY, Shen CY, Jin KX (1988) Mycoplasmas and fastidious vascular prokaryotes associated with tree disease in China. In: Tree Mycoplasmas and Mycoplasma Diseases. Ed Hiruki CJ, Alberta Press, Canada, 69–96 pp.

Uyemoto JK, Kirkpatrick BC (2011) X-disease phytoplasma In: Virus and virus like diseases of pome and stone fruits. Eds Hadidi A, Barba M, Candresse T, Jelkmann W, APS, St. Paul, Minnesota, USA, 243–245 pp.

Valiūnas D, Alminaite A, Jomantiene R, Davis RE, Maas JL (2004) Possible cause of European blueberry disease is related to North American milkweed yellows phytoplasma. *Journal of Plant Pathology* **86**, 135–140.

Valiūnas D, Staniulis J, Davis RE (2006) '*Candidatus* Phytoplasma fragariae', a novel phytoplasma taxon discovered in yellows diseased strawberry, *Fragaria* × *ananassa*. *International Journal of Systematic and Evolutionary Microbiology* **56**, 277–281.

Valiūnas D, Jomantiene R, Davis RE (2007) Phytoplasmas detected in cultivated fruit plants in Lithuania. *Bulletin of Insectology* **60**, 139–140.

Valiūnas D, Jomantiene R, Davis RE (2009a) Establishment of a new phytoplasma subgroup, 16SrI-Q, to accommodate a previously undescribed phytoplasma found in diseased cherry in Lithuania. *Journal of Plant Pathology* **91**, 71–75.

Valiūnas D, Jomantiene R, Ivanauskas A, Abraitis R, Staniene G, Zhao Y, Davis RE (2009b) First report of a new phytoplasma subgroup, 16SrIII-T, associated with decline disease affecting sweet and sour cherry trees in Lithuania. *Plant Disease* **93**, 550.

van der Meer FA (1987) Full blossom of red currant. United States Department of Agriculture Handbook 631, 155–157 pp.

Verdin E, Salar P, Danet J, Choueiri E, Jreijiri F, El Zammar S, Gélie B, Bové JM, Garnier M (2003) '*Candidatus* phytoplasma phoenicium' sp. nov., a novel phytoplasma associated with an emerging lethal disease of almond trees in Lebanon and Iran. *International Journal of Systematic and Evolutionary Microbiology* **53**, 833–838.

Verma R, Mungekar D, Gaikwad P, Tomer SPS, Datar VV (2012) First report of a phytoplasma associated with an axillary shoot proliferation disease in papaya in India. *New Disease Reports* **25**, 18.

Vindimian ME, Grassi A, Ciccotti A, Poggi Pollini C, Terlizzi F (2004) Epidemiological studies on rubus stunt (RS) in blackberry orchards located near Trento (Italy). *Acta Horticulturae* **656**, 177–180.

Westwood MN, Cameron HR (1978) Environment-induced remission of pear decline symptoms. *Plant Disease Reporter* **62**, 176–179.

White DT, Blackall LL, Scott PT, Walsh KB (1998) Phylogenetic positions of phytoplasmas associated with dieback, yellow crinkle and mosaic diseases of papaya, and their proposed inclusion in '*Candidatus* Phytoplasma australiense' and a new taxon, '*Candidatus* Phytoplasma australasia'. *International Journal of Systematic and Evolutionary Microbiology* **48**, 941–951.

Wilcox RB (1942) Blueberry stunt, a virus disease. *Plant Disease Reporter* **26**, 211–213.

Xu BY, Chen TA (1996) Cranberry false blossom disease in New Jersey. *Phytopathology* **86**, S125.

Yang I-L (1975) Studies on the peanut witches' broom disease in Taiwan. *T'ai-wan-Nung-Yeh-Chi-k'an* **11**, 93–108.

Yi CK, La YJ (1973) Mycoplasma-like bodies found in the phloem elements of jujube infected with witches' broom disease. *Journal of Korean Forerestry Society* **20**, 111–114.

Yvon M, Thebaud G, Alary R, Labonne G (2009) Specific detection and quantification of the phytopathogenic agent of '*Candidatus* Phytoplasma prunorum'. *Molecular and Cellular Probes* **23**, 227–234.

Zamharir MG (2014). Molecular study of a new disease of peach in Iran associated with a phytoplasma. *Advances in Microbiology* **4**, 20–24.

Zawadzka BJ (1976) Reaction of apple cultivars infection by apple proliferation disease. *Acta Horticulturae* **67**, 113–120.

Žežlina I, Rot M, Kač M, Trdan S (2016) Causal agents of stone fruit diseases in Slovenia and the potential for diminishing their economic impact: a review. *Plant Protection Science* **52**, 149–157.

Zhang L, Li ZN, Zhang HW, Tao Y, Wu YF (2013) Detection and identification of aster yellows group phytoplasma (16SrI-C) associated with peach red leaf disease. *Journal of Phytopathology* **161**, 359–362.

Zhu SF, Hadidi A, Lee I-M, Gundersen DE, Zhang CL (1997) Characterization of the phytoplasmas associated with cherry lethal yellows and jujube witches' broom disease in China. *Acta Horticulturae* **472**, 701–707.

Zirak L, Bahar M, Ahoonmanesh A (2009a) Molecular characterization of phytoplasmas related to peanut witches' broom and "stolbur" groups infecting plum in Iran. *Journal of Plant Pathology* **91**, 713–716.

Zirak L, Bahar M, Ahoonmanesh A (2009b) Characterization of phytoplasmas associated with almond diseases in Iran. *Journal of Phytopathology* **157**, 736–741.

Zreik Laj;aj, Carle P, Bové JM, Garnier M. (1995) Characterizationof the mycoplasmalike organism associated with witches' broom disease of lime and proposition of a '*Candidatus*' taxon for the organism, '*Candidatus* Phytoplasma aurantifolia'. *International Journal of Systematic Bacteriology* **45**, 449–453.

Zunnoon-Khan S, Michelutti R, Arocha-Rosete Y, Scott J, Crosby W, Bertaccini A (2010) First report of '*Candidatus* Phytoplasma asteris'- related strain associated with peach rosette in Canada. *Plant Disease* **94**, 916.

Chapter 7
Phytoplasma Diseases in Ornamental Crops

Maria Grazia Bellardi, Assunta Bertaccini, Madhupriya, and Govind Pratap Rao

Abstract An extensive and updated review of the literature reporting the phytoplasma associated diseases in a number of ornamental plants and their classification is presented with major emphasis to reports in the main floricultural areas. Symptomatology of phytoplasma diseases is described in the most relevant traditional species as well as in emerging species used in floriculture and gardening worldwide.

Keywords Ornamental species · Phytoplasma · Symptomatology · PCR · RFLP

7.1 Introduction

The phytoplasmas are an important group of pathogens which drastically damage growth and marketing parameters of ornamental plants and affect their commercial value (Chaturvedi et al. 2010a). Many ornamental plants are affected by phytoplasmas that are very often associated with significant economic impacts. The number of phytoplasmas identified in ornamental species has greatly increased over the last decades as a consequence of increased production and worldwide commercial distribution of plant material (consisting of cut flowers, foliage or flowering potted plants, shoots, seeds, bulbs, rhizomes, etc.). In addition, some species and new hybrids are becoming more economically important all over the world, but no information is available regarding susceptibility and/or tolerance to these pathogens. As a consequence, the incidence varies from overall infection due to phytoplasmas frequently found, to those observed only occasionally. The severity of symptoms differs considerably among ornamental species, hybrids, and varieties, ranging from

---

M. G. Bellardi (✉) · A. Bertaccini
Department of Agricultural and Food Sciences, *Alma Mater Studiorum* – University of Bologna, Bologna, Italy
e-mail: mariagrazia.bellardi@unibo.it

Madhupriya · G. P. Rao
Division of Plant Pathology, Advance Center for Plant Virology, Indian Agricultural Research Institute, New Delhi 110012, India

G. P. Rao et al. (eds.), *Phytoplasmas: Plant Pathogenic Bacteria - I*,
https://doi.org/10.1007/978-981-13-0119-3_7

malformations and yellowing causing little or no appreciable damage to severe virescence, phyllody, and growth reduction (Lee et al. 1998). Phytoplasma diseases of ornamentals have been described worldwide in a wide range of plant genera, and the associated phytoplasmas belong to 14 different 16S ribosomal groups and to about 30 ribosomal subgroups (Table 7.1).

Table 7.1 Symptomatology, identification, and distribution of phytoplasma diseases occurring in ornamental plant species worldwide

Host species/trivial name (family)	Diseases	Country	Phytoplasma group/subgroup	References
Adenium obesum (Apocynaceae)	Witches' brooms	Myanmar	16SrII	Win et al. (2012)
	Little leaf	India	16SrI	Raj et al. (2007a)
Allamanda cathartica (Apocynaceae)	Leaf yellowing	India	16SrVI-D	Khasa et al. (2016)
Alstroemeria (Alstroemeriaceae)	Virescence	Italy	16SrI-B	Bertaccini et al. (1996a)
	Deformation and dieback	Mexico		Cervantes-Diaz et al. (2004)
	Little leaf	India	16SrI	Singh et al. (2011)
Anemone coronaria (Ranunculaceae)	Virescence	Italy	16SrI-C	Vibio et al. (1995)
Aquilegia columbine (Ranunculaceae)	Yellows, stunting, virescence, and phyllody	Lithuania	16SrI	Samuitiene et al. (2004)
Aralia cordata/ Japanese spikenard (Araliaceae)	Yellow dwarf	South Korea	16SrIII	Lee et al. (2004)
Argyranthemum frutescens (Asteraceae)	Yellowing	Italy	16SrIX-C	Ferretti et al. (2015)
Asclepias physocarpa (Asclepiadaceae)	Yellowing and stunting	Italy	16SrXII-A	Bertaccini et al. (2006)
Begonia spp. (Begoniaceae)	Stunting and little leaf	Brazil	16SrIII	Ribeiro et al. (2006)
Brachyscome (Asteraceae)	Yellows and witches' broom	India	16SrI	Madhupriya et al. (2013a)
Brugmansia candida (Solanaceae)	Little leaf	Australia	–	Hiruki (1986)
Calendula officinalis (Asteraceae)	Phyllody, virescence, proliferation, and witches' broom	Iran	16SrII-D	Esmailzadeh Hosseini et al. (2016)
		Italy	16SrI	Marcone et al. (1997b)
		Serbia	16SrXII-A	Pavlovic et al. (2014)
Calendula arvensis (Asteraceae)	Stunting	Italy	16SrII-E	Tolu et al. (2006)

(continued)

Table 7.1 (continued)

Host species/trivial name (family)	Diseases	Country	Phytoplasma group/subgroup	References
Callistephus chinensis/ (China aster) (Asteraceae)	Virescence	Korea	16SrII-A	Win et al. (2011)
	Yellowing and stunting	Iran	16SrI-B	Sichani et al. (2014)
Carpobrotus edulis (Aizoaceae)	Leaf yellowing, little leaf, and reduced size of flowers	India	16SrIX	Shukla et al. (2014)
Cassia italica (Caesalpiniaceae)	Stunting	Oman	16SrXXIX-A	Al-Saady et al. (2008)
Cassia surattensis (Caesalpiniaceae)	Witches' broom	China	16SrV	Hong et al. (2005)
Catharanthus roseus (Apocynaceae)	Little leaf	USA	16SrI-A	Davis et al. (1990)
		Italy	16SrI-B	Parrella et al. (2014)
	Virescence	Argentina	16SrI	Torres et al. (2004)
		Egypt	16SrI	Omar et al. (2008)
		Mexico	16SrXIII-A	Gundersen et al. (1994)
	Witches' broom and virescence	Mexico	16SrIII-J; 16SrI	Poghosyan et al. (2015)
	Virescence and phyllody	Malaysia	16SrXXXII	Nejat et al. (2013)
	Phyllody	India	16SrI	Khurana et al. (1981)
	Yellowing and little leaf		16SrI-C	Chaturvedi et al. (2009b)
Celosia argentea (Amaranthaceae)	Phyllody	Lithuania	16SrI-L; 16SrI-M	Samuitiene and Navalinskiene (2006)
		Iran	16SrI-M	Aldaghi and Bertaccini (2015)
	Leaf malformation	Brazil	16SrIII-J	Eckstein et al. (2012)
	–	Iran	16SrVI	Babaie et al. (2007)
Celosia plumosa & *Celosia cristata* (Amaranthaceae)	Yellows	Israel	16SrI; 16SrIII	Tanne et al. (2000)
Chionanthus retusus/ snow flower fringe (Oleaceae)	Dwarfing	Korea	16SrI	Lee (2004)
Chrysanthemum morifolium (Asteraceae)	Yellowing, stunting, and dwarfing	South Korea	16SrXII; 16SrI	Chung and Kim (2005), Bongnam et al. (2007), Chung and Hun (2008)
	Flattened stem	China	16SrI	Min et al. (2009)
	Little leaf	India	16SrI	Raj et al. (2007b)

(continued)

Table 7.1 (continued)

Host species/trivial name (family)	Diseases	Country	Phytoplasma group/subgroup	References
Chrysanthemum coronarium (Asteraceae)	Witches' broom	Japan	16SrI-B	Okuda (1997)
		China		Zhong and Shen (2004)
Chrysanthemum frutescens (Asteraceae)	Yellows	Japan	16SrI-B	Okuda (1997)
		Italy	16SrI-B	Bertaccini et al. (1990a)
Chrysanthemum indicum hybridum (Asteraceae)	Virescence and abnormal growth	Serbia	16SrXII-A	Duduk et al. (2006)
Codiaeum variegatum (Euphorbiaceae)	Leaf yellows and witches' broom	India	16SrI	Tiwari et al. (2014)
		Colombia		Perilla-Henao et al. (2012)
Cosmos bipinnatus (Asteraceae)	Phyllody	Mexico	16SrI-B	Rojas-Martínez et al. (2003a)
Cycas revoluta (Cycadaceae)	Yellowing	India	16SrII	Kumar et al. (2012a)
Cyclamen persicum (Primulaceae)	Virescence	Italy	16SrI-B; 16SrI-C	Alma et al. (2000), Satta et al. (2013)
Cytisus scoparius (Fabaceae)	Witches' broom	Germany	16SrV-C; 16SrI-B	Contaldo et al. (2015b)
Dahlia cultorum (Asteraceae)	Shoot proliferation, narrow leaf, and flower bud deficiency	Poland	16SrI-B; 16SrX-A	Kaminska and Śliwa (2008b)
		Italy	16SrX-A	Marzachì et al. (1999)
Delphinium sp. (Ranunculaceae)	Yellows, stunting, phyllody, and virescence	United Kingdom	16SrIII	Harju et al. (2008)
Daucus carota/Queen Anne's lace (Plumbaginaceae)	Yellows	Canada	16SrI	Chang et al. (2004)
Dendranthema grandiflora (Asteraceae)	Yellow dwarf and witches' broom	South Korea	16SrI; 16SrXII	Chung and Kim (2005)
	Phyllody	India	.	Rani et al. (2014)
Dicentra spectabilis/bleeding heart (Fumariaceae)	Shoot proliferation	Poland	16SrI-B; 16SrI-A	Kaminska et al. (2004)
Dictamnus albus (Rutaceae)	Stunting	USA	16SrIII	Valiunas et al. (2007)
Dimorphotheca sinuata (Asteraceae)	Stunting, virescence, and phyllody	Italy	16SrIX	Marcone et al. (2001)
Duranta repens (Verbenaceae)	Virescence and stunting	India	16SrI	Singh et al. (2011)

(continued)

Table 7.1 (continued)

Host species/trivial name (family)	Diseases	Country	Phytoplasma group/subgroup	References
Echinacea purpurea/purple coneflower (Euphorbiaceae)	Phyllody	USA	16SrI-B	Lee et al. (2008)
	Yellowing, virescence, and phyllody	Slovenia	16SrI	Radisek et al. (2008)
	Virescence, phyllody, malformations, and stunting	Italy	16SrI-B; 16SrIX-C	Bertaccini et al. (2009)
	Flower deformation and pale coloration	Czech Republic	16SrI-C; 16SrIII-B	Franova et al. (2009, 2013)
Erysimum linifolium (Brassicaceae)	Stunting and malformation	Italy	16SrI-B	Paltrinieri et al. (2015)
Euphorbia pulcherrima (Euphorbiaceae)	Branch inducing and stem flat	Italy, USA	16SrIII-H	Bertaccini et al. (1996b), Lee et al. (1997)
		South Korea		Chung and Choi (2010)
Gaillardia sp. (Asteraceae)	–	Iran	16SrI-B	Sichani et al. (2014)
Gentiana scabra/ Japanese gentian (Gentianaceae)	Witches' broom	South Korea	Unclassified	Lee (2004)
Gerbera jamesonii (Asteraceae)	Phyllody	Australia	16SrII-D	Siddique (2006)
	Virescence and phyllody	Italy	16SrI	Spanò et al. (2011)
Gypsophila paniculata (Caryophyllaceae)	Stunting and witches' broom	Israel	16SrII	Gera et al. (2006)
Gladiolus sp. (Iridaceae)	Yellowing and malformation of flower and stunting	Italy, Portugal	16SrI-B	Bertaccini et al. (1990b, 1992b), Louro et al. (1996)
		India	16SrI	Raj et al. (2009)
Helianthus annuus (Asteraceae)	Virescence and phyllody	Argentina	16SrIII-J	Guzman et al. (2014)
	Virescence, phyllody, and	Iran	16SrII-D	Salehi et al. (2015)
	Witches' broom	Iran	16SrII; 16SrVI	Tazehkand et al. (2010)
Helianthus debilis (Asteraceae)	Little leaf	USA	16SrI	Harrison and Helmick (2008)
Hibiscus rosa-sinensis (Malvaceae)	Witches' broom	Brazil	16SrXV-A; 16SrXII-A	Montano et al. (2001, 2011)
	Yellow leaf	India	16SrI	Chaturvedi et al. (2010b)
	Leaf yellowing and phyllody	India	16SrVI-D	Khasa et al. (2016)

(continued)

Table 7.1 (continued)

Host species/trivial name (family)	Diseases	Country	Phytoplasma group/subgroup	References
Hydrangea macrophylla (Saxifragaceae)	Phyllody virescence	Italy Canada	16SrI-B	Bertaccini et al. (1992a), Duduk et al. (2013)
	Phyllody	Japan	16SrXII-C	Sawayanagi et al. (1999)
			16SrI	Takinami et al. (2013)
	Stunting, virescence, necrosis, and redness of the leaf edge	Italy Bulgaria	16SrI-B; 16SrXII-A	Bertaccini et al. (1995, 2015)
Iberis sempervirens (Brassicaceae)	Stunting	Italy	16SrX; 16SrI-B; 16SrI-A	Contaldo et al. (2015a)
Impatiens balsamina (Balsaminaceae)	Virescence, wrinkled leaves, and stunted internods	China	16SrV-B	Li et al. (2014)
Jasminum sambac/Arabian jasmine (Oleaceae)	Witches' broom	Oman	16SrII	Al-Zadjali et al. (2007)
		India	16SrXI	Madhupriya et al. (2015)
Koelreuteria paniculata/goldenrain tree (Koelreuteriae)	Stunting	Korea	16SrI	Kamala-Kannan et al. (2010)
Lachenalia aloides (Asparagaceae)	Flower malformation	Italy	16SrI-B	Bellardi et al. (2017)
Lilium longiflorum (Liliaceae)	Virescence	Czech Republic	16SrI	Bertaccini et al. (2005)
	Flattened stem	Korea	16SrXII	Chung and Jeong (2003)
	Zigzag line pattern in leaves	Mexico	16SrI	Cortes-Martinez et al. (2007)
	Flower dropping	Poland	16SrI-B; 16SrXII-A	Bertaccini et al. (2002)
	Leaf scorch	Poland	16SrX-A	Kaminska and Śliwa (2008a)
Ligustrum lucidum/Chinese privet (Oleaceae)	Witches' broom	South Korea	16SrV-B	Lee (2004)
Limonium sinuatum (Plumbaginaceae)	Witches' broom	South Korea	16SrI	Chung and Kim (2005)

(continued)

Table 7.1 (continued)

Host species/trivial name (family)	Diseases	Country	Phytoplasma group/subgroup	References
Lupinus polyphyllus	Malformations	Italy	16SrXII-A	Bellardi et al. (2013)
Magnolia grandiflora (Magnoliaceae)	Stunting, leaf necrosis, and witches' broom	Poland	16SrI-B; 16SrX-A	Kaminska et al. (2001b)
Matthiola incana (Brassicaceae)	Stunting, malformation, and virescence	Italy	16SrII	Davino et al. (2007)
Mirabilis jalapa (Nyctaginaceae)	Yellowing and general stuntings	Israel	16SrII	Sobolev et al. (2007)
	Little leaf	India	16SrII	Kumar et al. (2012b)
Opuntia ficus-indica (Cactaceae)	Proliferation of cladodes and lack of flowers and fruits	Italy	16SrII-C	Granata et al. (2006), Tessitori et al. (2006)
	Proliferation and stunting of cladodes	USA	16SrI-B; 16SrV-A	Bertaccini et al. (2007)
	Buds proliferation, thickening, and heart-shaping of cladodes	Mexico	16SrI	Zak et al. (2011)
Opuntia cylindrica (Cactaceae)	–	Egypt	16SrII	Omar et al. (2014)
Opuntia monacantha	Shoot proliferation	Lebanon	16SrII	Choueiri et al. (2005)
Pachysandra terminalis/Japanese spurge (Buxaceae)	Yellows	South Korea	16SrI-B	Back et al. (2010)
Petunia hybrida (Solanaceae)	Flat stem	South Korea	16SrI	Chung and Hun (2008)
	Yellows	India	16SrI	Singh et al. (2011)
		Iran	16SrII	Faghihi et al. (2014)
Phlox sp. (Polemoniaceae)	Malformation	Lithuania	16SrI-M	Navalinskiene and Samuitiene (2004)
Portulaca grandiflora (Portulacaceae)	Little leaf	India	16SrVI	Ajaykumar et al. (2007), Samad et al. (2008)
Psylliostachys suworowii (Plumbaginaceae)	Yellows	Canada	16SrI	Chang et al. (2004)

(continued)

Table 7.1 (continued)

Host species/trivial name (family)	Diseases	Country	Phytoplasma group/subgroup	References
Ranunculus sp. (Ranunculaceae)	Phyllody and virescence	Italy	16SrI-B; 16SrI-C	Bertaccini et al. (1988, 1990b), Parrella et al. (2008)
Rhododendron sp./azalea (Ericaceae)	Witches' broom	Czech Republic	16SrXII-A	Mertelik et al. (2004)
Rosa alba (Rosaceae)	Witches' broom and bud proliferation	Poland	16SrI-B; 16SrX-A	Kaminska and Sliwa (2004)
	Little leaf	India	16SrI	Chaturvedi et al. (2009a)
Saponaria officinalis (Caryophyllaceae)	Witches' broom	India	16SrVI-D	Khasa et al. (2016)
Silene nicaeensis (Caryophyllaceae)	Flower malformation	Italy	16SrI-B	Cozza et al. (2008)
Spartium junceum (Fabaceae)	Witches' broom	Italy, Spain	16SrX-D; 16SrV-C; 16SrI	Marcone et al. (1996), Torres et al. (2002), Contaldo et al. (2015a)
Spiraea salicifolia/ spiraea (Rosaceae)	Stunting	USA	16SrIII-E	Griffiths et al. (1994)
		China	16SrI-B; 16SrV-B	Gao et al. (2007), Li et al. (2010)
Spiraea bumalda (Rosaceae)	Witches' broom	USA	16SrIII	Lockhart et al. (2012)
Solidago virgaurea/ goldenrod (Asteraceae)	Witches' broom	Korea	16SrIII	Lee et al. (1996)
Streblus asper/singhore (Moraceae)	Chlorosis and yellows	India	16SrI	Maurya et al. (2014)
Syringa vulgaris/lilac (Oleaceae)	Witches' broom	USA	16SrVII-A	Griffiths et al. (1999)
		South Korea	16SrV-B	Lee et al. (1996)
Tabebuia pentaphylla (Bignoniaceae)	Witches' broom	Brazil	16SrII	Mafia et al. (2008)
Tagetes erecta (Asteraceae)	Phyllody	Mexico	16SrI-B	Rojas-Martínez et al. (2003b)
	Witches' broom	India	16SrI	Raj et al. (2011)
Trillium spp. (Trilliaceae)	Virescence	Canada	16SrIII-F	Arocha-Rosete et al. (2016)
Veronica scutellata (Scrophulariaceae)	Yellowing and little leaf	United Kingdom	16SrI-A	Jones and Arocha (2006)
Viola tricolor (Violaceae)	Leaf yellows and little leaf	India	16SrI	Shukla et al. (2015)

(continued)

Table 7.1 (continued)

Host species/trivial name (family)	Diseases	Country	Phytoplasma group/subgroup	References
Xanthoceras sorbifolia (Sapindaceae)	Leaf rolling	China	16SrI	Zhang et al. (2009)
Zamia furfuracea (Arecaceae)	Yellowing	India	16SrII	Kumar et al. (2012a)
Zinnia elegans (Asteraceae)	Little leaf, yellowing, and phyllody	India	16SrI	Rao et al. (2012)
Zygocactus truncatus (Cactaceae)	Witches' broom	China	16SrII	Cai et al. (2007)

Fig. 7.1 Phytoplasma-infected plant of *E. linifolium* showing symptoms in only a part of shoots. At flowering stage, this affected plant does not bloom

7.2 Aegean Wallflower (*Erysimum linifolium* L.; sin. *Cheiranthus linifolium* L.)

Aegean wallflower (Brassicaceae), native to the Mediterranean region, is an evergreen perennial ornamental shrub used in rock gardens or in mixed garden borders. In 2012, a phytoplasma-like disease was observed for the first time in pot plants by an ornamental grower in the Albenga area (Liguria region; northern Italy) (Paltrinieri et al. 2015). Symptomatic *E. linifolium* showed reduced leaf size, rosetting, and stunting; in some cases, shortening of internodes and growth reduction occur in only part of the plant (Fig. 7.1). An increasing percentage of symptomatic plants were found at the flowering stage, when affected plants did not bloom. Phytoplasmas belonging to subgroup 16SrI-B ('*Candidatus* Phytoplasma asteris') were detected by nested PCR followed by RFLP analyses on both the 16S rRNA and *tuf* genes.

Symptomatology associated with aster yellows (AY) presence in *E. linifolium* is very severe, probably due to becoming infected when the plants are in early growth stages. In 2010, phytoplasmas belonging to 16SrII group ('*Ca. P. aurantifolia*') were detected in *E. cheiri* (sin. *C. cheiri*), a different species cultivated in southeastern Iran, showing witches' broom and phyllody (Tazehkand et al. 2010).

7.3 *Allamanda cathartica* L.

A. cathartica (Apocynaceae), commonly called golden trumpet or common trumpet vine, was observed in India showing leaf yellowing symptoms, and phytoplasmas belonging to 16SrVI group ('*Ca.* P. trifolii'-related) were detected (Khasa et al. 2016).

7.4 *Alstroemeria* spp.

Aster yellows (16SrI) phytoplasmas were detected in *Alstroemeria* (Alstroemeriaceae) plants growing under greenhouse conditions and showing virescence symptoms in both Italy and the Netherlands (Bertaccini et al. 1996a). Phytoplasmas of the same ribosomal group were identified in malformed plants in Mexico (Cervantes-Diaz et al. 2004) and, more recently in plants showing a little leaf disease in India (Singh et al. 2011).

7.5 Bleeding Heart (*Dicentra spectabilis* L.)

The ornamental species *D. spectabilis* (bleeding heart, ladies locket) (Fumariaceae) produces fleshy tuberous roots, but it is more frequently listed as a perennial than bulbous crop. It is propagated by cuttings or by seeds. Symptoms of shoot proliferation, along with small reddened or chlorotic leaves, were reported in Poland and were associated with 16SrI-B or 16SrI-A phytoplasmas (Kaminska et al. 2004).

7.6 *Brachyscome* spp.

Madhupriya et al. (2013a) reported leaf yellows and witches' broom symptoms on *Brachyscome* spp. (Asteraceae) in India. Sequence analysis of amplified sequences revealed 99% identity with the 16S rRNA gene of strains belonging to '*Ca.* P. asteris' (16SrI group).

7.7 Burning Bush (*Dictamnus albus* L.)

Valiunas et al. (2007) identified phytoplasma symptoms represented by twisting and recumbent growth of stems, stunting, phyllody, leaf yellowing, and leaf crinkle in this ornamental shrub in Lithuania and found their association with 16SrIII-F phytoplasmas.

7.8 Calendula (*Calendula officinalis* L.)

Phytoplasmas belonging to subgroup 16SrII-E were detected in wild calendula (Asteraceae) showing malformed flowers in Sardinia, Italy (Tolu et al. 2006). Phytoplasma-infected *C. officinalis* was reported in India (Khurana et al. 1981; Rani et al. 2014) and in Iran (Esmailzadeh Hosseini et al. 2016); different phytoplasmas were identified in the various countries (Table 7.1).

7.9 Candytuft (*Iberis sempervirens* L.)

Iberis sempervirens (Brassicaceae) is one of the few flowering plants available in Europe for the market in wintertime. In winter/spring 2013, the two varieties Tahoe and Fish Back, produced by seed in Ligurian Riviera (Italy), showed symptoms of yellowing, stunting, and witches' broom (Fig. 7.2). Detection of 16SrX group phytoplasmas was obtained after nested PCR/RFLP analyses. Sequencing of the 16S rDNA gene confirmed that the phytoplasma infecting *I. sempervirens* showed 99% identity to '*Ca.* P. mali'. Moreover, RFLP analyses indicated the presence of aster yellows phytoplasmas (subgroups 16SrI-B and 16SrI-A) in mixed infection with 16SrX phytoplasmas in the Fish Back variety. This has been one of the few detections of phytoplasmas related to the 16SrX-A group from a herbaceous species worldwide (Contaldo et al. 2015a).

7.10 Chinese Aster (*Callistephus chinensis* L.)

The first reports of phytoplasmas in Chinese aster (Asteraceae) were by electron microscopy (Hemmati and Mc Lean 1980) in Canada and in some cases also by serology (Sinha and Benhamou 1983). Wang and Hiruki (2005) reported detection and estimation of genetic divergence of phytoplasmas associated with the Chinese aster yellows with the help of heteroduplex mobility assay (HMA). Flower virescence symptoms were recently observed also in Yezin, Myanmar. The symptoms usually start with the emergence of new yellow leaves during the vegetative growth

Fig. 7.2 *Iberis sempervirens* showing symptoms of yellowing, stunting, and witches' broom

stage, followed by the leaf petiole turning upright with the clustering of leaves; then, the affected plants stop growing and remain stunted. At the later stage, some flowers show green petals instead of normal color. The phytoplasma was identified as belonging to subgroup 16SrII-A (Win et al. 2011). Further reports of phytoplasma presence are from Lithuania (Navalinskiene et al. 2005).

7.11 Chinese Hibiscus (*Hibiscus rosa-sinensis* L.)

In Brazil, witches' broom disease was first reported in São Paulo State in plants of *H. rosa-sinensis* (Malvaceae); leaf yellowing and malformation as well as short internodes were also present in symptomatic plants (Vicente et al. 1974). Later, the disease was observed, in the State of Rio de Janeiro, in plants of the same species; they displayed similar symptoms and premature dropping of flowers (Kitajima et al. 1984; Kitajima 1994). The phytoplasma associated with hibiscus witches' broom disease in Brazil is reported as '*Ca.* P. brasiliense' belonging to subgroup 16SrXV-A that was in some cases associated with "stolbur" phytoplasmas (16SrXII-A)

(Montano et al. 2001, 2011). In Australia, an unidentified phytoplasma has been reported to be associated with a witches' broom disease of *H. heterophyllus*, an Australian native species that is also grown commercially (Hiruki 1987). Chaturvedi et al. (2010b) reported in India a little leaf disease of *H. rosa-sinensis* associated with 16SrI group of phytoplasmas, and Khasa et al. (2016) reported association with the same disease of clover proliferation (16SrVI) group phytoplasmas.

7.12 *Chrysanthemum* spp.

Phytoplasma diseases mainly represented by virescence (Fig. 7.3) in *Chrysanthemum* spp. (sin. *Dendranthema grandiflorum* L.) (Asteraceae) were firstly found in Sweden, Belgium, Brazil, and Japan (Pettersson and Tomenius 1979; Verhoyen et al. 1979; Kitajima and Costa 1979, Shiomi and Sugiura 1983). Conti et al. (1988) described typical yellows disease in *Chrysanthemum*, and this species was also used as a model plant to study several phytoplasma features in Italy. Galetto et al. (2007) produced polyclonal and monoclonal antibodies against membrane proteins of 'Ca. P. asteris'; Bosco et al. (2007) reported the multiplication rate of the same phytoplasma in three leafhopper vector species (*Euscelis incisus*, *Euscelidius variegatus*, and *Macrosteles quadripunctulatus*); and D'Amelio et al. (2007) described effects of elicitors of plant resistance on *Chrysanthemum* yellows-infected plants. In Serbia, 16SrXII-A phytoplasmas (Duduk et al. 2006) and in South Korea, aster yellows (16SrI) (Bongnam et al. 2007) were reported in *Chrysanthemum*, but in the latter case, two different symptoms were also described such as witches' broom and yellows, and the phytoplasmas 16SrI and 16SrXII-A were reported as being associated with these symptoms, respectively (Chung and Kim 2005). The identification of 16SrII group phytoplasmas has been reported in *Chrysanthemum* in Japan (Naito et al. 2007). Min et al. (2009) reported a 16SrI-B phytoplasma associated with flattened stems, shortening of internodes, yellowing of leaf margins, root death, and dwarfing of plants in China. They also indicated a significant loss in quality of

Fig. 7.3 Strong virescence in two *Chrysanthemum* plants due to phytoplasma presence

flowers due to phytoplasmas; in some affected plants, there is no flower at all. Bayat et al. (2013) reported the association of a '*Ca.* P. phoenicium'-related phytoplasma strain with a *Chrysanthemum* disease in Iran. Raj et al. (2007a) reported association of '*Ca.* P. asteris'-related phytoplasmas (16SrI) with little leaf disease on *Chrysanthemum morifolium*. Aido (2017) recorded symptoms of little leaf, yellowing, chlorosis, phyllody, witches' broom, and stunting in *Chrysanthemum* plants during 2015–2017 and verified the presence of four groups and five subgroups of phytoplasmas: 16SrI-B, 16SrII-A, 16SrII-D, 16SrVI-D, and 16SrXIV-A in India on the basis of RFLP analysis of 16S rDNA amplified sequences.

7.13 Cock's Comb (*Celosia argentea* L., sin. *C. cristata* L.)

Celosia argentea (Amaranthaceae) is grown in Western countries as an ornamental plant, either as potted or for cut flowers. Yellows diseases are common in Israel, and phytoplasma presence was detected in some of diseased plants. Commercial fields of *Celosia plumosa* and *C. cristata* exhibited yellows symptoms and even total crop failure where 16SrI and16SrIII phytoplasmas were detected (Tanne et al. 2000). Phytoplasmas belonging to 16SrI-M were identified on *C. argentea* from Lithuania (Samuitiene and Navalinskiene 2006), '*Ca.* P. asteris' (16SrI-B) and '*Ca.* P. australasia' (16SrII-D) were also detected in samples from India (Madhupriya et al. 2017) and Iran (Aldaghi and Bertaccini 2015), while clover proliferation phytoplasmas (16SrVI group) were identified in Iran (Babaie et al. 2007).

7.14 Cycads

Abnormal yellowing symptoms were observed in India on two species, *Cycas revoluta* (Cycadaceae) and *Zamia furfuracea* (Zamiaceae), of the order Cycadales both associated with the presence of 16SrII phytoplasmas (Kumar et al. 2012a).

7.15 Cyclamen (*Cyclamen persicum* Mill.)

Phytoplasmas were detected for the first time on cyclamen hybrids (Primulaceae) in Italy (Bertaccini 1990). During the first year, symptoms consisted of phyllody and virescence; during the second year, the plants stopped flower production, and the new leaves were dwarfed, very similar to those of wild cyclamens. Some years later, phytoplasmas associated with a cyclamen disease in Germany were found to be

Fig. 7.4 Symptoms related to phytoplasma presence in cyclamen pot and flowers. Little leaf and typical green petals are present together with flower malformations and virescence

identical, based on RFLP patterns, to the American aster yellows (AAY) strain, belonging to the 16SrI-B subgroup (Seemüller et al. 1998). In 2000, in Italy, phytoplasmas belonging to aster yellows 16SrI-B and I-C subgroups were found in five cyclamen plants with virescence and yellow stunted leaves and in one plant showing phyllody and rolled and thickened leaves. Two cyclamens, representing the two syndromes, were chosen as source plants for transmission trials in which three leafhopper species, known as vectors of 16SrI-B and 16SrI-C phytoplasmas, were used to inoculate healthy cyclamen and periwinkle plants. The extremely low level of transmission obtained, to both cyclamen and periwinkle, suggested that cyclamen is an unsuitable species for phytoplasma acquisition and can be regarded as a dead-end host plant for these phytoplasmas (Alma et al. 2000). Recently, cyclamen plants showing phytoplasma-associated symptoms were observed in a farm specialized in potted production in Liguria region (Italy). The symptomatology was represented mainly by strong modification of flowers enclosing virescence and phyllody; in several leaves stunting and rosetting were also present (Fig. 7.4). RFLP analyses allowed to identify phytoplasmas as belonging to 16SrI-B subgroup and further classified as rpI-B, SecYI-B, and GroElIII groups (Satta et al. 2013).

7.16 Desert Rose (*Adenium obesum* L.)

Desert rose (Apocynaceae) is an exotic ornamental plant from warm climates, grown for its attractive fleshy stem, leaves, and bright colorful flowers. It is a succulent plant originating from East Africa, commonly cultivated in humid, tropical areas such as India, the Philippines, and Thailand. Raj et al. (2007a) reported the association of '*Ca.* P. asteris' (group 16SrI) with a little leaf disease in India, while Win et al. (2012) indicated desert rose as a new host for '*Ca.* P. aurantifolia' (group 16SrII) in Myanmar.

7.17 Four O'Clock Flower (*Mirabilis jalapa* L.)

The four o'clock flower (Nyctaginaceae), a native of tropical South America, has been naturalized as an ornamental garden plant in many parts of the world. Plants with small yellow leaves and distorted flowers were observed in home gardens in the north of Israel (Sobolev et al. 2007). Sequence analysis of the PCR product from symptomatic *M. jalapa* clustered within those of phytoplasmas in group 16SrII. The same phytoplasma group was detected also in India (Kumar et al. 2012b) in stunted plants showing crowding of younger leaves, shortening of internodes, and small-sized leaves and flowers.

7.18 Garden Cosmos (*Cosmos bipinnatus* Cav.)

In 2003 16SrI-B phytoplasmas were detected in garden cosmos (Asteraceae) plants showing different symptoms in Mexico. It was suggested that a single type of phytoplasma was associated with the symptoms of phyllody, apical dwarfing, and yellowing (Rojas-Martínez et al. 2003a).

7.19 Garden Croton (*Codiaeum variegatum* L.)

Croton (Euphorbiaceae) with its amazing colors and leathery leaves has been popular in tropical gardens for centuries. Reports of phytoplasma diseases include 16SrII group in Uganda (Arocha et al. 2008) and 16SrI in Colombia (Perilla-Henao et al. 2012). Identification of subgroups 16SrI-B and 16SrVI-C in samples from India was also recently reported (Tiwari et al. 2014).

7.20 *Gladiolus* spp.

Aster yellows disease of *Gladiolus* spp. (Iridaceae) has been widely distributed throughout the USA, where it was first described (Magie et al. 1952). Originally, the disease, referred to as "grassy top," "hairy roots," or "green fin" (Albouy 1966), was reported to be widespread in Belgium, France (Cousin et al. 1968), Italy, Rumania, and Portugal (Bertaccini and Marani 1980; Bellardi et al. 1985; Bertaccini et al. 1990b, 1992b; Ploaie et al. 1981; Louro et al. 1996). Phytoplasmas infecting gladiolus were mainly identified in virescent flowers (Fig. 7.5) on the basis of histopathology, PCR assays, and dot hybridization, which were determined to be 16SrI-B and 16SrI-A phytoplasmas (Bertaccini et al. 1990b, 1992b; Rudzinska-Langwald

Fig. 7.5 Top, a gladiolus flower spike cv. Rose Supréme showing virescence due to aster yellows phytoplasmas compared to a normal flower spike below

and Kaminska 2003). In India Raj et al. (2009) identified '*Ca.* P. asteris' associated with malformation of floral spikes.

7.21 Hydrangea (*Hydrangea macrophylla* Thunb.)

The genus *Hydrangea* (Saxifragaceae) is composed of several deciduous shrubs; this genus encompasses almost 25 species; of these, *H. macrophylla*, native to Japan and China, is the most popular cultivated species, with over 600 recognized cultivars and hybrids, grown in both temperate and subtropical climates. Phytoplasma diseases may have a significant impact on the appearance, health, and market value of hydrangea. Cousin and Sharma (1986) and Pisi et al. (1990) detected phytoplasmas in TEM sections. Phytoplasmas belonging to subgroups 16SrI-A and 16SrI-B have been found infecting hydrangea worldwide, especially in Canada, Japan, and Europe (Bertaccini et al. 1992a; Hiruki et al. 1994; Marzachì et al. 1999; Alioto et al. 2000; Duduk et al. 2013). In Japan, the disease was called "Japanese hydrangea phyllody" (JHP) and identified phytoplasmas belong to '*Ca.* P. japonicum' (Sawayanagi et al. 1999). In addition, 16SrXII-A "stolbur" phytoplasmas were reported in Bulgaria more than 20 years ago in mixed infection with aster yellows (Bertaccini et al. 1995). In 2011 and 2012, an epidemiological survey was carried out in Liguria and Lazio regions (Italy) (Bertaccini et al. 2015). Hydrangea plants showing stunting, flower virescence and phyllody, yellowing, necrosis, and redness of the leaf edge were collected in commercial greenhouses and plants with flower virescence and red edges of leaves (Fig. 7.6) infected with phytoplasmas in the 16SrI-B subgroup. Further, RFLP analysis of the *GroEl* gene with *Tru1*I and *Alu*I allowed it to be assigned to the GroElI subgroup III. In Bolsena city (Lazio region), plants showing growth reduction, flower virescence, and phyllody, but with

Fig. 7.6 From left: hydrangea plants with flower virescence and red edges of leaves infected by 16SrI-B phytoplasmas; hydrangea plants, showing growth reduction, flower virescence, and phyllody infected by 16SrXII-A "stolbur"

asymptomatic leaves (Fig. 7.6), were infected by 16SrXII-A, "stolbur" phytoplasmas that were further characterized by their *tuf* gene. In these diseased hydrangea plants '*Ca.* P. asteris'-related phytoplasmas were identified in *L. striatellus*, while "stolbur" (16SrXII-A) phytoplasmas were present in *Anaceratogallia* sp. More recently, in Japan Kesumawati et al. (2006) studied the interaction between phytoplasma concentration and green-flowering stability in *H. macrophylla* and phylogenetic analyses based on multigene sequences, indicating that symptomatic hydrangea plants were associated with phytoplasmas belonging to '*Ca.* P. asteris' (Takinami et al. 2013).

7.22 Ice Plant (*Carpobrotus edulis* L.)

Shukla et al. (2014) reported extensive yellowing, little leaves, and reduced flower size in *C. edulis* (Aizoaceae) at Gorakhpur gardens, Eastern Uttar Pradesh, India. The associated phytoplasma was identified as belonging to the 16SrIX-C subgroup.

7.23 *Jasminum sambac* L.

Jasmine (Oleaceae) is a popular ornamental plant and is traditionally used for flowers and tea; it is cultivated commercially also for the perfume industry and herbal medicine. Group 16SrII phytoplasmas in Arabian jasmine were detected in Oman (Al-Zadjali et al. 2007), group 16SrI-B in Italy (Marzachì et al. 1999), and group 16SrXI in India (Madhupriya et al. 2015) associated with little leaf, yellows, and witches' broom symptoms.

Fig. 7.7 Lachenalia plants showing growth reduction and severe leaf and flower malformations

7.24 Lachenalia (*Lachenalia aloides* L.)

Lachenalia (Asparagaceae) is an elegant bulbous ornamental plant endemic to southern Africa. In 2016, several lachenalia pot plants at a flower bulbous grower in the Liguria region (Italy) were showing phytoplasma-like symptoms: growth reduction, severe flower malformation and virescence, yellow stripes, and variegation on the rolled leaves; in only one plant, some of the smaller flowers were normal in color (yellow and purple) (Fig. 7.7). Nested PCR and RFLP analyses using restriction enzyme *Tru1*I classified the phytoplasmas as belonging to ribosomal group 16SrI-B (Bellardi et al. 2017).

7.25 Lilac (*Syringa* spp.)

Hibben et al. (1986) detected phytoplasmas in the phloem sieve tubes of leaves of *Syringa vulgaris* (Oleaceae) plants showing lilac witches' broom by Dienes' stain. Electron microscopy observations and susceptibility of lilac to this disease were reported in the USA, and the phytoplasma detected was classified in subgroup 16SrVII-A (Hibben et al. 1986; Hibben and Franzen 1989, Griffiths et al. 1994, 1999).

7.26 Lily (*Lilium* spp.)

The earliest description of aster yellows-type diseases in lilies was that of Ogilvie and Guterman in 1929 in the USA. They described a disease on *Lilium longiflorum* cultivars characterized by severe leaf chlorosis and malformation, stunted growth, and flower distortion. In 1954, Brierley and Smith described symptoms of lily rosette in *L. longiflorum* cultivars Croft and Georgia. The affected plants showed stunting and rosette-like symptoms. Bertaccini and Marani (1982) described a flower and leaf malformation and discoloration in the lily hybrid Pink Perfection associated with multiple infections of lily mottle and lily symptomless viruses and the presence of phytoplasmas in Italy. Recently, PCR amplification of 16S rDNA and RFLP analysis indicated that stunting and flower bud deficiency symptoms in hybrid Casablanca were associated with infection by aster yellows phytoplasma and viruses (Kaminska et al. 1998; Kaminska and Korbin 1999; Bertaccini et al. 2005). In 1997–2000 phytoplasma infection was reported in plants of several lily cultivars with different symptoms (Poncarova-Vorackova et al. 1998; Kaminska and Korbin 2000; Bertaccini et al. 2002) in the Czech Republic and Poland. Moreover, in Poland, Kaminska and Śliwa (2008a) reported '*Ca.* P. mali' in oriental lilies. The presence of '*Ca.* P. solani' association with lily stem flattening disease was reported by Chung and Jeong (2003) in South Korea, while Cortes-Martinez et al. (2007) reported a 16SrI phytoplasma associated with a zigzag line pattern in leaves of *Lilium* sp. in Mexico.

7.27 Lupine (*Lupinus polyphyllus* Ltd.)

Ornamental lupine (Fabaceae) is an elegant herbaceous plant native to Northern America, used as pot plants, in home gardens, and for cut flower production. In May 2012, phytoplasma-like symptoms were observed on ornamental lupine pot plants, obtained by seed, at an ornamental grower in Ligurian Riviera (Italy) (Bellardi et al. 2013). Some plants at the blooming stage showed shortened internodes, the floral stems bent like an "S" shape, and younger leaves were smaller and rolled (Fig. 7.8). Nested PCR and RFLP analyses of the 16S rDNA from symptomatic leaves and flowers identified "stolbur" phytoplasmas (16SrXII-A) (Contaldo et al. 2013).

7.28 *Magnolia* spp.

The genus *Magnolia* spp. (Magnoliaceae) comprises about 80 species of trees and shrubs that are naturally distributed throughout eastern North America and Southeastern Asia. Magnolias are relatively free of pests and diseases; however, a severe phytoplasma disease designated as *Magnolia* stunt and yellows was observed

Fig. 7.8 Ornamental lupine flower stems: on the left, a normal one; on the right, plant infected by "stolbur" phytoplasmas

in plants growing in some gardens and nurseries and in imported plants in Poland. The identified phytoplasma was aster yellows (16SrI) (Kaminska et al. 2001b).

7.29 Marigold (*Tagetes erecta* L.)

Stunting and lack of flower production (Fig. 7.9) are the main symptoms exhibited by *Tagetes* spp. (Asteraceae) infected by phytoplasmas. Raj et al. (2011) reported little leaf and witches' broom symptoms on marigold from India. In Mexico, marigold phyllody was reported by Zavaleta-Mejia et al. (1993), and the phytoplasma which induces marigold phyllody belonged to the aster yellows group (16SrI-B) (Rojas-Martínez et al. 2003b). Pot marigold phyllody was observed in a botanical garden in Yazd Province of Iran (Esmailzadeh Hosseini et al. 2011). In Italy, witches' broom disease is frequent in small gardens.

7.30 Opuntia and Cactus Species

Ornamental cacti were reported to show several symptoms associated with phytoplasma presence. Some of these were ornamentals such as *Opuntia monacantha* Willd. (Cactaceae) in Lebanon (Choueiri et al. 2005), *Opuntia* sp., and *Zygocactus*

Fig. 7.9 Phytoplasma-infected marigold plants near healthy plants

truncatus Haw. in China (Cai et al. 2007, 2008). Phytoplasmas were reported to be associated with ornamental cacti such as *Echinopsis subdenudata* and *Opuntia* sp. also in Mexico (Hernández-Gutiérrez 1993; Leyva-Lopez et al. 1999; Avina-Padilla et al. 2009). Tessitori et al. (2006) and Granata et al. (2006) for the first time reported phytoplasmas associated with abnormal proliferation of cladodes in *Opuntia ficus-indica* Mill. in Italy. Bertaccini et al. (2007) developed a method for phytoplasma detection in *O. ficus-indica* in California (USA).

7.31 Pansy (*Viola tricolor* L.)

Leaf yellows and little leaf symptoms observed on *Viola tricolor* (Violaceae) plants at Indian Agricultural Research Institute campus, New Delhi, in 2012–2013 are associated with the presence of '*Ca.* P. asteris' (Shukla et al. 2015).

7.32 Paris Daisy [*Argyranthemum frutescens* (L.) Lch. Bip.]

Argyranthemum frutescens, known as Paris daisy, marguerite, or marguerite daisy (Asteraceae), is a perennial plant known for its lovely flowers. Phytoplasma infections in *A. frutescens* plants have already been reported and associated with phytoplasmas belonging to the aster yellows (16SrI) and elm yellows (16SrV) ribosomal groups (Bertaccini et al. 1990a, 1992a; Boarino et al. 2002). Ferretti et al. (2015) reported the association of 16SrIX-C phytoplasma group with plants showing general yellowing and stunting, little leaf, and/or abnormal proliferation of axillary shoots resulting in the appearance of witches' broom and reduced flower size.

7.33 Periwinkle [*Catharanthus roseus* (L.) G. Don.]

Periwinkle (Apocynaceae) is a perennial commonly used as an experimental host to maintain phytoplasmas since it is able to harbor the majority of known phytoplasmas (Shaw et al. 1993). Detection of phytoplasmas in periwinkle has been reported all around the world (Ploaie et al. 1977; Okuda 1977; Shishlova and Andreeva 1978; Mc Coy and Thomas 1980; Rao et al. 1983; Chen et al. 1984;Clark and Davies 1984; Grimaldi and Grasso 1988; Khurana et al. 1981; Chaturvedi et al. 2009b; Kumar and Byadgi 2012; Parrella et al. 2014; Madupriya et al. 2016). Using periwinkle as an experimental host has enabled the following research: (1) Chen and Hiruki (1978) reported the preservation of membranes of tubular bodies associated with phytoplasmas by tannic acid in *C. roseus* plants infected with aster yellows, (2) Carling and Millikan (1978) observed banded filaments associated with aster yellows in *C. roseus* in the USA, (3) Cousin and Abadie (1982) reported the action of phytoplasmas on the anther of *C. roseus* with the help of light and electron microscopy studies, (4) Schmitt et al. (1983) reported pleomorphism of phytoplasmas in periwinkle, (5) Moreno et al. (1985) described and compared several yellows diseases on *C. roseus* in Spain, (6) Rocha et al. (1986) detected phytoplasmas in *C. roseus* by indirect immunofluorescence microscopy, (7) Schmitt et al. (1987) observed freeze-fracture SEM of phytoplasmas in the phloem of *Catharanthus* sp., (8) a polyclonal antiserum was produced against a phytoplasma strain in periwinkle (Bellardi et al. 1992), (9) Davis et al. (1988) detected phytoplasmas in *C. roseus* using cloned nucleic acid hybridization probes, (10) Deng and Hiruki (1990) reported molecular cloning and detection of DNA of clover proliferation phytoplasmas in *C. roseus*, and (11) Davis et al. (1990) reported molecular cloning and detection of chromosomal and extrachromosomal DNA associated with little leaf disease in *C. roseus* in the USA.

The major aster yellows phytoplasma subgroups (16SrI-A, 16SrI-B, and 16SrI-C) have a wide host range and may also occur together in the same host. In periwinkle, subgroups 16SrI-A and 16SrI-B induce a wide variety of symptoms such as virescence, phyllody, small and light pink petals, flower malformations, shortening of internodes, elongation of internodes, plant yellowing, and small and deformed leaves. Periwinkles are known to be susceptible to the AY (16SrI) group phytoplasma in several countries, such as Malaysia (Khew et al. 1991) and Argentina (Torres et al. 2004). Omar et al. (2008) observed little leaves, shortened internodes, virescence, and witches' broom symptoms associated with infected periwinkle plants growing in Egypt, and the Egyptian phytoplasma virescence (EPV) detected in diseased periwinkle was identified as AY.

7.34 Persian Buttercup (*Ranunculus asiaticus* L.)

Phytoplasmas infecting Persian buttercup (Ranunculaceae) were first reported in France in 1968 (Devergne and Lovisolo 1969) and 20 years later in Italy (Bertaccini et al. 1988). The presence in diseased samples of a '*Ca.* P. asteris'-related strain, belonging to the 16SrI-B subgroup, was demonstrated by PCR/RFLP and phylogenetic analysis in Campania (southern Italy) (Parrella et al. 2008).

7.35 Petunia (*Petunia hybrida*)

Petunia (Solanaceae) is an economically important ornamental widely grown worldwide. In its hybrids, a phytoplasma disease was first reported in 1964 from plants showing stunting or yellowing (Doi et al. 1967). Natural occurrence of different groups of phytoplasmas was reported from several countries: a 16SrIII group was associated with a little leaf in Australia; 16SrI group was associated with flat stem, yellows, and witches' broom in Korea, Iran, and India and a 16SrXII group in Iran. Later, 16SrI phytoplasmas were associated with petunia flat stem in both China and South Korea (Chung and Hun 2008). Chung et al. (2013) reported "stolbur" phytoplasmas (16SrXII-A) in commercial petunias showing an unusual multiple plantlet sprouting from the lateral buds in a greenhouse in Gwacheon, Gyeonggi Province, China. Faghihi et al. (2014) reported the association of 16SrII phytoplasmas with plants showing witches' broom, yellowing, little leaf, phyllody, and virescence symptom in Sistan and Baluchestan Province of Iran. Leaf yellows symptoms were recorded on *Petunia* species associated with 16SrI phytoplasmas (Singh et al. 2011) and flattened stem and witches' broom symptoms (Madhupriya et al. 2014) in India.

7.36 *Phlox* spp.

Zajak (1979) observed by transmission electron microscopy phytoplasmas of 150–1200 nm diameter in sieve tubes of *Phlox paniculata* L. (Polemoniaceae) with flower virescence and Misra et al. (1985) detected them in phloem sieve tubes of *P. drummondii* Hook. plants in Rajasthan, India. Recently Navalinskiene and Samuitiene (2004) reported the same species infected by phytoplasmas belonging to subgroup 16SrI-M in Lithuania. Madhupriya et al. (2013b) reported extensive yellowing, stunting, proliferation of shoots, little leaves, and reduced size of flowers like symptoms on *P. drummondii* at New Delhi, India. Sequence analysis confirms 99% sequence identity with the 16S rRNA gene of strains belonging to the '*Ca.* P. phoenicium' (16SrIX group).

7.37 Purple Coneflower (*Echinacea* spp.)

Purple coneflower (Asteraceae) is native to North America. Plants showing general leaf yellowing, reddening, and stunting were described in Alberta (Canada) where phytoplasma-infected plants sometimes have extremely small, numerous, branched, axillary shoots coming from the stem nodes, giving them a bunchy or witches' broom appearance (Hwang et al. 1997). In the purple coneflower, phytoplasmas belonging to the aster yellows group have been identified (Stanosz et al. 1997; Khadhair et al. 1997; Chang et al. 2000; Lee et al. 2008); in particular, in the USA the subgroups 16SrI-A and 16SrI-B were identified. Radisek et al. (2008) also reported a 16SrI-C phytoplasma infecting purple coneflower in Slovenia, while Bertaccini et al. (2009) reported the association of 16SrIX-C and 16SrI-B phytoplasmas in *E. purpurea* (L.) Moench. plants showing yellowing, phyllody, and virescence symptoms (Fig. 7.10). In Serbia, *E. purpurea* and *E. angustifolia* L. were observed to show phytoplasma symptoms. The symptoms on *E. purpurea* were yellowing in the early stages of disease development; foliage reddening, plant stunting, and proliferation of axillary shoots appear as the disease progresses, and infected plants showed bunchy or witches' broom appearance. Symptoms on *E. angustifolia* were stunting, shortened internodes, and purplish-reddening leaves and stalks. Flowers on such plants were found smaller and did not bear seeds. Molecular identification confirmed the presence of 16SrXII-A ("stolbur") phytoplasmas in both species investigated (Pavlovic et al. 2010). Purple coneflower plants showing leaf reddening and flower abnormalities were observed in South Bohemia (Czech Republic) where phytoplasmas were observed by transmission electron microscopy and identified as belonging to 16SrI-C and 16SrIII-B subgroup. The identity of the latter was also confirmed after sequence analyses of the 16S–23S ribosomal operon, ribosomal protein L15, and protein translocase genes (Franova et al. 2009, 2013).

Fig. 7.10 Purple coneflower showing strong virescence and malformation symptoms due to the presence of mixed phytoplasma infection in Italy

7.38 Queen Anne's Lace (*Daucus carota* L.) and Poker Statice [*Psylliostachys suworowii* (Regel) Roshk]

Queen Anne's lace (Apiaceae) and poker statice (Poaceae) are widely cultivated ornamentals. They were found with a yellows-type disease with typical phytoplasma symptoms in an experimental farm near Brooks, Alberta, Canada, in 1996. Further study provided the first record of 16SrI phytoplasmas in Queen Anne's lace and poker statice (Chang et al. 2004).

7.39 Rose (*Rosa* spp.)

Rose (Rosaceae) is the most important and popular garden plant in the world and the most important commercial cut flower cultivated in glasshouses. Several virus-like diseases of uncertain etiology have been reported in rose throughout the world. Rose wilt was first recorded in Australia in 1908, but the symptoms were later described more fully by Grieve (1931) and Fry and Hammett (1971) in New Zealand. In the United Kingdom, Hollings (1961) recorded the occurrence of rose bud proliferation, rose wilt, and rose dieback, also called rose winter dieback (Thomas 1979). Some rose degeneration symptoms have been reported from Bulgaria (Hristova 1974), France (Devergne and Coujon 1975), the Netherlands (Bos and Perqin 1975), and Poland (Kaminska et al. 2001a, 2003, 2006; Kaminska and Sliwa 2004). Rose rosette disease, also referred as rose witches' broom, was first reported in Manitoba on wild rose species by Connors in 1941 and subsequently in other parts of Canada and the USA (Epstein and Hill 1995). The disease is endemic in much of the south-east, south-central, and north-central USA (Hindal et al. 1988; Tipping and Sindermann 2000). Similar symptoms known as rose dieback or rose wilt were recorded in garden roses by Cheo (1970) and Gumpf and Weathers (1974). In 1976, rose leaf curl (Slack et al. 1976a), which resembles rose wilt disease and rose spring dwarf, was more fully described in the USA (Slack et al. 1976b). Research in various parts of the world demonstrated the presence of an aster yellows (16SrI) phytoplasma associated with phyllody symptoms in Poland (Kaminska et al. 2003), India (Fig. 7.11) (Chaturvedi et al. 2009a), and China (Gao et al. 2008). Madhupriya et al. (2017) reported leaf chlorosis, phyllody, virescence, and little leaf symptoms in various rose cultivars associated with subgroups 16SrI-B and 16Sr II-D in India.

Fig. 7.11 Yellows and pale flower in rose affected by aster yellows phytoplasmas in India

7.40 Rose Balsam (*Impatiens balsamina* L.)

Rose balsam (Balsaminaceae) is an ornamental species cultivated in China, where the red flower is often used as nail polish in rural regions. Phytoplasmas of the 16SrI group were reported in plants showing phyllody in China (Li et al. 2011). More recently (Li et al. 2014), subgroup 16SrV-B phytoplasmas were detected in plants with wrinkled leaves, phyllody, deformed and shortened internodes, stunting, and no seed production.

7.41 Moss Rose (*Portulaca grandiflora* L.)

Ajaykumar et al. (2007) and Samad et al. (2008) reported in India a little leaf disease of *P. grandiflora* (Portulacaceae) known also as eleven o'clock, Mexican rose, moss rose, Vietnam rose, sun rose, and rockrose. The symptomatic plants displayed bud proliferation, downward curling, and diminished size of leaves, followed by overall stunted growth and general yellowing; some plants also show a witches' broom appearance and 16SrVI phytoplasmas were identified.

7.42 Soapwort (*Saponaria officinalis* L.)

S. officinalis (Caryophyllaceae) is a common perennial plant from the carnation family. This plant has many common names, including common soapwort, bouncing-bet, crow soap, wild sweet William, and soapweed. Khasa et al. (2016) observed witches' broom symptom associated with 16SrVI group phytoplasmas in India.

7.43 Scotch Broom (*Cytisus scoparius* L. sin. *Sarothamnus scoparius* L.)

A witches' broom and stunting were observed in a group of shrubs of *Cytisus scoparius*, better known as common broom or scotch broom growing in Berlin-Dahlem Botanical Garden and Botanical Museum in Berlin (Germany). This is a perennial shrub native to western and central Europe, but it is considered an invasive plant in North America and New Zealand. Symptomatic *C. scoparius* were analyzed by nested PCR/RFLP, and '*Ca.* P. spartii' in mixed infection with '*Ca.* P. asteris' was detected. In some samples of *C. scoparius*, "stolbur" phytoplasmas were also identified (Contaldo et al. 2015b).

7.44 Siamese Rough Bush (*Streblus asper* Lour)

Streblus asper (Moraceae) is an important medicinal and ornamental plant distributed in tropical countries such as India. Maurya et al. (2014) recorded association of '*Ca.* P. asteris' (16SrI) with *S. asper* plants showing chlorosis and little leaf symptoms growing in different gardens and nurseries of Gorakhpur, India.

7.45 Spanish Broom (*Spartium junceum* L.)

Spanish broom (Fabaceae) is a deciduous shrub with dark green, round stems and alternate leaves; inflorescences are terminal clusters of several bright yellow somewhat fragrant flowers. This ornamental shrub grows naturally especially in southern Italy where it is affected by spartium witches' broom (SpaWB) disease, characterized by proliferation of axillary buds and stem fasciation. Two phytoplasmas were associated with this disease: '*Ca.* P. spartii' a member of the apple proliferation phylogenetic group (16SrX-D), and a phytoplasma belonging to elm yellows group (16SrV-C) (Marcone et al. 1996, 1997a, 2004; Torres et al. 2002; Mancini et al. 2010). Both phytoplasmas were reported associated with SpaWB in Italy, while

Fig. 7.12 Typical SpaWB symptoms in a plant growing in the city of Ercolano (Campania region, Italy)

only '*Ca.* P. spartii' was reported in Spain. More recently typical SpaWB symptoms were observed in a plant up to 2 m tall growing in the city of Ercolano (Campania region, Italy) (Fig. 7.12). Nested PCR/RFLP analyses showed the presence of '*Ca.* P. spartii' in mixed infection with '*Ca.* P. asteris' (Contaldo et al. 2015b).

7.46 *Spiraea* spp.

Spireas (*Spiraea* spp.) (Rosaceae) are woody perennial ornamentals that are widely grown throughout Minnesota and in other US states due to their cold hardiness and adaptability to a variety of low-maintenance landscape settings. Phytoplasma-infected plants were recorded as showing severe stunting, small leaves, and shoot proliferation (witches' broom). Diverse phytoplasmas were detected in particular in the USA 16SrIII-E phytoplasmas were identified (Griffiths et al. 1994), while 16SrI-B and 16SrV-B subgroups were identified in China (Gao et al. 2007; Li et al. 2010).

7.47 Sunflower (*Helianthus annuus* L.)

Phyllody in sunflower (Asteraceae) was reported in Sudan by Nour (1962), while the first information regarding the presence of phytoplasmas in sunflower was revealed by transmission electron microscopy in southern France (Signoret et al. 1976). Guzman et al. (2014) reported virescence, phyllody, flower malformation, shortened internodes, and abnormal branches on sunflower associated with 16SrIII-J in Argentina. Previous studies have reported the infections caused by phytoplasmas from 16SrII and 16SrVI groups in sunflower in Iran (Tazehkand et al. 2010).

Fig. 7.13 Virescent and malformed flowers of *M. incana* infected by phytoplasmas

7.48 Tenweeks stock (*Matthiola incana* R. Br.)

Tenweeks stock (Brassicaceae) is a common herbaceous ornamental species cultivated in all temperate areas for flower production. In April 2007, a severe disease occurred in Sicily (southern Italy) in a glasshouse cultivation of tenweeks stock cultivar White-Beach. Plants were stunted and rosetted, but the main symptoms, appearing at the flowering stage, were malformation of white flowers and virescence (Fig. 7.13). The incidence of symptomatic plants was about 65%. PCR, RFLP, and sequencing carried out on the 16S rRNA gene, together with phylogenetic comparison of the 16S rRNA, confirmed that the phytoplasmas detected belonged to ribosomal subgroup 16SrII-A ('*Ca.* P. aurantifolia') (Davino et al. 2007).

7.49 Zinnia (*Zinnia elegans* L.)

Rao et al. (2012) reported typical little leaf, chlorosis, witches' broom, yellowing, and phyllody symptoms on *Zinnia elegans* (Asteraceae) at Gorakhpur, India. Sequence analysis revealed 99% sequence identity with the 16S rRNA gene of strains belonging to the '*Ca.* P. asteris' (16SrI) group.

References

Aido T (2017) Characterization and management of phytoplasma disease of *Chrysanthemum morifolium* R. PhD thesis, Indian Agricultural Institute. New Delhi, India, 91 pp.

Ajaykumar PV, Samad A, Shasany AK, Gupta M, Kalam M, Rastogi S (2007) First record of a '*Candidatus* Phytoplasma' associated with little leaf disease of Portulaca grandiflora. Australasian Plant Disease Notes **2**, 67–69.

Albouy J (1966) Le probleme des "germs fins" du glaeul. *Annals de l'Epiphytie* **17**, 81–91.

Aldaghi M, Bertaccini A (2015) Preliminary study on some ornamental plant phytoplasma diseases in north of Iran.*Phytopathogenic Mollicutes* **5**(1-Supplement), S67–S68.

Alioto D, Marcone C, Ragozzino E (2000) Virescence and phyllody in hydrangea (*Hydrangea macrophylla*). *Informatore Fitopatologico* **50**, 59–62.

Alma A, Marzachì C, d'Aquilio M, Bosco D (2000) Cyclamen (*Cyclamen persicum* L.): a dead-end host species for 16SrI-B and -C subgroup phytoplasmas. *Annals of Applied Biology* **136**, 173–178.

Al-Saady NA, Khan AJ, Calari A, Al-Subhi AM, Bertaccini A (2008) '*Candidatus* Phytoplasma omanense', a phytoplasma associated with witches' broom of *Cassia italica* (Mill.) Lam. in Oman. *International Journal of Systematic and Evolutionary Microbiology* **58**, 461–466.

Al-Zadjali AD, Natsuaki T, Okuda S (2007) Detection, identification and molecular characterization of a phytoplasma associated with Arabian jasmine (*Jasminum sambac* L.) witches' broom in Oman. *Journal of Phytopathology* **155**, 211–219.

Arocha Y, Singh A, Pandey M, Tripathi AN, Chandra B, Shukla SK, Singh Y, Kumar A, Srivastava RK, Zaidi NW, Arif M, Narwal S, Tewari AK, Gupta MK, Nath PD, Rabindran R, Khirbat SK, Byadgi AS, Singh G, Boa E (2008) New plant hosts for group 16SrII, '*Ca.* P. aurantifolia', in India. *Plant Pathology* **17**, 36.

Arocha-Rosete Y, Morales-Lizcano NP, Hasan A, Yoshioka K, Moeder W, Michelutti R, Satta E, Bertaccini A, Scott J (2016) First report of the identification of a '*Candidatus* Phytoplasma pruni'-related strain in *Trillium* species in Canada. *New Disease Reports* **34**, 19.

Avina-Padilla K, Parra-Cota F, Ochoa-Sanchez JC, Perales-Segovia C, Martinez-Soriano JP (2009) Phytoplasmas associated to diseases of ornamental cacti in Mexico. *Journal of Biological Sciences* **9**, 268–271.

Babaie G, Khatabi B, Bayat H, Rastgou M, Hosseini A, Salekdeh GH (2007) Detection and characterization of phytoplasmas infecting ornamental and weed plants in Iran. *Journal of Phytopathology* **155**, 368–372.

Back C-G, Chung H, Jung H-Y (2010) First report of '*Candidatus* Phytoplasma asteris' associated with yellowing and rosetting of Japanese spurge in Korea. *Journal of General Plant Pathology* **76**, 355–357.

Bayat H, Verhoeven JThJ, Botermans M, Peters D, Hassani-Mehraban A (2013) First report of a 16SrIX group ('*Ca.* P. phoenicium'-related) phytoplasma associated with a *Chrysanthemum* disease. *Plant Disease* **97**, 1110–1110.

Bellardi MG, Vicchi V, Bertaccini A (1985) Micoplasmosi del gladiolo. *Informatore Fitopatologico* **35**, 35–39.

Bellardi MG, Vibio M, Bertaccini A (1992) Production of a polyclonal antiserum to CY-MLO using infected *Catharanthus roseus*. *Phytopathologia Mediterranea* **31**, 53–55.

Bellardi MG, Cavicchi L, Bozzano G, Parrella G, Contaldo N, Bertaccini A (2013) Infezione mista da fitoplasmi e TSWV in lupino ornamentale in Liguria. *Clamer informa* **2**, 51–55.

Bellardi MG, Parmeggiani S, Bozzano G, Contaldo N, Bertaccini A (2017) Primo caso di un'infezione da fitoplasmi in *Lachenalia*. *Clamer informa* **2**, 72–80.

Bertaccini A (1990) Cyclamen: a new host of mycoplasma-like organisms. *Phytopathologia Mediterranea* **29**, 213–214.

Bertaccini A, Marani F (1980) Mycoplasma-like organisms in *Gladiolus* sp. with malformed and virescent flowers. *Phytopathologia Mediterranea* **19**, 121–128.

Bertaccini A, Marani F (1982) Electron microscopy of two viruses and mycoplasma-like organism in lilies with deformed flowers. *Phytopathologia Mediterranea* **21**, 8–14.
Bertaccini A, Marani F, Rapetti S (1988) Phyllody and virescence in ranunculus hybrids. *Acta Horticulturae* **234**, 123–128.
Bertaccini A, Davis RE, Lee I-M, Conti M, Dally EL, Douglas SM (1990a) Detection of chrysanthemum yellows mycoplasmalike organisms (MLO) by dot-hybridization and Southern blot analysis. *Plant Disease* **74**, 40–43.
Bertaccini A, Davis RE, Lee I-M (1990b) Distinctions among mycoplasma like organisms (MLOs) in *Gladiolus*, *Ranunculus*, *Brassica* and *Hydrangea* through detection with nonradioactive cloned DNA-probes. *Phytopathologia Mediterranea* **29**, 107–113.
Bertaccini A, Davis RE, Hammond RW, Vibio M, Bellardi MG, Lee I-M (1992a) Sensitive detection of mycoplasmalike organisms in field collected and *in vitro* propagated plants of *Brassica*, *Hydrangea* and *Chrysanthemum* by polymerase chain reaction. *Annals of Applied Biology* **121**, 593–599.
Bertaccini A, Bellardi MG, Vibio M, Lee I-M, Davis RE (1992b) Detection of mycoplasmalike organisms in gladiolus using microscopy and DNA-probes. *Acta Horticulturae* **325**, 703–708.
Bertaccini A, Vibio M, Sakalieva D, Guergov E, Karov S (1995) Molecular characterization of phytoplasmas infecting vegetable and flower crops in Bulgaria. I Congresso Nazionale Biotecnologia, Bologna, Italia, S.1.3.
Bertaccini A, Vibio M, Bellardi MG, Danielli A (1996a) Identification of phytoplasmas in alstroemeria. *Acta Horticulturae* **432**, 312–319.
Bertaccini A, Bellardi MG, Vibio M (1996b) Virus diseases of ornamental shrubs. X. *Euphorbia pulcherrima* Willd. Infected by viruses and phytoplasmas. *Phytopathologia Mediterranea* **35**, 129–132.
Bertaccini A, Kaminska M, Botti S, Martini M (2002) Molecular evidence for mixed phytoplasma infection in lily plants. *Acta Horticulturae* **568**, 35–41.
Bertaccini A, Franova J, Botti S, Tabanelli D (2005) Molecular characterization of phytoplasmas in lilies with fasciation in the Czech Republic. *FEMS Microbiology Letters* **249**, 79–85.
Bertaccini A, Bellardi MG, Botti S, Paltrinieri S, Restuccia P (2006) Phytoplasma infection in *Asclepias physocarpa*. *Acta Horticulturae* **722**, 229–234.
Bertaccini A, Calari A, Felker P (2007) Developing a method for phytoplasma identification in cactus pear samples from California. *Bulletin of Insectology* **60**, 257–258.
Bertaccini A, Paltrinieri S, Contaldo N, Duduk B, Nahdi S, Benni A, Bellardi MG (2009) Different phytoplasmas infecting purple coneflower in Italy. *Journal of Plant Pathology* **91**, S4.49.
Bertaccini A, Paltrinieri S, Contaldo N, Cavicchi L, Mori N, Bellardi MG (2015) Severe diseases induced by viruses and phytoplasmas in *Hydrangea* in Italy. *Acta Horticulturae* **1072**, 105–112.
Boarino A, Boccardo G, D'aquilio M, Veratti F, Marzachì C (2002) Detection and molecular characterization of group V phytoplasmas in several herbaceous hosts. *Petria* **12**, 391–392.
Bongnam C, Jeongsoo K, Myeongil J (2007) Phytoplasma disease in field grown *Chrysanthemum* (*Dendrathema grandiflorum*). *Horticulture Environment and Biotechnology* **48**, 81–84.
Bos L, Perqin FW (1975) Rose bud proliferation, a disorder of unknown etiology. *Netherlands Journal Plant Pathology* **81**, 187–198.
Bosco D, Galetto L, Leoncini P, Saracco P, Raccah B, Marzachì C (2007) Interrelationships between '*Ca.* P. asteris' and its leafhopper vectors (Homoptera: Cicadellidae). *Journal of Economic Entomology* **100**, 1504–1511
Brierley P, Smith FF (1954) New records of virus diseases of ornamental plants. *Plant Disease Reporter* **38**, 739–741.
Cai H, Chen HR, Kong BH, Yang GH, Liu T (2007) First report of a phytoplasma associated with Christmas cactus witches' broom. *Plant Pathology* **56**, 346.
Cai H, Wei W, Davis RE, Chen H, Zhao Y (2008) Genetic diversity among phytoplasmas infecting *Opuntia* species: virtual RFLP analysis identifies new subgroups in the peanut witches' broom phytoplasma group. *International Journal of Systematic and Evolutionary Microbiology* **58**, 1448–1457.

Carling DE, Millikan DF (1978) Banded filaments associated with the aster yellows MLO in *Vinca rosea*. *Canadian Journal of Microbiology* **24**, 1417–1418.

Cervantes-Diaz L, Zavaleta-Mejia E, Rojas-Martinez RI, Alanis-Martinez I, Ochoa-Martinez DL, Sanchez-Garcia P (2004) First report of phytoplasma occurrence in *Alstroemeria* sp. plants in Mexico. *Revista Mexicana de Fitopatologia* **22**, 134–139.

Chang KF, Howard RJ, Blade SF, Hwang SF (2000) Survey of aster yellows of *Echinacea* in Alberta in 1999. *Canadian Plant Disease Survey* **80**, 88–89.

Chang KF, Hwang SF, Khadair AH, Kawchuk L, Howard R (2004) Detection and molecular characterization of an aster yellows phytoplasma in poker statice and Queen Anne's lace in Alberta, Canada. *Microbiological Research* **159**, 43–50.

Chaturvedi Y, Singh M, Rao GP, Snehi SK, Raj SK (2009a) First report of association of '*Candidatus* Phytoplasma asteris' (16SrI group) with little leaf disease of rose (*Rosa alba*) in India. *Plant Pathology* **58**, 788.

Chaturvedi Y, Tewari AK, Upadhyaya PP, Prabhuji SK, Rao GP (2009b) Association of '*Candidatus* Phytoplasma asteris' with little leaf and phyllody disease of *Catharanthus roseus* in Eastern Uttar Pradesh, India. *Medicinal Plants* **1**, 103–108.

Chaturvedi Y, Rao GP, Tiwari AK, Duduk B, Bertaccini A (2010a) Phytoplasma on ornamentals: detection, diversity and management. *Acta Phytopathologica et Entomologica Hungarica* **45**, 31–69.

Chaturvedi Y, Singh M, Snehi SK, Raj SK, Rao GP (2010b) Association of '*Candidatus* Phytoplasma asteris' (16SrI group) with yellows and little leaf disease of *Hibiscus rosa-sinensis* in India. *Plant Pathology* **59**, 796.

Chen MH, Hiruki C (1978) The preservation of membranes of tubular bodies associated with mycoplasma like organisms by tannic acid. *Canadian Journal of Botany* **56**, 2878–2882.

Chen ZY, Shen JY, Peng BZ, Zheng GB, Chen MY (1984) Mycoplasma associated with periwinkle phyllody disease. *Acta Phytopathologica Sinica* **14**, 233–234.

Cheo P (1970) Rose wilt or dieback-a new virus disease attacks rose in California. *Lasca Leaves* **20**, 88–89.

Choueiri E, Massad R, Jreijiri F, Danet J-L, Salar P, Bové J-M, Foissac X (2005) First report of a 16SrII group phytoplasma associated with shoot proliferation of a cactus (*Opuntia monacantha*) in Lebanon. *Plant Disease* **89**, 1129–112.

Chung BN, Jeong MI (2003) Detection and molecular characterization of a "stolbur" phytoplasma in *Lilium* oriental hybrids. *The Plant Pathology Journal* **19**, 106–110.

Chung BN, Kim BD (2005) Two groups of phytoplasma from chrysanthemum (*Dendrathemum grandiflorum*) distinguished by symptoms and 16S rRNA gene sequence in Korea. *The Plant Pathology Journal* **21**, 132–136.

Chung BN, Hun KY (2008) Occurrence of petunia flattened stem caused by phytoplasma. *The Plant Pathology Journal* **24**, 279–282.

Chung BN, Choi GS (2010) Occurrence of poinsettia stem flat disease caused by phytoplasma in Korea. *Plant Disease* **94**, 792.

Chung BN, Jeong MI, Choi SK, Joa JH, Choi KS, Choi IM (2013) Occurrence of "stolbur" phytoplasma disease in spreading type *Petunia hybrida* cultivars in Korea. *The Plant Pathology Journal* **29**, 465–470.

Clark MF, Davies DL (1984) Mycoplasma detection and characterization. *Report East Malling Research Station*, 108–109.

Connors IL (1941) 20th Annual Report of the Canadian Plant Disease Survey, 1940, 1–7.

Contaldo N, Bertaccini A, Bozzano G, Parrella G, Cavicchi L, Bellardi MG (2013) Infezione mista da fitoplasmi "stolbur" e dal virus dell'avvizzimento maculato del pomodoro (TSWV) in *Lupinus polyphyllus* Ltd. *Petria* **23**, 37–34.

Contaldo N, Bellardi MG, Paltrinieri S, Bertaccini A (2015a) Phytoplasma identification in iberis showing stunting and witches' broom symptoms. *Phytopathogenic Mollicutes* **5**(1-Supplement), S85–S86.

Contaldo N, Paltrinieri S, Ardizzi S, Duduk B, Bertaccini A, Bellardi MG (2015b) Identification and molecular characterization of multiple phytoplasma infection in *Spartium junceum* and *Cytisus scoparius*. *Acta Horticulturae* **1072**, 113–116.

Conti M, D'Agostino G, Casetta A, Mela L (1988) Some characteristics of *Chrysanthemum* yellows disease. *Acta Horticulturae* **234**, 129–136.

Cortes-Martinez NE, Moctezuma EV, Molina LXZ, Aguilar-Rios J (2007) A new phytoplasma associated with a zigzag line pattern in leaves of *Lilum spp.* in Mexico. *Bulletin of Insectology* **60**, 283–284.

Cousin MT, Maillet PL, Gourret JP, Grison C, Staron T (1968) Presence de particules de type "mycoplasme" dans le phloeme de plantes infectées par trois maladies de type "jaunisse europeenne" le "stolbur" de la tomate, l'aster yellows du glaïeul, la phyllodie du bifli. Etude cytologique et ultrastructurale. Premiers essais a le lutte chemique. *Comptes Rendus Hebdomadaires des Seances de l'Academie des Sciences* **54**, 887–895.

Cousin MT, Abadie M (1982) Action of mycoplasmas on the anther. Light and electron microscope study. *Station de Pathologie Vegetale* **5**, 41–57.

Cousin MT, Sharma AK (1986) Association of mycoplasma like organisms (MLOs) with mild type of *Hydrangea virescence*: a study with 60–1000 nm thick sections. *Journal of Phytopathology* **115**, 274–282.

Cozza R, Bernardo L, Calari A, Silvestro G, Duduk B, Bertaccini A (2008) Molecular identification of '*Candidatus* Phytoplasma asteris' inducing histological anomalies in Silene nicaeensis. *Phytoparasitica* **36**, 290–293.

D'Amelio RD, Massa N., Gamalero E, D'Agostino G, Sampò S, Berta G, Faoro F, Iriti M, Bosco D, Marzachì C (2007) Preliminary results on the evaluation of the effects of elicitors of plant resistance on *Chrysanthemum* yellows phytoplasma infection. *Bulletin of Insectology* **60**, 317–318.

Davino S, Calari A, Davino M, Tessitori M, Bertaccini A, Bellardi MG (2007) Virescence of ten-weeks stocks associated to phytoplasma infection in Sicily. *Bulletin of Insectology* **60**, 279–280.

Davis RE, Lee I-M, Dally EL, Delwitt N, Douglas SM (1988) Cloned nucleic acid hybridization probes in detection and classification of mycoplasma-like organisms (MLOs). *Acta Horticulturae* **234**, 115–122.

Davis RE, Lee I-M, Douglas SM, Dally EL (1990) Molecular cloning and detection of chromosomal and extrachromosomal DNA of the mycoplasma like organism associated with little leaf disease in periwinkle (*Catharanthus roseus*). *Phytopathology* **80**, 789–793.

Deng SJ, Hiruki C (1990) Molecular cloning and detection of DNA of the mycoplasma like organism associated with clover proliferation. *Canadian Journal of Plant Pathology* **12**, 383–388.

Devergne JC, Lovisolo O (1969) Les viroses des plants ornamentales dans le bassin méditaerranéen. *Annales de Phytopathologie* **1**, 285–301.

Devergne JC, Coujon C (1975) Etude d'anomalies de croissance chez le rosier de serre. *Annales de Phytopathologies* **7**, 71–79.

Doi Y, Teranaka M, Yora K, Asuyama H (1967) Mycoplasma- or PLT group-like microorganisms found in the phloem elements of plants infected with mulberry dwarf, potato witches' broom, aster yellows, or paulownia witches' broom. *Annals of the Phytopathological Society of Japan* **33**, 259–266.

Duduk B, Dukić N, Bulajić A, Krstić B, Bertaccini A (2006). "Stolbur" phytoplasmas infecting chrysanthemum plants in Serbia. *Pesticides and Phytomedicine* **21**, 107–112.

Duduk B, Arocha Y, Mitrović J, Michelutti R, Benabid R, Scott J, Bertaccini A (2013) Caratterizzazione molecolare di '*Candidatus* Phytoplasma asteris' in ortensia in Canada. *Petria* **23**, 139–142.

Eckstein B, Gonçalves Da Silva E, Bedendo PI (2012) Shoot proliferation and leaf malformation of *Celosia argentea* and *Celosia spicata* caused by a phytoplasma of the 16SrIII-J group. *Journal of Phytopathology* **160**, 206–208.

Epstein AH, Hill J (1995) The biology of rose rosette disease: a mite-associated disease of uncertain aetiology. *Journal of Phytopathology* **143**, 353–360.

Esmailzadeh Hosseini SA, Salehi M, Khanchezar A, Shamszadeh M (2011) The first report of a phytoplasma associated with pot marigold phyllody in Iran. *Bulletin of Insectology* **64**(Supplement), S109–S110.

Esmailzadeh Hosseini SA, Salehi M, Mirchenari SM, Bertaccini A (2016) First report of a 16SrII-D phytoplasma associated with *Calendula officinalis* phyllody in Iran. *New Disease Reports* **34**, 22.

Faghihi MM, Taghavi SM, Salehi M, Sadeghi MS, Samavi S, Siampour M (2014) Characterisation of a phytoplasma associated with *Petunia* witches' broom disease in Iran. *New Disease Reports* **30**, 21.

Ferretti L, Costantini E, Pasquini G (2015) Identification of 16SrIX-C phytoplasmas in *Argyranthemum frutescens* in Italy *Phytopathologia Mediterranea* **54**, 21–27.

Franova J, Pribylova J, Petrzik K (2009) Purple coneflower with reddening and phyllody: a new host of clover phyllody phytoplasma. *European Journal of Plant Pathology* **123**, 85–90.

Franova J, Špak J, Šimková M (2013) First report of a 16SrIII-B subgroup phytoplasma associated with leaf reddening, virescence and phyllody of purple coneflower. *European Journal of Plant Pathology* **136**, 7–12.

Fry PR, Hammett KRW (1971) Rose wilt virus in New Zealand. *Journal of Agriculture Research* **14**, 735–743.

Galetto L, Bosco D, Fletcher J, Marzachì C (2007) Production of polyclonal and monoclonal antibodies specific against membrane proteins of '*Candidatus* Phytoplasma asteris', *Chrysanthemum* yellows isolate. *Bulletin of Insectology* **60**, 211–212.

Gao R, Wang J, Li XD, Zhu XP, Tian GZ (2007) First report of *Spirea* witches' broom disease in China. *Plant Disease* **91**, 635.

Gao R, Zhang G-M, Lan Y-F, Zhu T-S, Yu X-Q, Zhu X-P, Li X-D (2008) Molecular characterization of phytoplasma associated with rose witches' broom in China. *Journal of Phytopathology* **156**, 93–98.

Gera A, Maslenin L, Rosner A, Zeidan M, Weintraub PG (2006) Phytoplasma diseases in ornamental crops in Israel. *Acta Horticulturae* **722**, 155–162.

Granata G, Paltrinieri S, Botti S, Bertaccini A (2006) Aetiology of *Opuntia ficus-indica* malformations and stunting disease. *Annals of Applied Biology* **149**, 317–325.

Grieve BJ (1931) Rose wilt and dieback. A virus disease occurring in Australia. *The Australian Journal of Experimental Biology and Medical Science* **8**, 107–121.

Griffiths HM, Gundersen DE, Sinclair WA, Lee I-M, Davis RE (1994) Mycoplasmalike organisms from milkweed, goldenrod, and spirea represent two new 16S rRNA subgroups and three new strain subclusters related to peach X-disease MLOs. *Canadian Journal of Plant Pathology* **16**, 255–260.

Griffiths HM, Sinclair WA, Smart CD, Davis RE (1999) The phytoplasma associated with ash yellows and lilac witches broom: '*Candidatus* Phytoplasma fraxini'. *International Journal of Systematic Bacteriology* **49**, 1605–1614.

Grimaldi V, Grasso S (1988) Mycoplasma disease of *Dimophotheca sinuata* and *Catharanthus roseus* in Sicily. *Acta Horticulturae* **234**, 137–143.

Gumpf DJ, Weathers LG (1974) Detection of a new infectious disease of roses in California. *Acta Horticulture* **36**, 53–57.

Gundersen DE, Lee I-M, Rehner SA, Davis RE, Kingsbury DT (1994) Phylogeny of mycoplasma-like organisms (phytoplasmas): a basis for their classification. *Journal of Bacteriology* **176**, 5244–5254.

Guzman F, Giolitti F, Fernandez F, Nome C, Lenardon S, Conci L (2014) Identification and molecular characterization of a phytoplasma associated with sunflower in Argentina. *European Journal of Plant Pathology* **138**, 679–683.

Harrison NA, Helmick E E (2008) First report of a '*Candidatus* Phytoplasma asteris'-related strain associated with little leaf disease of *Helianthus debilis* in Florida, USA. *Plant Pathology* **57**, 772.

Harju VA, Skelton AL, Monger WA, Jarvis B, Mumford RA (2008) Identification of an X-disease (16SrIII) group phytoplasma ('*Candidatus* Phytoplasma pruni') infecting delphiniums in the UK. *Plant Pathology* **57**, 769–769.

Hemmati K, Mc Lean DL (1980) Ultrastructure and morphological characteristics of mycoplasma-like organisms associated with tulelake aster yellows. *Phytopathology Zeitschrift* **99**, 146–154.

Hernández-Gutiérrez L (1993) Plagas y enfermedades del nopal en México. Reporte de investigación No. 11. Ciestaam, Universidad Autonoma Chapingo, Chapingo, Ed. de México. México, pp. 52.

Hibben CR, Lewis CA, Castello JD (1986) Mycoplasma like organisms, cause of lilac witches' broom. *Plant Disease* **70**, 342–345.

Hibben CR, Franzen LM (1989) Susceptibility of lilacs to mycoplasma like organisms. *Journal of Environmental Horticulture* **7**, 163–167.

Hindal DF, Amrine JW, Williams RL, Stasny TA (1988) Rose rosette disease on multiflora rose (*Rosa multiflora*) in Indian and Kentucky. *Weed Technology* **2**, 442–444.

Hiruki C (1986) Little-leaf disease of *Brugmansia candida* a new disease associated with mycoplasma-like organisms. *Annals of Phytopathological Society of Japan* **52**, 675–682.

Hiruki C (1987) Witches' broom of *Hibiscus heterophyllus*, a mycoplasma disease occurring in Australia. *Annals of Phytopathological Society of Japan* **53**, 1–6.

Hiruki C, Romg XD, Deng SJ (1994) *Hydrangea* virescence, a disease associated with mycoplasma like organisms in Canada. *Acta Horticulturae* **377**, 325–333.

Hollings M (1961) Virus and its effect on some ornamental shrubs. Nurseryman, Seedsman and Glasshouse Grower, 930–931.

Hong C, Aolin LX, Baohoa K, Hairu C (2005) Detection and identification of phytoplasma associated with sunshine tree witches' broom. *Acta Phytopathologica Sinica* **35**, 19–23.

Hristova M (1974) Rose virus diseases. *Plant Science* **10**, 121–129.

Hwang SF, Chang KF, Howard RJ, Khadhair AH, Gaudiel RG, Hiruki C (1997) First report of a yellows phytoplasma disease in purple coneflower (*Echinacea* spp.) in Canada. *Journal of Plant Disease & Protection* **104**, 182–192.

Kamala-Kannan S, Park S-M, Oh B-T, Kim H-M, Lee K-J (2010) First report of aster yellows phytoplasma in goldenrain tree (*Koelreuteria paniculata*) in Korea. *Journal of Phytopathology* **158**, 197–199.

Kaminska M, Korbin M, Komorowska B, Pulawska J (1998) Stunting and flower buds defiency of *Lilium* sp.: a new phytoplasma associated disease. *Acta Physiologiae Plantarum* **20**, 49–53.

Kaminska M, Korbin M (1999) Graft and dodder transmission of phytoplasma affecting lily to experimental hosts. *Acta Physiologiae Plantarum* **21**, 21–26.

Kaminska M, Korbin M (2000) Phytoplasma infection in *Lilium* sp. plants. *Phytopathologia Polonica* **20**, 45–57.

Kaminska M, Dziekanowska D, Rudzinska-Langwald A (2001a) Detection of phytoplasma infection in rose, with degeneration symptoms. *Journal of Phytopathology* **149**, 3–10.

Kaminska M, Sliwa H, Rudzinska-Langwald A (2001b) The association of phytoplasma with stunting, leaf necrosis and witches' broom symptoms in magnolia plants. *Journal of Phytopathology* **149**, 719–724.

Kaminska M, Sliwa H, Malinowski T, Skrzypczak CZ (2003) The association of aster yellows phytoplasma with rose dieback disease in Poland. *Journal of Phytopathology* **151**, 469–476.

Kaminska M, Sliwa H (2004) First report of phytoplasma belonging to apple proliferation group in roses in Poland. *Plant Disease* **88**, 1283.

Kaminska M, Śliwa H, Rudzińska-Langwald A (2004) First report of shoot proliferation of bleeding heart (*Dicentra spectabilis*) in Poland, associated with phytoplasma infection. *Plant Pathology* **53**, 801.

Kaminska M, Sliwa H, Malinowski T (2006) Occurrence of rose degeneration symptoms in rose plant in Poland; presumed phytoplasmal aetiology of rose proliferation and dieback. *Acta Horticulturae* **722,** 163–174.

Kaminska M, Śliwa H (2008a). First report of '*Candidatus* Phytoplasma mali' in oriental lilies and its association with leaf scorch in Poland. *Plant Pathology* **57**, 363.

Kaminska M, Śliwa H (2008b) Mixed infection of dahlia plants in Poland with apple proliferation and aster yellows phytoplasmas. *Plant Pathology* **57**, 363.

Kesumawati E, Kimata T, Uemachi T, Hosokawa M, Yazawa S (2006) Correlation of phytoplasma concentration in *Hydrangea macrophylla* with green-flowering stability. *Scientia Horticulturae* **108**, 74–78.

Khadhair AH, Hwang SF, Chang KF, Howard RJ (1997) Molecular identification of aster yellows phytoplasmas in purple coneflower and monarda based on PCR amplification and RFLP analysis of 16SrDNA sequences. *Zeitschrift für Pflanzenkrankheiten und Pflanzenschutz* **104**, 403–410.

Khasa E, Taloh G, Prabha AK, Madhupriya, Rao GP (2016) Molecular characterization of phytoplasmas of clover proliferation group associated with three ornamental plant species in India. *3 Biotech* **6**, 252.

Khew KL, Davis RE, Ong CA, Lee I-M, Su HJ, Tsai MC (1991) Detection of a Malaysian mycoplasmalike organism (MLO) and its differentiation from other Asian, European, and North American MLOs by use of cloned chromosomal and extrachromosomal MLO DNA probes. *Journal of Plant Protection in the Tropics* **8**, 167–180.

Khurana SMP, Singh V, Nagaich BB (1981) Floral abnormalities in periwinkle and calendula induced by potato yellows pathogens. *Indian Journal of Horticulture* **38**, 282–284.

Kitajima EW (1994) Enfermidades de plantas associadas a organismos do tipo micoplasma. *Annual Review of Plant Pathology* **2**, 153–174.

Kitajima EW, Costa AS (1979) Mycoplasma-like organisms associated with yellow-type diseases in cultivated plants and ornamentals in São Paulo state and the Federal district. *Fitopatologia Brasileira* **4**, 317–327.

Kitajima EW, Ribeiro RLD, Lin MR, Ribeiro MISD, Kimura O, Costa CL, Pimentel JP (1984) Lista comentada de vírus e organismos do tipo micoplasma em plantas cultivadas esilvestres do Estado do Rio de Janeiro. *Fitopatologia Brasileira* **9**, 607–625.

Kumar S, Byadgi AS (2012) Molecular detection and characterization of phytoplasma associated with periwinkle. *Vegetos* **25**, 127–135.

Kumar S, Singh V, Lakhanpaul S (2012a) Molecular detection and phylogeny of phytoplasmas affecting cycads in India. *New Disease Reports* **26**, 27.

Kumar S, Singh V, Lakhanpaul S (2012b) First report of *Mirabilis* and *Chrysanthemum* little leaf associated with '*Candidatus* Phytoplasma aurantifolia' in India. *Australasian Plant Disease Notes* **7**, 71–73.

Jones P, Arocha Y (2006) A natural infection of Hebe is associated with an isolate of '*Candidatus* Phytoplasma asteris' causing a yellowing and little leaf disease in the UK. *Plant Pathology* **55**, 821.

Lee I-M, Klopmeyer M, Bartoszyk IM, Gundrrse-Rindal DE, Chou TS, Thomson KL, Eisenreich R (1997) Phytoplasma induced free-branching in commercial poinsettia cultivars. *Nature Biotechnology* **15**, 178–182.

Lee I-M, Gundersen-Rindal DE, Bertaccini A (1998) Phytoplasma: ecology and genomic diversity. *Phytopathology* **88**, 1359–1366.

Lee I-M, Bottner KD, Dally EL, Davis RE (2008) First report of purple coneflower phyllody associated with a 16SrI-B phytoplasma in Maryland. *Plant Disease* **92**, 654.

Lee JT, Kim EH, Yea MC, Kwon OY (1996) Detection of phytoplasmas from *Paulownia tomentosa*, *Syringa vulgaris* and *Solidago virgaurea* var. *gigantea* using polymerase chain reaction (PCR) and their relationships. *Korean Journal of Plant Pathology* **12**, 191–196.

Lee JT (2004) Phytoplasmal diseases in Korea. National Institute of Agricultural Science and Technology, Suwan, Korea, 226 pp.

Leyva-Lopez NE, Aguilar-Rojas OI, Leal-Klevezas DS, Martinez-Soriano JP (1999) Presence of phytoplasmas in Mexican cacti. *Phytopathology* **89**, S45.

Li Z, Wu Z, Liu H, Hao X, Zhang C, Wu H-F (2010) *Spiraea salicifolia*: a new plant host of '*Candidatus* Phytoplasma ziziphi'-related phytoplasma. *Journal of General Plant Pathology* **76**, 299–301.

Li Z, Zhang L, Liu P, Bai Y, Wu Y (2011) First report of an aster yellows phytoplasma as the cause of rose balsam phyllody in China. *Journal of Phytopathology* **159**, 799–801.

Li ZY, Hao ZM, Dong JG, Wu D, Cao ZY (2014) First report of elm yellows subgroup 16SrV-B phytoplasma as the cause of rose balsam phyllody in China. *Disease Notes* **98**, 565.

Lockhart B, Mollov D, Voth-Hulshout J (2012) Association of spirea stunt phytoplasma with a disease of *Spiraea* spp. in Minnesota. *Plant Health Progress* https://www.plantmanagement-network.org/pub/php/brief/2012/spirea/

Louro D, Vibio M, Bertaccini A (1996) Deteccao e caracterizacao molecular de fitoplasmas associados a doencas de culturas ornamentais em Portugal. 1 Meeting of the Portuguese Society of Plant Pathology, Vila Real, Portugal, 166–168.

Mafia RG, Barreto RW, Vanetti CA, Hodgetts J, Dickinson M, Alfenas AC (2008) A phytoplasma associated with witches' broom disease of *Tabebuia pentaphylla* in Brazil. *Plant Pathology* **57**, 365.

Mancini E, Marcone C, De Feo V, Senatore F, Formisano C (2010) Changes in the composition of volatile compounds of *Spartium junceum* induced by the phytoplasmal disease, *Spartium* witches' broom. *Plant Biosystems* **144**, 568–572.

Marcone C, Ragozzino A, Schneider B, Lauer U, Smart CD, Seemüller E (1996) Genetic characterization and classification of two phytoplasmas associated with *Spartium* witches' broom disease. *Plant Diseases* **80**, 365–371.

Marcone C, Ragozzino A, Seemüller E (1997a) Witches' broom of *Sarothamnus scoparius*: a new disease associated with a phytoplasma related to the spartium witches' broom agent. *Journal of Phytopathology* **145**, 159–161.

Marcone C, Ragozzino A, Seemüller E (1997b) Detection and identification of phytoplasmas in yellows-diseased weeds in Italy. *Plant Pathology* **46**, 530–537.

Marcone C, Ragozzino A, Camele I, Rana GL, Seemüller E (2001) Updating and extending genetic characterization and classification of phytoplasmas from wild and cultivated plants in southern Italy. *Journal of Plant Pathology* **83**, 133–138.

Marcone C, Gibb KS, Streten C, Schneider B (2004) '*Candidatus* Phytoplasma spartii', '*Candidatus* Phytoplasma rhamni' and '*Candidatus* Phytoplasma allocasuarinae', respectively associated with spartium witches' broom, buckthorn witches' broom and allocasuarina yellows diseases. *International Journal of Systematic and Evolutionary Microbiology* **54**, 1025–1029.

Marzachì C, Alma M, d'Aquilio M, Minuto G, Boccardo G (1999) Detection and identification of phytoplasmas infecting cultivated and wild plants in Liguria (Italian Riviera). *Journal of Plant Pathology* **2**, 127–136.

Madhupriya, Rao GP, Khurana SMP (2013a) '*Candidatus* Phytoplasma asteris' association with leaf yellows and witches' broom symptoms of *Brachyscome* species in India. *Phytopathogenic Mollicutes* **3**, 91–94.

Madhupriya, Rao GP, Khurana SMP (2013b) Association of pigeon pea witches' broom phytoplasma (16SrIX) infecting *Phlox drummondii* in India. *New Disease Reports* **27**, 15.

Madhupriya, Rao GP, Khurana SMP (2014) Association of '*Candidatus* Phytoplasma asteris' (16SrI group) with flattened stem and witches' broom symptoms of *Petunia hybrida* in India. *Indian Phytopathology* **67**, 106–108.

Madhupriya, Rao GP, Khurana SMP (2015) Rice yellow dwarf phytoplasma (16SrXI-B subgroup) infecting *Jasminum sambac* in India. *Phytoparasitica* **43**, 77–80.

Madhupriya, Tiwari AK, Rao GP (2016) First molecular identification of '*Candidatus* Phytoplasma trifolii' (16SrVI-C) in *Croton bonplandianum* from India. *Journal of Plant Pathology* **98**, 178.

Madhupriya, Banyal N, Dantuluri RVS, Manimekalai R, Rao GP, Khurana SMP (2017) Association of different groups of phytoplasma in flower malformation, phyllody, foliar yellowing, and little leaf disease of rose (*Rosa* sp.). *The Journal of Horticultural Science & Biotechnology* **94**, 424-431.

Magie RO, Smith FF, Brierly P (1952) Occurrence of western aster yellows virus infection in gladioulus in eastern United States. *Plant Disease Reporter* **36**, 468–470.

Maurya R, Madhupriya, Chaturvedi Y, Upadhaya PP, Rao GP (2014) Association of '*Candidatus* Phytoplasma asteris' (16SrI group) with chlorosis and little leaf disease of *Streblus asper*. *Medicinal Plants* **6**, 68–70.

Mc Coy RE, Thomas DL (1980) Periwinkle witches' broom on South Florida. *Proceedings of the Florida State Horticultural Society* **93**, 179–181.

Mertelik J, Kloudova K, Vanc P, Mokra V, Sediva J (2004) First detection of phytoplasmas in rhododendron in the Czech Republic. *Plant Disease* **88**, 906.

Min H, Hu SB, Li ZN, Wu YF, Zhang CP, Wei T (2009) A phytoplasma associated with an outbreak of an unusual disease of chrysanthemum in China in 2008. *Plant Disease* **93**, 840.

Misra S, Sharma AK, Cousin MT (1985) Phyllody disease of *Phlox drummondii* Hook in Rajasthan, India, associated with mycoplasma-like organism-study on thin-thick sections. *International Journal of Tropical Plant Diseases* **3**, 7–13.

Montano HG, Davis RE, Dally EL, Hogenhout S, Pimentel JP, Brioso PS (2001) '*Candidatus* Phytoplasma brasiliense', a new phytoplasma taxon associated with hibiscus witches' broom disease. *International Journal of Systematic and Evolutionary Microbiology* **51**, 1109–1118.

Montano HG, Contaldo N, David TVA, Silva IB, Paltrinieri S, Bertaccini A (2011) Hibiscus witches' broom disease associated with different phytoplasma taxa in Brazil. *Bulletin of Insectology* **64**(Supplement), S249–S250.

Moreno P, Llacer G, Medina V (1985) Description and comparison of several yellow disease on *Vinca rosea*. *Agricola* **28**, 287–310.

Naito T, Tanaka M, Taba S, Toyosata T, Oshiro A, Takaesu K, Hokama K, Usugi T, Kawano S (2007) Occurrence of *Chrysanthemum* virescence caused by '*Ca.* P. aurantifolia' in Okinawa. *Journal of General Plant Pathology* **73**, 139–141.

Navalinskiene M, Samuitiene M (2004) Identification of *Tobacco mosaic tobamovirus* and 16SrI-M phytoplasma sub group in *Phlox drummondii* plants. *Botanica Lithuanica* **10**, 223–231.

Navalinskiene M, Samuitienė M, Jomantienė R (2005) Molecular detection and characterization of phytoplasma infecting *Callistephus chinensis* plants in Lithuania. *Phytopathologia Polonica* **35**, 109–112.

Nejat N, Vadamalai G, Davis RE, Harrison NA, Sijam K, Dickinson M, Abdullah SN and ZhaoY (2013) '*Candidatus* Phytoplasma malaysianum', a novel taxon associated with virescence and phyllody of Madagascar periwinkle (*Catharanthus roseus*). *International Journal of Systematic and Evolutinary Microbiology* **63**, 540–548.

Nour MA (1962) Witches' broom and phyllody in some plants in Khatoum province Sudan. *FAO Plant Protection Bulletin* **10**, 49.

Ogilvic L, Guterman CEF (1929) A mosaic disease of the aster lily. *Phytopathology* **19**, 311–316.

Okuda S (1977) Studies on the causal agents of yellows or witches' broom diseases of plants. *Special Bulletin of the College of Agriculture Utsunomiya University* **32**, 70.

Omar AF, Emeran AA, Abass JM (2008) Detection of phytoplasma associated with periwinkle virescence in Egypt. *Plant Pathology Journal* **7**, 92–97.

Omar AF, Dewir YH, El-Mahrouk ME (2014) Molecular identification of phytoplasmas in fasciated cacti and succulent species and associated hormonal perturbation. *Journal of Plant Interactions* **9**, 632–639.

Paltrinieri S, Contaldo N, Bertaccini A, Bellardi MG, Cavicchi L (2015) First report of phytoplasmas associated with *Erysimum linifolium* L. stunting. *Acta Horticulturae* **1072**, 117–121.

Parrella G, Paltrinieri S, Botti S, Bertaccini A (2008) Molecular identification of phytoplasmas from virescent ranunculus plants and from leafhoppers in southern Italian crops. *Journal of Plant Pathology* **90**, 537–543.

Parrella G, Paltrinieri S, Contaldo N, Vitale MR, Bertaccini A (2014) Characterization of '*Candidatus* Phytoplasma asteris' strains associated with periwinkle virescence in Southern Italy. *Phytopathogenic Mollicutes* **4**, 53–58.

Pavlovic S, Ivanovic Z, Stojanovic S, Starovic M, Josic D, Martini M (2010) Identification of phytoplasma of 16SrXIIA group infecting two *Echinacea* species in Serbia. In: Current status and perspectives of phytoplasma disease research and management. Eds Bertaccini A, Lavina A, Torres E. COST FA0807, Sitges, Spain, 32.

Pavlovic S, Starovic M, Stojanovic S, Aleksic G, Zdravkovic M, Josic D (2014) The first report of "stolbur" phytoplasma associated with phyllody of *Calendula officinalis* in Serbia. *Plant Disease* **98**, 1152.

Perilla-Henao ML, Dickinson M, Franco-Lara L (2012) First report of '*Ca.* P. asteris' affecting woody hosts (*Fraxinus uhdei*, *Populus nigra*, *Pittosporum undulatum*, and *Croton spp.*) in Colombia. *Plant Disease* **96**, 1372.

Pettersson ML, Tomenius K (1979) Mycoplasma-like organisms in marguerite, *Chrysanthemum frutescens* hybr. *Vaxtskyddsnotiser* **43940**, 95–99.

Pisi A, Bellardi MG, Rizzi E, Scala C, Filippini G (1990) Ultrastructural study of *Hydrangea* leaf tissue infected by mycoplasma-like organisms. *Phytopathologia Mediterranea* **29**, 168–174.

Ploaie PG, Petre Z, Ionica M (1977) Identification of some plant diseases with symptoms of proliferation, chlorosis and dwarfing associated with mycoplasma in Romania. *Analele Institutulve de Cercetari pentree Protectia Plantelor* **12**, 17–25.

Ploaie PG, Nicolaescu M, Ionica M, Avram M (1981) Chlorosis and virescence of gladiolus, a new disease caused by organisms of the class Mollicutes-mycoplasmatales. *Analele Institutului de Cercetari pentru Protectia Plantelor* **16**, 35–39.

Poghosyan A, Hernandez-Gonzalez J, Lebsky V (2015) Phytoplasmas associated with naturally infected *Catharanthus roseus* in La Paz, Baja California Sur, Mexico. *Phytopathogenic Mollicutes* **5**(1-Supplement), S25–S26.

Poncarova-Vorackova Z, Franova J, Valova P, Mertelik J, Navratil M, Nebesarova J (1998) Identification of phytoplasma infecting *Lilium martagon* in the Czech Republic. *Journal of Phytopathology* **146**, 609–612.

Radisek S, Ferant N, Jakse J, Javornik B (2008) Identification of a phytoplasma from the aster yellows group infecting purple coneflower (*Echinacea purpurea*) in Slovenia. *New Disease Reports* **18**, 13.

Raj SK, Khan MS, Snehi SK (2007a) Association of '*Candidatus* Phytoplasma asteris' with little leaf disease of desert rose. *Plant Pathology* **56**, 1040.

Raj SK, Khan MS, Kumar S (2007b) Molecular identification of '*Candidatus* Phytoplasma asteris' associated with little leaf disease of *Chrysanthemum morifolium*. *Plant Pathology* **2**, 21–22.

Raj SK, Snehi SK, Kumar S, Banerji BK, Dwivedi AK, Roy RK, Goel AK (2009) First report of '*Candidatus* Phytoplasma asteris' (16SrI group) associated with colour-breaking and malformation of floral spikes of gladiolus in India. *Plant Pathology* **19**, 23.

Raj SK, Snehi SK, Khan MS, Singh M, Chaturvedi Y, Tiwari AK, Rao GP (2011) Molecular detection and identification of phytoplasmas associated with little leaf and witches' broom disease of marigold (*Tagetes erecta* L.) in India. *Phytopathogenic Mollicutes* **1**, 41–46.

Rani A, Misra P, Singh J, Kumar P, Rani R, Shukla P (2014) PCR based detection of phytoplasma association in pot marigold (*Calendula officinalis* L.) and guldawari (*Dendranthema grandiflora* L.). *Asian Journal of Biological Science* **l**, 238–241.

Rao BLS, Husain A, Manohar SK (1983) Association of mycoplasma-like bodies with little leaf disease of periwinkle (*Catharanthus roseus* G.) in India. *Indian Journal of Plant Pathology* **1**, 122–124.

Rao GP, Tiwari AK, Singh M, Chaturvedi Y, Madhupriya (2012) Molecular characterization and phylogenetic relationships of an isolate of '*Candidatus* Phytoplasma asteris' affecting *Zinnia elegans* in India. *Phytopathogenic Mollicutes* **2**, 33–36.

Ribeiro CFL, Amaral Mello APO, Bedendo IP, Gioria R (2006) Phytoplasma associated with shoot proliferation in begonia. *Piracicaba Brazil* **63**, 475–477.

Rocha DA, Okhi ST, Hiruki C (1986) Detection of mycoplasma like organisms *in situ* by indirect immunofluorescence microscopy. *Phytopathology* **76**, 864–868.

Rojas-Martínez RI, Zavaleta-Mejía E, Martínez-Soriano JP, Lee I-M (2003a) Identification of the phytoplasma associated with cosmos (*Cosmos bipinnatus* Cav.) phyllody and classification by RLFP analysis of 16S rDNA. *Revista Mexicana de Fitopatologia* **21**, 83–86.
Rojas-Martínez RI, Zavaleta-Mejia E, Lee I-M, Martini M, Aspiros HS (2003b) Detection and characterization of the phytoplasma associated with marigold phyllody in Mexico. *Journal of Plant Pathology* **85**, 81–86.
Rudzinska-Langwald A, Kaminska M (2003) Changes of ultrastructure and cytoplasmic free calcium in *Gladiolus* x hybridus van Houtte roots infected by aster yellows phytoplasma. *Acta Societatis Botanicorum Poloniae* **72**, 269–282.
Salehi M, Esmailzadeh Hosseini SA, Salehi E (2015) Characterisation of a phytoplasma associated with sunflower phyllody in Fars, Isfahan and Yazd provinces of Iran. *New Disease Reports* **31**, 6.
Samad A, Ajaykumar PV, Shasany AK, Gupta MK, Alam M, Rastogi S (2008) Occurrence of a clover proliferation (16SrVI) group phytoplasma associated with little leaf disease of *Portulaca grandiflora* in India. *Plant Disease* **92**, 832.
Samuitiene M, Navalinskiene M, Jomantiene R, Davis RE (2004) Molecular characterization of phytoplasmas infecting columbine (*Aquilegia* sp. L.) plants. *Biologija* **2**, 15–17.
Samuitiene M, Navalinskiene M (2006) Molecular detection and characterization of phytoplasma infecting *Celosia argentea* L. plants in Lithuania. *Agronomy Research* **4**, 345–348.
Satta E, Contaldo N, Mejia JF, Paltrinieri S, Bertaccini A, Bellardi MG (2013) Characterization on six genes of '*Candidatus* Phytoplasma asteris'- related phytoplasmas infecting cyclamen. *Phytopathogenic Mollicutes* **3**, 72–76.
Sawayanagi T, Horikoshi N, Kanheira T, Shinhoara M, Bertaccini A, Cousin MT, Hiruki C, Namba S (1999) '*Candidatus* Phytoplasma japonicum', a new phytoplasma taxon associated with Japanese hydrangea phyllody. *International Journal of Systematic Bacteriology* **49**, 1275–1285.
Schmitt U, Peterzold H, Marwitz R (1983) On the pleomorphism of MLO in *Catharanthus roseus* (*Vinca rosea*). *Phytopathology Zeitschrift* **108**, 314–326.
Schmitt U, Peterzold H, Martwitz R (1987) Bud like structure of mycoplasma-like organisms (MLO) demonstrated by freeze-fracturing. *Naturwissenschaften* **74**, 396–397.
Seemüller E, Marcone C, Lauer U, Ragozzino A, Göschl M (1998) Current status of molecular classification of the phytoplasmas. *Journal of Plant Pathology* **80**, 3–26.
Shaw ME, Kirkpatrick BC, Golino DA (1993) The beet leafhopper-transmitted virescence agent causes tomato big bud disease in California. *Plant Disease* **77**, 290–295.
Shiomi T, Sugiura M (1983) Water dropwort yellows and *Chrysanthemum* witches' broom occurred in Ishikawa prefecture. *Annals of Phytopathology Society Japan* **49**, 367–370.
Shishlova ZHN, Andreeva LG (1978) Some biological features and diseases of *Catharanthus roseus* in geographic investigations in the Rostov Botanic Garden. *Referativnyi Zhurnal* **3**, 108–110.
Shukla K, Madhupriya, Dubey D, Upadhyaya PP, Rao GP (2014) Association of pigeon pea witches' broom phytoplasma infecting *Carpobrotus edulis* (L.) N.E.Br in India. *Phytopathogenic Mollicutes* **4**, 27–32.
Shukla K, Madhupriya, Upadhyaya PP, Rao GP (2015) '*Candidatus* Phytoplasma asteris' association with little leaf and leaf yellows symptoms of *Viola tricolor* in India. *Indian Phytopathology* **68**, 3.
Sichani FV, Bahar M, Zirak L (2014) Characterization of phytoplasmas related to aster yellows group infecting annual plants in Iran, based on the studies of 16S rRNA and rp genes. *Journal of Plant Protection Research* **54**, 1–8.
Siddique ABM (2006) Phytoplasma associated with gerbera phyllody in Australia. *Jounal of Phytopathology* **154**, 730–732.
Signoret PA, Louis C, Alliot B (1976) Mycoplasma-like organisms associated with sunflower phyllody in France. *Phytopathologische Zeitschrift* **86**, 186–189.

Singh M, Chaturvedi Y, Tiwari AK, Rao GP, Snehi SK, Raj SK, Khan MS (2011) Diversity among phytoplasmas infecting ornamental plants grown in India. *Bulletin of Insectology* **64**(Supplement), S69–S70.

Sinha RC, Benhamou N (1983) Detection of mycoplasma-like organism antigens from aster yellows-diseased plants by two serological procedures. *Phytopathology* **73**, 1199–1202.

Slack SA, Traylor JA, Williams HE, Nyland G (1976a) Rose leaf curl, a distinct component of a disease complex which resembles rose wilt. *Plant Disease Reporter* **60**, 178–182.

Slack SA, Traylor JA, Nyland G, Williams HE (1976b) Symptoms, indexing and transmission of rose spring dwarf disease. *Plant Disease Reporter* **60**, 183–187.

Sobolev I, Weintrab PG, Gera A, Tam Y, Spiegel S (2007) Phytoplasma infection in the four o'clock flower (*Mirabilis jalapa*). *Bulletin of Insectology* **60**, 281–282.

Spanò R, Marzachì C, Mascia T, Bubici G, Gallitelli D (2011) First report of 'Candidatus Phytoplasma asteris' from *Gerbera jamesonii* in Apulia, southern Italy. *Journal of Plant Pathology* **93**, S4.63.

Stanosz GR, Heimann MF, Lee I-M (1997) Purple coneflower is a host of the aster yellows phytoplasma. *Plant Disease* **81**, 424.

Takinami Y, Maejima K, Takahashi A, Keima T, Shiraishi T, Okano Y, Komatsu K, Oshima K, Namba S (2013) First report of '*Candidatus* Phytoplasma asteris' infecting hydrangea showing phyllody in Japan. *Journal of General Plant Pathology* **79**, 209–213.

Tanne E, Kuznetsova L, Cohen J, Alexandrova S, Gera A (2000) Phytoplasmas as causal agents of *Celosia* disease in Israel. *HortScience* **35**, 1103–1106.

Tazehkand S, Akbar Hosseini P, Heydarnejad J, Varsani A, Massumi H (2010) Identification of phytoplasmsa associated with cultivated and ornamental plants in Kerman province, Iran. *Journal of Phytopathology* **158**, 713–720.

Tessitori M, Masenga V, Marzachì C (2006) First report of phytoplasma associated with abnormal proliferation of cladodes in cactus pear (*Opuntia ficus-indica*) in Italy. *Plant Pathology* **55**, 292.

Thomas BJ (1979) Some degeneration and dieback diseases of the rose. Annual Report -Glasshouse Crops Research Institute, 178–190.

Tipping PW, Sindermann AB (2000) Natural and augmented spread of rose rosette disease of multiflora rose in Maryland. *Plant Disease* **12**, 1344.

Tiwari AK, Shukla K, Kumar S, Madhupriya, Rao GP (2014) '*Candidatus* Phytoplasma asteris' association with leaf yellows and witches' broom symptoms of *Codiaeum variegatum* in India. *Indian Journal of Plant Pathology* **96**, S4.114.

Tolu G, Botti S, Garau R, Prota VA, Sechi A, Prota U, Bertaccini A (2006) Identification of 16SrII-E phytoplasmas in *Calendula arvensis*, *Solanum nigrum*, and *Chenopodium* spp. *Plant Disease* **90**, 325–330.

Torres E, Botti S, Paltrinieri S, Martin MP, Bertaccini A (2002) First report of *Spartium* witches' broom disease in Spain. *Plant Pathology* **51**, 807.

Torres L, Galdeano E, Docampo D, Conci L (2004) Characterization of aster yellows phytoplasma associated with *Catharanthus* little leaf in Argentina. *Journal of Plant Pathology* **86**, 209–214.

Valiunas D, Samuitiene M, Rasomavicius V, Staniulis J, Davis RE (2007) Subgroup 16SrIII-F phytoplasma strains in an invasive plant, *Heracleum sosnowskyi*, and an ornamental, *Dictamnus albus*. *Journal of Plant Pathology* **88**, 137–140.

Verhoyen M, Genot M, Colin J, Horvat F (1979) Chrysanthemum yellows etiology. *Phytopathologische Zeitschrift* **96**, 59–64.

Vibio M, Bertaccini A, Schilirò E, Rapetti S (1995) Caratterizzazione molecolare di fitoplasmi associati a virescenza in anemone. *Informatore fitopatologico* **11**, 60–62.

Vicente M, Caner J, July JR (1974) Corpúsculos do tipo micoplasma em *Hibiscus rosa sinensis*. *Arquivos do Instituto Biológico Saho Paulo* **41**, 53–58.

Wang K, Hiruki C (2005) Distinctions between phytoplasmas at the subgroup level detected by heteroduplex mobility assay. *Plant Pathology* **54**, 625–633.

Win NKK, Kim Y-H, Chung H, Jung H-Y (2011) Detection of 16SrII group phytoplasma in China aster (*Callistephus chinensis*). *Tropical Plant Pathology* **36**, 186–189.

Win NKK, Back CG, Kim YH, Jung HY (2012) Desert rose witches' broom disease associated with '*Candidatus* Phytoplasma aurantifolia'. *Journal of General Plant Pathology* **78**, 73–76.

Zajak Z (1979) Mycoplasma-like organisms in phloem of (*Phlox paniculata* L.) with symptoms of flower greening. *Referativny Zhurnal* **3**, 79–106.

Zak LF, Yáñz-Morales MJ, Alanis-Martínez I, González-Pérez E (2011) New hosts of 16SrI phytoplasma group associated with edible *Opuntia ficus-indica* crop and its pests in Mexico. *African Journal of Microbiology Research* **5**, 910–918.

Zavaleta-Mejia E, Cardenas SE, Gomez RO (1993) Organismos tipo micoplasma asociados con la filodia del cempazuchil. XX Congreso Nacional de Fitopatologia, Zacatecas, Mexico, 15.

Zhang CP, Wu KK, Li ZN, Zhang J, Wang WL, Wu YF (2009) Occurrence of an aster yellows (16SrI) group phytoplasma associated with a leaf roll disease of shiny leaf yellow horn in China. *Plant Pathology* **58**, 790.

Zhong B-X, Shen Y-W (2004) Accumulation of pathogenesis-related type-5 like proteins in phytoplasma infected garland chrysanthemum *Chrysanthemum coronarium*. *Acta Biochimica Biophysica Sinica (Shanghai)* **36**, 773–779.

Chapter 8
Phytoplasma Diseases of Medicinal Crops

Govind Pratap Rao, Carmine Marcone, Maria Grazia Bellardi, and Madhupriya

Abstract Phytoplasma diseases of medicinal plants occur worldwide and are of great concern. So far 19 different phytoplasma ribosomal groups encompassing various subgroups have been reported. The subgroup 16SrI-B phytoplasmas are the prevalent agents mainly detected in Europe, North America and Asia. Phytoplasma diseases of medicinal plants severely reduce yield and quality of crops along with the longevity of the plants. Changes in the composition of secondary metabolites are induced, while the levels of valuable phytochemicals are greatly affected. In contrast, an accumulation of pharmaceutically important compounds such as vinblastine and vincristine is reported in periwinkle upon phytoplasma infections. Important phytoplasma diseases of several medicinal plants with special reference to their impact on active biological constituents and secondary metabolites are reviewed. General information on geographic distribution, diagnostics, genetic diversity, natural transmission and management aspects of phytoplasmas infecting medicinal plants are also discussed.

Keywords Medicinal species · Phytoplasmas · Symptomatology · Diagnosis · Diversity · Secondary metabolites

8.1 Introduction

Medicinal plants are used in official and traditional medicine because they contain many biologically active compounds, known as phytochemicals, which are able to provide human and animal health benefits. Because of their increasing application

G. P. Rao (✉) · Madhupriya
Division of Plant Pathology, Advance Center for Plant Virology, Indian Agricultural Research Institute, New Delhi 110012, India
e-mail: gprao_gor@rediffmail.com

C. Marcone (✉)
Department of Pharmacy, University of Salerno, Fisciano (Salerno), Italy

M. G. Bellardi
Department of Agricultural and Food Sciences, *Alma Mater Studiorum* – University of Bologna, Bologna, Italy

G. P. Rao et al. (eds.), *Phytoplasmas: Plant Pathogenic Bacteria - I*,
https://doi.org/10.1007/978-981-13-0119-3_8

in the pharmaceutical and cosmetic industries, medicinal plants are currently recognized as an important element of global economy. The diseases and pests affecting medicinal crops are increasing at appreciable levels (Carrubba et al. 2015). Yellows, witches' broom and decline diseases associated with phytoplasma infections severely impair productivity, phytochemical content and longevity of the affected plants. Phytoplasmas are wall-less bacterial plant pathogens of the class *Mollicutes* that colonize the phloem sieve tubes of diseased plants and are transmitted by phloem-feeding homopteran insects (Weintraub and Beanland 2006). The main effect of phytoplasma infection is the impairment of the sieve tube functions. Inhibition of phloem transport along with other impaired physiological functions including reduced photosynthesis, altered secondary metabolism and disturbed plant hormone balance could account for symptoms exhibited by phytoplasma-infected plants (Marcone 2014). Phytoplasmas are currently classified within the provisional genus '*Candidatus* Phytoplasma' based on 16S rDNA sequence analysis (IRPCM 2004; Martini et al. 2014). Approximately 33 major phylogenetic groups were identified. This figure is broadly in accordance with the number of phytoplasma groups established by restriction fragment length polymorphism (RFLP) analysis of PCR-amplified rDNA. Within the majority of the phytoplasma groups, several distinct subgroups have been delineated, based on the RFLP analysis of 16S rDNA sequences (Bertaccini et al. 2014).

Phytoplasma diseases of medicinal plants occur worldwide; however, the majority of the reports are from Europe and south eastern Asian countries. Many of these diseases either were previously of unknown aetiologies or mistakenly presumed to be induced by virus. Phytoplasma diseases of medicinal plants affect plant species belonging to over 70 botanical families, mostly Apiaceae and Asteraceae (Marcone et al. 2016). They differ considerably in geographic distribution and belong to diverse phytoplasma ribosomal groups and subgroups. The 16SrI-B phytoplasmas are the prevalent subgroup occurring mainly in Europe, North America and Asia. Phytoplasmas of the 16SrXII-A subgroup with a few exceptions, e.g. *Artemisia scoparia* witches' broom and *Salvia miltiorrhiza* red leaf agents, both occur in China, as well as phytoplasmas of the 16SrX and 16SrXX groups are restricted to Europe, whereas phytoplasmas of the 16SrII group are prevalent in Asia, Australia and South Africa. Also, phytoplasmas of the 16SrIV, 16SrXIII, 16SrXV, 16SrXXIX, 16SrXXX and 16SrXXXII groups have only been identified in diseased medicinal plants in Asia and South America. Based on the sequences retrieved from GenBank, identified phytoplasmas mainly belong to the 16SrI, 16SrII, 16SrIII, 16SrIV, 16SrV, 16SrVI, 16SrVII, 16SrVIII, 16SrIX, 16SrX, 16SrXI, 16SrXII, 16SrXIII, 16SrXIV, 16SrXV, 16SrXX, 16SrXXIV, 16SrXXIX, 16SrXXX and 16SrXXXII groups (Marcone et al. 2016).

Like other phytoplasma diseases, a number of diseases of medicinal plants are associated with genetically different phytoplasmas, up to seven in some cases, which induce similar symptoms in a given plant host and occur either within the same geographic areas or different continents. Among these, there are yellows of dill, celery, *Cirsium*, dandelion, mallow, Chinaberry, myrtle, rose, purple coneflower and grapevine; periwinkle yellows and phyllody; and alfalfa witches' broom. The phytoplasma diseases of medicinal plants are associated either with one of more ribosomal groups or a single plant host may be singly, doubly or multiply infected with distinctly different groups of phytoplasmas. The diversity of their potential disease reservoir has been increased with the discovery of new phytoplasma host species.

8.2 Medicinal Plants

Plants infected by phytoplasmas exhibit a variety of symptoms indicative of profound disturbances in plant growth (Lee et al. 2000). A range of symptoms are induced by phytoplasmas on different medicinal plant species. The most characteristic symptoms associated with phytoplasmas in medicinal plants are virescence, phyllody, witches' broom, abnormal internodes, elongation, stunting, foliar yellowing, little leaf, phloem necrosis, crinkling, leaf burn and death of the plant (Fig. 8.1) (McCoy et al. 1989; Chaube et al. 2015; Marcone et al. 2016; Samad et al. 2006). Application of PCR to the detection of phytoplasmas associated with medicinal plant diseases has greatly facilitated the identification of a wide array of phytoplasmas in medicinal plants (Marcone and Rao 2008; Favali et al. 2008; Raj et al. 2008; Madhupriya et al. 2014; Chaube et al. 2015).

Phytoplasma diseases have been reported in approximately 200 medicinal plants worldwide, but here 35 important medicinal plant species are listed with information on distribution, identity, diversity, epidemiology and effect of phytoplasma infection on secondary metabolites.

8.3 Alfalfa (*Medicago sativa* L.)

Alfalfa (Fabaceae) is an important medicinal herb used as an antiscorbutic, aperient, diuretic, oxytocic, haemostatic, nutritive, stimulant and tonic. Its expressed juice is emetic and is also anodyne in the treatment of gravel. The plant helps in convalescence or anaemia, haemorrhage, menopausal complaints, premenstrual tension and fibroids. A poultice of the heated leaves has been applied to the ear in the treatment of earache. The leaves are rich in vitamin K. Alfalfa witches' broom has been recorded in Asian countries as well as in Europe, Americas and Australia and is associated with genetically different phytoplasmas including members of the 16SrI, 16SrII-C and 16SrII-D, 16SrIII-B, 16SrV-B, 16SrVI, 16SrVII-C and 16SrXII-A groups and subgroups (for review see Marcone et al. 2016).

Fig. 8.1 Century plant infected by a subgroup 16SrI-B phytoplasma showing symptoms of virescence and phyllody (**a**). Symptoms related to phytoplasma presence in rosemary: little leaves and typical witches' brooms are visible (**b**). Gum plant phytoplasma-infected plants showing malformed flowers and virescence (H, healthy plant) (**c**). Hyssop crop located in hilly areas of central Italy: diseased plants show stunting, extreme reduction in leaf size, rosetting, yellowing and lack of blossoms (**d**). Phytoplasma-infected purple coneflower plants showing virescence and phyllody (**e**). Wild bergamot plants show stunting, yellowing, flower malformations and virescence (**f**).

8.4 *Andrographis paniculata* (Burm. f.) Wall. ex Nees

Kalmegh, also known as Indian Echinacea or king of bitters (Acanthaceae), is an annual or biennial herbaceous plant, native to India and Sri Lanka, which is traditionally used for the treatment of a wide array of diseases in Asia, America and Africa continents. It possesses antimicrobial, antiviral, anti-inflammatory, hepatoprotective, antidiabetic, antihyperglycaemic and antioxidant properties. A yellows and witches' broom disease of kalmegh was observed in 2014 in the experimental fields of the Central Institute of Medicinal and Aromatic Plants, Lucknow, India. The disease incidence ranged from 7 to 10%. Symptoms shown by affected plants included virescence, shoot proliferation, witches' broom, little leaf and stunted growth. A subgroup 16SrII-D phytoplasma was identified in the diseased plants (Saeed et al. 2015).

8.5 Bermuda Grass [*Cynodon dactylon* (L.) Pers.]

Bermuda grass (Poaceae) is a hardy, evergreen, perennial grass with long rapid-growing, creeping runners or stolons. Native to the Mediterranean region, it is now widespread throughout warm temperate, subtropical and tropical regions. Bermuda grass has a variety of medicinal properties including antiviral and antimicrobial activities and is widely used in the traditional medicine to treat various ailments such as anasarca, cancer, convulsions, cough, cramps, diarrhoea, dropsy, epilepsy, headache, haemorrhage, hypertension, hysteria, measles, rubella, snakebite, sores, stones, tumours, urogenital disorders, warts and wounds (Nagori and Solanki 2011). This species is affected by Bermuda grass white leaf disease (BGWL) associated with the presence of '*Ca.* P. cynodontis', a member of the phytoplasma group 16SrXIV, subgroups 16SrXIV-A and 16SrXIV-C (Marcone et al. 2004a). BGWL disease is known to occur in Asia, Europe, Turkey, Sudan, Australia, and Cuba (Marcone et al. 1997a, 2004a; Arocha et al. 2005; Rao et al. 2007; Salehi et al. 2009; Çağlar et al. 2013; Khanna et al. 2015; Mitrović et al. 2015). However, a subgroup 16SrXII-A phytoplasma was also detected in BGWL-affected plants in Serbia (Mitrović et al. 2012). The most characteristic symptoms of the disease are extensive leaf whitening, proliferation of axillary shoots, bushy growing habit, shortened

Fig. 8.1 (continued) Stunting and yellowing related to phytoplasma presence in an Italian crop of *Parietaria* spp. (pellitory) (**g**). Phytoplasma-infected thyme plants showing little leaves, yellowing and reddening of the lamina and abnormal root production from the aerial part (**h**). Experimental field of St. John's worth where plants show severe dwarfing, reduction in leaf size, yellow leaves, witches' broom and lack of flower production (**i**). Woolly foxglove plants infected by a subgroup 16SrI-B phytoplasma showing stunting, virescence and red edges of leaves; flower spikes can be only partially virescent (**j**). Little leaf and witches' broom in periwinkle infected by '*Ca.* P. asteris' (**k**). Witches' broom disease of hemp associated with the presence of '*Ca.* P. trifolii' (**l**). Little leaf disease of *Tylophora asthmatica* (**m**). Witches' broom of *S. officinalis* (**n**). Leaf chlorosis in lemon grass (**o**)

stolons and rhizomes, stunting and death of the plants. The leafhopper *Exitianus capicola* has been reported as a vector of '*Ca*. P. cynodontis' in Iran (Salehi et al. 2009).

8.6 Bitter Gourd (*Momordica charantia* L.)

Bitter gourd (Cucurbitaceae) is one of the most popular vegetables with numerous medicinal properties such as stomachic and carminative and is used for diabetes treatment (Grover and Yadav 2004). Jiménez and Montano (2010) reported a phytoplasma disease of bitter gourd in Brazil: infected plants exhibited yellowing, reduced leaf size and witches' broom. The associated phytoplasma was identified as an X-disease phytoplasma subgroup 16SrIII-J. A little leaf and phyllody disease of *M. charantia* was reported in Myanmar, Indonesia and Thailand and found to be associated with 16SrI-B and 16SrII phytoplasmas (Sdoodee et al. 1999; Davis et al. 2003; Win et al. 2014).

8.7 Centaury (*Centaurium erythraea* L.)

The centaury (Gentianaceae) is a herbaceous ornamental and medicinal plant first reported as a phytoplasma host in 2016 in Italy (Paltrinieri et al. 2016). The affected plants showed virescence and phyllody; only a few flowers had a pink colour, but were strongly malformed (Figure 8.1a). The associated phytoplasma was identified by PCR assays with primers amplifying the *tuf* gene in a cocktail nested PCR reaction (Contaldo et al. 2011a). Centaury was found to be infected by the aster yellows agent '*Ca*. P. asteris', subgroup 16SrI-B (Paltrinieri et al. 2016).

8.8 *Datura* spp.

Datura inoxia Mill. (Solanaceae) is an annual invasive weed used as a medicinal plant for therapeutic purposes. *Datura* leaves and flowers are the source of a drug used for asthma treatment. Raj et al. (2009) reported little leaf disease of *D. inoxia* in India. The diseased plants exhibited proliferation of branches with shortened internodes and reduced leaf size which give rise to the little leaf appearance. Analysis of the partial 16S rRNA gene sequence of the phytoplasma identified in little leaf-affected *D. inoxia* revealed its highest identity (97%) with '*Ca*. P. trifolii' (16SrVI group).

D. stramonium L. is another species used in medicine to treat asthma, gastrointestinal problems, aches, abscesses, arthritis, boils, headaches, haemorrhoids, rattlesnake bites, sprains, swellings and tumours. *D. stramonium* contains active compounds such as hyoscine, as well as atropine, hyoscyamine, apohyoscine and meteloidine. Thus it is poisonous and hallucinogenic as well as a pain killer (Duke 1985). The 16SrVI and 16SrXII-A phytoplasmas have been detected in diseased *D. stramonium* plants showing mainly witches' broom symptoms in India, Czech Republic, Greece and Turkey (Fialová et al. 2009; Özdemir et al. 2009; Lotos et al. 2013; Singh et al. 2013; Singh and Upadhyaya 2015).

8.9 *Eucalyptus* spp.

Eucalyptus (Myrtaceae) is a fast-growing evergreen tree, native to Australia, which is now grown all over the world. The essential oils extracted from eucalyptus leaves have powerful medicinal properties such as anti-inflammatory, antimicrobial, astringent, tonic, antispasmodic, deodorant, expectorant, stimulant, rubefacient, febrifuge and hypoglycaemic. Several species of eucalyptus are affected by eucalyptus little leaf (ELL), a disease first described in India in 1971 where it was thought to be caused by a virus (Sastry et al. 1971). Later, phytoplasmas were detected in ELL-affected trees by electron and light microscopy (Nayar 1973; Nayar and Ananthapadmanabha 1977; Sharma et al. 1983; Ghosh et al. 1984). The disease has also been reported from Italy, China, Sudan, Iran and Brazil (Marcone et al. 1996b, 1997b; Camele et al. 1999; de Souza et al. 2015; Marcone 2015; Salehi et al. 2016). The symptoms include abnormally minute leaves, yellowing, shortened internodes and premature growth from axillary buds, both giving the shoots a bushy and broom-like appearance. A phytoplasma of the elm yellows or 16SrV group was detected in ELL-affected trees in Italy (Marcone et al. 1996b). Also, phytoplasmas of 16SrI group, subgroups 16SrI-B and 16SrI-C, were identified in diseased eucalyptus trees in Italy (Camele et al. 1999), whereas 16SrI-B and 16SrII phytoplasmas were identified in diseased eucalyptus trees in Iran and Brazil, respectively (de Souza et al. 2015; Salehi et al. 2016). However, the identity of phytoplasmas infecting eucalyptus in the other countries where the disease was detected has never been determined by molecular methods.

8.10 Garlic (*Allium sativum* L.)

Garlic (Liliaceae) is a potent medicinal plant, native to China, but now cultivated and used throughout the world. The most important constituents of this plant which are responsible for pharmacological effects are organosulfur compounds such as allicin, diallyl disulphide, S-allyl cysteine and diallyl trisulphide. Pharmacological effects of garlic include antifungal, antibacterial, antiprotozoal, antiviral, antihypertensive, antiatherosclerotic, antithrombotic, anticancer, anti-inflammatory and immunomodulatory activities. Phytoplasma diseases of garlic have been observed in Argentina and Canada (Conci et al. 1998; Khadhair et al. 2002). The disease occurring in Argentina was termed 'tristeza del ajo' (garlic decline). Symptoms are evident during the second half of the growth cycle. Affected plants die and have spongy bulbs with undeveloped cloves. However, if symptoms appear towards the end of the cycle, plants show only changes in the leaf colour, but bulbs seem normal. Symptomatic plants were highly colonized by phytoplasmas as shown by transmission electron microscopy (Conci et al. 1998). The phytoplasma infecting garlic in Argentina was identified as a member of the X-disease or 16SrIII group, subgroup 16SrIII-J (Conci et al. 1998; Galdeano et al. 2013). A subgroup 16SrI-A phytoplasma was detected in diseased garlic plants in Canada. Affected plants had elongated hollow stems terminating in proliferated, silky, round heads. Recently, Goel et al. (2017) reported the association of '*Ca.* P. phoenicium' (16SrIX) with a yellows and decline disease of garlic in India.

8.11 Gum Plant (*Grindelia robusta* Nutt.)

Gum plant (Asteraceae), native of California, is a medicinal herb recommended for its expectorant and sedative action. This perennial plant has a smooth, round, striate stem, much divided into ascending branches, each of which ends in a large, yellow flower head surrounded by clusters of three or four leaves. The names gum weed or gum plant come from the sticky white sap (oleo-resin) exuded from the flower buds. In spring-summer 2007, a phytoplasma disease affecting *G. robusta* was observed in the Herb Garden "Augusto Rinaldi Ceroni" of Casola Valsenio (northern Italy) (Bertaccini et al. 2008). First symptoms appeared only at blooming (in May), and an increasing percentage of symptomatic plants was found in the following month. In fact, the foliage was asymptomatic: all leaves were normal in colour and size. Almost all symptoms involved only the flowers that showed malformations and reduction of size, virescence and phyllody and partial or complete rosetting (Figure 8.1c). The phytoplasma strain infecting *G. robusta* plants was identified as '*Ca.* P. asteris', subgroup 16SrI-B. The essential oil components from healthy and infected plants were compared. The healthy plants had 50% higher concentration of monoterpenes, especially limonene and bornyl acetate compared to diseased plants (Fraternale et al. 2007; Bellardi et al. 2009).

8.12 Hemp (*Cannabis sativa* L.)

Hemp (Cannabinaceae) is an annual herbaceous plant native to Central Asia which is now grown worldwide. The flowers and to a lesser extent the leaves, stems and seeds of hemp contain psychoactive chemical compounds known as cannabinoids which are used for several purposes including medicinal ones. This plant has been used for treatment of HIV/AIDS, glaucoma, eye problems, cancer and cachexia and treatment of pain, muscle spasticity, insomnia, asthma, hypertension and depression. Moreover, it is used for cosmetic purposes. A witches' broom disease of hemp has been recorded in India (Figure 8.11) and Iran. Main symptoms included shoot proliferation, shortened internodes, little leaves and witches' brooms. Several phytoplasmas were associated with this disease in groups/subgroups 16SrI, 16SrVI-D and 16SrXIV-A (Raj et al. 2008; Sichani et al. 2014; Chaube et al. 2015; Kumar et al. 2017).

8.13 *Hibiscus rosa-sinensis* L.

Chinese hibiscus (Malvaceae) is a sweet, astringent, cooling herb that checks bleeding, soothes irritated tissues and relaxes spasms (Bown 1995). The flowers are aphrodisiac, demulcent, emmenagogue, emollient and refrigerant (Chopra et al. 1986). They are used in the treatment of excessive and painful menstruation, cystitis, venereal diseases, feverish illnesses, bronchial catarrh and coughs and to promote hair growth (Bown 1995; Chopra et al. 1986). In Brazil, witches' broom disease of *Hibiscus* was first reported in São Paulo State in plants of *H. rosa-sinensis* (Vicente et al. 1974). Later, the disease was observed, in the State of Rio de Janeiro, in plants of the same species, and they displayed similar symptoms and premature dropping of flowers (Kitajima et al. 1984; Davis 1995). In Australia, an unidentified phytoplasma has been reported to be associated with a witches' broom disease of *H. heterophyllus*, an Australian native species that is also grown commercially (Hiruki 1987). Many *Hibiscus* plants are affected by witches' broom disease in Brazil, which is characterized by excessive axillary branching, small leaves and deformed flowers. '*Ca.* P. brasiliense' (16SrXV-A) and a member of the 'stolbur' group (16SrXII) are associated with this disease in Brazil (Montano et al. 2001 2011). Chaturvedi et al. (2010) reported a yellows and little leaf disease on *Hibiscus* in India which showed symptoms of excessive yellowing, vein banding, little leaf, curling, puckering and stunting. Phylogenetic analysis of the Indian *Hibiscus* phytoplasma 16S rDNA sequences (GenBank accession numbers FJ939287 and FJ939288) showed close relationship with strains of aster yellows '*Ca.* P. asteris'.

8.14 Hyssop (*Hyssopus officinalis* L.)

Hyssop (Lamiaceae) is a herbaceous perennial herb known for the medicinal properties of its aerial parts which contains rather large amounts of bitter and antioxidative tannins (phenols, depsides of coffeic acid and several triterpenoid acids). During a survey carried out in 2002–2003 in Italy, a phytoplasma-related symptomatology was observed in a crop of *H. officinalis* located in hilly areas of the Marches-Tuscany

Apennines (central Italy). Almost 10% of the plants showed symptoms of yellowing and stunting (Figure 8.1d). Starting in May 2003, increasing percentages (30%) of hyssop plants growing in the same crop were found showing more severe symptoms of dwarfing, extreme reduction in leaf size, rosetting, yellows and lack of blossoms (Bertaccini et al. 2005a). The presence of a phytoplasma belonging to the "stolbur" group (16SrXII), subgroup 16SrXII-A, was detected. Only in a few cases (in 2002) phytoplasmas belonging to 16SrI and 16SrII groups were also detected. The infected samples used for essential oil extraction were all infected by the "stolbur" phytoplasma alone. A study was carried out to evaluate the effects of this phytoplasma on content and quality of essential oil of cultivated *H. officinalis*. With regards to essential oil extraction, due to rosette symptoms and lack of blossoms as a result of phytoplasma infection, the infected material had greater leaf content, and the essential oil obtained from healthy hyssop was approximately 1/2 less than that obtained from the infected plants. However, considering the stunting of infected hyssop, the number of infected plants necessary to reach 2 kg of material was almost three times that of the healthy ones. From GC/MS analysis, the identified components represented about 93% of the total essential oil. The results were similar in terms of qualitative pattern; however, significant differences were found in the percentage of the most characteristic components. In the essential oil from healthy plants, the carbonyl compounds pinocamphone and isopinocamphone were dominant, and they were found to be in accordance with the literature data which established levels within 5.5–17.5% of pinocamphone and 34.5–50.0% of isopinocamphone in hyssop (Salvatore et al. 1998). The essential oil from infected plants contained significantly lower levels of both these terpenic ketones in comparison to that from healthy hyssop. Conversely, that from infected hyssop was found to contain higher percentages of some components such as (E)-β-ocimene, γ-terpinene, sabinene hydrate, terpinolene, cis-thujone, germacrene-D, germacrene-B, bicyclogermacrene, caryophyllene oxide and limonene, in comparison to the essential oil obtained from the healthy plants. Finally, the most typical feature of the analysed commercial essential oil was the presence of high levels of camphor, bornyl acetate and cis-thujone (Mazzanti et al. 1997). These results suggest that phytoplasma infection could be responsible for significant alteration of the composition of essential oil as a consequence of secondary metabolism alteration in the infected plant (Bertaccini et al. 2005a).

8.15 Indian Ginseng [*Withania somnifera* (L.) Dunal]

Indian ginseng (Solanaceae) is an evergreen shrub widely cultivated as a medicinal crop in several Asian and African countries. The overall medicinal properties of Indian ginseng make it a valuable therapeutic agent for addressing anxiety, cancer, microbial infections, immunomodulation and neurodegenerative disorders. A witches' broom disease of *W. somnifera* was first observed in 1988 in the experimental field of the Central Institute of Medicinal and Aromatic Plants, Lucknow,

India (Zaim and Samad 1995). Affected plants showed symptoms of little leaf, shortened internodes, excessive branching resulting in a witches' broom appearance and death of affected twigs and leaves. The phytoplasma presence was confirmed on the basis of transmission electron microscopy observations and tetracycline treatment (Zaim and Samad 1995). In subsequent years the disease has spread to other parts of the country, reaching an incidence of 70% in some instances, and phytoplasmas of 16SrI and 16SrVI groups were identified in diseased plants (Khan et al. 2006b; Samad et al. 2006).

8.16 Lemongrass (*Cymbopogon citratus* Stapf.)

Lemongrass is a member of Poaceae family and native to warm temperate and tropical regions. It is a tall perennial grass having tremendous medicinal value in relieving cough and nasal congestion. This species is used for the production of citronella oil, which is used as an insect repellent especially against mosquitoes and in aromatherapy (Jayasinha 1999; Shah et al. 2011). Leaf chlorosis symptoms in *C. citratus* (Figure 8.1o) were recorded in 2014 in a herbal garden at the Amity University campus, Manesar, Haryana, India. The associated phytoplasma was identified as a member of 16SrII group (Madhupriya et al. 2014).

8.17 *Ocimum* sp.

Basil (Lamiaceae) plants are used in cold, cough fever, sore throat, hoarseness, earache, eye troubles, sinus, congestion, pneumonia, respiratory disorder like bronchitis and asthma, acidity, flatulence and indigestion. Basil oil obtained from *O. basilicum* is used in a variety of medicine and perfumery. It has antibacterial activity against *Staphylococcus aureus* and *Mycobacterium tuberculosis* in vitro as well as other pathogens including fungi. Arocha et al. (2006) reported a little leaf and witches' broom disease of *O. basilicum* in Cuba and identified the associated phytoplasma as belonging to 16SrI group. Rao et al. (2017) reported the association of 16SrI-B subgroup on *O. canum* plants in India.

8.18 Onion (*Allium cepa* L.)

Onion is a well-known nutraceutical and medicinal plant that is cultivated and used around the world. This plant has a wide range of beneficial properties including antiasthmatic, diuretic, antibiotic, anticarcinogenic, anticholesterol and antioxidant activities. Onion is affected by onion yellows (OY), a phytoplasma disease which was first described in 1945 in California, USA (Severin and Frazier 1945). Later, this disease has been recorded in Michigan, Texas, Canada, several European countries (France, Germany, Poland, Romania, Lithuania and Italy), Japan and Pakistan (McCoy et al. 1989; Seemüller et al. 1998; Marcone et al. 2000; Khadhair et al. 2002; Lee et al. 2003, 2004a; Jomantiene et al. 2010; Maejima et al. 2014; Ahmad et al. 2015). In Italy, OY is known to occur in all onion-growing areas, and a disease incidence of 10% was recorded in crops for seed production in Apulia (southern Italy) (Vibio et al. 1995; Marcone et al. 2000, 2016). The most characteristic symptoms of OY include virescence, phyllody, flower proliferation and other flower abnormalities, all resulting in sterility, witches' brooms, internode elongation and etiolation, shortened internodes, stunting and yellowing. OY is mainly associated with phytoplasmas of the 16SrI group, subgroups 16SrI-B, 16SrI-A, 16SrI-L and 16SrI-M (Marcone et al. 2000; Khadhair et al. 2002; Lee et al. 2003, 2004a; Jomantiene et al. 2010; Maejima et al. 2014). However, a 16SrVI group phytoplasma was also detected in a single diseased onion plant showing symptoms of stunting in Texas (Lee et al. 2003). Also, a subgroup 16SrI-B phytoplasma was recorded in diseased plants of other *Allium* species such as *A. altyncolicum*, *A. ampeloprasum* and *A. fistulosum* showing typical OY symptoms in Czech Republic, Italy and Japan. In some diseased *A. ampeloprasum* plants in Italy, the additional presence of a 16SrXII-A phytoplasma was detected (Bertaccini et al. 1999; Navrátil et al. 2000). OY is mainly spread in nature by infected plant propagating material and insect vectors. In North America and Europe, the main vectors of the 16SrI phytoplasmas including those infecting onion are polyphagous leafhoppers such as *Macrosteles* spp., *Euscelis* spp., *Scaphytopius* spp. and *Aphrodes* spp. (Lee et al. 2004a), whereas *M. striifrons* and *Hishimonus sellatus* are vectors of the OY agent in Japan (Nishimura et al. 1998). Goel et al. (2017) reported the association of '*Ca.* P. oryzae' (16SrXI) and '*Ca.* P. phoenicium' (16SrIX) causing a yellows disease of onion in India.

8.19 *Parietaria* spp.

Pellitory plants (Urticaceae) are annual or perennial herbaceous plants characterized by diuretic, laxative, refrigerant and demulcent activities. In 2002–2003, plants of pellitory (*Parietaria officinalis* L., sin. *Parietaria erecta* M. et K.; *Parietaria judaica* Auct. An. L., sin. *P. diffusa* M. et K., *P. ramiflora* Moench.) were found showing severe stunting, little leaves and yellows symptoms (Figure 8.1g) in Emilia-Romagna region (Northern Italy). Samples were collected and tested to verify the presence of phytoplasmas associated with yellows symptoms. The phytoplasma was determined to belong to the 16SrXII group, subgroup 16SrXII-A, by nested PCR/RFLP assays (Bertaccini et al. 2005b).

8.20 Periwinkle [*Catharanthus roseus* (L.) G. Don]

Periwinkle (Apocynaceae) is a medicinal plant of pharmaceutical interest for its ability to biosynthesize more than 130 alkaloids, most of which have pharmacological activities. Among these, there are terpenoid indole alkaloids, including vinblastine and vincristine, which are used for their antineoplastic activity in the treatment of many cancers, and ajmalicine and serpentine, which are used as antihypertensive and sedative, respectively (Zhou et al. 2009; Pan et al. 2016). However, terpenoid indole alkaloids are produced in periwinkle only in very small amounts and usually have some chiral centres. They are therefore difficult to synthesize chemically, making them expensive to produce on a large scale (Zhou et al. 2009). Phytoplasma infections in periwinkle cause an increase of metabolites related to the biosynthetic pathways of terpenoid indole alkaloids (loganic acid, secologanic, vindoline) and phenylpropanoids (chlorogenic acid and polyphenols) (Choi et al. 2004; Favali et al. 2004). In particular, it has been shown that as a consequence of the infection of periwinkle plants with clover phyllody phytoplasma (16SrIII-B subgroup), the content of terpenoid indole alkaloids such as vindoline, ajmalicine, serpentine, vinblastine and vincristine increased. The total alkaloid content of dry weight was 550.96 μg/g, whereas that of healthy plants was 452.48 μg/g (Favali et al. 2004). However, the level reached by each compound in diseased plants differed greatly among the plant organs. Vinblastine reached its highest concentration in the roots of infected plants (148.44 μg/g), whereas it was not detectable or present at a very low level (< 0.05 μg/g) in those of healthy plants. The content of vindoline (precursor of vinblastine) increased in diseased leaves (222.00 versus 131.90 μg/g of healthy plants) and stems (13.22 versus <0.05 μg/g of healthy plants) but decreased in diseased roots (< 0.05 versus 142.95 μg/g of healthy plants). Also, phytoplasma infections led to a significant two- to fourfold increase in chlorogenic acid in periwinkle, compared to healthy plants (Choi et al. 2004). Srivastava et al. (2014) have shown that in flowers of periwinkle plants infected with a 16SrI-B phytoplasma, the contents of vindoline, catharanthine and vincristine plus vinblastine were two-, ten- and fourfold higher, respectively, than those of healthy flowers, whereas the contents of serpentine plus ajmalicine were lower than those of healthy flowers. However, in the leaves there were no significant differences between healthy and infected plants. Also, several genes including some involved in mevalonate pathway of the terpenoid indole alkaloid biosynthesis system, namely, geranyl geranyl pyrophosphate synthase, geraniol 10 hydroxylase, desacetoxyvindoline-4-hydroxylase, strictosidine synthase, secologanin synthase and deacetylvindoline-4-O-acetyltransferase, were expressed at higher levels in flowers of diseased periwinkles than in those of healthy plants (Srivastava et al. 2014). Over the last few years, many attempts have been made to boost the yield of terpenoid indole alkaloids in periwinkle. Despite the progress made, they still cannot be successfully produced on an industrial scale, owing to several biological and technological constraints. The various means by which terpenoid indole alkaloid accumulation can be increased include biotic elicitors such as phytoplasmas (Zhou et al. 2009). Therefore, phytoplasma infections in periwinkle, in spite of the detrimental effects on plant host, could be considered as beneficial from the pharmaceutical point of view. Genetically different phytoplasmas belonging to at least eight 16Sr groups (16SrI, 16SrII, 16SrIII, 16SrVI, 16SrIX, 16SrXII, 16SrXIII and 16SrXXXII) have been reported to occur in naturally infected periwinkle plants in several areas from Europe, Americas and Asia. Among these, there are two recently described '*Candidatus* Phytoplasma' taxa, namely, '*Ca.* P. malasianum' (subgroup 16SrXXXII-A) and '*Ca.* P. hispanicum' (subgroup 16SrXIII-A), associated with Malaysian and Mexican periwinkle virescence diseases, respectively (Seemüller et al. 1998; Lee et al. 2000, 2004a, 2012; Torres et al. 2004; Chaturvedi et al. 2009; Ayman et al. 2010; Chen et al. 2011; Kumar and Byadgi 2012; Galdeano et al. 2013; Mitrović et al. 2013; Nejat et al. 2013; Caicedo et al. 2015; Khanna et al. 2015; Pérez-López et al. 2016; Davis et al. 2016; Marcone et al. 2016; Madhupriya 2016). Naturally infected periwinkle plants showed symptoms of yellowing, stunting, little leaf (Figure 8.1k), virescence, phyllody and witches' broom. Also, many phytoplasmas have been experimentally transmitted from naturally infected, field-grown plants of various species to periwinkle in which they are routinely maintained by periodic grafting (Dickinson et al. 2013) or by micropropagation (Bertaccini et al. 1992).

8.21 *Plantago* spp.

Fleaworts or plantain (Plantaginaceae) is a low-growing perennial plant, native to Europe and temperate parts of Asia, which is distributed in all temperate regions of the world. It has astringent, antibacterial, anti-inflammatory, expectorant, haemostatic, ophthalmic and laxative effects. Subgroup 16SrI-B and 16SrXII-A phytoplasmas were detected in diseased plants of *P. lanceolata* in Czech Republic and Italy (Fránová and Šimková 2009; Marcone 2011; Mori et al. 2015) and of *P. major* in Germany, Italy, Serbia and Hawaii (USA) (Marcone et al. 2000; Lee et al. 2004a; Bellardi and Bertaccini 2005; Borth et al. 2006; Josic et al. 2012; Mori et al. 2015), whereas a subgroup 16SrI-A phytoplasma was identified in diseased *P. coronopus* plants in Germany (Marcone et al. 2000; Lee et al. 2004a). Affected *P. lanceolata* plants showed symptoms of reduced growth, narrow and chlorotic leaves, necrosis and flower abnormalities. Symptoms of reddening, purpling, little leaf, crinkling and flower abnormalities were observed on diseased *P. major* plants. On this species, a disease incidence of 35% was recorded in a commercial field located in Pancevo, Serbia (Josic et al. 2012).

8.22 Port Royal Senna [*Cassia italica* (Mill.) Spreng. Sin. *Senna italica* Mill.]

C. italica (Leguminosae) originated in the drier regions of tropical Africa and has adapted to warm temperatures and may grow throughout the year. The leaves, pods and seeds are mostly used in traditional medicine. Samples showing witches' broom symptoms were observed and studied in Oman, and the affected plants yielded positive results for the presence of '*Ca*. P. omanense' (Al-Saady et al. 2008) affiliated to ribosomal group 16SrXXIX-A (Esmailzadeh Hosseini et al. 2016a).

8.23 Pot Marigold (*Calendula officinalis* L.)

Pot marigold, also known as calendula, is a member of the aster family (Asteraceae). It is native to Mediterranean countries, but is now grown as a medicinal and ornamental plant throughout the world. Extracts of this plant have antiviral, antigenotoxic, antifungal, antibacterial, anti-inflammatory and astringent properties and are also present in many topical preparations for cosmetic uses. Phytoplasma diseases of pot marigold have been reported from Italy, Canada, Iran and Serbia (Hwang et al. 1977a; Marcone et al. 2000; Lee et al. 2004a; Wang and Hiruki 2005; Esmailzadeh-Hosseini et al. 2011, 2016a, b; Pavlović et al. 2014). Typical symptoms associated with these diseases included virescence, phyllody, proliferation of axillary buds, stunting and yellowing. Disease incidences of 12 and 20% were recorded in Iran and Serbia, respectively (Esmailzadeh-Hosseini et al. 2011; Pavlović et al. 2014). A subgroup 16SrI-B phytoplasma was identified in diseased pot marigold plants in Canada and Italy (Marcone et al. 2000; Lee et al. 2004a; Wang and Hiruki 2005), whereas subgroup 16SrII-D and 16SrXII-A phytoplasmas were detected in diseased pot marigold plants in Iran and Serbia, respectively (Esmailzadeh-Hosseini et al. 2011, 2016a, b; Pavlović et al. 2014). Also, the pot marigold-infecting phytoplasma occurring in Iran was transmitted from naturally infected plants to periwinkle through dodder (*Cuscuta* spp.) bridges (Esmailzadeh-Hosseini et al. 2011).

8.24 Purple Coneflower [*Echinacea purpurea* (L.) Moench.]

Purple coneflower (Asteraceae) has shown many potential pharmacological activities and is traditionally employed in the formulation of dietary supplements used as immunostimulants mainly in the prevention and treatment of inflammatory and viral diseases. The first report of a yellows disease in purple coneflower was described in 1977 in Canada (Hwang et al. 1977b). In 2008, a phytoplasma from the aster yellows group infecting this species was reported in Slovenia (Radisek et al. 2009). Symptoms consisted of virescence and phyllody (Figure 8.1e). In 2010, almost 60% of the plants cultivated in a crop in northern Italy showed severe phyllody symptoms and, in some cases, purplish reddening of the basal leaves. Diseased purple coneflower plants showed a yield of essential oils which differed slightly from that of healthy plants (0.026 versus 0.029% in healthy plants). These plants were infected with a subgroup 16SrIX-C phytoplasma. Significant differences between diseased and healthy plants in the components of essential oils were observed for limonene (4.4% versus 2.2% in healthy plants), *cis*-verbenol (5.7% versus 1.8% in healthy plants), verbenone (11.6% versus 2.7% in healthy plants), carvone (2.5% versus 0.8% in healthy plants), germacrene-D (8.5% versus 10.8% in healthy plants), caryophyllene oxide (3.3% versus 4.5% in healthy plants) and spathulenol (3.2% versus 4.4% in healthy plants) (Pellati et al. 2011). *E. purpurea* affected by a witches' broom disease was reported infected with a subgroup 16SrXII-A phytoplasma in Serbia (Pavlović et al. 2011) and in Taiwan (Tseng et al. 2012), whereas diseased *E. purpurea* plants showing symptoms of leaf reddening, virescence and phyllody were found to be infected by 16SrI-C and 16SrIII-B phytoplasmas in South Bohemia, Czech Republic (Fránová et al. 2009, 2013).

8.25 Red Sage (*Salvia miltiorrhiza* Bunge)

Red sage, also known as Chinese sage (Lamiaceae), is a perennial plant highly valued for its roots in traditional Chinese medicine. This species has been widely used in China and, to a lesser extent, in Japan, the USA and European countries for the treatment of cardiovascular and cerebrovascular diseases. In China, red sage alone or combined with other Chinese herbal medicines was applied for the treatment of a variety of diseases such as angina pectoris, myocardial infarction, hypertension, hyperlipidaemia and acute ischaemic stroke. Recently, a phytoplasma disease of red sage, the *Salvia miltiorrhiza* red leaf, has been recorded in the major cultivation areas of China (Yang et al. 2016). Disease incidence ranged from 30 to 70%. Main symptoms shown by affected plants were stunting, inflorescence malformation, leaf reddening, root browning and root rot. A subgroup 16SrXII-A phytoplasma ('*Ca.* P. solani'-related) was identified in all symptomatic plants (Yang et al. 2016).

8.26 *Rehmannia glutinosa* (Gaertn.) Steud.

Rehmannia glutinosa (Scrophulariaceae) is used for loss of blood, lower back pain, lumbago, cough, hectic fever, diabetes, urinary incontinence, deafness, uterine bleeding, vertigo and tinnitus and for regulating menstrual flow. *R. glutinosa* has also astringent properties and helps to protect and support the liver and adrenal glands. Přibylová et al. (2001) reported a little leaf disease of *R. glutinosa* in Czech Republic, Europe. Infected plants showed proliferating shoots, leaves reduced in size with vein clearing and chlorosis, shortened internodes and virescent petals and died in advanced stages of the disease. RFLP analysis of phytoplasma 16S rDNA confirmed the presence of an aster yellows-related phytoplasma (16SrI-B) in the diseased plants.

8.27 Rosemary (*Rosmarinus officinalis* L.)

Rosemary (Lamiaceae) is a common and attractive evergreen shrubby species growing wild in the Mediterranean area. Essential oils extracted from leaves and flowers of this plant have a variety of medicinal properties including carminative, stomachic and nervine activities, thereby curing many cases of headache. In the spring of 2011, symptoms of yellowing, witches' broom and stunting were observed on rosemary plants cultivated at Liguria region, Italy (Figure 8.1b); phytoplasma presence was registered and identified as subgroup 16SrXII-A in 2011 samples and 16SrI-B in 2012 samples (Contaldo et al. 2012).

8.28 Sandal (*Santalum album* L.)

Sandal (Santalaceae) is a hemi-root parasitic tree, which commonly occurs in the dry regions of peninsular India, particularly in Karnataka and Tamil Nadu states. Sandal wood oil is useful in supporting the lymphatic, nervous and cardiovascular systems. It gives relief in itching, inflammation, nausea, vomiting, sunstroke, healing wounds, scars and acne. Sandal spike is a major disease of sandal associated with phytoplasma presence. Khan et al. (2006a) reported spike disease of sandal in India associated with witches' broom symptoms consisting of small, narrow leaves which turn pale-green or yellow on branches acquiring a spike-like appearance. BLAST and phylogenetic analyses of 16S rDNA sequences revealed that *Santalum* phytoplasma (GenBank accession number DQ0932357) is most similar (99%) to '*Ca.* P. asteris'-related strains, subgroup 16SrI-B members (Khan et al. 2006a, 2008).

8.29 Soapwort (*Saponaria officinalis* L.)

Saponaria officinalis (Caryophyllaceae), also known as bouncing-bet or soapwort, is a perennial medicinal plant important for the pharmaceutical industry and used as an expectorant, alterative, laxative and ointment for some skin diseases and arthritic conditions. *S. officinalis* plants with typical symptoms (23% in 2011 and 47% in 2012) of phytoplasma infection were observed in Pancevo plantation, Serbia. The symptoms appeared in May with leaves changing colour from green to brown with severe reddening and necrosis. Severely diseased plants died. The infected plants had a significant reduction in biomass and quality. The association of 16SrXII-A was confirmed by PCR/RFLP analyses of the 16S rDNA gene (Josic et al. 2013). A 16SrI-M phytoplasma was detected in diseased *S. officinalis* plants in Lithuania (Lee et al. 2004a), whereas Khasa et al. (2016) reported the presence of a 16SrVI-D subgroup phytoplasma in diseased *S. officinalis* showing witches' broom disease in India (Figure 8.1n).

8.30 Spanish Broom (*Spartium junceum* L.)

Spanish broom (Fabaceae) is a woody shrub, with deep golden-yellow flowers that is common in Mediterranean areas. This rapidly growing plant has several medicinal properties. The young herbaceous tips of flowering shoots, harvested in spring, are cardiotonics, emetics, diuretic and purgatives. Flowers possess a potent anti-ulcerogenic activity (Yesilada and Takaishi 1999). Spanish broom is severely affected by a lethal disease, the spartium witches' broom (SpaWB). The most characteristic symptoms of the disease are pronounced witches' brooms, shortened internodes, off-season growth and death of the plants. SpaWB occurs in Italy and Spain (Marcone et al. 1996a, 2004b, 2016; Torres et al. 2002; Spallino et al. 2013; Contaldo et al. 2015). Two genetically different phytoplasmas which induce the same symptoms were identical in Italy, as '*Ca.* P. spartii' (16SrX-D) and a member of the elm yellows group, subgroup 16SrV-C (Lee et al. 2004b). However, in Spain only '*Ca.* P. spartii' has been reported to be associated with SpaWB (Torres et al. 2002). The yield of essential oils extracted from flowers of SpaWB-affected Spanish broom plants is lower than that of healthy plants. Also, substantial amounts of sesquiterpenes and a marked decrease in the amount of *n*-alkanes and aliphatic compounds occur in the essential oils from flowers of diseased plants. Sesquiterpenes could not be detected in the volatile fraction of healthy plants (Mancini et al. 2010a). Great differences were also identified in the content of alkaloid compounds between healthy and diseased Spanish broom plants, the latter containing seven different alkaloids that were not present in healthy plants (Mancini et al. 2010b).

8.31 St. John's Worth (*Hypericum perforatum* L.)

St. John's worth (Hypericaceae, sin. Guttiferae) has completed the transition from noxious weed to wild collected resource. The drug consists of the dried flowering tops or aerial parts of the plant. Its major constituents include naphthodianthrones, acylphloroglucinols, flavonoids, biflavones, phenylpropanes and an essential oil rich in sesquiterpenes. Its most important, carefully validated use is in the treatment of mild to moderate depression (Barnes et al. 2001). During the spring and summer of 2002, a phytoplasma infection was observed in an experimental field of *H. perforatum* (cvs. Godet Derborance and Zorzi) located in Ozzano (Bologna, northern Italy) in its second year of cultivation (Bellardi et al. 2002). Severe symptoms consisted of dwarfing, extreme reduction in leaf size, yellow leaves, witches' broom, reduced internodal length and lack of flower production (Figure 8.1i). The majority of the symptoms (more than 50%) were observed in 'Godet Derborance'. This disease was associated with a 16SrXII-A phytoplasma (Paltrinieri et al. 2002). In the spring of 2003, during further investigations in the same Italian experimental field, plants of cv Godet Derborance failed to sprout, while almost 20% of cv Zorzi showed symptoms similar to those previously noticed. A phytoplasma belonging to ash yellows group (16SrVII group) was identified in the diseased plants (Bruni et al. 2005). Qualitative/quantitative phytochemical variations were observed in dried flowering tops of cv Zorzi infected by 16SrVII group phytoplasma. The affected plants exhibited decreased amounts of rutin, hyperoside, isoquercitrin, amentoflavone and pseudohypericin, whereas the chlorogenic acid content was doubled. Hypericin, quercitrin and quercetin contents were not severely affected. The essential oil yield was drastically reduced in infected material (0.11% vs 0.75% in healthy material) and revealed an increased abundance of sesquiterpenes (β-caryophyllene, δ-elemene and germacrene-D, in particular), and a matching decrease in monoterpene hydrocarbons and aliphatics was noticed (Bruni et al. 2005). A subgroup 16SrXII-A phytoplasma has been recorded in yellows-diseased *H. perforatum* and *H. barbatum* plants in Serbia (Pavlović et al. 2012).

8.32 Thyme (*Thymus vulgaris* L.)

Thyme (Lamiaceae), known as an important aromatic pot plant, often used as an ornamental in the garden, has been found infected by phytoplasmas in Italy and Spain. In the former country, the first cases, quite sporadic, occurred in 2008 and again in 2009 on plants propagated from cuttings purchased from local companies in the Albenga area (Liguria region; northern Italy). The disease (appearance in the fall) consisted of dwarfing and redness of the shoot tip (the percentage of infected plants was 20%). Phytoplasmas identified belonged to the subgroup 16SrX-A (apple proliferation group, '*Ca.* P. mali') (in 2008) and to 16SrV-A subgroup (elm yellows group, '*Ca.* P. ulmi') (in 2009) (A. Bertaccini and M.G. Bellardi, unpublished data). Symptoms such as proliferation of axillary shoots, a mass of adventitious roots of orange-brown colour and herbaceous consistency, which flowed from the bark, were present. The branches were twisted in the upper portions corresponding to the adventitious roots. The foliage was dense with rosetting symptoms. Overall, the plants were reduced in dimensions both the canopy volume and the stem diameter. The leaf colour was lighter green in the two asymptomatic plants and green and/or silver powdery in the symptomatic ones (Figure 8.1h). The associated phytoplasma was identified as a member of group 16SrI (Bellardi et al. 2016). A subgroup 16SrXII-A phytoplasma has been detected in diseased *T. vulgaris* plants in Spain (Battle et al. 2000).

8.33 *Tylophora asthmatica* (L.F.) Wight and Arn.

Tylophora asthmatica, also known as Indian ipecac and "anantmool", is a dark copper-coloured delicate creeper found growing wild in the plains of India and other subtropical regions of the world (Bhavan 1992). It has been traditionally used as a folk medicine in certain regions of India for the treatment of bronchial asthma, inflammation, bronchitis, allergies, rheumatism and dermatitis (Gore et al. 1980; Khare 2004). '*Ca.* P. aurantifolia' strain was reported to be associated with the little leaf and leaf yellowing disease of "anantmool" in India (Fig. 8.1m) (Madhupriya et al. 2014).

8.34 Valerian (*Valeriana officinalis* L.)

Valerian (Valerianaceae) is a perennial plant native to North America, temperate Asia and some European countries that have been used as a medicinal herb since at least the time of ancient Greece and Rome. It has antispasmodic, carminative, diuretic, hypnotic, hypotensive, relaxant, sedative, stimulant, warming and powerfully nervine properties. A yellows phytoplasma disease of valerian has been reported from Canada, Lithuania and Serbia (Hwang et al. 1977a; Khadhair et al. 2008; Lee et al. 2004a; Mitrović et al. 2013). Symptoms observed in affected plants included leaf reddening, yellowing, stunting, small leaves which had a stiff texture, rosetting and proliferation of axillary shoots. Subgroup 16SrI-A, 16SrI-M and 16SrXII-A phytoplasmas were found to be associated with the disease in Canada, Lithuania and Serbia, respectively (Khadhair et al. 2008; Lee et al. 2004a; Mitrović et al. 2013). Phytoplasma infections in diseased valerian plants in Canada were also detected by transmission electron microscopy (Hwang et al. 1977a). This study revealed a very large number of phytoplasma cells in the sieve tubes of affected plants.

8.35 *Vernonia cinerea* Less.

Vernonia cinerea (Compositae) has sweet, cold, tonic, stomachic and astringent properties. It is also used to cure asthma, bronchitis and fevers; it is one of the best remedies for the treatment of typhoid. In case of female diseases particularly in leucorrhoea, its juice is used and given to the patients with gud (jaggery). The juice of *V. cinerea* is a good remedy for intestinal worms also. Brown (2008) observed typical lethal yellowing symptoms in *V. cinerea* in Jamaica. Blast analysis determined *Vernonia* phytoplasma sequence (GenBank accession number EU057983) to be most similar (99%) to that of CLY phytoplasma in Jamaica (GenBank accession number AF49807) and Florida (GenBank accession number AF498309), member of the 16SrIV group. Therefore, *Vernonia* phytoplasma is identified as a member of the 16SrIV group.

8.36 Wild Bergamot (*Monarda fistulosa* L.)

Wild bergamot (Lamiaceae), native to North America, is an annual or perennial medicinal plant known for its strong therapeutic effects: its essential oil is characterized by high antibacterial, antimycotic and anti-inflammatory activities. During a survey carried out in 2009 in Italy, wild bergamot showing yellows, stunting, virescence and flower buds proliferation (Figure 8.1e) was found to be infected by a subgroup 16SrXII-A phytoplasma (Bellardi et al. 2011). In 2010, a study to verify the correlation among phytoplasma presence, symptom expression and the effects of these prokaryotes on essential oil composition was carried out. One essential oil was extracted from stunted plants with virescence and phyllody and infected by both aster yellows and "stolbur" group (16SrXII-A) phytoplasmas. The essential oil extracted from asymptomatic plants infected by the "stolbur" phytoplasma was reported to have a marked decrease in monoterpene compounds (Contaldo et al. 2011b).

8.37 Woolly Foxglove (*Digitalis lanata* Ehrh.)

Woolly foxglove (Scrophulariaceae) is a well-known medicinal plant, used as a commercial source of cardiac glycosides, compounds that are therapeutically relevant for the treatment of heart diseases. In particular, this species is known to contain primary glycosides, such as lanatoside A, lanatoside B and lanatoside C. *Digitalis* glycosides, which are difficult to synthesize, so it is easier to extract them from foxglove leaves. During the spring-summer of 2007–2008 at the Herb Garden of Casola Valsenio (Ravenna, northern Italy), plants grown in an experimental field of *D. lanata* were found to be infected with '*Ca.* P. asteris', subgroup 16SrI-B, as confirmed by the amplification with primers for several genes (Bellardi et al. 2007). The major symptoms associated with the disease were stunting, witches' broom, virescence, little leaf, reddish lamina, severe phyllody and partial or complete rosetting of flower spikes with a production of axillary shoots. In addition, a suitable phytochemical investigation was undertaken to evaluate the effect of this pathogen on the content of *D. lanata* secondary metabolites. These analyses demonstrated that the secondary metabolite mainly affected by the phytoplasma was lanatoside C (153.2 μg/100 mg in healthy material versus 76.1 μg/100 mg in the infected one) (Pellati et al. 2009). Several years before a 16SrI-B phytoplasma was also identified in *D. lutea* L. (yellow foxglove) showing stunting end yellow leaves, cultivated in the same Herb Garden of Casola Valsenio (Bellardi et al. 1999).

8.38 Transmission and Management

Little is known about the transmission of phytoplasma diseases of medicinal plants. Phytoplasmas infecting these plants are known to be spread by grafting, cutting, micropropagation or other ways of asexual plant propagation. A number of insect vectors are responsible for transmission of phytoplasmas on medicinal plants (Arocha et al. 2005; Favali et al. 2008; Harrison and Oropeza 2008). One of the phytoplasma diseases that occur on *C. roseus* plants in nature is periwinkle yellows reported to be transmitted by the leafhopper *Macrosteles quadripunctulatus* (Bosco et al. 1997). Natural infections of periwinkle are reported in gardens in different Italian regions. The transmission of phytoplasmas to periwinkle by insects was largely used to study the transmission characteristics and the spread of these pathogens (Carraro et al. 1991); however, transmission of phytoplasmas from host plants to *C. roseus* and back to host plants can be obtained by grafting and by dodder (Marwitz et al. 1974; Carraro et al. 1991). In particular, it has been demonstrated that dodder transmits different phytoplasmas to periwinkle with different efficiency, which may be due to their pathogenic effects on the vector plant (Carraro et al. 1991; Musetti et al. 1992).

Control of epidemic outbreaks of phytoplasma diseases could be performed either by controlling the vector or by eliminating the pathogen from the infected plants by meristem tip culture or antibiotic treatments (Bertaccini 2007). At present, insect vector control using pesticides is the tool of choice for limiting disease outbreaks; however, removal of sources of inoculum is also an efficient tool in the case of phytoplasmas spread by monophagous insect vectors. Vector control is difficult to achieve when wild reservoir plants are sources of inoculum for polyphagous leafhoppers. Similarly, it is easier to control monophagous insects reproducing on the affected crop than insects completing their life cycle on wild plants.

Phytoplasmas can be eliminated from their plant hosts, as they are generally thermolabile and are not present in the shoot meristem (Lee and Davis 1992). Furthermore they are sensitive to some antibiotics such as tetracycline (Ishiie et al. 1967; Heintz 1989). Several methods have been applied to clean plant material from phytoplasmas; these include in vitro tissue culture such as shoot tip (Dale and Cheyne 1993) or micropropagation (Davies and Clark 1994) sometimes in combination with heat or antibiotic treatment.

Symptoms of periwinkle little leaf disease of *C. roseus* associated with phytoplasmas were reduced by treatment with gibberellic acid or tetracycline (Kar et al. 1983). Phytoplasmas could be eliminated in shoot-tip cultures of *C. roseus* employing oxytetracycline (Singh et al. 2007). Foliar spray of oxytetracycline, tetracycline hydrochloride or penicillin and dipping of cuttings of infected plants in indole-3-acetic acid (IAA) or indole-3-butyric acid (IBA) have been attempted by various workers for remission of symptoms in various plants or their control to some extent.

Therefore, a real way to control phytoplasma infection is to prevent outbreaks by producing clean material or by finding phytoplasma-resistant varieties. Knowledge about the mechanisms of plant host resistance to phytoplasmas is little, but the paucity of effective disease management strategies for these diseases lends a high priority to these questions. Efforts continue to identify germplasm encoding natural resistance to phytoplasmas and to incorporate such genes via selection and breeding programmes that may involve resistance to either the pathogen itself or the insect vector. Plant defence-related proteins, known to be active in host responses to invasion by other types of pathogens, might occur in response to mollicute infection; confirmation of this hypothesis would require demonstration that the compound is in the right place at the right time and is present in effective concentrations (Garnier et al. 2001).

8.39 Conclusions and Perspectives

Medicinal plants constitute a group of industrially important crops which are of great value. Plant-based drugs are being increasingly preferred in medicinal science. Phytoplasmas are associated with diseases in many medicinal plants inducing serious economic losses. Therefore, phytoplasma diseases are major constraints in profitable cultivation of medicinal plants and lower the quantity and quality of desirable compounds. Epidemics of these diseases have disturbed the natural active ingredients resulting in the withdrawal of many medicinal plant varieties from cultivation. Also, some diseases of such crops are associated each with several genetically different phytoplasmas which induce similar symptoms in a given plant host and occur either within the same geographic areas or different continents. Newly discovered diseases of medicinal plants are increasingly being attributed to phytoplasma infections. This rapid increase of infection may be related to the fact that currently the use of insecticides to control insect vectors of phytoplasmas is prohibited on medicinal plants, whereas insect control and/or management by other, environmentally friendly means is not fully effective (Marcone et al. 2016). Phytoplasma diseases of medicinal plants severely reduce the yield and quality of crops and longevity of plants. Changes in the composition of secondary metabolites occurring in diseased plants can be related to the role of phytoplasma infections in triggering plant defence responses in which, however, the levels of valuable phytochemicals are greatly affected. An exception is represented by phytoplasma diseases of periwinkle in which an accumulation of pharmaceutically important compounds such as vinblastine and vincristine occurs upon phytoplasma infections. The impact of phytoplasma infections on medicinal plants should be taken into account in promoting good agricultural practices for cultivation and propagation of these plants.

References

Ahmad SJN, Ahmad JN, Irfan M, Ahmad M, Aslam M (2015) Report on phytoplasma new host plants in Pakistan. *Phytopathogenic Mollicutes* **5**(1-Supplement), S71–S72.

Al-Saady NA, Khan AJ, Calari A, Al-Subhi AM, Bertaccini A (2008) '*Candidatus* Phytoplasma omanense', a phytoplasma associated with witches' broom of *Cassia italica* (Mill.) Lam. in Oman. *International Journal of Systematic and Evolutionary Microbiology* **58**, 461–466.

Arocha Y, Horta D, Piñol B, Palenzuela I, Picornell S, Almeida R, Jones P (2005) First report of phytoplasma associated with Bermuda grass white leaf disease in Cuba. *Plant Pathology* **54**, 233.

Arocha Y, Piñol B, Picornell B, Almeida R, Jones P, Boa E (2006) Basil little leaf: a new disease associated with a phytoplasma of the 16SrI (aster yellows) group in Cuba. *Plant Pathology* **55**, 822.

Ayman OF, Kumar Y, Hallan V, Zaidi AA (2010) Molecular characterization of the phytoplasmas associated with toon trees and periwinkle in India. *Journal of General Plant Pathology* **76**, 351–354.

Barnes J, Anderson LA, Phillipson JD (2001) St. John's wort (*Hypericum perforatum* L.): a review of its chemistry, pharmacology and clinical properties. *Journal of Pharmacy and Pharmacology* **53**, 583–600.

Battle A, Martínez MA, Laviña A (2000) Occurrence, distribution and epidemiology of grapevine yellows in Spain. *European Journal of Plan Pathology* **106**, 811–816.

Bellardi MG, Bertaccini A (2005) Malattie delle piante officinali – virosi e fitoplasmosi. Ed. l'Informatore Agrario, Verona, Italia, 197 pp.

Bellardi MG, Rubies-Autonell C, Bertaccini A (1999) Infezioni da virus e fitoplasmi in piante officinali in Emilia- Romagna. III Contributo. *Informatore Fitopatologico* **49**, 47–53.

Bellardi MG, Benni A, Bertaccini A (2002) Deperimento dell'iperico infetto da fitoplasmi. *Natural 1* **26**, 42–46.

Bellardi MG, Benni A, Paltrinieri S, Bertaccini A (2007) A severe disease induced by '*Candidatus* Phytoplasma asteris' in *Digitalis lanata*. *Bullettin of Insectology* **60**, 275–276.

Bellardi MG, Contaldo N, Benni A, Curini M, Epifano F, Genovese S, Bertaccini A (2009) Effects of phytoplasma infection on the quality of *Grindelia robusta* essential oil. *Journal of Plant Pathology* **91**, 231–240.

Bellardi MG, Contaldo N, Cavicchi L, Bertaccini A, Biffi S (2011) Infezioni da fitoplasmi in *Monarda fistulosa*. *Natural 1* **99**, 31–33.

Bellardi MG, Contaldo N, Bozzano G, Parodi C, Cavicchi L, Bertaccini A (2016) Infezioni da fitoplasmi in timo. *Clamer informa* **2**, 71–76.

Bertaccini A (2007) Phytoplasmas: diversity, taxonomy, and epidemiology. *Frontiers in Bioscience* **12**, 673–689.

Bertaccini A, Davis RE, Lee I-M (1992) *In vitro* micropropagation for maintenance of mycoplasma-like organisms in infected plant tissues. *HortScience* **27**, 1041–1043.

Bertaccini A, Fránová J, Paltrinieri S, Martini M, Navrátil M, Lugaresi C, Nebesářová J, Simková M (1999) Leek proliferation: a new phytoplasma disease in Czech Republic and Italy. *European Journal of Plant Pathology* **105**, 487–493.

Bertaccini A, Vallesi P, Gotti R, Benni A, Bellardi MG (2005a) Infezioni da fitoplasmi in *Hyssopus officinalis* L. - Analisi dell'olio essenziale. *Petria* **15**, 31–32.

Bertaccini A, Vallesi P, Bellardi MG (2005b) Infezioni da fitoplasmi in *Parietaria officinalis* L. e P. *judaica* Auct. An. L. Incontri fitoiatrici Difesa delle colture ortoflorofrutticole Torino, Italia, 36–37.

Bertaccini A, Paltrinieri S, Benni A, Bellardi MG (2008) Una grave malattia associata alla presenza del fitoplasma del giallume dell'astro in *Grindelia robusta* Nutt. *Petria* **18**, 338–340.

Bertaccini A, Duduk B, Paltrinieri S, Contaldo N (2014) Phytoplasmas and phytoplasma diseases: a severe threat to agriculture. *American Journal of Plant Sciences* **5**, 1763–1788.

Bhavan BV (1992) Selected medicinal plants of India. Tata Press. Bombay, India, 333–336 pp.

Borth WB, Fukuda SK, Hamasaki RT, Hu JS, Almeida RPP (2006) Detection, characterisation and transmission by *Macrosteles* leafhoppers of watercress yellows phytoplasma in Hawaii. *Annals of Applied Biology* **149**, 357–363.
Bosco D, Minucci C, Boccardo G, Conti M (1997) Differential acquisition of chrysanthemum yellows phytoplasma by three leafhopper species. *Entomologia Experimentalis et Applicata* **83**, 219–224.
Bown D (1995) Encyclopedia of herbs and their uses. Dorling Kindersley, London.
Brown SE (2008) First report of lethal yellowing group (16Sr IV) of phytoplasmas in *Vernonia cinerea* in Jamaica. *Plant Disease* **92**, 1132.
Bruni R, Pellati F, Bellardi MG, Benvenuti S, Paltrinieri S, Bertaccini A, Bianchi A (2005) Herbal drug quality and phytochemical composition of *Hypericum perforatum* L. affected by ash yellows phytoplasma infection. *Journal of Agricultural and Food Chemistry* **53**, 964–968.
Çağlar BK, Satar S, Elbeaino T (2013) Detection and molecular characterization of Bermuda grass (*Cynodon dactylon*) white leaf phytoplasma from Turkey. *International Journal of Agriculture and Biology* **15**, 90–94.
Caicedo JD, Rivera-Vargas LI, Segarra AE, Davis RE (2015) Detection and molecular characterisation of a group 16SrIX phytoplasma infecting citrus (*Citrus sinensis* and *C. limon*), coffee (*Coffea arabica*), periwinkle (*Catharanthus roseus*), and tabebuia (*Tabebuia heterophylla*) in Puerto Rico. *Australasian Plant Disease Notes* **10**, 28.
Camele I, Bertaccini A, Martini M, Rana GL (1999) Identificazione molecolare di fitoplasmi in piante di eucalipto e mirto con scopazzi. *Petria* **9**, 259–268.
Carraro L, Osler R, Loi N, Favali MA (1991) Transmission characteristics of the clover phyllody agent by dodder. *Journal of Phytopathology* **133**, 15–22.
Carrubba A, Lo Verde G, Salamone A (2015) Sustainable weed, disease and pest management in medicinal and aromatic plants. In: Medicinal and Aromatic Plants of the World. Ed Máthé Á. Springer Science+Business Media Dordrecht, Heidelberg, Germany, 2015-235 pp.
Chaturvedi Y, Tiwari AK, Upadhyaya PP, Prabhuji SK, Rao GP (2009) Association of '*Candidatus* phytoplasma asteris' with little leaf and phyllody disease of *Catharanthus roseus* in Eastern Uttar Pradesh, India. *Medicinal Plant* **1**, 103–108.
Chaturvedi Y, Singh M, Snehi SK, Raj SK, Rao GP (2010) First report of '*Candidatus* Phytoplasma asteris' (16SrI group) associated with yellows and little leaf diseases of *Hibiscus rosa-sinensis* in India. *Plant Pathology* **59**, 796.
Chaube S, Kumar S, Dubey D, Tiwari AK, Upadhyaya PP, Rao GP (2015) Identification of a novel phytoplasma (16SrXIV-A subgroup) associated with little leaf and witches' broom of *Cannabis sativa* L. ssp. *sativa* and *C. sativa* L. ssp. *indica* in India. *Phytoparasitica* **43**, 275–279.
Chen W-Y, Huang Y-C, Tsai M-L, Lin C-P (2011) Detection and identification of a new phytoplasma associated with periwinkle leaf yellowing disease in Taiwan. *Australasian Plant Pathology* **40**, 476–483.
Choi YH, Tapias EC, Kim HK, Lefeber AWM, Erkelens C, Verhoeven JThJ, Brzin J, Zel J, Verpoorte R (2004) Metabolic discrimination of *Catharanthus roseus* leaves infected by phytoplasma using ^{1}H-NMR spectroscopy and multivariate data analysis. *Plant Physiology* **135**, 2398–2410.
Chopra RN, Nayar SL, Chopra IC (1986) Glossary of Indian medicinal plants. Council of Scientific and Industrial Research, New Delhi, India.
Conci VC, Gomez GG, Docampo DM, Conci LR (1998) Phytoplasma associated with symptoms of "tristeza del ajo" (garlic decline) in garlic (*Allium sativum* L.). *Journal of Phytopathology* **146**, 473–477.
Contaldo N, Canel A, Makarova O, Paltrinieri S, Bertaccini A, Nicolaisen M (2011a) Use of a fragment of the *tuf* gene for phytoplasma 16Sr group/subgroup differentiation. *Bulletin of Insectology* **64**(Supplement), S45–S46.
Contaldo N, Bellardi MG, Cavicchi L, Epifano F, Genovese S, Curini M, Bertaccini A (2011b) Phytochemical effects of phytoplasma infections on essential oil of *Monarda fistulosa* L. *Bulletin of Insectology* **64**(Supplement), S177-S178.

Contaldo LN, Bertaccini A, Bozzano G, Cavicchi L, Bellardi MG (2012) Detection and molecular characterization of phytoplasmas infecting *Rosmarinus officinalis*. *Journal of Plant Pathology* **94**, S4.60.

Contaldo N, Paltrinieri S, Ardizzi S, Duduk B, Bertaccini A, Bellardi MG (2015) Identification and molecular characterization of multiple phytoplasma infection in *Spartium junceum* and *Cytisus scoparius*. *Acta Horticulturae* **1072**, 113–116.

Dale PJ, Cheyne VA (1993) The elimination of clover diseases by shoot tip culture. *Annals of Applied Biology* **123**, 25–32.

Davies DL, Clark MF (1994) Maintenance of mycoplasma-like organisms occurring in *Pyrus* species by micropropagation and their elimination by tetracycline therapy. *Plant Pathology* **43**, 819–823.

Davis RE (1995) Fitoplasmas: fitopatogenos procarioticos sem parede celular, habitantes de floema e transmitidos por artrópodes. *Revue Annual Patologie Plant* **3**, 1–27.

Davis RE, Harrison NA, Zhao Y, Wei W, Dally EL (2016) '*Candidatus* Phytoplasma hispanicum', a novel taxon associated with Mexican periwinkle virescence disease of *Catharanthusroseus*. *International Journal of Systematic and Evolutionary Microbiology* **66**, 3463–3467.

Davis RI, Jacobson SC, De La Rue SJ, Tran-Nguyen L, Gunua TG, Rahamma S (2003) Phytoplasma disease surveys in the extreme north of Queensland, Australia, and the island New Guinea. *Australasian Plant Pathology* **32**, 269–277.

de Souza AN, de Carvalho SL, da Silva FN, Alfenas AC, Valverde Zauza EA, Carvalho CM (2015) First report of phytoplasma associated with *Eucalyptus urophylla* showing witches' broom in Brazil. *Phytopathogenic Mollicutes* **5**(1-Supplement), S83–S84.

Dickinson M, Tuffen M, Hodgetts J (2013) The phytoplasmas: an introduction. *Methods in Molecular Biology* **938**, 1–14.

Duke J (1985) CRC Handbook of medicinal herbs. CRC Press, Inc., Boca Raton, USA.

Esmailzadeh-Hosseini S, Salehi M, Khanchezar A, Shamszadeh M (2011) The first report of a phytoplasma associated with pot marigold phyllody in Iran. *Bulletin of Insectology* **64**(Supplement), S109–S110.

Esmailzadeh Hosseini SA, Salehi M, Mirchenari SM, Contaldo N, Paltrinieri S, Bertaccini A (2016a) Occurrence of a '*Candidatus* Phytoplasma omanense'-related strain in bindweed showing a witches' broom disease in Iran. *Phytopathogenic Mollicutes* **6**, 63–68.

Esmailzadeh Hosseini SA, Salehi M, Mirchenari SM, Bertaccini A (2016b) First report of a 16SrII-D phytoplasma associated with *Calendula* officinalis phyllody in Iran. *New Disease Reports* **34**, 22.

Favali MA, Musetti R, Benvenuti S, Bianchi A, Pressacco L (2004) *Catharanthus roseus* L. plants and explants infected with phytoplasmas: alkaloid production and structural observations. *Protoplasma* **223**, 45–51.

Favali MA, Fossati F, La Sanità Toppi D, Musetti R (2008) *Catharanthus roseus* phytoplasmas. In: Characterization, Diagnosis and Management of phytoplasmas. Eds Harrison NA, Rao GP, Marcone C. Studium Press LLC, Texas, USA, 195-218 pp.

Fialová R, Válová P, Balakishiyeva G, Danet J-L, Šafárová D, Foissac X, Navrátil M (2009) Genetic variability of "stolbur" phytoplasma in annual crop and wild plant species in south Moravia. *Journal of Plant Pathology* **91**, 411–416.

Fránová J, Šimková M (2009) Association of '*Candidatus* Phytoplasma asteris' with yellowing and phyllody of *Plantago lanceolata*. *Folia Microbiologica* **54,** 469–472.

Fránová J, Přibylová J, Petrzik K (2009) Purple coneflower with reddening and phyllody: a new host of clover phyllody phytoplasma. *European Journal of Plant Pathology* **123**, 85–90.

Fránová J, Špak J, Šimková M, (2013) First report of a 16SrIII-B subgroup phytoplasma associated with leaf reddening, virescence and phyllody of purple coneflower. *European Journal of Plant Pathology* **136**, 7–12.

Fraternale D, Giampieri L, Bucchini A, Ricci D (2007) Essential oil composition and antioxidant activity of aerial parts of *Grindelia robusta* from Central Italy. *Fitoterapia* **78**, 443–445.

Galdeano E, Guzmán FA, Fernández F, Conci LR (2013) Genetic diversity of 16SrIII group phytoplasmas in Argentina. Predominance of subgroups 16SrIII-W and -X. *European Journal of Plant Pathology* **137**, 753–764.

Garnier M, Foissac X, Gaurivaud P, Laigret F, Renaudin J, Saillard C, Bové JM (2001) Mycoplasmas, plants, insect vectors: a matrimonial triangle. *Compte Rendu Academie Scientifique* **324**, 923–928.

Ghosh SK, Mohamed Ali MI, Balasundaran M (1984) Light and fluorescent microscopic studies on little leaf disease of eucalypts. *Phytopathologische Zeitschrift* **110**, 207–212.

Goel S, Madhupriya, Yadav A, Rao GP (2017) Identification and characterization of 16SrIX and 16SrXI groups of phytoplasmas associated with leaf yellows and declining disease of garlic and onion in India. *Indian Phytopathology* **70**, 368–372.

Gore KV, Rao K, Guruswamy MN (1980) Physiological studies with *Tylophora asthmatica* in bronchial asthma. *Indian Journal of Medical Research* **71**, 144–148.

Grover JK, Yadav SP (2004) Pharmacological actions and potential uses of *Momordica charantia*: a review. *Journal of Ethnopharmacology* **93**, 123–132.

Harrison NA, Oropeza C (2008) Coconut lethal yellowing. In: Characterization, Diagnosis and Management of Phytoplasmas. Eds Harrison NA, Rao GP, Marcone C. Plant Pathogens Series-5. Studium Press LLC, USA, 219-248 pp.

Heintz W (1989) Transmission of a new mycoplasma-like organism (MLO) from *Cuscuta odorata* (Ruiz et Pav.) to herbaceous plants and attempts to its elimination in the vector. *Journal of Phytopathology* **125**, 171–186.

Hiruki C (1987) Witches' broom of *Hibiscus heterophyllus*, a mycoplasma disease occurring in Australia. *Annals of Phytopathological Sociaty Japan* **53**, 1–6.

Hwang SF, Chang KF, Howard RJ, Blade SF (1977a) Yellows diseases of calendula (*Calendula officinalis*) and valerian (*Valeriana officinalis*) in Alberta, Canada, associated with phytoplasma infection. *Zeitschrift für Pflanzenkrankheiten und Pflanzenschutz* **104**, 452–458.

Hwang SF, Chang KF, Howard RJ, Khadhair AH, Gaudiel RG, Hiruki C (1977b) First report of a yellows disease in purple coneflower (*Echinacea* spp. L.) in Canada caused by a phytoplasma. *Zeitschrift für Pflanzenkrankheiten und Pflanzenschutz* **104**, 182–192.

IRPCM (2004) '*Candidatus* Phytoplasma', a taxon for the wall-less, non-helical prokaryotes that colonize plant phloem and insects. *International Journal of Systematic and Evolutionary Microbiology* **54**, 1243–1255.

Ishiie T, Doi Y, Yora K, Asuyama H (1967) Suppressive effects of antibiotics of tetracycline group on symptom development of mulberry dwarf disease. *Annals of Phytopathological Society of Japan* **33**, 267–275.

Jayasinha P (1999) Lemongrass - A literature review. Sri Lanka: Industrial Technology Institute.

Jiménez NZA, Montano HG (2010) Detection of phytoplasma in desiccated tissue of *Momordica charantia*, *Catharanthus roseus* and *Sechium edule*. *Tropical Plant Pathology* **35**, 381–384.

Jomantiene R, Davis RE, Lee I-M, Zhao Y, Bottner-Parker K, Valiunas D, Petkauskaite R (2010) Onion is host for two phytoplasma lineages, subgroups 16SrI-A and 16SrI-(B/L)L, in Lithuania: a *Hinf*I site revealed a snp marking divergent branches of evolution. *Journal of Plant Pathology* **92**, 461–470.

Josic D, Pavlovic S, Pivic R, Kuzmanovic S, Stojanovic S, Popovic T, Starovic M (2012) Cultivated and wild plantain (*Plantago major*) as a host of "stolbur" phytoplasma in Serbia. *Journal of Medicinal Plants Research* **6**, 284–288.

Josic D, Starovic M, Stojanovic S, Popovic T, Dolovac N, Zdravkovic J, Pavlovic S (2013) First report of group 16SrXII-A phytoplasma causing "stolbur" disease in *Saponaria officinalis* plants in Serbia. *Plant Disease* **97**, 420.

Kar RK, Kabi T, Pattnaik H (1983) Chemotherapy of periwinkle little leaf disease. *Current Science* **52**, 496–497.

Khadhair A-H, Evans IR, Choban B (2002) Identification of aster yellows phytoplasma in garlic and green onion by PCR-based methods. *Microbiological Research* **157**, 161–167.

Khadhair A-H, Hiruki C, Deyholos M (2008) Molecular characterization of aster yellows phytoplasma associated with valerian and sowthistle plants by PCR–RFLP analyses. *Journal of Phytopathology* **156**, 326–331.

Khan JA, Srivastava P, Sing SK (2006a) Identification of a '*Candidatus* Phytoplasma asteris'-related strain associated with spike disease of sandal (*Santalum album*) in India. *Plant Pathology* **55**, 572.

Khan JA, Srivastava P, Singh SK (2006b) Sensitive detection of a phytoplasma associated with little leaf symptoms in *Withania somnifera*. *European Journal of Plant Pathology* **115**, 401–408.

Khan JA, Sing SK, Ahmad J (2008) Characterisation and phylogeny of a phytoplasma inducing sandal spike disease in sandal (*Santalum album*). *Annals of Applied Biology* **153**, 365–372.

Khanna S, Singh J, Singh R, Kumar P, Rani T, Baranwal VK, Sirohi A, Bertaccini A (2015) Evidence of association of a '*Candidatus* Phytoplasma cynodontis' with bermuda grass (*Cynodon dactylon*) and '*Candidatus* Phytoplasma asteris' with periwinkle (*Catharanthus roseus*) from western Uttar Pradesh, India. *Crop Protection* **74**, 138–144.

Khare CP (2004) Enclyclopedia of Indian medicinal plants. Springer-Verlag, Berlin-Heidelberg, Germany, 523 pp.

Khasa E, Taloh A, Gopala, Prabha T, Rao GP (2016) Molecular characterization of phytoplasmas of clover proliferation group associated with three ornamental plant species in India. *3 Biotech* **6**, 237.

Kitajima EW, Ribeiro RLD, Lin MR, Ribeiro MISD, Kimura O, Costa CL, Pimentel JP (1984) Lista comentada de vırus e organismos do tipo micoplasma em plantas cultivadas e silvestres do Estado do Rio de Janeiro. *Fitopatologia Brasileira* **9**, 607–625.

Kumar M, Madhupriya, Rao GP (2017) Molecular characterization, vector identification and sources of phytoplasmas associated with brinjal little leaf disease in India. *3 Biotech* **7**, 7.

Kumar S, Byadgi AS (2012) Molecular detection and characterization of phytoplasma associated with periwinkle. *Vegetos* **25**, 127–135.

Lee I-M, Davis RE (1992) Mycoplasmas which infect plants and insects. In: Mycoplasmas: Molecular Biology and Pathogenesis. Eds Maniloff J, McElhaney RN, Finch LR, Baseman JB. American Society for Microbiology, Washington DC, USA, 379-390 pp.

Lee I-M, Davis RE, Gundersen-Rindal DE (2000) Phytoplasma: phytopathogenic mollicutes. *Annual Review of Microbiology* **54**, 221–255.

Lee I-M, Martini M, Bottner KD, Dane RA, Black MC, Troxclair NB (2003) Ecological implications from molecular analysis of phytoplasmas involved in an aster yellows epidemic in various crops in Texas. *Phytopathology* **93**, 1368–1377.

Lee I-M, Gundersen-Rindal DE, Davis RE, Bottner KD, Marcone C, Seemüller E (2004a) '*Candidatus* Phytoplasma asteris', a novel phytoplasma taxon associated with aster yellows and related diseases. *International Journal of Systematic and Evolutionary Microbiology* **54**, 1037–1048.

Lee I-M, Martini M, Marcone C, Zhu SF (2004b) Classification of phytoplasma strains in the elm yellows group (16SrV) and proposal of '*Candidatus* Phytoplasma ulmi' for the phytoplasma associated with elm yellows. *International Journal of Systematic and Evolutionary Microbiology* **54**, 337–347.

Lee I-M, Bottner-Parker KD, Zhao Y, Bertaccini A, Davis RE (2012) Differentiation and classification of phytoplasmas in the pigeon pea witches' broom group (16SrIX): an update based on multiple gene sequence analysis. *International Journal of Systematic and Evolutionary Microbiology* **62**, 2279–2285.

Lotos L, Tsialtas JT, Maliogka VI, Kaloumenos N, Eleftherohorinos IG, Katis NI (2013) First report of '*Candidatus* Phytoplasma solani' related strain associated with a disease of *Datura stramonium* in Greece. *Journal of Plant Pathology* **95,** 447.

Madhupriya (2016) Molecular characterization of phytoplasmas associated with important ornamental plant species in Northern India. PhD Thesis, Amity University Haryana, India, 301 pp.

Madhupriya, Maurya R, Rao GP (2014) '*Candidatus* Phytoplasma aurantifolia' related strains affecting two important medicinal plants (*Cymbopogon citratus* and *Tylophora asthmatica*) in India. *Indian Phytopathology* **67**, 303–307.

Maejima K, Oshima K, Namba S (2014) Exploring the phytoplasmas, plant pathogenic bacteria. *Journal of General Plant Pathology* **80**, 210–221.

Mancini E, Marcone C, De Feo V, Senatore F, Formisano C (2010a) Changes in the composition of volatile compounds of *Spartium junceum* induced by the phytoplasmal disease, spartium witches' broom. *Plant Biosystems* **144**, 568–572.

Mancini E, Marcone C, De Martino L, De Feo V (2010b) Profili alcaloidei di piante di ginestra affette da spartium witches' broom. *Petria* **20**, 718–721.

Marcone C (2011) Current status of phytoplasma diseases of medicinal and nutraceutical plants in southern Italy. *Bulletin of Insectology* **64**(Supplement), S233–S234.

Marcone C (2014) Molecular biology and pathogenicity of phytoplasmas. *Annals of Applied Biology* **165**, 199–221.

Marcone C (2015) Current status of phytoplasma diseases of forest and landscape trees and shrubs. *Journal of Plant Pathology* **97**, 9–36.

Marcone C, Rao GP (2008) '*Candidatus* Phytoplasma cynodontis': the causal agent of Bermudagrass white leaf disease. In: Characterization, Diagnosis and Managemen of Phytoplasma. Eds Harrison NA, Rao GP, Marcone C. Studium Press LLC, USA, 353–364 pp.

Marcone C, Ragozzino A, Schneider B, Lauer U, Smart CD, Seemüller E (1996a) Genetic characterization and classification of two phytoplasmas associated with spartium witches' broom disease. *Plant Disease* **80**, 365–371.

Marcone C, Ragozzino A, Seemüller E (1996b) Detection of an elm yellows-related phytoplasma in eucalyptus trees affected by little leaf disease in Italy. *Plant Disease* **80**, 669–673.

Marcone C, Ragozzino A, Seemüller E (1997a) Detection of Bermuda grass white leaf disease in Italy and genetic characterization of the associated phytoplasma by RFLP analysis. *Plant Disease* **81**, 862–866.

Marcone C, Ragozzino A, Seemüller E (1997b) Identification and characterization of the phytoplasma associated with elm yellows in southern Italy and its relatedness to other phytoplasmas of the elm yellows group. *European Journal of Forest Pathology* **27**, 45–54.

Marcone C, Lee I-M, Davis RE, Ragozzino A, Seemüller E (2000) Classification of aster yellows-group phytoplasmas based on combined analyses of ribosomal RNA and *tuf* gene sequences. *International Journal of Systematic and Evolutionary Microbiology* **50**, 1703–1713.

Marcone C, Schneider B, Seemüller E (2004a) '*Candidatus* Phytoplasma cynodontis', the phytoplasma associated with Bermuda grass white leaf disease. *International Journal of Systematic and Evolutionary Microbiology* **54**, 1077–1082.

Marcone C, Gibb KS, Streten C, Schneider B (2004b) '*Candidatus* Phytoplasma spartii', '*Candidatus* Phytoplasma rhamni' and *Candidatus* Phytoplasma allocasuarinae', respectively associated with spartium witches' broom, buckthorn witches' broom and allocasuarina yellows diseases. *International Journal of Systematic and Evolutionary Microbiology* **54**, 1025–1029.

Marcone C, Bellardi MG, Bertaccini A (2016) Phytoplasma diseases of medicinal and aromatic plants. *Journal of Plant Pathology* **98**, 379–404.

Martini M, Marcone C, Lee I-M, Firrao G (2014) The family *Acholeplasmataceae* (including phytoplasmas). In: The Prokaryotes - Firmicutes and Tenericutes, 4th edition, XXII. Eds Rosenberg E, DeLong EF, Lory S, Stackebrandt E, Thompson F. Springer-Verlag, Berlin, Germany, 469–504 pp.

Marwitz R, Petzold H, Ozel M (1974) Studies on the transfer of the possible causal agent of apple proliferation to a herbaceous host. *Phytopathology* **81**, 85–91.

Mazzanti G, Lu M, Salvatore G, Capasso F, Basso F, Pasquale R, Evans FJ (1997) Spasmolytic action of the essential oil from *Hyssopus officinalis* var. decumbens and its major components. *Phytotherapy Research* **12**, 92–94.

McCoy RE, Caudwell A, Chang CJ, Chen TA, Chiykowski LN, Cousin MT, Dale JL, de Leeuw GTN, Golino DA, Hackett KJ, Kirkpatrick BC, Marwitz R, Petzold H, Sinha RC, Sugiura M, Whitcomb RF, Yang IL, Zhu BM, Seemüller E (1989) Plant diseases associated with mycoplasma-like organisms. In: The Mycoplasmas.V. Eds Whitcomb RF, Tully JG. Academic Press, Inc., San Diego, California, USA. 545-640 pp.

Mitrović J, Pavlović S, Duduk B (2013) Survey and multigene characterization of "stolbur" phytoplasmas on various plant species in Serbia. *Phytopathologia Mediterranea* **52**, 434–441.

Mitrović J, Smiljković M, Seemüller E, Reinhardt R, Hüttel B, Büttner C, Bertaccini A, Kube M, Duduk B (2015) Differentiation of '*Candidatus* Phytoplasma cynodontis' based on 16S rRNA and *groEl* genes and identification of a new subgroup, 16SrXIV-C. *Plant Disease* **99**, 1578–1583.

Mitrović M, Jović J, Cvrković T, Krstić O, Trkulja N, Toševski I (2012) Characterisation of a 16SrII phytoplasma strain associated with bushy stunt of hawkweed oxtongue (*Picris hieracioides*) in south-eastern Serbia and the role of the leafhopper *Neoaliturus fenestratus* (Deltocephalinae) as a natural vector. *European Journal of Plant Pathology* **134**, 647–660.

Montano HG, Davis RE, Dally EL, Hogenhout S, Pimentel JP, Brioso PST (2001) '*Candidatus* Phytoplasma brasiliense', a new phytoplasma taxon associated with hibiscus witches' broom disease. *International Journal of Systematic and Evolutionary Microbiology* **51**, 1109–1118.

Montano HG, Contaldo N, David TVA, Silva IB, Paltrinieri S, Bertaccini A (2011) Hibiscus witches' broom disease associated with different phytoplasma taxa in Brazil. *Bulletin of Insectology* **64**(Supplement), S249–S250.

Mori N, Quaglino F, Tessari F, Pozzebon A, Bulgari D, Casati P, Bianco PA (2015) Investigation on "bois noir" epidemiology in north-eastern Italian vineyards through a multidisciplinary approach. *Annals of Applied Biology* **166**, 75–89.

Musetti R, Favali MA, Carraro L, Osler R (1992) An attempt to differentiate by microscopic methods two plant mycoplasma-like organisms. *Cytobios* **72**, 71–82.

Nagori BP, Solanki R (2011) *Cynodon dactylon* (L.) Pers.: A valuable medicinal plant. *Research Journal of Medicinal Plants* **5**, 508–514.

Navrátil M, Fránová J, Válová P, Fialová R, Šinková M, Nebesářová J (2000) First report on a yellows disease of *Allium altyncolicum* associated with phytoplasma infection. *Zeitschrift für Pflanzenkrankheiten und Pflanzenschutz* **107**, 380–386.

Nayar R (1973) Spike disease of sandal and associated hosts of sandal. *Journal of Indian Academy of Wood Science* **4**, 96–102.

Nayar R, Ananthapadmanabha HS (1977) Little-leaf disease in collateral hosts of sandal (*Santalum album* L.). *European Journal of Forest Pathology* **7**, 152–158.

Nejat N, Vadamalai G, Davis RE, Harrison NA, Sijam K, Dickinson M, Akmar Abdullah SN, Zhao Y (2013) '*Candidatus* Phytoplasma malaysianum', a novel taxon associated with virescence and phyllody of Madagascar periwinkle (*Catharanthus roseus*). *International Journal of Systematic and Evolutionary Microbiology* **63**, 540–548.

Nishimura N, Nakajima S, Sawayanagi T, Namba S, Shiomi T, Matsuda I, Tsuchizaki T (1998) Transmission of *Cryptotaenia japonica* witches' broom and onion yellows phytoplasmas by *Hishimonus sellatus* Uhler. *Annals of the Phytopathological Society of Japan* **64**, 474–477.

Özdemir Z, Saygili H, Sahin F, Karsavuran Y, Bayrak OF, Oral B (2009) Host range and genetic characterization of a phytoplasma causing tomato stolbur disease in Turkey. *Acta Horticulturae* **808**, 255–262.

Paltrinieri S, Tasso L, Bellardi MG, Bertaccini A (2002) Identificazione molecolare di fitoplasmi associati a nanismo e scopazzi in iperico. *Petria* **12**, 393–394.

Paltrinieri S, Bellardi MG, Lesi F, Satta E, Davino S, Parrella G, Contaldo N, Bertaccini A (2018) Multilocus typing for characterization of '*Candidatus* Phytoplasma asteris'-related strains in several ornamental species in Italy. *Acta Horticulturae* **1193**, 55-61.

Pan Q, Mustafa NR, Tang K, Choi YH, Verpoorte R (2016) Monoterpenoid indole alkaloids biosynthesis and its regulation in *Catharanthus roseus*: a literature review from genes to metabolites. *Phytochemistry Reviews* **15**, 221–250.

Pavlović S, Starović M, Stojanović S, Popović T, Aleksić G, Dražić S, Jošić D (2011) *Echinacea purpurea* - a host of 16SrXII-A phytoplasma group in Serbia. *Phytopathogenic Mollicutes* **1**, 35–39.

Pavlović S, Jošić D, Starović M, Stojanović S, Aleksić G, Stojišin V, Radanović D (2012) The first "stolbur" phytoplasma occurrence on two St. John's worth species (*Hypericum perforatum* L. and *Hypericum barbatum* L.) in Serbia. *Journal of Medicinal Plant Research* **6**, 906–911.

Pavlović S, Starovic M, Stojanovic S, Aleksic G, Kojic S, Zdravkovic M, Josic D (2014) The first report of "stolbur" phytoplasma associated with phyllody of *Calendula officinalis* in Serbia. *Plant Disease* **98**, 1152.

Pellati F, Bruni R, Bellardi MG, Bertaccini A (2009) Optimization and validation of high-performance liquid chromatography method for the analysis of cardiac glycosides in *Digitalis lanata*. *Journal of Chromatography A* **1216**, 3260–3269.

Pellati F, Epifano F, Contaldo N, Orlandini G, Cavicchi L, Genovese S, Bertelli D, Benvenuti S, Curini M, Bertaccini A, Bellardi MG (2011) Chromatographic methods for metabolite profiling of virus- and phytoplasma-infected plants of *Echinacea purpurea*. *Journal of Agricultural and Food Chemistry* **59**, 10425–10434.

Pérez-López E, Luna-Rodríguez M, Olivier CY, Dumonceaux TJ (2016) The underestimated diversity of phytoplasmas in Latin America. *International Journal of Systematic and Evolutionary Microbiology* **66**, 492–513.

Přibylová J, Špak J, Fránová J, Petrzik K (2001) Association of aster yellows subgroup 16SrI-B phytoplasmas with a disease of *Rehmannia glutinosa* var. *purpurea*. *Plant Pathology* **50**, 776–781.

Radisek S, Ferant N, Jakse J, Javornik B (2009) Identification of a phytoplasma from the aster yellows group infecting purple coneflower (*Echinacea purpurea*) in Slovenia. *Plant Pathology* **58**, 392.

Raj SK, Snehi SK, Khan MS, Kumar S (2008) '*Candidatus* Phytoplasma asteris' (group 16SrI) associated with a witches' broom disease of *Cannabis sativa* in India. *Plant Pathology* **57**, 1173.

Raj SK, Snehi SK, Kumar S, Khan MS (2009) First finding of '*Candidatus* Phytoplasma trifolii' (16SrVI group) associated with little leaf disease of *Datura inoxia* in India. *Plant Pathology* **58**, 791.

Rao GP, Raj SK, Snehi SK, Mall S, Singh M, Marcone C (2007) Molecular evidence for the presence of '*Candidatus* Phytoplasma cynodontis', the Bermuda grass white leaf agent, in India. *Bulletin of Insectology* **60**, 145–146.

Rao GP, Madhupriya, Tiwari AK, Dubey DK (2017) Identification of *Ocimum canum* as possible alternative host of sesame phyllody phytoplasma strain 16SrI-B in India. *Phytopathogenic Mollicutes* **7**, 77–81.

Saeed ST, Khan A, Samad A (2015) First report on the molecular identification of phytoplasma (16SrII-D) associated with witches' broom of kalmegh (*Andrographis paniculata*) in India. *Plant Disease* **99**, 155.

Salehi M, Izadpanah K, Siampour M, Taghizadeh M (2009) Molecular characterization and transmission of Bermuda grass white leaf phytoplasma in Iran. *Journal of Plant Pathology* **91**, 655–661.

Salehi M, Esmailzadeh Hosseini SA, Salehi E (2016) First report of a '*Candidatus* Phytoplasma asteris' related phytoplasma associated with eucalyptus little leaf disease in Iran. *Journal of Plant Pathology* **98**, 175.

Salvatore G, D'Andrea A, Nicoletti M (1998) A pinocamphone poor oil of *Hyssopus officinalis* L. var. decumbens from France (Banon). *Journal of Essential Oil Research* **10**, 563–567.

Samad A, Shasany AK, Gupta S, Ajayakuar PV, Darokar MP, Khanuja SPS (2006) First report of a 16SrVI group phytoplasma associated with witches' broom disease of *Withania somnifera*. *Plant Disease* **90**, 248.

Sastry KSM, Thakur RN, Gupta JH, Pandotra VR (1971) Three virus disease of *Eucalyptus citriodora*. *Indian Phytopathology* **24**, 123–126.

Sdoodee R, Schneider B, Padovan AC, Gibb KS (1999) Detection and genetic relatedness of phytoplasmas associated with plant diseases in Thailand. *Journal of Biochemistry, Molecular Biology and Biophysics* **3**, 133–140.

Seemüller E, Marcone C, Lauer U, Ragozzino A, Göschl M (1998) Current status of molecular classification of the phytoplasmas. *Journal of Plant Pathology* **80**, 3–26.

Severin HHP, Frazier NW (1945) California aster yellows on vegetable and seed crops. *Hilgardia* **16**, 573–596.

Shah G, Shri R, Panchal V, Sharma N, Singh B, Mann AS (2011) Scientific basis for the therapeutic use of *Cymbopogon citratus*, stapf (lemongrass). *Journal of Advanced Pharmacological Technological Research* **2**, 3–8.

Sharma JK, Mohanan C, Florence EJM (1983) A little-leaf disease of eucalyptus in Kerala, India. *European Journal of Forest Pathology* **13**, 385–388.

Sichani FV, Bahar M, Zirak L (2014) Characterization of phytoplasmas related to aster yellows group infecting annual plants in Iran, based on the studies of 16S rRNA and *rp* genes. *Journal of Plant Protection Research* **54**, 1–8.

Singh N, Upadhyaya PP (2015) Weeds diversity in Eastern U.P. (Gorakhpur) associated with phytoplasma. *World Journal of Pharmaceutical Research* **4**, 968–974.

Singh SK, Aminuddin, Srivastava P, Singh BR, Khan AJ (2007) Production of phytoplasma-free plants from yellow leaf diseased *Catharanthus roseus* L. (G.) Don. *Journal of Plant Diseases and Protection* **114**, 2–5.

Singh N, Tiwari R, Upadhyaya PP (2013) A strain of phytoplasma related to 16SrVI group in *Datura stramonium* in India. *Greener Journal of Biological Sciences* **3**, 253–257.

Spallino RE, Rizza S, Marzachì C, Tessitori M (2013) Scopazzi della ginestra in Sicilia. *Petria* **23**, 41–44.

Srivastava S, Pandey R, Kumar S, Nautiyal CS (2014) Correspondence between flowers and leaves in terpenoid indole alkaloid metabolism of the phytoplasma-infected *Catharanthus roseus* plants. *Protoplasma* **251**, 1307–1320.

Torres E, Botti S, Paltrinieri S, Martin MP, Bertaccini A (2002) First report of spartium witches' broom disease in Spain. *Plant Pathology* **51**, 807.

Torres L, Galdeano E, Docampo D, Conci L (2004) Characterization of an aster yellows phytoplasma associated with *Catharanthus* little leaf in Argentina. *Journal of Plant Pathology* **86**, 209–214.

Tseng Y-W, Deng W-L, Chang C-J, Su C-C, Chen C-L, Jan FJ (2012) First report of a 16SrII-A subgroup phytoplasma associated with purple coneflower (*Echinacea purpurea*) witches' broom disease in Taiwan. *Plant Disease* **96**, 582.

Vibio M, Camele I, Bertaccini A, Rana GL, D' Aloisio V, Benni A (1995) Metodi diagnostici molecolari per l'individuazione di infezioni da fitoplasmi in cipolla. *L'Informatore Agrario* **35**, 75–77.

Vicente M, Caner J, July JR (1974) Corpusculos do tipo micoplasma em *Hibiscus rosa-sinensis*. *Arquives Instituto Biologico Saho Paulo* **41**, 53–58.

Wang K, Hiruki C (2005) Distinctions between phytoplasmas at the subgroup level detected by heteroduplex mobility assay. *Plant Pathology* **54**, 625–633.

Weintraub PG, Beanland L (2006) Insect vectors of phytoplasmas. *Annual Review of Entomology* **51**, 91–111.

Win NKK, Kim Y-H, Jung H-Y (2014) Bitter gourd little leaf disease associated to '*Candidatus* Phytoplasma asteris'. *Tropical Plant Pathology* **39**, 82–88.

Yang R, Wang G, Wang S, Zhang D, Wei L, Chen H, Li O, Hu X (2016) Molecular identification and diversity of '*Candidatus* Phytoplasma solani' associated with red-leaf disease of *Salvia miltiorrhiza* in China. *Journal of Phytopathology* **164**, 882–889.

Yesilada E, Takaishi Y (1999) A saponin with anti-ulcerogenic effect from the flowers of *Spartium junceum*. *Phytochemistry* **51**, 903–908.

Zaim M, Samad A (1995) Association of phytoplasmas with a witches-broom disease of *Withania somnifera* (L.) Dunal in India. *Plant Science* **109**, 225–229.

Zhou M-L, Shao J-R, Tang Y-X (2009) Production and metabolic engineering of terpenoid indole alkaloids in cell cultures of the medicinal plant *Catharanthus roseus* (L.) G. Don (Madagascar periwinkle). *Biotechnology and Applied Biochemistry* **52**, 313–323.

Chapter 9
Phytoplasma Diseases of Palms

Egya Ndede Yankey, Joao Bila, Yaima Arocha Rosete, Carlos Oropeza, and Fabian Pilet

Abstract Palms are native to tropical and subtropical areas of the world and contribute to the ecology and economy of many countries. Phytoplasmas which are cell-wall-less prokaryotes are associated with diseases of significant economic importance to palms worldwide. The diseases are vectored by insects, such as *Haplaxius crudus*, which transmits the lethal yellowing phytoplasma in Florida. In other places, such as in Africa, the vectors have so far been elusive, although putative vectors have been identified. Management of these palm diseases relies on integrated approaches involving the use of resistant materials and cultural methods such as early detection and removal of affected palms. The gaps in the present knowledge of the diseases provide excellent opportunities for further research on these enigmatic plant pathogens.

Keywords Palms · Phytoplasmas · Coconut · Lethal yellowing · Taxonomy · Vectors

---

E. N. Yankey (✉)
Oil Palm Research Institute, Coconut Programme, Council for Scientific and Industrial Research, Sekondi, Ghana
e-mail: ndedeyankey@yahoo.com

J. Bila
Departamento de ProtecçãoVegetal, Faculdade de Agronomia e Engenharia Florestal, Universidade Eduardo Mondlane, Maputo, Mozambique

Y. A. Rosete
Sporometrics, Toronto, Canada

C. Oropeza
Centro de Investigación Científica de Yucatan, Merida, México

F. Pilet
CIRAD, UMR PVBMT, Saint-Pierre, La Réunion, France

G. P. Rao et al. (eds.), *Phytoplasmas: Plant Pathogenic Bacteria - I*,
https://doi.org/10.1007/978-981-13-0119-3_9

9.1 Introduction

Palms belong to the Arecaceae family (formerly Palmaceae) which has about 180 genera and 2600 species. Palms are found in tropical and subtropical areas of the world and have significant economic and ecological importance. They provide man with food, medicine, construction materials, ornamentals, and fuel, particularly in rural communities. Palms of significant economic importance include the coconut (*Cocos nucifera* L.), the African oil palm (*Elaeis guineensis* Jacq.), and the date palm (*Phoenix dactylifera* L.). Phytoplasmas are cell-wall-less prokaryotes that until recently could not be cultivated in cell-free media (Lee et al. 2000; Razin 2007). Success in the work of Contaldo and co-workers (Contaldo et al. 2016) has, however, begun to overcome this hurdle. Phytoplasmas are associated with diseases in several plant species including grasses, vegetables, food crops, ornamental plants, fiber plants, and palms (Lee et al. 2000; Bertaccini and Duduk 2009; Gasparich 2010). In palms, phytoplasmas are associated with diseases in over 50 palm species (Harrison et al. 1999; Cordova et al. 2017); however, their devastative effect has been more prominent in coconut than in the other palms. In coconut, the most important phytoplasma disease is lethal yellowing (LY), as it is referred to in the Americas and the Caribbean region. Diseases with similar symptoms as LY occur in Africa and are collectively called lethal yellowing-like diseases (LYD). In Asia, diseases with similar symptoms as LY have also been reported. However, in these regions the phytoplasmas involved are different from those associated with diseases in Africa and the Caribbean region. The first scientific report of the disease was made at the close of the nineteenth century in Jamaica. The disease has wreaked great havoc in Jamaica, Ghana, Tanzania, Togo, Nigeria, Mozambique, Mexico, and very recently Cote d'Ivoire (Eden-Green 1997; Dollet et al. 2009; Arocha-Rosete et al. 2014). The disease is now the number one threat to coconut cultivation in many countries where it has killed millions of trees.

9.2 History and Distribution

The Americas and Caribbean Region LY has killed millions of palm trees in the Americas, mostly coconuts of the highly susceptible variety Atlantic Tall (Zizumbo-Villarreal et al. 2008).The infected susceptible coconut palms develop a syndrome that starts with nut drop within a few days, followed by necrosis of inflorescences and yellowing of younger leaves and then of the older leaves until all are affected and the palm dies (Zizumbo-Villarreal et al. 2008). Palms with such symptoms were originally reported in the nineteenth century in the Caribbean (Fawcett 1891 in Eden-Green 1997), but the first epidemic with extensive loss of palms occurred in Jamaica during the 1960s (Eden-Green 1997). Further spread of the disease has occurred in the USA (McCoy et al. 1983), México (Oropeza and Zizumbo 1997), Belize (Eden-Green 1997), Honduras (Ashburner et al. 1996), and Guatemala

(Mejía et al. 2004). However, LY has not yet moved south of Honduras into Nicaragua or any other country within Central America. On the other hand, it has spread through the Caribbean more rapidly, and it is currently present as far as Antigua (Ntushelo et al. 2013), threatening to spread to South America.

Africa Diseases with symptoms resembling those of LY were first reported in Africa at the beginning of the twentieth century. Nigeria appears to be the first country affected by LYD in West Africa in 1913 where it was called "awka wilt" disease (Johnson 1918). Similar diseases were reported in Togo ("kaincope" disease), Cameroon ("kribi" disease), and Ghana ("Cape St. Paul wilt" disease) by the early 1930s (Eden-Green 1997). Equatorial Guinea is also reported to be affected by the disease (Dollet et al. 2009). In East Africa, the disease has been reported in Tanzania (Lethal Disease: LD) (Schuiling et al. 1981), Kenya, and Mozambique (Eden-Green 1997; Mpunami et al. 1999; Bila et al. 2015a, b). Reports across coconut-growing regions of Mozambique, Nigeria, and Tanzania reveal that lethal yellowing-like diseases (LYD) remains a serious issue (Eziashi and Omar 2010; Munguambe et al. 2013; Arocha-Rosete et al. 2014; Bila et al. 2015b). In Nigeria, LYD has nearly wiped out the palm plantations in south-east and south-south and is now spreading toward the south-west (Eziashi and Omar 2010). In Mozambique, outbreaks of LYD have caused successions of epidemics and losses of millions of coconut palms, threatening the livelihood of a significant part of the Mozambican people, mainly for those living by the coastal belt. Likewise, in Tanzania the lethal disease (LD) has caused extensive losses to coconut plantations and is now the main limiting factor of coconut production throughout the coastal belt of the mainland (Schuiling et al. 1992; Mpunami et al. 2000; Eziashi and Omar 2010). Similar devastation occurred in Ghana, causing the near collapse of the once vibrant coconut industry in the Volta Region, one of the main coconut-producing areas of Ghana, as well as destroying thousands of hectares of palms in the Western and Central Regions of Ghana (Ofori and Nkansah-Poku 1997; Nkansah-Poku et al. 2009). The Cote d'Ivoire lethal yellowing (CILY) phytoplasma disease has destroyed hundreds of hectares of palms in a recent outbreak, with over 7000 more hectares at risk in the Grand-Lahou region of Cote d'Ivoire (Arocha-Rosete et al. 2014). The country hosts the International Coconut Genebank for Africa and the Indian Ocean, and the spread of the disease to new coconut-growing areas is of great concern.

Oceania The main production zone of coconut in Asia/Oceania was not affected by lethal yellowing diseases associated with one of the phytoplasmas of the palm groups (i.e., 16SrIV and 16SrXXII) during the twentieth century. In 2008, lethal yellowing-like symptoms were observed on both old and young coconut trees in the Madang Province of Papua-New Guinea (PNG) (Kelly et al. 2011). The disease was called "bogia" coconut syndrome (BCS). The Madang Province of PNG hosts the International Coconut Genebank for the South Pacific (ICG-SP), and because of its geographic proximity to Indonesia, a major global producer of coconut, the situation is of particular concern.

Asia A range of lethal yellowing-type diseases have been reported from Asia. However, the phytoplasmas involved in these diseases are different from those that cause LYD/LY in Africa and the Caribbean region, and the diseases are generally not lethal.

Coconut yellowing disease. In Malaysia, coconut lethal yellowing-like symptoms were observed about 90 years ago, but it was only reported as a serious threat after 2000 and referred to as coconut yellowing disease (CYD) attributed to a phytoplasma (Nejat et al. 2009a). Symptoms of CYD start with the yellowing of the canopy or eventually browning and chlorosis of the spear leaf. The coconut trees usually die within 5 months after infection (Nejat et al. 2009b). Similar symptoms of CYD have been observed on other palms such as foxtail palm (*Wodyetia bifurcata*) in 2012 (Naderali et al. 2013) and lipstick palm (*Cyrtostachys renda*) (Naderali et al. 2014). In 2013, a yellow decline was observed also on royal palms (*Roystonea regia*) (Naderali et al. 2014). The phytoplasma associated with CYD shared 99% 16S rRNA identity with Bermuda grass white leaf phytoplasma related to '*Ca.* P. cynodontis' (16SrXIV) (Nejat et al. 2009a, b).

"Weligama" coconut leaf decline (WCLWD) is a phytoplasma associated with lethal yellowing disease of coconut occurring in the Southern part of Sri Lanka (Perera et al. 2012). The first report of the disease was made in 2006 in the Weligama area, and the disease has since spread to other coconut-growing areas in Southern Sri Lanka (Wejisekara et al. 2008; Perera et al. 2012).

Symptoms of the disease include flattening and downward bending of leaflets giving the frond a ribbed or flaccid appearance. This is followed by yellowing and drying up of leaf margins. Dried fronds hang on the crown for some time before eventually falling off. Unlike lethal yellowing-like diseases, necrosis in unopened inflorescences and premature nut drop have not been observed for WCLWD. With fronds falling off, the crown becomes smaller and the trunk begins to taper. Female flower production declines and the palm becomes unproductive. WCLWD is associated with a phytoplasma belonging to the 16SrXI group (Perera et al. 2012) that is similar to the sugarcane white leaf phytoplasma (99% identity at the 16S rRNA gene level), the sugarcane grassy shoot phytoplasma (99% identity), as well as Kerala wilt phytoplasma. The vector of the disease is unknown, but the plant hopper *Proutista moesta* and the lace bug *Stephanitis typica* are considered the likely candidates (Wejisekara et al. 2008).

Kalimantan wilt disease of coconut was first reported in Indonesia by farmers in 1978, and an outbreak was observed in 1988 (Warokka et al. 2006). The involvement of phytoplasmas is not clear because of the detection of phytoplasmas in both symptomatic and asymptomatic coconut trees using universal phytoplasma primers, and the phytoplasma involved was not fully characterized (Warokka et al. 2006).

Kerala/root wilt disease of coconut is a devastating disease of coconut found in Southern India. The disease is known internationally as Kerala wilt disease; it does not kill the palms but causes massive yield losses. The phytoplasma involved was identified as a '*Ca.* P. oryzae'-related strain and belongs to 16SrXI-B subgroup

(Manimekalai et al. 2014). The insect *P. moesta* has been identified as the vector of the disease (Edwin and Mohankumar 2007a, b).

9.3 LY Phytoplasma Detection and Taxonomy

The identification of the phytoplasmas agent of LY was based on (a) observations by transmission electron microscopy (TEM) of their presence in phloem vessels of affected palms (Beakbane et al. 1972; Heinze et al. 1972; Plavsic-Banjac et al. 1972) and (b) differential responses of infected palms to the antibiotics penicillin and oxytetracycline (Hunt et al. 1974; McCoy 1972). Detection of the presence of phytoplasmas in the plant species showing LY symptoms or similar syndromes during the 1970s to the early 1990s was determined only by TEM (McCoy et al. 1983; Howard 1995), but more recently molecular techniques, particularly PCR, were developed for detection of group 16SrIV phytoplasmas (Harrison et al. 1999). The use of PCR, sequencing and phylogenetic analysis techniques has allowed the identification of groups and subgroups of the phytoplasmas associated with some of the syndromes reported for palms and non-palm species (Harrison and Oropeza 2008). Based on phylogenetic analysis of the 16S rRNA gene, the phytoplasmas associated with LY in the Americas have been classified within group 16SrIV (Lee et al. 1998). There are five subgroups identified: 16SrIV-A, 16SrIV-B, 16SrIV-D, 16SrIV-E, and 16SrIV-F (Ntushelo et al. 2013). Subgroup 16SrIV-A has been found in coconut and other palm species and has killed millions of coconut palms of the highly susceptible Atlantic Tall variety in different countries in the Americas, including Antigua (Myrie et al. 2014), Mexico (Narvaez et al. 2016), Belize, Cuba, Honduras, Jamaica, St. Kitts and Nevis (Myrie et al. 2012), and Florida (USA) (Ntushelo et al. 2013). Subgroup16SrIV-B was found in coconuts in Mexico and in coconuts and *Acrocomia aculeata* palms in Honduras (Roca et al. 2006). Subgroup 16SrIV-D has been found in a non-palm species *Carludovica palmata* (Cordova et al. 2000) and in *Sabal mexicana*, *Pseudophoenix sargentii*, and *Thrinax radiata* palms in Mexico (Vázquez-Euán et al. 2011) and in *Phoenix canariensis*, *Phoenix dactylifera*, *P. reclinata*, *P. roebelenii*, *P. sylvestris*, *Syagrus romanzoffiana*, *Butiagrus "nabonnandii"* (*Butia odorata* × *S. romanzoffiana*), *and Sabal palmetto* palms in the USA. Subgroup 16SrIV-E has only been found in coconut palms in the Dominican Republic (Martinez et al. 2008). Subgroup16SrIV-F was found in *Washingtonia robusta* and *Phoenix dactylifera* palms in the USA (Ntushelo et al. 2013). Based on 16S rDNA RFLP and sequence analyses of the 16S rRNA gene, Mpunami et al. (1999) showed that the Mozambican and Ghanaian samples were more similar to each other than those from Kenya and Tanzania. Tymon et al. (1998) showed that the phytoplasmas causing "awka" disease in Nigeria (LDN) and Cape St. Paul wilt disease (CSPWD) in Ghana were associated with either the same or very closely related strains. Later on, with advances in molecular tools, finer differentiation of phytoplasma groups associated with LYD in Africa has been established. The Mozambican and Nigerian LYD phytoplasma groups have been described as '*Ca.* P. palmicola' and assigned to 16SrXXII-A subgroup, whereas the closest group, from Ghana and Côte d'Ivoire, was assigned to a different ribosomal subgroup 16SrXXII-B and has been named as a '*Ca.* P. palmicola'-related strain (Harrison et al. 2014). Apart from '*Ca.* P. palmicola', the other group associated with LYD in Mozambique is the Tanzanian LD phytoplasma (Bila et al. 2015a). Until the end of the 1990s, phytoplasmas associated with the "maladie de Kaincopé" in Togo, "awka wilt" in Nigeria, and CSPWD in Ghana were listed in the 16SrIV group (Lee et al. 1998; Harrison and Oropeza 2008). Later, they were included in a new group, 16SrXXII (Wei et al. 2007). Based on their 16S rRNA gene sequences, phytoplasmas from West Africa (Nigeria and Ghana) and East Africa (Tanzania) were confirmed as two different subclades (Tymon et al. 1998), separate from the Caribbean coconut phytoplasmas (Gundersen et al. 1994). Such subclades were informally proposed as three separate 'Candidatus Phytoplasma' species (IRPCM 2004), which were further supported by phylogenetic analysis of secA gene sequences (Hodgetts et al. 2008). Based on the near full-length 16S rRNA gene, 16S–23S rRNA intergenic spacer region, and partial 23S rRNA gene sequences, Harrison et al. (2014) confirmed that the LY phytoplasma strain from Mozambique shared 100% identity with that of the "awka wilt" phytoplasma strain or LDN from Nigeria (GenBank accession number Y14175). The sequences from the CSPW phytoplasma were observed to be identical to those of the CILY phytoplasma. Sequences from the Mozambique LY phytoplasma shared 99.0–99.6% identity with those of CSPWD and CILY phytoplasmas but less than 97.5% identity with reference strains from all previously described '*Candidatus* Phytoplasma' species. Further analysis of virtual 16S rRNA RFLP profiles and similarity coefficients delineated the Mozambique LY phytoplasma strains as members of group 16SrXXII-A (Harrison et al. 2014).

The "bogia" coconut syndrome (BCS) phytoplasma is related to but distinct from all phytoplasmas described above; it appears to be closer to the lethal disease phytoplasma (Tanzania) (formally group 16SrIV) with only 96% identity (Kelly et al. 2011) and group 16SrXXII, suggesting that this phytoplasma is part of the lethal yellowing disease cluster.

9.4 Symptoms of LY in Coconut

There is no one symptom which predicts the presence of LYD but rather the symptoms can be considered as a syndrome involving a series of symptoms. The symptoms of LYD are typically very similar and include fruit abortion, necrosis of inflorescences, progressive yellowing of the leaves, rotting of the stem apical tissues, and wilting and collapse of the palm crown leading to coconut mortality (Dollet et al. 2009; Bertaccini et al. 2014; Harrison et al. 2014) (Figs. 9.1 and 9.2). Though the diseases are very similar, there are some differences in the evolution of the symptoms. The diseases are usually grouped into stages, but these are not

Fig. 9.1 Typical LYD symptoms progression in Mozambique: premature nut drop (**a**); progressive yellow discoloration from the oldest to the youngest leaves followed by skirt-shaped brown discoloration (necrosis) of the older leaves (**b, c**); rotting and death of the apical meristem/spear leaves (**d**) followed by wilting and collapse of the entire crown (**e**) leaving an empty stem (**f**)

consistent across countries. For the Cape St. Paul wilt disease (CSPWD) in Ghana, five stages have been suggested by Dery and Philippe (1997): stage 1, premature nut drop, blackening of the inflorescence with or without one or two yellow leaves; stage 2, more than two leaves, but less than half the canopy, yellow; stage 3, more than half canopy yellow but with some green leaves still present; stage 4, all leaves in canopy yellow; and stage 5, telephone pole.

In the case of the Cote d'Ivoire LY (CILY) and LY, three stages of the disease have been outlined but still with differences. Stage 1 of LY is recognized solely by premature nut fall (Harrison and Elliot 2005). CILY stage 1, on the other hand, includes the starting of yellowing of the older leaves as well as the beginning of inflorescence blackening, in addition to premature nut fall (Arocha-Rosete et al. 2017). LY stage 2 involves only necrosis of the inflorescences (Harrison and Elliot 2005), while coconut palms at CILY stage 2 additionally exhibit yellowing of the older leaves, progressing to the younger leaves. Yellowing of LY-affected palms does not show yellowing of fronds until stage 3 of the disease. For CSPWD, leaf yellowing is present in more than two leaves but less than half the canopy at disease stage 2, in more than half of the canopy at stage 3, and in all leaves in the canopy at

Fig. 9.2 Sequence of symptom development of CSPWD in Ghana. Premature dropping of nuts (**a**); toppling of crown (**c**); blackening of inflorescence (**b**); progressive yellowing of fronds (**d**–**e**); bare trunks or telephone poles (**g**–**h**); a severely devastated farm (**i**) (Courtesy of J. Nkansah-Poku)

stage 4 (Dery and Philippe 1997). For both LY and CILY at stage 3, the yellowed leaves eventually turn brown, desiccate, and hang down, forming a skirt around the trunk before falling and turning into telephone poles.

9.5 Host Range of LYD Phytoplasmas

Symptoms similar to those of LY in coconuts have been observed in more than 50 other palm species (McCoy et al. 1983; Harrison et al. 1999; Howard 1995; Ntushelo et al. 2013; Myrie et al. 2014; Vázquez-Euán et al. 2011; Narvaez et al. 2006, 2016; Roca et al. 2006; Cordova et al. 2017). McCoy et al. (1983) classified palms according to their level of susceptibility, and the most susceptible palms include *Cocos nucifera*, *Phoenix dactylifera*, and different *Pritchardia* species. They also reported a list of palm species that were not known to be affected by LY, including *Elaeis guineensis*, *Roystonea regia*, *Sabal palmetto*, *Thrinax radiata*, and *Washingtonia robusta*. However, more recent reports have shown that for *R. regia* (Narvaez et al.

2016), *S. palmetto* (Harrison et al. 2009), and *T. radiata* (Narvaez et al. 2006), some individuals can be affected by LY. Non-palm species have also been reported to be associated with LY phytoplasmas. That is the case of *Pandanus utilis* (Thomas and Donselman 1979) and *Carludovica palmata* (Cordova et al. 2000). In both cases LY-type syndromes were observed. In addition, five species of plants (*Emilia fosbergii*, *Synedrella nodiflora*, *Stachytarpheta jamaicensis*, *Macroptilium lathyroides*, and *Cleome rutidosperma*) growing below the coconut palms in groves in Jamaica have been found positive with group 16SrIV phytoplasma, although they were symptomless (Brown et al. 2008; Brown and McLaughlin 2011). In Africa, the oil palm (*Elaeis guineensis*) and fan palm (*Borassus aethiopum*) were found harboring the LYD phytoplasma in Mozambique and assigned as alternate hosts for '*Ca*. P. palmicola' (Bila et al. 2015b).

In Côte d'Ivoire, the CILY phytoplasma was found in six plant species from five botanical families based on secA and 16S rDNA sequence analysis from nested PCR (Arocha-Rosete et al. 2016). The plant species included Poaceae (*Paspalum vaginatum*, *Pennisetum pedicellatum*), Verbenaceae (*Stachytarpheta indica*), Plantaginaceae (*Scoparia dulcis*), Phyllanthaceae (*Phyllanthus muellerianus*), and Cyperaceae (*Diplacrum capitatum*). No symptoms of phytoplasma diseases were observed on these alternative plant hosts, suggesting that asymptomatic reservoirs are present for the CILY phytoplasmas. Recently, cassava (*Manihot esculenta*) has also been described as an alternative host plant of the CILY phytoplasma (Kra et al. 2017). In Ghana, Yankey et al. (2009) and Yankey (Yankey 2012) performed extensive surveys during both rainy and dry season in two different areas, Asebu and Cape Three Points in the Central and Western Regions, respectively, sampling and analyzing 57 plant species. PCR amplicons using CSPWD-specific primers were produced from *Desmodium adscendens*; however, sequence analysis confirmed the presence of *Bacillus megaterium* and *Rhodobacter sphaeroides*. These bacteria have been previously isolated from trunks of date palms (*Phoenix canariensis*, Chabaud) affected by the lethal decline phytoplasma using a universal phytoplasma primer pair (P1/P7) in Texas (Harrison et al. 2002). Three of the species identified as alternate hosts in Côte d'Ivoire (*S. indica*, *P. pedicellatum*, and *M. esculenta*) had been assessed in Ghana using direct PCR, but they were not found to harbor the phytoplasma even though the phytoplasmas implicated in both diseases are in the same subgroup (16SrXXII-B). The phytoplasma recently described as BCS (Bogia coconut syndrome), phylogenetically related to the palm phytoplasma, is homologous with an emerging phytoplasma of another monocot: the banana wilt associated phytoplasma (BWAP), observed on cooking banana plants (Davis et al. 2012).

9.6 Transmission of LYD Phytoplasmas

Using different approaches evidence has been obtained that supports *Haplaxius crudus* (Hemiptera: Auchenorrhyncha: Cixiidae) (previously *Myndus crudus*) as an insect vectoring LY phytoplasmas in Florida (McCoy et al. 1983). Most notable are the studies by Howard (1995) demonstrating that *H. crudus* insects were able to transmit phytoplasmas to different palm species that developed LY symptoms and died within insect-proof cages. Later studies using PCR have shown that phytoplasmas in *H. crudus* belong to the 16SrIV group (Harrison and Oropeza 1997). Transmission studies similar to those by Howard (1995) testing *H. crudus* as a vector have been performed outside Florida without success so far. In addition, occurrence of LY (16SrIV phytoplasmas) has been found in *Cedusa* sp. (Hemiptera: Auchenorrhyncha: Derbidae) in Jamaica (Brown et al. 2006). Although no transmission studies have been carried out testing *Cedusa* insects, this occurrence supports the potential existence of other vectors of LY phytoplasmas.

Based on molecular screening, different plant hoppers and leafhoppers were associated with LYD in different parts of the world (Mpunami et al. 2000; Bila et al. 2017), but this does not prove their ability to transmit the phytoplasma, since vector status can only be established after successful transmission trials. Nevertheless, distinction between resistance to the vector transmission ability or to the phytoplasma host infection cannot be easily distinguished (Jarausch et al. 2013). Based on molecular screening, Pentatomidae specimens of *Platacantha lutea* were revealed as a potential insect vectoring LYD phytoplasma in northern Mozambique (Dollet et al. 2011). A further search, including specimens from all over the country, has shown *Diostrombus mkurangai* Wilson (Wilson 1987) as a potential LYD phytoplasma insect vector in Mozambique (Bila et al. 2017). However, the same study failed to confirm any of the pantatomids previously tested as harboring LYD phytoplasma. In Tanzania, LD phytoplasma is associated with the plant hoppers *D. mkurangai* and *Meenoplus* sp. (Mpunami et al. 2000), but transmission studies have not been successful. The vector for CSPWD phytoplasma in Ghana is still unknown despite massive surveys that resulted in over 12,500 specimens representing 203 species of 19 families (Pilet et al. 2009), as well as cage transmission trials with more than 70,000 *Myndus adiopodoumeensis* for 28 months (520 adults/seedling/month). Although two species, *Diostrombus* sp. (Derbidae) and *Myndodus adiopodoumeensis* (Synave) (Cixiidae), formerly placed in the genus *Myndus* (*M. adiopodoumeensis*) were found to carry the CSPWD phytoplasma, transmission trials were inconclusive (Philippe et al. 2009). In Côte d'Ivoire, surveys of coconut farms in Grand-Lahou revealed the presence of eight major Hemiptera families: Aphrophoridae, Achilidae, Derbidae, Flatidae, Membracidae, Pentatomidae, Tropiduchidae, and Cicadellidae. Specimens from the families Cicadellidae and Derbidae were the most abundantly collected. The family Cicadellidae was represented by a recently described genus and species of the tribe Erythroneurini, *Nedotepa curta* Dmitriev (Cicadellidae: Typhlocybinae: Erythroneurini) (Dmitriev 2016). Dmitriev (2016) provided a morphological description and detailed illustrations of *N. curta* specimens collected from coconut palms in the Western Region of Ghana. *N. curta* collected in Grand-Lahou in Côte d'Ivoire appears to be morphologically identical to specimens from Ghana. Nested PCR and sequence analyses of the 16S rRNA and the *sec*A genes confirmed the presence of the CILY phytoplasma in 216 out of 296 (73%) of the *N. curta* specimens, which suggested *N. curta* as a potential vector for the CILY phytoplasma (Kwadjo et al. 2018).

9.7 Management of LYD

Integrated pest management (IPM) is an approach to crop production and protection that combines different management strategies and practices to grow healthy crops and minimize the use of pesticides (FAO 2017). Successful IPM against LYD is through strict quarantine and disease surveillance, prompt detection followed by immediate removal and destruction of LYD infected trees, proper weeding of alternative plant hosts, replanting with resistant varieties, and control of the insect vector (Brown et al. 2008; Eziashi and Omar 2010; Myrie et al. 2011; Munguambe et al. 2013; Gurr et al. 2016).

In Jamaica it is believed that there is a LY variation that might involve a phytoplasma different to that found before that is affecting palms that were previously considered resistant to LY. So far a coconut variety able to resist this disease has not been found. However by applying an IPM program, it was possible to manage LY in Michael Black Farm, a coconut plantation located in the parish of St. Thomas, an area affected by the LY variant. This IPM program includes (a) monitoring to detect palms with early LY symptoms, (b) elimination of these palms as soon as they are detected, (c) immediate replanting, and (d) weeding. There are approximately 62,000 coconuts in Michael Black Farm, and within the period of 2002–2013, only 915 palms or 1.5% have been lost (Coconut Industry Board 2013). In Mexico, management of LY is based on the use of coconut germplasm known to be resistant to LY (Zizumbo-Villarreal et al. 2008) as the main approach, and such management is working for replanting programs in different regions of Mexico that have been affected by LY.

In Mozambique, the widely endorsed management approach for LYD is felling and burning of symptomatic coconut trees, which are replaced by another variety believed to be tolerant to the disease (Munguambe et al. 2013). Finding germplasm that is resistant to LYD in Mozambique has so far been a challenge. In Ghana, a field practice called "slow down" has been implemented in an integrated control strategy based on early detection and prompt removal of diseased palms to control CSPWD spread (Danyo 2011). This was based on previous observations that regularly cutting out all diseased palms slows down the rate of spread of the disease (Nkansah-Poku et al. 2009). Moreover, Nkansah-Poku et al. (2005) showed that the treatment of infected farms with insecticide by hot fogging followed by felling diseased and contact palms, immediately upon detection, slows down CSPWD and in some cases completely holds the disease for few years.

Intercropping is widely used as an alternative means for managing a range of diseases. Intercropping coconut with other crops have failed to lower the disease incidence, but provided an alternative source of income as insurance against CSPWD in Ghana (Andoh-Mensah and Ofosu-Budu 2012) and LD in Tanzania (Oleke et al. 2012). In Mozambique, Bila et al. (2016) observed that intercropped coconut farms were more vulnerable to LYD infection than single cropping. Hence, the effects of intercropping on LYD epidemiology require more effort to be elucidated.

In Jamaica, the heavy losses suffered by the Maypan hybrid which was planted extensively as a control of LY were partly explained by the genetic contamination of the Panama Tall (PNT), the pollen parent, with pollen from the susceptible Jamaican Tall ecotype and a large percentage of off-types observed in the MYD mother palms (Baudouin et al. 2008; Lebrun et al. 2008). The use of marker-assisted breeding in ensuring genetic purity of breeding materials is therefore an important factor to be considered. Swarbrick et al. (2013) developed a marker based on coconut receptor-like kinase genes and a high-throughput genotyping system based on high-resolution melt curve analysis to validate the genetic purity of breeding material resistant to CSPWD in Ghana, as well as to identify infected breeding material before it is provided to growers, and to prevent resistance breakdown. Such technologies need to be encouraged in the breeding programs of the various countries.

9.8 Non-lethal Yellowing Diseases of Palms

Phytoplasmas are also associated with other palm diseases which do not belong to the lethal yellowing type diseases. These diseases are associated with phytoplasmas belonging to 16Sr groups other than those in 16SrIV and 16SrXXII. The host plants involved in such diseases include the oil palm, date palm, and arecanut palm (Table 9.1). Although these diseases can cause significant economic losses, they do not cause the death of the palms. Yield losses of up to 50% have been reported for the yellow leaf disease of arecanut in India (Muddumadiah et al. 2014). The "Alwijam" disease of date palms in Saudi Arabia and Kuwait is also known to cause significant fruit failure at harvest (Alhudaib et al. 2007; Gurr et al. 2015).

Table 9.1 List of palm phytoplasma diseases associated with nonlethal yellowing phytoplasmas

Disease	Host species	Ribosomal group	country	Reference
Oil palm lethal wilt	Oil palm	16SrI-B	Colombia	Alvarez et al. (2014)
Oil palm stunting	Oil palm	16SrI-B	India	Medhi et al. (2012)
	Arecanut palm	16SrXI	India	Nair et al. (2014)
Yellow leaf	Arecanut palm	16SrI-B	India	Muddumadiah et al. (2014)
"Alwijam"	Date palm	16SrI	Saudi Arabia; Kuwait	Alhudaib et al. (2007, 2002)
White tip dieback	Date palm	16SXIV-A	Sudan	Cronjé et al. (2000)

9.9 Conclusions

Phytoplasma-associated palm diseases have significant economic impact, particularly, for poor families that depend on the palms for their sustenance. The ability of the diseases to cause unrestrained destruction in coconut, for example, is a major concern that requires international concerted effort to address. The many gaps that exist in the present understanding of the diseases provide exciting opportunities for further research. It is hoped that the recent breakthrough in the culture of phytoplasmas will help to speed up research on these enigmatic plant pathogens and their diseases.

Both LY and CSPWD phytoplasma DNA have been detected in coconut embryos in Mexico (Cordova et al. 2003) and Ghana (Nipah et al. 2007), respectively. Oropeza et al. (2017) confirmed the occurrence of the transmission of LY phytoplasma from coconut embryos to plantlets obtained from in vitro germination of zygotic embryos from the seeds of LY phytoplasma-infected coconut palms. This raises concerns about possible seed transmission of LY, a development which requires appropriate quarantine and seed movement policies to forestall intra and inter county spread of the disease.

The epidemiology of these diseases needs to be elucidated. In many countries, the insect vectors of the disease are unknown and the search for the vectors must continue. In Mozambique, field testing of the transmission ability of *D. mkurangai* to confirm its LYD phytoplasma vector status would have to be done. Conducting a transmission field trial for putative LYD insect vector is regarded as logistically difficult. Indirect potential insect vector screening through membrane-feeding assay followed by PCR testing for phytoplasma detection on the media (Tanne et al. 2001) maybe an alternative option. Similarly, fast and more effective transmission trials would have to be developed to prove the vectoring capacity of *N. curta* in both Côte d'Ivoire and Ghana.

References

Al-Awadhi H, Hanif A, Suleman P, Montasser M (2002) Molecular and microscopical detection of phytoplasma associated with yellowing disease of date palm *Phoenix dactilifera* L. in Kuwait. *Kuwait Journal of Science Engeneering* **29**, 87–109.

Alhudaib K, Arocha Y, Wilson M, Jones P (2007) "Al-Wijam", a new phytoplasma disease of date palm in Saudi Arabia. *Bulletin of Insectology* **60**, 285–286.

Alvarez E, Mejía JF, Contaldo N, Paltrinieri S, Duduk B, Bertaccini A (2014) '*Candidatus* Phytoplasma asteris' strains associated with oil palm lethal wilt in Colombia. *Plant Disease* **98**, 311–318.

Andoh-Mensah E, Ofosu-Budu GK (2012) Evaluation of coconut citrus intercropping systems in context of coconut lethal yellowing in Ghana. *African Journal of Food Agricultural Nutrition and Development* **12**, 6945–6962.

Arocha-Rosete Y, Konan Konan JL, Diallo HA, Allou K, Scott JA (2014) Identification and molecular characterization of the phytoplasma associated with a lethal yellowing-type disease of coconut in Côte d'Ivoire. *Canadian Journal of Plant Pathology* **36**, 141–150.

Arocha-Rosete Y, Diallo HA, Konan Konan JL, Kouamé AEP, Séka K, Kra KD, Toualy MN, Kwadjo KE, Daramcoum WAMP, Beugré NI, Ouattara BWM, Kouadjo-Zaka CG, Allou K, Fursy-Rodelec ND, Doudjo-Ouattara ON, Yankey N, Dery S, Maharaj A, Saleh M, Summerbell, R, Contaldo N, Paltrinieri S, Bertaccini A, Scott J (2016) Detection and identification of the coconut lethal yellowing phytoplasma in weeds growing in coconut farms in Côte d'Ivoire. *Canadian Journal of Plant Pathology* **38**, 164–173.

Arocha-Rosete Y, Diallo HA, Konan Konan JL, Yankey N, Saleh M, Pilet F, Contaldo N, Paltrinieri S, Bertaccini A, Scott J (2017) Detection and differentiation of the coconut lethal yellowing phytoplasma in coconut-growing villages of Grand-Lahou, Côte d'Ivoire. *Annals of Applied Biology* **170**, 333–347.

Ashburner GR, Cordova I, Oropeza C, Illingworth R, Harrison NA (1996) First report of coconut lethal yellowing in Honduras. *Plant Disease* **80**, 960.

Baudouin L, Lebrun P, Berger A, Myrie W, Been B, Dollet M (2008) The Panama Tall and the Maypan hybrid coconut in Jamaica: did genetic contamination cause a loss of resistance to lethal yellowing? *Euphytica* **161**, 353-360.

Beakbane AB, Slater CHW, Posnette AC (1972) Mycoplasmas in the phloem of coconut, *Cocos nucifera* L., with lethal yellowing disease. *Journal of Horticultural Science* **47**, 265.

Bertaccini A, Duduk B (2009) Phytoplasma and phytoplasma disease: a review of recent research. *Phytopathologia Mediterranea* **48**, 355–378.

Bertaccini A, Duduk B, Paltrinieri S, Contaldo N (2014) Phytoplasmas and phytoplasma diseases: a severe threat to agriculture. *American Journal of Plant Sciences* **5**, 1763–1788.

Bila J, Mondjana A, Samils B, Högberg N (2015a) High diversity, expanding populationsand purifying selection in phytoplasmas causing coconut lethal yellowing in Mozambique. *Plant Pathology* **64**, 597–604.

Bila J, Högberg N, Mondjana A, Samils B (2015b). African fan palm (*Borassus aethiopum*) and oil palm (*Elaeis guineensis*) are alternate hosts of coconut lethal yellowing phytoplasma in Mozambique. *African Journal of Biotechnology* **14**, 3359–3367.

Bila J, Persson I, Högberg N, Mondjana A, Manuel L, Tostao E, Johansson N, Soderlind J, Santos L (2016) Effect of farming practices and farm history on the incidence of coconut lethal yellowing in Mozambique. *African Crop Science Journal* **24**, 167–178.

Bila J, Högberg N, Mondjana A, Berit S, Michael W, Santos L (2017) First report of '*Candidatus* Phytoplasma palmicola' detection in the planthopper *Distrombus mkurangai* in Mozambique. *Bulletin of Insectology* **70**, 45–48.

Brown SE, McLaughlin WA (2011) Identification of lethal yellowing group (16SrIV) of phytoplasmas in the weeds *Stachytarphetaja maicensis*, *Macroptilium lathyroides* and *Cleome rutidosperma* in Jamaica. *Phytopathogenic Mollicutes* **1**, 27–34.

Brown SE, Been BO, McLaughlin WA (2006) Detection and variability of the lethal yellowing group (16Sr IV) phytoplasmas in the *Cedusa* sp. (Hemiptera: Auchenorrhyncha: Derbidae) in Jamaica. *Annals of Applied Biology* **149**, 53–62.

Brown SE, Been BO, McLaughlin WA (2008) First report of the presence of the lethal yellowing group (16SrIV) of phytoplasmas in the weeds *Emilia fosbergii* and *Synedrella nodiflora* in Jamaica. *Annals of Applied Biology* **57**, 770.

Coconut Industry Board (2013) Final report of project sustainable coconut production through control of Coconut Lethal Yellowing (CFC/FIGOOF/22). CFC, Amsterdam, The Netherlands.

Contaldo N, Satta E, Zambon S Paltrinieri S, Bertaccini A (2016) Development and evaluation of different complex media for phytoplasma isolation and growth. *Journal of Microbiological Methods* **127**, 105–110.

Cordova I, Oropeza C, Almeyda H, Harrison NA (2000) First report of a phytoplasma-associated leaf yellowing syndrome of palma jipi plants in southern México. *Plant Disease* **84**, 807.

Cordova I, Jones P, Harrison NA, Oropeza C (2003) *In situ* detection of phytoplasma DNA in embryos from coconut palms with lethal yellowing disease. *Molecular Plant Pathology* **4**, 99–108.

Cordova LI, Mota NL, Puch HC, Oropeza C, Senz L (2017) Detection and identification of lethal yellowing phytoplasma 16SrIV-A and D associated with *Adonidia merrillii* palms in Mexico. *Australasian Plant Pathology* **46**, 389–396.

Cronjé P, Dabek AJ, Jones P, Tymon AM (2000) Slow decline: a new disease of mature date palms in North Africa associated with phytoplasma. *Plant Pathology* **49**, 804.

Danyo G (2011) Review of scientific research into the Cape Saint Paul wilt disease (CSPWD) of coconut in Ghana. *African Journal of Agricultural Research* **6**, 4567–4578.

Davis RI, Kokoa P, Jones LM, Mackie J, Constable FE, Rodoni BC, Gunua TG, Rossel JB (2012) A new wilt disease of banana plants associated with phytoplasmas in Papua New Guinea (PNG). *Australasian Plant Disease Notes* **7**, 91.

Dery SK, Philippe R (1997) Preliminary study on the epidemiology of Cape St Paul wilt disease of coconut in Ghana. In: Proceedings of an International workshop on lethal yellowing-like diseases of coconut. Eds Eden-Green SJ, Ofori F, Natural Resources Institute, Chatham, United Kingdom, 255–260 pp.

Dmitriev DA (2016) Three unusual species of Erythroneurini (Hemiptera: Cicadellidae) with descriptions of new genera. *Zootaxa* **4173**, 183–191.

Dollet M, Quaicoe R, Pilet F (2009) Review of coconut "lethal yellowing" type diseases: diversity, variability and diagnosis. *Oleagineux* **16**, 97–101.

Dollet M, Macome F, Vaz A, Fabre S (2011) Phytoplasmas identical to coconut lethal yellowing phytoplasmas from Zambesia (Mozambique) found in a pentatomide bug in Cabo Delgado province. *Bulletin of Insectology* **64**(Supplement), S139–S140.

Eden-Green SJ (1997) History, distribution and research on coconut lethal yellowing-like diseases of palms. In: Proceedings of the International Workshop on Lethal Yellowing-like Diseases of Coconut. Ed Eden-Green SJ, Ofori F, Natural Resources Institute, Chatham, United Kingdom, 9–25 pp.

Edwin BT, Mohankumar C (2007a) Kerala wilt disease phytoplasma: phylogenetic analysis and identification of a vector *Proutista moesta*. *Physiological and Molecular Plant Pathology* **71**, 41–47.

Edwin BT, Mohankumar C (2007b) Molecular identification of *Proutista moesta* as the vector and the phylogenetic analysis of KWD phytoplasma. *Indian Journal of Biotechnology* **6**, 560–563.

Eziashi E, Omar I (2010) Lethal yellowing disease of the coconut palms (*Cocos nucifera* L.): an overview of the crises. *African Journal of Biotechnology* **9**, 9122–9127.

FAO (2017) AGP-Integrated Pest Management (http://www.fao.org/agriculture/crops/thematic-sitemap/theme/pests/ipm/en/).

Gasparich GE (2010) Spiroplasmas and phytoplasmas: Microbes associated with plant hosts. *Biologicals* **38**, 193–203.

Gundersen DE, Lee I-M, Rehner SA, Davis RE, Kingsbury DT (1994) Phylogeny of mycoplasma-like organisms (phytoplasmas): a basis for their classification. *Journal of Bacteriology* **176**, 5244–5254.

Gurr GM, Bertaccini A, Gopurenko D, Krueger R, Alhudaib K, Liu J, Fletcher MJ (2015) Phytoplasmas and their insect vectors: implications for date palm. In: Sustainable Pest Management in Date Palm: Current Status and Emerging Challenges. Eds Waqas W, Faleiro R, Miller J, Thomas A. Springer, 287–314 pp.

Gurr GM, Johnson AC, Ash GJ, Wilson BAL, Ero MM, Pilotti CA, Dewhurst CF, You MS (2016) Coconut lethal yellowing diseases: a phytoplasma threat to palms of global economic and social significance. *Frontieres in Plant Sciences* **7**, 1521.

Harrison NA, Oropeza C (1997) Recent studies on detection of lethal yellowing disease phytoplasmas in the Americas. In: Proceedings of the International Workshop on Lethal Yellowing-like Diseases of Coconut. Eds Eden-Green SJ, Ofori F, Natural Resources Institute, Chatham, United Kingdom, 221–234 pp.

Harrison NA, Elliot M (2005) Lethal yellowing (LY) of palm. Plant Pathology Department, University of Florida, IFAS Extension, 222 pp. URL http://edis.ifas.ufl.edu.

Harrison NA, Oropeza C (2008) Coconut lethal yellowing. In: Characterization, Diagnosis and Management of Phytoplasmas. Eds Harrison NA, Rao GP, Marcone C. Studium Press. Houston, USA, 219–248 pp.

Harrison NA, Cordova I, Richardson P, Dibonito R (1999) Detection and diagnosis of lethal yellowing. In: Current Advances in Coconut Biotechnology. Eds Oropeza C, Verdeil JL, Ashburner GR, Cardeña R, Santamaría JM. Kluwer Academic Publishers, Dordrecht, The Netherlands, 183–196 pp.

Harrison NA, Womack M, Carpio ML (2002) Detection and characterization of a lethal yellowing (165rIV) group phytoplasma in Canary Island date palms affected by lethal decline in Texas. *Plant Disease* **86**, 676–681.

Harrison NA, Helmick EE, Elliot ML (2009) First report of a phytoplasma-associated lethal decline of *Sabal palmetto* in Florida, USA. *Plant Pathology* **58**, 792.

Harrison NA, Davis RE, Oropeza C, Helmick E, Narvaez M, Eden-Green S, Dollet M, Dickinson M (2014) '*Candidatus* Phytoplasma palmicola', a novel taxon associated with a lethal yellowing-type disease (LYD) of coconut (*Cocos nucifera* L.) in Mozambique. *International Journal of Systematic and Evolutionary Microbiology* **64**, 1890–1899.

Heinze KG, Schuiling M, Romney DH (1972) The possible cause of lethal yellowing disease of coconut. *FAO Plant Protection Bulletin* **20**, 58–68.

Hodgetts J, Boonham N, Mumford R, Harrison N, Dickinson M. 2008. Phytoplasma phylogenetics based on analysis of secA and 23S rRNA gene sequences for improved resolution of candidate species of '*Candidatus* Phytoplasma'. *International Journal of Systematic and Evolutionary Microbiology* **58**, 1826–1837.

Howard FW (1995) Lethal yellowing vector studies. I. Methods of experimental transmission. In: Lethal Yellowing Research and Practical Aspects. Eds Oropeza C, Howard FW, Ashburner GR. Kluwer Academic Publishers, Dordrecht, The Netherlands, 43–57 pp.

Hunt P, Dabek AJ, Schuiling M (1974) Remission of symptoms following tetracycline treatment of lethal yellowing infected coconut palms. *Phytopathology* **63**, 307–312.

IRPCM (2004) '*Candidatus* Phytoplasma', a taxon for the wall-less, nonhelical prokaryotes that colonize plant phloem and insects. *International Journal of Systematic and Evolutionary Microbiology* **54**, 1243–1255.

Jarausch W, Angelini E, Eveillard S, Malembic-Maher S (2013) Management of fruit tree and grapevine phytoplasma disease through genetic resistance. *Phytopathogenic Mollicutes* **3**, 16–24.

Johnson WH (1918) Annual Report of the Agricultural Department of Southern Provinces, Nigeria for the year 1917. Government Publication.

Kelly PL, Reeder R, Kokoa P, Arocha Y, Nixon T, Fox A (2011) First report of a phytoplasma identified in coconut palms (*Cocos nucifera*) with lethal yellowing like symptoms in Papua New Guinea. *New Disease Reports* **23**, 9.

Kra KD, Toualy YMN, Kouamé AC, Diallo HA, Arocha Rosete Y (2017) First report of a phytoplasma affecting cassava orchards in Côte d'Ivoire. *New Disease Reports* **35**, 21.

Kwadjo KF, Beugré NI, Dietrich CH, Kodjo ATT, Atta-Diallo H, Yankey N, Dery S, Wilson M, Konan-Konan JL, Contaldo N, Paltrinieri S, Bertaccini A Rosete YA (2018) Identification of *Nedotepa curta* Dmitriev as a potential vector of the Côte d'Ivoire lethal yellowing phytoplasma in coconut palms sole or in mixed infection with a '*Candidatus* Phytoplasma asteris'-related strain. *Crop Protection* **110**, 48-56.

Lebrun P, Baudouin L, Myrie W, Berger A, Dollet M (2008) Recent lethal yellowing outbreak: why is the Malayan yellow dwarf coconut no longer resistant in Jamaica. *Tree Genetics and Genomes* **4**, 125–131.

Lee I-M, Gundersen-Rindal DE, Davis RE, Bartoszyk IM (1998) Revised classification scheme of phytoplasmas based on RFLP analyses of 16SrRNA and ribosomal protein gene sequences. *International Journal of Systematic Bacteriology* **48**, 1153–1169.

Lee I-M, Davis RE, Gundersen-Rindal, DE (2000) Phytoplasmas: phytopathogenic mollicutes. *Annual Review of Microbiology* **54**, 221–255.

Manimekalai R, Soumya VP, Nair S, Thomas GV, Baranwal VK (2014) Molecular characterization identifies 16SrXI-B group phytoplasma ('*Candidatus* Phytoplasma oryzae'-related strain) associated with root wilt disease of coconut in India. *Science Horticulturae* **165**, 288–294.

Martinez RT, Narvaez M, Fabre S, Harrison NA, Oropeza C, Dollet M, Hichez E (2008) Coconut lethal yellowing on the southern coast of the Dominican Republic is associated with a new 16SrIV group phytoplasma. *Plant Pathology* **57**, 366.

McCoy RE (1972) Remission of lethal yellowing in coconut palms treated with tetracycline antibiotics. *Plant Disease Reports* **56,** 1019–1021.

McCoy RE, Howard FW, Tsai JH, Donselman HM, Thomas DL, Basham RA, Atilano RA, Eskafi FM, Britt L, Collins ME (1983) Lethal yellowing of palms. *University of Florida Agricultural Experiment Stations Bulletin* **834**.

Medhi A, Baranwal VK, Babu MK, Deepthi P (2012) Sequence analysis of 16S rRNA and secA genes confirms the association of 16SrI-B subgroup phytoplasma with oil palm (*Elaeis guineensis* Jacq.) stunting disease in India. *Journal of Phytopathology* **160**, 6–12.

Mejía F, Palmieri M, Oropeza C (2004) First report of coconut lethal yellowing disease in Guatemala. *Plant Pathology* **53**, 80.

Mpunami A, Tymon A, Jones P, Dickinson M (1999) Genetic diversity in the coconut lethal yellowing disease phytoplasma of East Africa. *Plant Pathology* **48**, 109–114.

Mpunami A, Tymon A, Jones P, Dickinson MJ (2000) Identification of potential vectors of the coconut lethal disease phytoplasma. *Plant Pathology* **49**, 355–361.

Muddumadiah C, Madhupriya, Kumar S, Manimekalai R, Rao GP (2014) Detection and characterisation of 16SrI-B phytoplasmas associated with yellow leaf disease of arecanut palm in India. *Phytopathogenic Mollicutes* **4**, 77–82.

Munguambe N, Timbrine R, Freire M, Gadaga S, Dias P, Do Rosario B, Pudivitr J, Pilet F (2013) Large scale management of coconut lethal yellowing disease in Mozambique. *Acta Phytopathologica Sinica* **43**, 205.

Myrie WA, Douglas L, Harrison NA, McLaughlin W, James M (2012) First report of lethal yellowing disease associated with subgroup 16SrIV, a phytoplasma on St. Kitts in the Lesser Antilles. *New Disease Reports* **26**, 25.

Myrie W, Oropeza C, Sáenz L, Harrison N, Roca MM, Córdova I, Ku S, Douglas L (2011) Reliable improved molecular detection of coconut lethal yellowing phytoplasma and reduction of associated disease through field management strategies. *Bulletin of Insectology* **64**(Supplement), S203–S204.

Myrie W, Harrison N, Douglas L, Helmick E, Gore-Francis J, Oropeza C, McLaughlin WA (2014) Identification of lethal yellowing disease of palms associated with infection by subgroup 16SrIV-A phytoplasmas in Antigua, West Indies. *New Disease Reports* **29**, 12.

Naderali N, Nejat N, Vadamalai G, Tan YH (2013) First report of two distinct phytoplasma species, '*Candidatus* Phytoplasma cynodontis' and '*Candidatus* Phytoplasma asteris', simultaneously associated with yellow decline of *Wodyetia bifurcate* (foxtail palm) in Malaysia. *Plant Disease* **97**, 1504.

Naderali N, Vadamalai G, Tan YH, Nejat N (2014) Detection and identification of aster yellows phytoplasma associated with lipstick yellow frond disease in Malaysia. *Journal of Phytopathology* **162**, 264–268.

Nair S, Roshna OM, Soumya VP, Hegde V, Suresh Kumar M, Manimekalai R, Thomas GV (2014) Real-time PCR technique for detection of arecanut yellow leaf disease. *Australasian Plant Pathology* **43**, 527–529.

Narvaez M, Cordova I, Orellana R, Harrison NA, Oropeza C (2006) First report of a lethal yellowing phytoplasma in *Thrinax radiata* and *Coccothrinax readii* palms in the Yucatan peninsula of Mexico. *Plant Pathology* **55**, 292.

Narvaez M, Córdova-Lara I, Reyes-Martínez C, Puch-Hau C, Mota-Narvaez L, Collí A, Caamal G, Harrison N, Sáenz L, Oropeza C (2016) Occurrence of 16SrIV subgroup-A phytoplasmas in *Roystonea regia* and *Acrocomia mexicana* palms with lethal yellowing-like syndromes in Yucatán, Mexico. *Journal of Phytopathology* **164**, 1111–1115.

Nejat N, Sijam K, Abdullah SNA, Dickinson M (2009a) First report of a 16SrXIV, '*Candidatus* Phytoplasma cynodontis' group phytoplasma associated with coconut yellow decline in Malaysia. *Plant Pathology* **58**, 389.

Nejat N, Sijam K, Abdullah SNA Vadamalai G., Dickinson M (2009b) Phytoplasmas associated with disease of coconut in Malaysia are from two different phylogenetic groups and occur in other plant species. *Plant Pathology* **58**, 1152–1160.

Nipah JO, Jones P, Dickinson MJ (2007) Detection of lethal yellowing phytoplasma in embryos from coconut palms infected with Cape St Paul wilt disease in Ghana. *Plant Pathology* **56**, 777–784.

Nkansah-Poku J, Dery SK, Philippe R (2005) Reduction of spread of CSPWD of coconut by insecticidal hot-fogging and removal of diseased palms. *Ghana Journal of Agricultural Science* **1**, 193–198.

Nkansah-Poku J, Philippe R, Quaicoe RN (2009) Cape Saint Paul wilt disease of coconut in Ghana: surveillance and management of disease spread. *Oleagineux* **16**, 111–115.

Ntushelo K, Harrison NA, Elliott ML (2013) Palm phytoplasmas in the Caribbean basin. *Palms* **57**, 93–100.

Ofori F and Nkansah-Poku J (1997) Cape Saint Paul wilt disease of coconut in Ghana: history of its occurrence and spread. In: Proceedings of an International workshop on lethal yellowing–like diseases of coconut. Eds Eden-Green SJ, Ofori F, Natural Resources Institute, Chatham, United Kingdom, 27–32 pp.

Oleke JM, Isinika A, Manyong V (2012) Farmers' perception of coconut mite damage and crop diversification alternatives in the coastal belt of Tanzania. *International Journal of Acarology* **38**, 471–479.

Oropeza C, Cordova I, Puch-Hau C, Castillo R, Chan JL, Sáenz L (2017) Detection of lethal yellowing phytoplasma in coconut plantlets obtained through *in vitro* germination of zygotic embryos from the seeds of infected palms. *Annals of Applied Biology* **171**, 28–36.

Oropeza C, Zizumbo D (1997) History of lethal yellowing in Mexico. In: Proceedings of an International workshop on lethal yellowing–like diseases of coconut. Eds Eden-Green SJ, Ofori F, Natural Resources Institute, Chatham, United Kingdom, 69–76 pp.

Perera L, Meegahakumbura MK, Wijesekara HRT, Fernando WBS, Dickinson MJ (2012) A phytoplasma is associated with the Weligama coconut leaf wilt disease in Sri Lanka. *Journal of Plant Pathology* **94**, 205–209.

Philippe R, Simon R, Deschamp S, Nkansak-Poku J, Pilet F, Fabre S, Dollet M (2009) Study on the transmission of lethal yellowing in Ghana. *OCL* **16**, 102–106.

Pilet F, Philippe R, Reignard S, Descamps S, Quaicoe RN, Nkansah-Poku J, Fabre S, Dollet M (2009) Identification of potential insect vectors of the Cape Saint Paul wilt disease of coconut in Ghana by PCR. *OCL* **16**, 107–110.

Plavsic-Banjac B, Hunt P, Maramorosch K (1972) Mycoplasmalike bodies associated with lethal yellowing disease of coconut palms. *Phytopathology* **62**, 298–299.

Razin S (2007) Molecular biology and genomics of *Mollicutes*. *Bulletin of Insectology* **60**, 101–103.

Roca MM, Castillo MG, Harrison NA Oropeza C (2006) First report of a 16SrIV group phytoplasma associated with declining coyol palms in Honduras. *Plant Disease* **90**, 526.

Schuiling M, Mpunami A, Kaiza DA (1992) Lethal disease of coconut palm in Tanzania. 2. History, distribution and epidemiology. *Oléagineux* **47**, 516–522.

Schuiling M, Nienhaus F, Kaiza DA (1981) The syndrome in coconut palms affected by a lethal disease in Tanzania. *Zeitschift fur Pflanzenkranheiten und Pflanzenschutz* **88**, 665–677.

Swarbrick JP, Yankey NE, Nipah OJ, Quaicoe R, Dickinson MJ (2013) Identification of receptor like kinase genes in coconut and development of a marker for validation of breeding materials resistant to a phytoplasma disease in Ghana. *African Journal of Biotechnology* **12**, 6347–6357.

Tanne E, Boudon-Padieu E, Clair D, Davidovich M, Melamed S, Klein M (2001) Detection of phytoplasma by polymerase chain reaction of insect feeding medium and its use in determining vectoring ability. *Phytopathology* **91**, 741–746.

Thomas DL, Donselman HM (1979) Mycoplasmalike bodies and phloem degeneration associated with declining *Pandanus* in Florida. *Plant Disease Reporter* **63**, 911–916.

Tymon AM, Jones P, Harrison NA (1998) Phylogenetic relationships of coconut phytoplasmas and the development of specific oligonucleotide PCR primers. *Annals of Applied Biology* **132**, 437–452.

Vázquez-Euán R, Harrison NA, Narvaez M, Oropeza C (2011) Occurrence of a 16SrIV group phytoplasma not previously associated with palm species in Yucatan, Mexico. *Plant Disease* **95**, 256–262.

Warokka WA, Jones P, Dickinson M (2006) Detection of phytoplasmas associated with Kalimanthan wilt disease of coconut by polymerase chain reaction. *Jurnal Littri* **12**,154–160.

Wei W, Davis RE, Lee I-M, Zhao Y (2007) Computer-simulated RFLP analysis of 16S rRNA genes: identification of ten new phytoplasma groups. *International Journal of Systematic and Evolutionary Microbiology* **57**, 1855–1867.

Wejisekara HRT, Perera L, Wickramananda IR, Herath I, Meegahakumbura MK, Fernando WBS, de Silva PHPR (2008) Preliminary investigation on "weligama" coconut leaf wilt disease: a new disease in Southern Sri Lanka. Second Plantation Crop Symposium. Samayawardana Printers, Colombo, Sri Lanka, 336–341 pp.

Wilson MR (1987) African Derbidae (Homoptera, Fulgoroidea): taxonomic notes with descriptions of new species collected mainly from coconut. *Journal of Natural History* **21**, 567–595.

Yankey EN (2012) The lethal disease of coconut in Ghana: developing markers and pathogen quantification techniques for the breeding of resistant or tolerant varieties. PhD thesis, University of Nottingham. Access from the University of Nottingham repository: http://eprints.nottingham.ac.uk/13953/1/575157.pdf.

Yankey EN, Pilet F, Quaicoe RN, Dery SK, Dollet M, Dzogbefia VP (2009) Search for alternate hosts of the coconut Cape Saint Paul wilt disease pathogen. *OCL* **16**, 123–126.

Zizumbo-Villarreal D, Colunga-GarcíaMarín P, Fernández-Barrera M, Torres-Hernández N, Oropeza C (2008) Mortality of Mexican coconut germplasm due to lethal yellowing. *Plant Genetic Resources Newsletter, FAO-Bioversity* **156**, 23–33.

Chapter 10
Major Phytoplasma Diseases of Forest and Urban Trees

Carmine Marcone, Liliana Franco-Lara, and Ivo Toševski

Abstract In the northern hemisphere, yellows, witches' broom, and decline diseases of several forest and urban tree species are widespread and of considerable economic and ecological significance. Elm (*Ulmus* spp.) and alder (*Alnus* spp.) are affected by elm yellows (EY) and alder yellows (ALY), respectively. These diseases are mainly associated with the presence of closely related phytoplasmas, the EY agent '*Candidatus* Phytoplasma ulmi' and the ALY agent, which are members of the EY or 16SrV group, subgroups 16SrV-A and 16SrV-C, respectively. Ash (*Fraxinus* spp.) is affected by ash yellows, a disease which occurs mainly in North America and is associated with the presence of '*Candidatus* Phytoplasma fraxini', a member of subgroup 16SrVII-A. Poplar (*Populus* spp.), sandal (*Santalum album*), paulownia (*Paulownia* spp.), and mulberry (*Morus* spp.) are affected by yellows diseases associated with phytoplasmas of different 16SrI subgroups. Several species of conifers are affected by yellows and witches' broom diseases associated with phytoplasmas belonging to at least five taxonomic groups (16SrI, 16SrIII, 16SrVI, 16SrIX, and 16SrXXI) and several different subgroups. A number of urban tree species grown in the Sabana de Bogotà (Colombia) are affected by decline diseases which are primarily associated with 16SrI and 16SrVII phytoplasmas. This chapter summarizes the current knowledge of major phytoplasma diseases of forest and urban trees grown in the northern hemisphere.

Keywords Elm yellows · Alder yellows · Ash yellows · '*Candidatus* Phytoplasma' species · 16Sr group/subgroups · Decline

---

C. Marcone (✉)
Department of Pharmacy, University of Salerno, Fisciano (Salerno), Italy
e-mail: cmarcone@unisa.it

L. Franco-Lara
Facultad de Ciencias Básicas y Aplicadas, Universidad Militar Nueva Granada, Cajicá, Colombia

I. Toševski
Department of Plant Pests, Institute for Plant Protection and Environment, Zemun, Serbia

CABI Europe – Switzerland, Delémont, Switzerland

G. P. Rao et al. (eds.), *Phytoplasmas: Plant Pathogenic Bacteria - I*,
https://doi.org/10.1007/978-981-13-0119-3_10

10.1 Introduction

Like all plants, forest and urban trees are constantly besieged by problems brought by adverse environmental factors, insects and other animals, and diseases. Of the diseases, those associated with phytoplasma presence are of considerable economic and ecological significance, either because of their local impact or widespread distribution (Marcone 2015). Phytoplasmas are wall-less bacterial plant pathogens of the class *Mollicutes*. In the host plant, they reside in the sieve tube elements and are transmitted from plant to plant by phloem-feeding homopteran insects, mainly leafhoppers, planthoppers, and psyllids (Weintraub and Beanland 2006). Phytoplasmas cannot be transmitted mechanically; however, they can be spread by the use of infected vegetative propagating material (Lee et al. 2000; Dickinson et al. 2013). Phytoplasmas induce a wide range of symptoms that are either specific (such as virescence, phyllody, witches' brooms, rosetting, off-season growth, and brown discoloration of phloem tissue) or largely nonspecific (such as foliar yellowing and reddening, small leaves, leaf roll, premature autumn coloration, premature defoliation, reduced growth, dieback, and decline). However, symptom expression is also highly variable for a given host plant combination.

Phytoplasmas are currently classified within the provisional genus '*Candidatus* Phytoplasma' based primarily on 16S rDNA sequence analysis (IRPCM 2004; Martini et al. 2014). Approximately 33 major phylogenetic groups were identified within the phytoplasma clade (Bertaccini et al. 2014). This figure is broadly in accordance with the number of phytoplasma groups established by restriction fragment length polymorphism (RFLP) analysis of PCR-amplified rDNA. Within the majority of the phytoplasma groups, several distinct subgroups have been delineated, based on RFLP analysis of 16S rDNA sequences (Bertaccini et al. 2014; Martini et al. 2014). In addition, multi-locus sequence typing (MLST) employing genes with varying degrees of genetic variability has provided insights into the genetic diversity of phytoplasmas that are relatively homogeneous at the 16S rDNA sequence level. MLST proved to be the most useful molecular tool for the distinction of genetically closed but pathologically and/or ecologically distinct strains that are essential and highly relevant for epidemiological studies (Lee et al. 2010; Davis et al. 2013; Zhao and Davis 2016).

10.2 Alder

Alder yellows (ALY) is a decline disease that affects several *Alnus* (alder) species including *A. glutinosa*, *A. cordata*, *A. incana*, *A. hirsuta*, *A. rugosa*, *A. subcordata*, *A. tenuifolia*, and *A. rubra*. The disease has been reported from several European countries (for reviews see Seemüller et al. 1998; Marcone and Rao 2016). It has also been observed in Washington State (USA) on *A. rubra* (red alder) (Lederer and Seemüller 1991; Mäurer et al. 1993). However, molecular identification and

Fig. 10.1 Alder yellows. Symptoms of foliar yellowing in *Alnus glutinosa* (European alder). Left, healthy branch (**a**). Stunting and small leaves in *A. cordata* (**b**)

classification of the phytoplasma(s) infecting *A. rubra* in Washington State have not been accomplished.

In all affected species, the symptoms are similar. These include yellowing, sparse foliage, premature autumn coloration, small leaves, reduced terminal growth, dieback, and decline. Sometimes these symptoms occur in the crown, while the sprouts at the base of trunks show normal growth (Fig. 10.1). Latent infections are also common (Lederer and Seemüller 1991; Maixner and Reinert 1999; Holz et al. 2016). Of 500 European alder trees examined by Lederer and Seemüller (1991) using DAPI fluorescence methods, phytoplasma presence could be detected in all the trees older than approximately 5 years, irrespective of symptom expression. About 80% of infected trees were non-symptomatic, whereas only 20% showed ALY symptoms. Usually, healthy-appearing trees were heavier colonized than symptomatic trees. The phytoplasma population was always higher in petioles and young twigs than in several-year-old branches, trunks, and roots. Colonization was characterized by an uneven phytoplasma distribution in adjacent sieve tubes, with some tubes packed by phytoplasma cells while others contained a considerably lower number of them. A rather similar situation was observed in *A. incana* trees of which, with the exception of a few relatively young trees, all samples proved to be phytoplasma positive and showed very high numbers of the colonizing agents irrespective of symptom expression (Lederer and Seemüller 1991). Under natural infection conditions, a detection rate of more than 85% of ALY phytoplasma infections in *A. glutinosa* was recorded by PCR assays (Malembic-Maher et al. 2009).

The ALY phytoplasma is a member of EY group, subgroup 16SrV-C (Lee et al. 2004a). It shares 99.7–100% 16S rDNA sequence identity with "flavescence dorée" (FD), Palatinate grapevine yellows (PGY), and spartium witches' broom (SpaWB) phytoplasmas, which are other members of the EY group (Lee et al. 2004a). On the basis of 16S rDNA and 16S–23S rDNA spacer region sequences, ALY phytoplasma is a homogeneous pathogen throughout Europe. However, typing based on RFLP and sequence analyses of ribosomal protein genes (*rplV*, *rpsC*, *rplF*, and *rplR*) and non-ribosomal loci, including *secY*, *map*, *tuf*, *uvrB*, and *degV* genes, revealed a considerable molecular variability within this phytoplasma. Some strains proved to be very closely related or identical to either FD or PGY phytoplasma strains (Lee et al. 2004a; Arnaud et al. 2007; Malembic-Maher et al. 2007, 2009; Mehle et al. 2011; Ember et al. 2011; Holz et al. 2016).

The ALY phytoplasma is transmitted by *Oncopsis alni* Schrank, a univoltine leafhopper extremely abundant on alder, with oligophagous feeding behavior. The distribution of *O. alni* is clearly correlated with that of ALY disease in Europe (Maixner and Reinert 1999). ALY phytoplasma was also detected in specimens of the psyllid, *Psylla alni* L., collected from ALY-diseased alder trees. However, in spite of the high numbers of infected individuals that were recorded, no successful transmission of the ALY phytoplasma to healthy alder seedlings was achieved with transmission trials using ALY-infected *P. alni* psyllids (Maixner and Reinert 1999). Molecular and epidemiological evidence indicate that ALY phytoplasma strains infecting alder trees in the Palatinate region of Germany also infect grapevines growing in neighboring vineyards, inducing the disease described as Palatinate grapevine yellows. *O. alni*, which feeds occasionally on grapevine, can transmit ALY strains from alder to grapevine (Maixner et al. 1995, 2000; Reinert and Maixner 1997). Therefore, diseased alder trees residing in close proximity to vineyards may serve as pathogen reservoirs.

10.3 Ash

Ash yellows (AshY) is a disease of *Fraxinus* (ash) species known to occur in North America (Sinclair et al. 1996). This disease is associated with the presence of the AshY agent '*Ca.* P. fraxini,' a member of the AshY phytoplasma group, subgroup 16SrVII-A (Griffiths et al. 1999a). AshY occurs naturally in at least 12 indigenous and exotic ash species in North America. These species include *F. americana* (white ash), *F. pennsylvanica* (green ash), *F. velutina* (velvet ash), *F. angustifolia* (syn. *F. oxyacarpa*), *F. bungeana*, *F. excelsior* (European ash), *F. nigra* (black ash), *F. latifolia* (syn. *F. oregona*, Oregon ash), *F. ornus* (flowering ash), *F. potamophila*, *F. profunda* (syn. *F. tomentosa*), and *F. quadrangulata* (blue ash) (Sinclair et al. 1996). Infections by a 16SrI group phytoplasma have also been detected in declining trees of *F. excelsior* in Poland and *F. uhdei* in Colombia (Kamińska and Berniak 2009; Perilla-Henao et al. 2012). In the latter country, diseased *F. uhdei* trees proved to be infected also by a 16SrVII group phytoplasma (Griffiths et al. 2001; Filgueira et al. 2004; Perilla-Henao et al. 2012).

AshY causes reduced apical and radial growth, loss of apical dominance or deliquescent branching, suppressed root development, premature flowering and shoot growth, and witches' broom. Subnormal greenness of the foliage and leaf malformations is common, while chlorosis may occur occasionally. Highly susceptible taxa show dieback of branches and roots, shoot proliferation, and stunted deliquescent branches and die prematurely. Affected white ash trees produce abnormally short, bushy roots or show necrosis of the rootlets that may lead to sudden wilting and death. Infected ash trees are highly sensitive to frost injuries. Cambial damage by frost, appearing as vertical cracks of the bark and/or tangential separation of bark and wood at the base of the trunk, is common in diseased white ash (Matteoni and Sinclair 1985). AshY incidence greater than 50% has been recorded in some white ash populations in northern USA states and in a velvet ash population in Utah. Incidence rates of 3–27% were found in green ash shade trees in Iowa and Wisconsin cities (for review see Sinclair et al. 1996). '*Ca.* P. fraxini' infections were also detected in healthy-appearing plants of ash. Possible explanations provided for these observations are either that the plants were tolerant, harbored strains of low virulence, or were in the early stages of colonization (Sinclair and Griffiths 1995; Sinclair et al. 1996).

Differences in virulence among strains of '*Ca.* P. fraxini' have been observed under natural infection conditions in which six cultivars of green ash and five cultivars of white ash were graft-inoculated with six distinctly different strains at two different locations of Iowa and New York, where the inoculated plants were maintained under observation for 3 years. The strains greatly differed in aggressiveness as indicated by host growth suppression, reduction of foliar greenness, and frequency of witches' broom. However, strain-cultivar interactions were not identified (Sinclair et al. 2000b).

The vectors of AshY are unknown. Field-collected *Paraphlepsius irroratus* and *P. spumarius* transmitted naturally acquired phytoplasma infections to caged ash seedlings (Matteoni and Sinclair 1988). However, the identities of the associated phytoplasma(s) were unknown, and laboratory colonies of these insect species failed to transmit '*Ca.* P. fraxini' under controlled conditions. Hill and Sinclair (2000) collected 33 taxa of leafhoppers, including members of 13 genera known to contain phytoplasma vectors, from two sites of high AshY incidence in New York State and tested by PCR assays for the presence of phytoplasmas. The most abundant genus was *Scaphoideus*, with numbers about six times greater than any other genus. '*Ca.* P. fraxini' was detected in 19 of the 812 individuals of *Scaphoideus* spp. and in 1 of 87 of *Colladonus clitellarius*. Phytoplasmas of the X-disease group or 16SrIII group were detected in one *Scaphoideus* sp., four *C. clitellarius*, and 4 out of 83 *Scaphytopius acutus* individuals. Phytoplasmas of the 16SrI group were identified in one out of 68 individuals of *Gyponana* spp. and 5 of *S. acutus*. Therefore, '*Ca.* P. fraxini'-infected leafhoppers should be regarded as potential vectors of this pathogen and should be checked experimentally for their transmission ability (Hill and Sinclair 2000).

10.4 Conifers

Over the last years, phytoplasmas were detected and identified at the molecular level in several conifer species. Schneider et al. (2005) reported the occurrence of a novel taxon, '*Ca.* P. pini', in *Pinus sylvestris* (Scots pine) and *P. halepensis* (Aleppo pine) trees grown in Germany and Spain, respectively. Symptoms included yellowing, stunted growth, dwarfed needles, and proliferation of shoots. In *P. sylvestris*, conspicuous shoot proliferation combined with dwarfed needles gives some affected branches a dense, ball-like appearance, whereas other branches are symptomless. These ball-like structures of affected branches were not observed in *P. sylvestris*. '*Ca*. P. pini' infections were detected in symptomatic and symptomless plant parts of *P. sylvestris* and *P. halepensis* trees. Also, some neighboring non-symptomatic trees proved to be infected by '*Ca*. P. pini'.

Following the report of Schneider et al. (2005), '*Ca*. P. pini' and related strains were detected in *P. sylvestris*, *P. nigra* (Austrian pine), *P. banksiana* (Jack pine), *P. tabuliformis* (Chinese pine), *P. mugo* (mountain pine), *Abies procera* (noble fir), *Tsuga canadensis* (Canadian hemlock), and *Picea pungens* (Colorado blue spruce) trees grown in Poland and the Czech Republic, in *Pinus* spp. and *P. mugo* in Croatia, and in *P. sylvestris* and *P. mugo* in Lithuania. Most of the affected trees showed symptoms of shoot proliferation, stunted growth, and dwarfed needles, while ball-like structures similar to those observed in Germany on *P. sylvestris* were observed on the same species in Poland and Czech Republic. Although rare, ball-like structures were also observed on diseased *P. sylvestris* and *P. mugo* trees in Lithuania (Śliwa et al. 2008; Valiunas et al. 2010, 2015; Kamińska and Berniak 2011; Kamińska et al. 2011; Ježić et al. 2012). Furthermore, '*Ca*. P. pini' was identified in China (Huang et al. 2011) in *Taxodium distichum* var. Imbricarium (pond cypress) showing abnormal shoot proliferation, little leaf and leaf necrosis, and Mozambique strains related to '*Ca*. P. pini' were also detected in coconut palms showing lethal yellowing (Bila et al. 2015).

'*Ca*. P. pini' is a member of the pine shoot proliferation phytoplasma group, subgroup 16SrXXI-A. Phytoplasmas of other taxonomic groups have also been detected in conifers. A member of the X-disease phytoplasma group (16SrIII) was identified in Italy in *Cupressus* sp. (cypress) trees showing witches' broom, stunting, and fasciation (Paltrinieri et al. 1998) and in *Picea abies* (Norway spruce) and *P. glauca* (white spruce) with symptoms similar to those associated with the above- mentioned '*Ca*. P. pini' infections in Poland (Kamińska and Berniak 2011). In the latter country, infections by a phytoplasma of the aster yellows group were also detected in diseased *P. pungens* trees, whereas a 16SrI-B phytoplasma was identified in diseased *Larix* sp. (larch) in Ukraine (Jomantiene et al. 2011; Kamińska and Berniak 2011). The affected larch trees showed symptoms of yellowing, dwarfing, proliferation, and necrosis of the needles. Hence, the larch-infecting agent was named larch dwarfed needle proliferation (LDNP) phytoplasma (Jomantiene et al. 2011). Of nearly 300 *P. sylvestris* and *P. mugo* trees in Lithuania, examined by Valiunas et al. (2015) using PCR assays, 80% were phytoplasma positive. Of these, 98% harbored *Ca*. P. pini'-related strains, whereas the remaining trees proved to be infected by a subgroup 16SrI-A phytoplasma. A '*Ca*. P. phoenicium'-related strain, subgroup 16SrIX-E, was reported to be associated with the juniper witches' broom (JunWB) disease of *Juniperus occidentalis* (western juniper) in Oregon (USA). This disease is characterized by abnormal proliferation of shoots, little leaves, shortened internodes, and ball-like structures. The incidence of JunWB disease was about 1% (Davis et al. 2010). A '*Ca*. P. trifolii'-related strain, member of the clover proliferation phytoplasma or 16SrVI group, has been identified in diseased trees of

Araucaria heterophylla (Norfolk Island pine) showing yellowing, little leaves, witches' broom, and shoot proliferation in India with percentage of affected trees that was about 15% (Gupta et al. 2010).

The suitability of gymnosperms as phytoplasma hosts has for a long time been questioned mainly because of the small pore sizes in the sieve cells, which may hinder phytoplasma movement within plants (Schneider et al. 2005). However, data obtained using PCR technology suggests that phytoplasma infections in conifers usually occur at low titer and are more common than previously thought. Although phytoplasma detection is often associated with different kinds of symptoms including shoot proliferation and dwarfed needles, the phytopathological relevance of these infections in conifers is not always clear, mainly because the same phytoplasma were sometimes detected also in symptomless trees. Thus, further research based mainly on graft inoculation studies is needed to validate the effect of phytoplasma presence in conifers.

10.5 Elm

Several *Ulmus* (elm) species and hybrids are affected by elm yellows (EY), a lethal and decline disease which is widespread in North America and Europe (Marcone 2017). The symptoms vary among the elm species. In those native to North America such as *U. americana* (American or white elm), *U. rubra* (red or slippery elm), *U. alata* (winged elm), *U. serotina* (September elm), and *U. crassifolia* (cedar elm), symptoms include leaf epinasty, leaf curl, chlorosis, premature casting of the leaves, a yellow to brown discoloration of the phloem in the roots and stem, and tree death that usually occurs within 1 or 2 years from the appearance of foliar symptoms. Red elm, which usually dies in the second year of symptom expression, often shows witches' broom that occurs over the entire crown and progressively increases in severity giving the tree a starved appearance. Discolored living phloem tissue of American, winged, September, and cedar elms has a characteristic odor of oil of wintergreen (methyl salicylate), whereas a pleasant odor resembling that of maple syrup is released by the inner bark of red elm shortly after the bark is dead. In *U. minor* (*U. carpinifolia*, European field elm), the most characteristic symptoms are pronounced witches' broom present at the tips of twigs and branches and at the root level (Fig. 10.2). For this reason, the disease of European field elm is often called elm witches' broom. Other symptoms include leaf epinasty, yellowing, stunting, small leaves, and premature leaf shedding. Witches' broom and stunting are also the typical symptoms of *U. glabra* (Scots elm) and *U. parvifolia* (Chinese elm) (Braun and Sinclair 1979; Pisi et al. 1981; Lee et al. 1993; Sinclair 2000). Leaf yellowing or reddening, reduced terminal growth, witches' broom formation, dieback, and decline have also been observed in several other European and Asian elm species including *U. pumila* (Siberian elm), *U. chenmoui* (Chenmou elm), *U. villosa* (cherry-bark elm), *U. laevis* (European white elm), *U. wallichiana* (Himalayan elm), *U. wilsoniana* (Wilson elm), *U. japonica* (Japanese elm), and *U.* x *hollandica*

Fig. 10.2 Elm yellows symptoms. Foliar yellowing, small leaves in *Ulmus chenmoui* (Courtesy of A. Bertaccini)

(Dutch elm) (Conti et al. 1987; Lee et al. 1995; Mittempergher 2000; Sfalanga et al. 2002). However, phloem discoloration is not known to occur in any of the European or Asian species. Symptomless trees of some European and Asian elm genotypes, in which phytoplasma infections were detected by nested PCR assays, have also been recorded (Lee et al. 1995, Sinclair et al. 2000a; Mittempergher 2000; Sfalanga et al. 2002; Carraro et al. 2004; Katanić et al. 2016).

EY is associated with the presence of '*Ca.* P. ulmi,' a member of the EY phytoplasma group, subgroup 16SrV-A (Lee et al. 2004a). A phytoplasma of the clover proliferation group (16SrVI group), subgroup 16SrVI-C, the Illinois elm yellows (ILEY) agent, was identified in nine American elm trees with symptoms similar to those of EY disease in a suburb west of Chicago, Illinois, USA. However, double infections with the ILEY agent and a phytoplasma of the aster yellows group (16SrI group) were also detected in two diseased elm trees although the ILEY agent was predominant (Jacobs et al. 2003). Double or multiple infections with the EY agent and a phytoplasma of the 16SrI group and/or the "stolbur" group (16SrXII-A subgroup) were also detected in diseased elm and elm hybrids in northern and central Italy (Lee et al. 1995). Single infections with 16SrI and 16SrXII-A phytoplasmas were detected in two diseased elm trees showing symptoms of yellowing in Croatia (Katanić et al. 2016), whereas subgroup 16SrV-B and 16SrI-B phytoplasmas were identified in diseased elm trees in China (Zhu et al. 2008; Gao et al. 2011).

EY is widespread in the eastern USA where several epidemics have killed hundreds to tens of thousands of American and red elm trees (Sinclair 2000). In Italy, significant EY outbreaks have occurred in three experimental fields established in the 1980s in northern and central Italy to test the adaptability of a number of elm species and 33 hybrid clones resistant to the Dutch elm disease to local environmental conditions (Conti et al. 1987; Lee et al. 1995; Mittempergher 2000). EY epidemics in European field elm have been observed in the Po valley and Friuli-Venezia Giulia region (Northern Italy) (Conti et al. 1987; Mittempergher 2000; Carraro et al. 2004) and in the Agri Valley (Basilicata region, southern Italy) where EY incidence greater than 80% has been recorded (Marcone 2015, 2017). In Friuli-Venezia Giulia, widespread occurrence of EY on Siberian elm has also been observed (Carraro et al. 2004), whereas a detection rate of approximately 75% of EY phytoplasma infections in European white elm trees has been recorded in Croatia using nested PCR assays. However, more than half of the infected trees were symptomless (Katanić et al. 2016).

The white-banded elm leafhopper, *Scaphoideus luteolus*, is the only known vector of '*Ca*. P. ulmi' in North America, although other vectors are likely to be involved in its natural spread. This likelihood is supported by the fact that numerous homopteran insects belonging to the genera that comprise known phytoplasma vectors have been found on elm and probably feed on it to some extent and that *S. luteolus* is rare or absent in some areas where severe EY outbreaks occur (for reviews see Sinclair 2000; Marcone 2017). In New York State, of the various leafhoppers and other homopteran insects collected from sites of EY occurrence and tested for their ability to transmit the EY agent to American elm seedlings, single transmissions were recorded for the leafhopper *Allygus atomarius* and the spittlebug *Philaenus spumarius* (Matteoni and Sinclair 1988). By real-time PCR assays, '*Ca*. P. ulmi' was also detected in several leafhopper taxa belonging to the genera *Allygus*, *Colladonus*, *Empoasca*, *Erythroneura*, *Graphocephala*, *Homalodisca*, *Orientus*, *Scaphoideus*, and *Typhlocyba*, which were captured in the University Park Campus of Pennsylvania State University, USA (Herath et al. 2010). However, it remains to be demonstrated if the above leafhopper taxa can transmit the pathogen. Rosa et al. (2014) reported that 3 out of 30 American elm seedlings exposed to individuals of the spittlebugs *Lepyronia quadrangularis* and *P. spumarius* and the leafhopper *Latalus* sp., collected from an EY-infected red elm tree in the Pennsylvania State University campus, harbored '*Ca*. P. ulmi.' *S. luteolus* is not known to occur in Europe, and Carraro et al. (2004) showed that *Macropsis mendax* vectors the EY agent in Friuli-Venezia Giulia (Northern Italy). This leafhopper is strictly monophagous, completes one generation per year, and overwinters as eggs on elm. It is unknown whether *M. mendax* is involved in the spread of the EY agent in other parts of Europe, and there is no information on its transmission efficiency. More recently by PCR assays, '*Ca*. P. ulmi' was detected in individuals of *Hyalesthes luteipes* collected from EY-affected elm trees in Serbia (Jović et al. 2010). EY phytoplasma seems to spread only from elm to elm, because plants from other genera growing on sites of EY occurrence have not been found to harbor the phytoplasma. There are a few cases of EY detection in grapevine from vineyards in Northern Italy where elm was traditionally used as grapevine support (Botti and Bertaccini 2007). The EY agent is also reported to spread among closely spaced trees of the same species through natural root grafts. This kind of transmission was considered relevant in shade tree losses in urban epidemics in North America (Sinclair 1981).

10.6 Mulberry

Mulberry dwarf (MD) is one of the most serious diseases of *Morus* (mulberry) species, including *M. alba*, *M. bombycis*, and *M. multicaulis*, which is known to occur in Japan and Korea (Sato et al. 1996; Ji et al. 2009). The phytoplasma presence in association with this disease was established in 1967 when Doi et al. (1967) observed numerous wall-less, pleomorphic bodies in the phloem sieve tube elements of yellows-diseased mulberry plants. Morphologically and ultrastructurally, these bodies resembled mycoplasmas, organisms known to cause animal and human diseases. At the same time, Ishiie et al. (1967) achieved remission of symptoms of diseased mulberry plants by applying tetracyclines. The most characteristic symptoms of MD are yellowing of the leaves, phyllody, stunting, proliferation, and witches' brooms. The disease is associated with the presence of the aster yellows agent '*Ca.* P. asteris,' subgroup 16SrI-B (Namba et al. 1993; Lee et al. 2004b; Ji et al. 2009). The MD phytoplasma is spread in nature by the leafhoppers *Hishimonoides sellatiformis* and *H. sellatus*, the first of these two species being a more efficient vector. Although phytoplasmas were not believed to be vertically transmitted to the progeny of the vectors for many years, PCR-based and electron microscopy investigations provided indications for transovarial transmission of MD phytoplasma by *H. sellatiformis* (Kawakita et al. 2000). *H. sellatus* experimentally transmitted the MD phytoplasma to five herbaceous species, i.e., periwinkle, white clover, Ladino clover, red clover, and Chinese milk vetch (Kim et al. 1985). Natural resistance to the MD phytoplasma identified in some *M. alba* cultivars is associated with the phytoalexin concentration of the mulberry tree cortex. The amount of a small group of compounds isolated by thin-layer chromatography was four times higher in the resistant than in susceptible cultivars (Kuai et al. 1999).

10.7 Paulownia

Several *Paulownia* (paulownia) species are affected by paulownia witches' broom (PaWB), one of the first described phytoplasma diseases (Doi et al. 1967). This disease is widespread in East Asia and has not been reported from other geographic areas. In PaWB-affected paulownia trees, growth and vigor are greatly reduced and the leaves on affected shoots are yellowish, malformed, and reduced in size, while flower clusters, when they are produced, show different degrees of distortion and virescence, along with sterility. Proliferation of slender shoots which arise from the main branches gives rise to typical witches' brooms. Affected shoots show phloem necrosis and an irregular cell arrangement in the woody cylinder, mainly around the vessel. Severely affected trees die prematurely. The wood of infected trees is of poor quality and often commercially unfit. In China, the expansion of paulownia plantations during the 1970s contributed significantly to the rapid spread of PaWB due to the large use of PaWB-infected root cuttings for propagation (Hiruki 1999). In northern China, the disease incidence was 10–20% in the early 1970s and more than 70% in the 1980s. A separate survey in China registered an incidence of 5–10% at the seedling stage, 10–20% in 1-year-old saplings, 50% at the middle age stage, and 70–100% in several-year-old trees.

PaWB is associated with a distinct member of the aster yellows phytoplasma group, subgroup 16SrI-D (Marcone et al. 2000; Lee et al. 2004b; Yue et al. 2008). Paulownia is the only natural host of PaWB and was experimentally transmitted to periwinkle by dodder (Doi and Asuyama 1981); a preliminary study showed the presence of very little polymorphisms in genes other than the 16S rRNA gene (Zhao et al. 2016). Three species of heteropteran insects of the family Pentatomidae (stinkbugs), namely, *Halyomorpha mista*, *H. halys*, and *H. picus*, are reported as vectors of the PaWB agent (Hiruki 1999; Weintraub and Beanland 2006). Transgenic resistance to PaWB phytoplasma has been obtained by expressing an antibacterial peptide encoded by the *shiva-1* gene in *P. tomentosa* x *P. fortunei* plants. Both symptom development and phytoplasma titer were significantly reduced in the developed transgenic plants (Du et al. 2005).

10.8 Poplar

Trees affected by phytoplasmas include several *Populus* (poplar) species. A phytoplasma disease, the poplar witches' broom (PopWB), was first observed in 1973 in Bulgaria on *P. nigra* cv Italica (Lombardy or black poplar) and *P. canadensis* (Canadian poplar) (Atanasoff 1973). Later, the disease was first reported from the Netherlands on *P. alba* (white poplar), *P. canescens* (gray poplar), and Lombardy poplar (van der Meer 1980) and then from France on Lombardy and white poplar, Germany on *P. tremula* (aspen) and Lombardy and white poplar, Hungary on white poplar, and Croatia and Serbia on Lombardy poplar (Sharma and Cousin 1986; Seemüller and Lederer 1988; Cousin 1996; Berges et al. 1997; Cousin et al. 1999; Šeruga et al. 2003; Mitrović et al. 2011).

The disease is characterized mainly by witches' broom, although in some instances, especially on *P. nigra* cv Italica trees, only nonspecific symptoms such as yellowing and undersized leaves, sparse foliage, stunting, dieback, and decline may be present (Fig. 10.3). On aspen trees, the disease is frequently associated with foliar reddening, yellowing, and decline symptoms, whereas witches' broom may occasionally develop on vigorous shoots. PopWB is associated with the aster yellows agent '*Ca.* P. asteris,' subgroup 16SrI-B (Berges et al. 1997; Marcone et al. 2000; Lee et al. 2004b). However, subgroup 16SrI-A and 16SrI-P phytoplasma strains have also been identified in PoWB-affected *P. nigra* cv Italica trees in Croatia and Serbia (Šeruga et al. 2003; Mitrović et al. 2011). Work based on RFLP analyses

Fig. 10.3 Poplar witches' broom showing dieback of declining *Populus nigra* 'Italica' (Lombardy poplar) in a tree (healthy at right) and reduced leaf size in a small branch compared with healthy branch in the right (Courtesy of B. Duduk)

of 16S rDNA and 16S–23S rDNA spacer region sequences revealed the presence of three different RFLP profiles among strains infecting *P. nigra* cv Italica, *P. alba*, and *P. tremula* in Germany, France, and Hungary, following digestion with *Alu*I restriction enzyme (Berges et al. 1997). Differences between French and German strains infecting *P. nigra* cv Italica trees were also obtained by heteroduplex mobility assays using 16S rDNA and 16S–23S rDNA spacer region sequences (Cousin et al. 1998). Phytoplasmas of 16SrI and 16SrVII groups were found in diseased *P. nigra* trees in Colombia, as mentioned below (Perilla-Henao et al. 2012). The strictly oligophagous leafhoppers *Rhytidodus decimusquartus* and *Tremulicerus vitreus*, which are particularly abundant on *Populus* species, were identified as vectors of the PopWB agent in France (Cousin 1996; Cousin et al. 1999).

10.9 Sandal

Santalum album L. (sandal) is affected by sandal spike (SAS), a serious disease which is widespread in southern India, mainly in the states of Karnataka, Tamil Nadu, and Kerala, but does not occur elsewhere (Thomas and Balasundaran 1999; Khan et al. 2006, 2008). The disease was first observed in the late 1890s and was thought to be caused by virus. Subsequent studies based on fluorescence and electron microscopy observations and symptom remission obtained by tetracycline treatments clearly demonstrated the phytoplasma etiology of this disease (Dijkstra and Ie 1969; Hull et al. 1969; Raychaudhuri et al. 1972). SAS disease has spread progressively over the years, devastating large forest tracts and threatening the sandal industry of southern India, where production of sandalwood oil is of major importance. The most characteristic symptoms of SAS are shortened internodes, small and extremely narrow leaves, and phyllody. The affected leaves, which are pale green or yellow, stand out stiffly from the shoots, with a spikelike appearance. The most severely affected trees die within 2 or 3 years from the appearance of the first symptoms. A disease incidence reaching up to 55% was recorded in southern Karnataka (Rao and Muniyappa 1988). SAS is associated with the aster yellows agent '*Ca.* P. asteris,' subgroup 16SrI-B (Marcone et al. 2000; Lee et al. 2004b; Khan et al. 2008). SAS phytoplasma was transmitted from diseased sandal trees to periwinkle and from this host back to sandal trees via dodder (*C. subinclusa*) bridges. The dodder-inoculated periwinkle plants developed witches' broom symptoms (Dijkstra and Lee 1972; Hiruki and Dijkstra 1973). It was also experimentally transmitted from sandal to sandal by grafting. The leafhopper *Coelidia indica*, originally identified as *Jassus indicus*, was reported as a natural vector of SAS (Rangaswami and Griffith 1941; Weintraub and Beanland 2006).

10.10 Phytoplasma Diseases of Urban Trees in the Sabana de Bogota: A Case Study

In the last 15 years, phytoplasmas were detected in several urban tree species in Bogotà, Colombia, and surrounding rural areas (Sabana de Bogotà). Affected species included *Fraxinus uhdei*, previously referred to as *F. chinensis* (Griffiths et al. 2001), *P. nigra*, *Acacia melanoxylon*, *Eugenia neomyrtifolia*, *Liquidambar styraciflua*, *Magnolia grandiflora*, *Pittosporum undulatum*, *Quercus humboldtii*, and *Croton* spp. Among these, *F. uhdei* and *P. undulatum*, which are native to Mexico and Australia, respectively, are the major street trees of Bogotà and are also well established in rural areas. Affected trees showed symptoms of tufted foliage, deliquescent branching, witches' brooms, epicormic shoots, dieback, decline, small leaves, and foliar yellowing (Fig. 10.4).

Phytoplasmas of 16SrI and 16SrVII groups were identified in diseased trees of each species through RFLP and sequence analyses of PCR-amplified 16S rDNA sequences (Griffiths et al. 2001; Filgueira et al. 2004; Perilla-Henao et al. 2012; Perilla-Henao and Franco-Lara 2013; Franco-Lara and Perilla Henao 2014). Additionally, phytoplasmas of 16SrV, 16SrIX, and 16SrXII groups were also recorded in diseased *L. styraciflua* trees (Franco-Lara et al. 2017). Extensive visual symptom inspections carried out in December 2013 revealed that the percentage of affected trees varied from 93% to 100% for *F. uhdei*, *L. styraciflua*, and *P. nigra* and from 85% to 66% for *E. neomyrtifolia*, *P. undulatum*, *M. grandiflora*, and *Q. humboldtii*, whereas the least affected species were *A. melanoxylon* and *Croton* spp. with percentages of 47 and 36%, respectively (Franco-Lara and Perilla Henao 2014).

Fig. 10.4 Symptoms of witches' broom, mild yellowing, and decline in trees of *Liquidambar styraciflua* infected with 16SrVII phytoplasmas in the Sabana de Bogotà (Colombia)

In the work by Perilla-Henao et al. (2016), nine leafhopper morphospecies belonging to subfamilies Typhlocybinae, Deltocephalinae, and Xestocephalinae were collected in two sites with a high percentage of diseased *F. uhdei*, *Croton* spp., *P. undulatum*, and *P. nigra* trees in Bogotà and tested by PCR. The phytoplasma detection rate was 87% (21 out of 24 tested leafhopper individuals were phytoplasma positive). Phytoplasmas of 16SrI and 16SrVII groups were detected in *Amplicephalus funzaensis*, *Exitianus atratus*, *Scaphytopius* sp., and two Typhlocybinae morphospecies including *Empoasca* sp. Single infections with either 16SrI or 16SrVII phytoplasmas were detected in *Haldorus* sp., *Xestocephalus desertorum*, and two Typhlocybinae morphospecies including *Empoasca* sp. Since *A. funzaensis* and *E. atratus* were the most abundant leafhoppers recorded in the inspected sites, they were also examined for their ability to transmit phytoplasmas under controlled conditions (Perilla-Henao et al. 2016). These leafhoppers, captured from symptomless grass (*Pennisetum clandestinum*) plants growing around the mentioned diseased trees, transmitted the naturally acquired phytoplasmas to caged common bean (*Phaseolus vulgaris*) seedlings (Perilla-Henao et al. 2016).

The reasons for the widespread occurrence of 16SrI and 16SrVII phytoplasmas in the Sabana de Bogotà may be complex. It is important to take into account that it is located in the tropical zone, at 2,600 m above sea level where the approximate day temperatures are 13–23 °C and night temperatures −3 to 10°C. Freezing temperatures are rare and usually last for minutes. There are no proper seasons, just rainy and dry periods. The Sabana de Bogotà is a continuous flat terrain delimited by mountains, with interspersed urban and rural locations, in which agricultural and grass lands intermix. Introduced species coexist with native species, and there are small areas of relatively wild native forests. Since there is no winter, the insects do not overwinter, and although their population dynamics has not been studied, it is possible to collect *A. funzaensis* and *E. atratus* all year around. These leafhoppers complete their life cycle in the grass *P. clandestinum* which covers large areas inside Bogotà and in its surroundings and may feed on many other plant species, including urban trees and crops, thereby facilitating the spread of phytoplasmas. This phenomenon is enhanced by the fact that phytoplasmas of groups 16SrI and 16SrVII seem to infect many plant families. In South America, the climatic patterns are cyclically affected by the ENSO ("El Niño Southern Oscillation"). ENSO is a climatic phenomenon rooted in the Pacific Ocean, characterized by irregular fluctuations between its warm ("El Niño") and cold ("La Niña") phases with a periodicity ranging from 2 to 7 years (Garreaud et al. 2009). It has been hypothesized that during extreme drought periods, when plants hosts are under water stress, insect vectors may change hosts temporarily looking for better resources. If this is the case, they may carry phytoplasmas to new unrelated plant species, expanding their host range (Franco-Lara and Perilla-Henao 2014). Another consideration is the fact that urban trees in Bogotà are under stress conditions due to air pollution and poor water and nutrient access. For example, it is evident that, depending on the part of the city in which the trees grow, they may be covered by soot affecting their photosynthetic area and the opening and closure of stomata. Studies show that changes in environmental variables such as CO_2, O_3, drought, and temperature affect the response of plants to pathogens in complex ways depending on the pathogen and circumstances of the disease (Eastburn et al. 2011). There is little information about this subject regarding trees, but it is likely that defense mechanisms against insects and pathogens may be altered in the trees of Bogotà. More studies are needed to understand this phenomenon; however all these factors may contribute to the spread of phytoplasmas in the Sabana de Bogotà.

10.11 Diversity of Phytoplasmas Infecting Forest and Urban Trees

Genetic and bioecological diversity of '*Ca*. Phytoplasma' taxa are especially pronounced in phytoplasmas associated with trees, where diversity of these obligate bacteria are an integral part of the forest ecosystem (Fernández et al. 2007). The presence of phytoplasmas in urban areas are most commonly related with human activities (propagation, horticulture, drafting, trading with plants), thus representing a permanent source for spreading of these pathogens. Unfortunately, many aspects of phytoplasma presence in forest and urban ecosystems are poorly studied, because more attention of the practitioners in past years has been given to phytoplasma diseases causing significant losses in economically important crops primarily related to food production (potato, rice, corn, and grapevine).

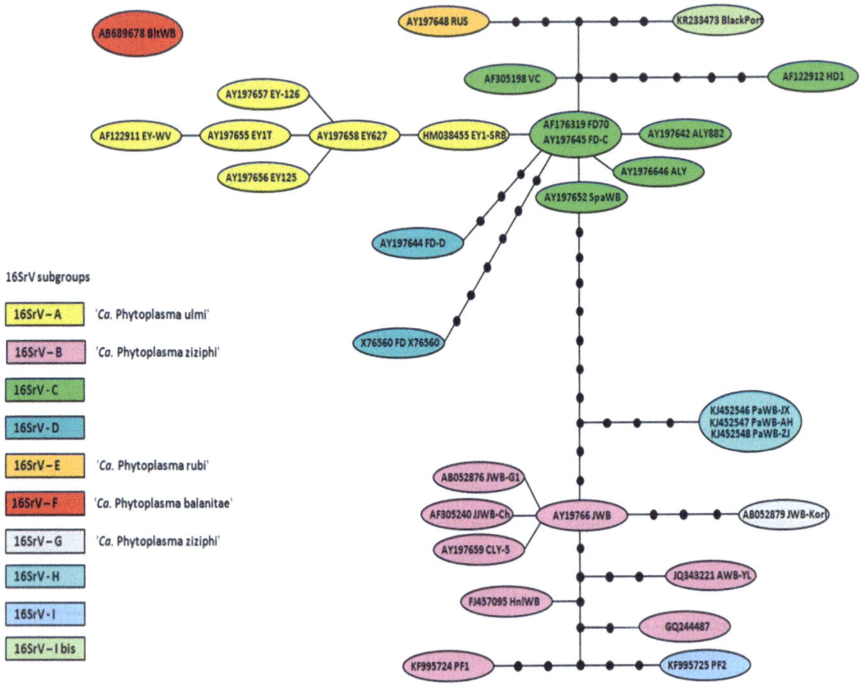

Fig. 10.5 Statistical parsimony network of 16S rRNA gene genealogy constructed using phytoplasma strains belonging to the 16SrV group. Colors correspond to the 16SrV subgroups. GenBank accession numbers are inside elliptical cycles. Black dots represent mutational differences separating different strains. '*Ca.* P. balanitae' was not connected with the other strains in the network at a confidence limit of 95%

The diversity of phytoplasmas associated with tree species shows important epidemiological aspects, varying from severe impacts on infected host plants on a large regional scale (Griffiths et al. 1999b), to discrete symptoms of discoloration or even absence of any visible symptoms. This is true for the members of the 16SrV group which share high 16S rRNA gene sequence similarity of 98.6–99.0% among subgroups (Lee et al. 2004a), while associated with a variety of plant species, especially woody perennial hosts. The 16SrV group consists of nine subgroups annotated as -A to -I (Fig. 10.5) associated with diseases in diverse plant genera and families in various geographical regions across the three biogeographic realms, i.e., Palearctic, Nearctic, and Indomalaya. As mentioned above, '*Ca.* P. ulmi' induces symptoms which vary among the elm species (Marcone 2017). However, some strains of this taxon recorded on large scales on some European elm genotypes in Europe do not exhibit visible pathological properties on the host plant (Jović et al. 2011; Katanić et al. 2016).

Genetic and ecological properties of the 16SrV group express all complexity associated with its diversity and consequential problems in phytoplasma classification according to rules proposed by IRPCM (2004). To date, four putative taxa among the 16SrV group members are described. They are '*Ca*. P. ulmi' (16SrV-A), '*Ca*. P. rubi' (16SrV-E), '*Ca*. P. ziziphi' (16SrV-B), and '*Ca*. P. balanitae' (16SrV-F), associated with *Ulmus* spp. (Ulmaceae), *Rubus* spp. (Rosaceae), *Ziziphus jujuba* (Rhamnaceae), and *Balanites triflora* (Zygophyllaceae), respectively (Fig. 10.5) (Jung et al. 2003; Lee et al. 2004a; Malembic-Maher et al. 2011; Win et al. 2013). In addition, members of the 16SrV-C and 16SrV-D subgroups including FD, PGY, ALY, and *Clematis* phytoplasmas did not meet all the proposed criteria for the description of '*Ca*. Phytoplasma' species in regard to common specific oligonucleotides in their 16S rRNA gene, and thus taxonomically are not recognized as valid species (Malembic-Maher et al. 2011). Various 16SrV-C phytoplasma strains do not induce symptoms or induce only mild symptoms of yellowing in *A. glutinosa* and *A. incana* (Betulaceae) (Arnaud et al. 2007; Cvrković et al. 2008), *Ailanthus altissima* (Simaroubaceae) (Filippin et al. 2011), and *Clematis vitalba* (Ranunculaceae) (Angelini et al. 2004; Filippin et al. 2009), while when present in *Vitis vinifera* (Vitaceae), 16SrV-C and 16SrV-D subgroup strains are responsible for the most severe and destructive disease, namely, FD. Discrepancy between genetic and ecological variation in phenotypic properties could be overcome by additional knowledge, especially in cases where '*Ca*. Phytoplasma' species could be represented by ecologically separated populations characterized by different host plant ranges, different insect vectors, different ecological niches, and evidence of further molecular diversity (IRPCM 2004; Malembic-Maher et al. 2009; Win et al. 2013; Fernández et al. 2016).

The specific ecological, epidemiological, and molecular properties of individual '*Ca*. Phytoplasma' species, as well as their economic importance and geographic distribution, have driven the effort to achieve proper species description by more complex molecular characterization and phylogenetic reconstruction. The same is true for '*Ca*. P. ulmi,' an economically important species in North America and in several European areas. This phytoplasma species shares high 16S rRNA gene sequence similarity with other members of the 16SrV group, especially with 16SrV-C subgroup, but has specific and unique biological properties, including specific host plant associations (*Ulmus* spp.) and two confirmed specific vector species, *S. luteolus* in North America (Baker 1949) and *Macropsis mendax* in Italy (Carraro et al. 2004). A description of species with such diverse bioecological characteristics could be expected to be based on many strains of different geographic origins not included in the original description. The number of characterized genomic markers and/or different strains required for the description of each phytoplasma taxon differs significantly among different taxa. This is a common case among those that are clearly differentiated on the basis of 16S rRNA gene sequences and those from more diverse subgroups or strains.

In general, the limited number of strains analyzed in the studies for species descriptions is a common event. This may cause a problem in subsequent species identifications especially for strains within described species that have not been previously elaborated with respect to geographic variability. The description of '*Ca*. P. ulmi' is a good example of such a case. This description is based on multigene characterization of only four isolates from North America and South Italy, i.e., regions where EY disease has a severe impact (Lee et al. 2004a). Analyzed strains were homogeneous in their 16S rRNA and ribosomal protein genes, with only minor diversity present in the *secY* gene. However, strains characterized in some subsequent studies, originating from France, Czech Republic, Serbia, and Croatia, suggested the presence of high genetic variability in EY phytoplasma and diverse multigene characteristics (Malembic-Maher et al. 2009; Navrátil et al. 2009; Jović et al. 2011; Katanić et al. 2016). The high genetic and ecological divergence was also recorded in other 16SrV phytoplasma subgroups suggesting the existence of more '*Ca.* Phytoplasma' species. The relationships between the 16S rRNA gene content of different strains are better visualized in reticulate graphs or networks (Posada and Crandall 2001) which additionally provide quantifications and qualitative patterns in phylogeny and genealogy among strains. Network analysis using statistical parsimony (Clement et al. 2000) strongly suggests segregation of several strains within the 16SrV group, i.e., strain HD1 associated with *Apocynum cannabinum* (Apocynaceae), strain of FD (X76560) associated with *V. vinifera* (Vitaceae), and strains PF1 and PF2 associated with *Diospyros kaki* (Ebenaceae) (Fig. 10.5) if genetic data are combined with well-determined ecological properties.

Similar to the phytoplasmas from the 16SrV group, phytoplasmas from the 16SrI group are relatively homogeneous at the 16S rRNA gene level while infecting different tree species. '*Ca.* P. asteris,' widely associated with diverse plants from the family Asteraceae, can induce serious disease when associated with various tree species. Subgroup 16SrI-B and 16SrI-C phytoplasmas were reported to be associated with a decline disease of European hackberry (*Celtis australis*, Cannabaceae) in Italy (Bertaccini et al. 1996), while a 16SrI-A phytoplasma was found to be associated with a stunt disease of gray dogwood (*Cornus racemosa*, Cornaceae) in North America (Lee et al. 2006). Several other forest and ornamental trees were also recorded as hosts for the various 16SrI subgroup phytoplasmas. Among them there are Indian mulberry (*Morinda citrifolia*, Rubiaceae), poplar, sandal, paulownia, and mulberry trees as well as the several urban trees in Sabana de Bogotà as mentioned above (Berges et al. 1997; Šeruga et al. 2003; Lee et al. 2004b; Davis et al. 2005; Khan et al. 2008; Yue et al. 2008; Ji et al. 2009; Mitrović et al. 2011; Perilla-Henao and Franco-Lara 2013; Franco-Lara and Perilla Henao 2014). In the Chinese scholar tree (*Sophora japonica* L.) showing severe stunting and yellowing, the presence of "stolbur" (16SrXII-A subgroup) and '*Ca.* P. japonicum'-related strains (16SrXII-D subgroup) was detected in China (Duduk et al. 2010). '*Ca.* P. solani' (16SrXII-A subgroup) was detected also in various forest and urban trees in Europe with or without mild symptoms of yellowing (Fernández et al. 2007; Katanić et al. 2016; J. Jović, personal comunication). Presence of 16SrXII-A subgroup phytoplasmas in such trees may have significant epidemiological relevance because these prokaryotes also infect economically important agricultural crops. Phytoplasmas belonging to at least five taxonomic groups and several different subgroups have been identified in conifers in Europe, North America, and Asia (Schneider et al. 2005; Davis et al. 2010; Gupta et al. 2010; Jomantiene et al. 2011; Kamińska and Berniak 2011; Valiunas et al. 2015).

Accurate identification and classification of phytoplasmas in affected plants is essential for understanding their diversity from a molecular and ecological standpoint. The accuracy in phytoplasma classification is complicated not only by existing difficulties in establishing their culture, but also by the interactions in which they are involved in the environment. Current knowledge regarding these bacteria is almost completely limited to their genetic peculiarities, and the traditional use of the 16S rRNA gene represents a primary tool for their detection and identification. The presence of phytoplasmas is conventionally correlated with existence of symptoms exhibited by affected plants. Symptom expression extends from very severe impact that leads to rapid death of the host plant to chronic metabolic misbalance which, if prolonged, may also lead to the death of the host. In contrast, there is growing data which describe latent presence of phytoplasmas in plants. Accordingly, detection of phytoplasmas in the plant is almost exclusively a matter of their pathogenic manifestation, while researchers chronically lack information regarding hosts where these bacteria occur in more or less equilibrium with the host plant. This is a very important issue, because interaction between plants and phytoplasmas without pathogenic effects seems not to be rare events.

10.12 Conclusions

Due to the detection and classification methods currently available, both the number of phytoplasma diseases of forest and urban trees and the number of the associated phytoplasma taxa are rapidly increasing. It is likely that there are still many unknown phytoplasma diseases of such kind of trees and that their economic and ecological significance is greater than currently known. However, for some of the mentioned diseases, there is still a lack of information on several aspects including insect vectors, phytoplasma-insect vector relationships, phytoplasma-host plant interactions, strain virulence, strain interference, host tolerance, host plant range, and impact of phytoplasma infections on growth and yield of affected plants. Also, in several instances, these diseases escape observation because they show nonspecific symptoms only, such as yellowing, stunting, and/or decline. The population of the phytoplasmas in diseased plants, especially in those with nonspecific symptoms, is often so low that it can only be detected with highly sensitive molecular methods. Latent phytoplasma infections, which are common in some forest and urban trees, can serve as inoculum reservoirs for further spread, including to agricultural crops. At present, the only approaches to control such diseases are (i) use of healthy plant material, (ii) elimination of infected trees, and (iii) growing resistant plants. Application of tetracycline antibiotics may be appropriate for the treatment of particularly valuable trees but is not allowed in several countries.

References

Angelini E, Squizzato F, Lucchetta G, Borgo M (2004) Detection of a phytoplasma associated with grapevine "flavescence dorée" in *Clematis vitalba*. *European Journal of Plant Pathology* **110**, 193–201.

Arnaud G, Malembic-Maher S, Salar P, Bonnet P, Maixner M, Marcone C, Boudon-Padieu E, Foissac X (2007) Multilocus sequence typing confirms the close genetic interrelatedness of three distinct "flavescence dorée" phytoplasma strain clusters and group 16SrV phytoplasmas infecting grapevine and alder in Europe. *Applied and Environmental Microbiology* **73**, 4001–4010.

Atanasoff D (1973) Witches' brooms on stems of elms and other trees. *Archiv für Phytopathologie und Pflanzenschutz* **9,** 241–243.

Baker WL (1949) Studies on the transmission of the virus causing phloem necrosis of American elm, with notes on the biology of its insect vector. *Journal of Economic Entomology* **42**, 729–732.

Berges R, Cousin MT, Roux J, Seemüller E (1997) Detection of phytoplasma infections in declining *Populus nigra* Italica trees and molecular differentiation of the aster yellows phytoplasmas identified in various *Populus species*. *European Journal of Forest Pathology* **27**, 33–43.

Bertaccini M, Mittempergher L, Vibio M (1996) Identification of phytoplasmas associated with a decline of European hackberry (*Celtis australis*). *Annals of Applied Biology* **128**, 245–253.

Bertaccini A, Duduk B, Paltrinieri S, Contaldo N (2014) Phytoplasmas and phytoplasma diseases: a severe threat to agriculture. *American Journal of Plant Sciences* **5**, 1763–1788.

Bila J, Mondjana A, Samils B, Högberg N (2015) Potential novel '*Candidatus* Phytoplasma pini'-related strain associated with coconut lethal yellowing in Mozambique. *Phytopathogenic Mollicutes* **5**(1-Supplement), S59–S60.

Botti S, Bertaccini A (2007) Grapevine yellows in Northern Italy: molecular identification of "flavescence dorée" phytoplasma strains and of "bois noir" phytoplasmas. *Journal of Applied Microbiology* **103**, 2325–2330.

Braun EJ, Sinclair WA (1979) Phloem necrosis of elms: symptoms and histopathological observations in tolerant hosts. *Phytopathology* **69**, 354–358.

Carraro L, Ferrini F, Ermacora P, Loi N, Martini M, Osler R (2004) *Macropsis mendax* as a vector of elm yellows phytoplasma of *Ulmus* species. *Plant Pathology* **53**, 90–95.

Clement M, Posada D, Crandall KA (2000) TCS: a computer program to estimate gene genealogies. *Molecular Ecology* **9**, 1657–1659.

Conti M, D'Agostino G, Mittempergher L (1987) A recent epiphytotic of elm yellows in Italy. 7th Congress of the Mediterranean Phytopathological Union, Granada, Spain, 208–209.

Cousin MT (1996) Witches' broom, a phytoplasma disease of poplar. In: Forest Trees and Palm Diseases and Control. Eds Maramorosch K, Raychaudhuri SP. Oxford and IBH Publishing Company, New Delhi, India, 267-283 pp.

Cousin M-T, Roux J, Boudon-Padieu E, Berges R, Seemüller E, Hiruki C (1998) Use of heteroduplex mobility analysis (HMA) for differentiating phytoplasma isolates causing witches' broom disease on *Populus nigra* cv Italica and "stolbur" or big bud symptoms on tomato. *Journal of Phytopathology* **146**, 97–102.

Cousin M-T, Berges R, Roux J, Moreau J-P, Hiruki C, Seemüller E (1999) *Populus nigra* L. Italica decline in France. Variability of the phytoplasma responsible for the disease in Europe. Results and perspectives. *Acta Horticulturae* **496**, 77–86.

Cvrković T, Jović J, Mitrović M, Petrović A, Krnjajić S, Malembic-Maher S, Toševski I (2008) First report of alder yellows phytoplasma on common alder (*Alnus glutinosa*) in Serbia. *Plant Pathology* **57**, 773.

Davis RI, Arocha Y, Jones P, Malau A (2005) First report of the association of phytoplasmas with plant diseases in the territory of Wallis and Futuna. *Australasian Plant Pathology* **34**, 417–418.

Davis RE, Dally EL, Zhao Y, Lee I-M, Jomantiene R, Detweiler AJ, Putnam ML (2010) First report of a new subgroup 16SrIX-E ('*Candidatus* Phytoplasma phoenicium'-related) phytoplasma associated with juniper witches' broom disease in Oregon, USA. *Plant Pathology* **59**, 1161.

Davis RE, Zhao Y, Dally EL, Lee I-M, Jomantiene R, Douglas SM (2013) '*Candidatus* Phytoplasma pruni', a novel taxon associated with X-disease of stone fruits, *Prunus* spp.: multilocus characterization based on 16S rRNA, *secY*, and ribosomal protein genes. *International Journal of Systematic and Evolutionary Microbiology* **63**, 766–776.

Dickinson M, Tuffen M, Hodgetts J (2013) The phytoplasmas: an introduction. *Methods in Molecular Biology* **938**, 1–14.

Dijkstra J, Ie TS (1969) Presence of mycoplasma-like bodies in the phloem of sandal affected with spike disease. *Netherlands Journal of Plant Pathology* **75**, 374–378.

Dijkstra J, Lee PE (1972) Transmission by dodder of sandal spike disease and the accompanying mycoplasma-like organisms via *Vinca rosea*. *Netherlands Journal of Plant Pathology* **78**, 218–228.

Doi Y, Asuyama H (1981) Paulownia witches' broom disease. In: Mycoplasma Diseases of Trees and Shrubs. Eds Maramorosch K, Raychaudhuri SP. Academic Press, New York, New York, USA, 135-146 pp.

Doi Y, Teranaka M, Yora K, Asuyama H (1967) Mycoplasma- or PLT group-like microorganisms found in the phloem elements of plants infected with mulberry dwarf, potato witches' broom, aster yellows, or paulownia witches' broom. *Annals of the Phytopathological Society of Japan* **33**, 259–266.

Du T, Wang Y, Hu Q-X, Chen J, Liu S, Huang W-J, Lin M-L (2005) Transgenic *Paulownia* expressing *shiva-1* gene has increased resistance to paulownia witches' broom disease. *Journal of Integrative Plant Biology* **47**, 1500–1506.

Duduk B, Tian JB, Contaldo N, Fan XP, Paltrinieri S, Chen QF, Zhao QF, Bertaccini A (2010) Occurrence of phytoplasmas related to "stolbur" and to '*Candidatus* Phytoplasma japonicum' in woody host plants in China. *Journal of Phytopathology* **158**, 100–104.

Ember I, Acs Z, Salar P, Danet J-L, Foissac X, Kölber M, Malembic-Maher S (2011) Survey and genetic diversity of phytoplasmas from the 16SrVC and -D subgroups in Hungary. *Bulletin of Insectology* **64**(Supplement), S33–S34.

Eastburn DM, McElrone AJ, Bilgin DD (2011) Influence of atmospheric and climatic change on plant-pathogen interactions. *Plant Pathology* **60**, 54–69.

Fernández EGB, Calari A, Hanzer V, Katinger H, Bertaccini A, Laimer M (2007) Phytoplasma infected plants in Austrian forests: role as a reservoir? *Bulletin of Insectology* **60**, 391–392.

Fernández FD, Galdeano E, Kornowski MV, Arneodo JD, Conci LR (2016) Description of '*Candidatus* Phytoplasma meliae', a phytoplasma associated with Chinaberry (*Melia azedarach* L.) yellowing in South America. *International Journal of Systematic and Evolutionary Microbiology* **66**, 5244–5251.

Filgueira JJ, Franco-Lara L, Salcedo JE, Gaitan SL, Boa ER (2004) Urapan (*Fraxinus udhei*) dieback, a new disease associated with a phytoplasma in Colombia. *Plant Pathology* **53**, 520.

Filippin L, Jović J, Cvrković T, Forte V, Clair D, Toševski I, Boudon-Padieu E, Borgo M, Angelini E (2009) Molecular characteristics of phytoplasmas associated with "flavescence dorée" in clematis and grapevine and preliminary results on the role of *Dictyophara europaea* as a vector. *Plant Pathology* **58**, 826–837.

Filippin L, De Pra V, Zottini M, Borgo M, Angelini E (2011) Nucleotide sequencing of imp gene in phytoplasmas associated to "flavescence dorée" from *Ailanthus altissima*. *Bulletin of Insectology* **64**, 49–50.

Franco-Lara L, Perilla Henao LM (2014) Phytoplasma diseases in trees of Bogotà, Colombia: a serious risk for urban trees and crops. In: Phytoplasmas and phytoplasma disease management: how to reduce their economic impact. Ed Bertaccini A. COST FA0807, Bologna, Italy, 90–100 pp.

Franco-Lara L, Contaldo N, Mejia JF, Paltrinieri S, Duduk B, Bertaccini A (2017) Detection and identification of phytoplasmas associated with declining *Liquidambar styraciflua* trees in Colombia. *Tropical Plant Pathology* **42**, 352–361.

Gao R, Wang J, Shao Y-H, Li X-D, Yang B-H, Chang WC, Zhao W-J, Zhu S-F (2011) Molecular identification of a phytoplasma associated with elm witches' broom in China. *Forest Pathology* **41**, 355–360.

Garreaud RD, Vuille M, Compagnucci R, Marengo J (2009) Present-day South American climate. *Palaeogeography, Palaeoclimatology, Palaeoecology* **281**, 180–195.

Griffiths HM, Sinclair WA, Smart CD, Davis RE (1999a) The phytoplasma associated with ash yellows and lilac witches' broom: '*Candidatus* Phytoplasma fraxini'. *International Journal of Systematic and Evolutionary Microbiology* **49**, 1605–1614.

Griffiths HM, Sinclair WA, Boudon-Padieu E, Daire X, Lee I-M, Sfalanga A, Bertaccini A (1999b) Phytoplasmas associated with elm yellows: molecular variability and differentiation from related organisms. *Plant Disease* **83**, 1101–1104.

Griffiths HM, Boa ER, Filgueira JJ (2001) Ash yellows: a new disease of *Fraxinus chinensis* in Colombia. *Phytopathology* **91**, S33–S34.

Gupta MK, Samad A, Shasany AK, Ajayakumar PV, Alam M (2010) First report of a 16SrVI '*Candidatus* Phytoplasma trifolii' isolate infecting Norfolk Island pine (*Auraucaria heterophylla*) in India. *Plant Pathology* **59**, 399.

Huang S, Tiwari AK, Rao GP (2011) '*Candidatus* Phytoplasma pini' affecting *Taxodium distichum* var. *imbricarium* in China. *Phytopathogenic Mollicutes* **1**, 91–94.

Herath P, Hoover GA, Angelini E, Moorman GW (2010) Detection of elm yellows phytoplasma in elms and insects using real-time PCR. *Plant Disease* **94**, 1355–1360.

Hill GT, Sinclair WA (2000) Taxa of leafhoppers carrying phytoplasmas at sites of ash yellows occurrence in New York State. *Plant Disease* **84**, 134–138.

Hiruki C (1999) Paulownia witches' broom disease important in East Asia. *Acta Horticulturae* **496**, 63–68.

Hiruki C, Dijkstra J (1973) Light and electron microscopy of vinca plants infected with mycoplasma-like bodies of the sandal spike disease. *Netherlands Journal of Plant Pathology* **79**, 207–217.

Holz S, Duduk B, Büttner C, Kube M (2016) Genetic variability of alder yellows phytoplasma in *Alnus glutinosa* in its natural Spreewald habitat. *Forest Pathology* **46**, 11–21.

Hull R, Horne RW, Nayar RM (1969) Mycoplasma-like bodies associated with sandal spike disease. *Nature* **224**, 1121–1122.

IRPCM (2004) '*Candidatus* Phytoplasma', a taxon for the wall-less, non-helical prokaryotes that colonize plant phloem and insects. *International Journal of Systematic and Evolutionary Microbiology* **54**, 1243–1255.

Ishiie T, Doi Y, Yora K, Asuyama H (1967) Suppressive effects of antibiotics of tetracycline group on symptoms development of mulberry dwarf disease. *Annals of the Phytopathological Society of Japan* **33**, 267–275.

Jacobs KA, Lee I-M, Griffiths HM, Miller FD Jr, Bottner KD (2003) A new member of the clover proliferation phytoplasma group (16SrVI) associated with elm yellows in Illinois. *Plant Disease* **87**, 241–246.

Ježić M, Poljak I, Idžojtić M, Ćurković-Perica M (2012) First report of '*Candidatus* Phytoplasma pini' in Croatia. 19th Congress of the International Organisation for Mycoplasmology, Toulouse, France, 158–159.

Ji X, Gai Y, Zheng C, Mu Z (2009) Comparative proteomic analysis provides new insights into mulberry dwarf responses in mulberry (*Morus alba* L.). *Proteomics* **9**, 5328–5339.

Jomantiene R, Valiunas D, Ivanauskas A, Urbanaviciene L, Staniulis J, Davis RE (2011) Larch is a new host for a group 16SrI, subgroup B, phytoplasma in Ukraine. *Bulletin of Insectology* **64**(Supplement), S101–S102.

Jović J, Cvrković T, Mitrović M, Petrović A, Krstić O, Krnjajić S and Toševski I (2010) Genetic variability among '*Candidatus* Phytoplasma ulmi' strains infecting elms in Serbia and survey of potential vectors. In: Current Status and Perspectives of Phytoplasma Disease Research and Management. Eds Bertaccini A, Laviña A, Torres E. COST Action FA0807, Sitges, Spain, 18.

Jović J, Cvrković T, Mitrović M, Petrović A, Krstić O, Krnjajić S, Toševski I (2011) Multigene sequence data and genetic diversity among '*Candidatus* Phytoplasma ulmi'strains infecting *Ulmus* spp. in Serbia. *Plant Pathology* **60**, 356–368.

Jung HY, Sawayanagi T, Kakizawa S, Nishigawa H, Wei W, Oshima K, Miyata S, Ugaki M, Hibi T, Namba S (2003) '*Candidatus* Phytoplasma ziziphi', a novel phytoplasma taxon associated with jujube witches' broom disease. *International Journal of Systematic and Evolutionary Microbiology* **53**, 1037–1041.

Kamińska M, Berniak H (2009) '*Candidatus* Phytoplasma asteris' in *Fraxinus excelsior* and its association with ash yellows newly reported in Poland. *Plant Pathology* **58**, 788.

Kamińska M, Berniak H (2011) Detection and identification of three '*Candidatus* Phytoplasma' species in *Picea* spp. trees in Poland. *Journal of Phytopathology* **159**, 796–798.

Kamińska M, Berniak H, Obdrzalek J (2011) New natural host plants of '*Candidatus* Phytoplasma pini' in Poland and the Czech Republic. *Plant Pathology* **60**, 1023–1029.

Katanić Z, Krstin L, Ježić M, Zebec M, Ćurković-Perica M (2016) Molecular characterization of elm yellows phytoplasmas in Croatia and their impact on *Ulmus* spp. *Plant Pathology* **65**, 1430–1440.

Kawakita H, Saiki T, Wei W, Mitsuhashi W, Watanabe K, Sato M (2000) Identification of mulberry dwarf phytoplasmas in the genital organs and eggs of leafhopper *Hishimonoides sellatiformis*. *Phytopathology* **90**, 909–914.

Khan JA, Srivastava P, Sing SK (2006) Identification of a '*Candidatus* Phytoplasma asteris'-related strain associated with spike disease of sandal (*Santalum album*) in India. *Plant Pathology* **55**, 572.

Khan JA, Sing SK, Ahmad J (2008) Characterisation and phylogeny of a phytoplasma inducing sandal spike disease in sandal (*Santalum album*). *Annals of Applied Biology* **153**, 365–372.

Kim YH, La YJ, Kim YT (1985) Transmission and histochemical detection of mulberry dwarf mycoplasma in several herbaceous plants. *Korean Journal of Plant Pathology* **1**, 184–189.

Kuai YZ, Zhu FP, Xia ZS, Chen PG (1999) Studies on the mechanism of mulberry varietal resistance to mulberry yellow dwarf disease. *International Journal of Tropical Plant Disease* **17**, 103–114.

Lederer W, Seemüller E (1991) Occurrence of mycoplasma-like organisms in diseased and non-symptomatic alder trees (*Alnus* spp.). *European Journal of Forest Pathology* **21**, 90–96.

Lee I-M, Davis RE, Sinclair WA, De Witt ND, Conti M (1993) Genetic relatedness of mycoplasmalike organisms detected in *Ulmus* spp. in the United States and Italy by means of DNA probes and polymerase chain reactions. *Phytopathology* **83**, 829–833.

Lee I-M, Bertaccini A, Vibio M, Gundersen DE, Davis RE, Mittempergher L, Conti M, Gennari F, (1995) Detection and characterization of phytoplasmas associated with disease in *Ulmus* and *Rubus* in northern and central Italy. *Phytopathologia Mediterranea* **34**, 174–183.

Lee I-M, Davis RE, Gundersen-Rindal DE (2000) Phytoplasma: phytopathogenic mollicutes. *Annual Review of Microbiology* **54**, 221–255.

Lee I-M, Martini M, Marcone C, Zhu SF (2004a) Classification of phytoplasma strains in the elm yellows group (16SrV) and proposal of '*Candidatus* Phytoplasma ulmi' for the phytoplasma associated with elm yellows. *International Journal of Systematic and Evolutionary Microbiology* **54**, 337–347.

Lee I-M, Gundersen-Rindal DE, Davis RE, Bottner KD, Marcone C, Seemüller E (2004b) '*Candidatus* Phytoplasma asteris', a novel phytoplasma taxon associated with aster yellows and related diseases. *International Journal of Systematic and Evolutionary Microbiology* **54**, 1037–1048.

Lee I-M, Zhao Y, Bottner KD (2006) *SecY* gene sequence analysis for finer differentiation of diverse strains in the aster yellows phytoplasma group. *Molecular and Cellular Probes* **20**, 87–91.

Lee I-M, Zhao Y, Davis RE (2010) Prospects of multiple gene-based systems for differentiation and classification of phytoplasmas. In: Phytoplasmas: Genomes, Plant Hosts and Vectors. Eds Weintraub PG, Jones P. CAB International, Wallingford, Oxfordshire, United Kingdom, 51–63 pp.

Maixner M, Reinert W (1999) *Oncopsis alni* (Schrank) (Auchenorrhyncha: Cicadellidae) as a vector of the alder yellows phytoplasma of *Alnus glutinosa* (L.) Gaertn. *European Journal of Plant Pathology* **105**, 87–94.

Maixner M, Rüdel M, Daire X, Boudon-Padieu E (1995) Diversity of grapevine yellows in Germany. *Vitis* **34,** 235–236.

Maixner M, Reinert W, Darimont H (2000). Transmission of grapevine yellows by *Oncopsis alni* (Schrank) (Auchenorrhynca: Macropsinae). *Vitis* **39**, 83–84.

Malembic-Maher S, Salar P, Vergnes D, Foissac X (2007) Detection and diversity of "flavescence dorée"-related phytoplasmas in alders surrounding infected vineyards in Aquitaine (France). *Bulletin of Insectology* **60**, 329–330.

Malembic-Maher S, Salar P, Carle P, Foissac X (2009) Ecology and taxonomy of "flavescence dorée" phytoplasmas: the contribution of genetic diversity studies. *Le Progrés Agricole et Viticole* **HS**, 132–134.

Malembic-Maher S, Salar P, Filippin L, Carle P, Angelini E, Foissac X (2011) Genetic diversity of European phytoplasmas of the 16SrV taxonomic group and proposal of '*Candidatus* Phytoplasma rubi'. *International Journal of Systematic and Evolutionary Microbiology* **61**, 2129–2134.

Marcone C (2015) Current status of phytoplasma diseases of forest and landscape trees and shrubs. *Journal of Plant Pathology* **97**, 9–36.

Marcone C (2017) Elm yellows: a phytoplasma disease of concern in forest and landscape ecosystems. *Forest Pathology* **47**, e12324.

Marcone C, Rao GP (2016) Elm and alder yellows: etiology, symptomatology and epidemiology. *Phytopathogenic Mollicutes* **6**, 63–70.

Marcone C, Lee I-M, Davis RE, Ragozzino A, Seemüller E (2000) Classification of aster yellows-group phytoplasmas based on combined analyses of ribosomal RNA and *tuf* gene sequences. *International Journal of Systematic and Evolutionary Microbiology* **50**, 1703–1713.

Martini M, Marcone C, Lee I-M, Firrao G (2014). The family *Acholeplasmataceae* (including phytoplasmas). In: The Prokaryotes – Firmicutes and Tenericutes, 4th edition, XXII. Eds Rosenberg E, DeLong EF, Lory S, Stackebrandt E, Thompson F. Springer-Verlag, Berlin, Germany, 469–504 pp.

Matteoni JA, Sinclair WA (1985) Role of the mycoplasmal disease, ash yellows, in decline of white ash in New York State. *Phytopathology* **75**, 355–360.

Matteoni JA, Sinclair WA (1988) Elm yellows and ash yellows. In: Tree Mycoplasmas and Mycoplasma Diseases. Ed Hiruki C. University of Alberta Press, Edmonton, Alberta, Canada, 19-31 pp.

Mäurer R, Seemüller E, Sinclair WA (1993) Genetic relatedness of mycoplasmalike organisms affecting elm, alder, and ash in Europe and North America. *Phytopathology* **83**, 971–976.

Mehle N, Rupar M, Seljak G, Ravnikar M, Dermastia M (2011) Molecular diversity of "flavescence dorée" phytoplasma strains in Slovenia. *Bulletin of Insectology* **64**(Supplement), S29–S30.

Mitrović J, Paltrinieri S, Contaldo N, Bertaccini A, Duduk B (2011) Occurrence of two '*Candidatus* Phytoplasma asteris'-related phytoplasmas in poplar trees in Serbia. *Bulletin of Insectology* **64**(Supplement), S57–S58.

Mittempergher L (2000) Elm yellows in Europe. In: The elms: breeding, conservation, and disease management. Ed Dunn CP. Kluwer Academic Publisher, Boston, Massachusetts, USA, 103-119 pp.

Namba S, Oyaizu S, Kato S, Iwanami S, Tsuchizaki T (1993) Phylogenetic diversity of phytopathogenic mycoplasmalike organisms. *International Journal of Systematic Bacteriology* **43**, 461–467.

Navrátil M, Šafářová D, Válová P, Fránová J, Šimková M (2009) Phytoplasma associated with witches' broom disease of *Ulmus minor* Mill. in the Czech Republic: electron microscopy and molecular characterization. *Folia Microbiologica* **54**, 37–42.

Paltrinieri S, Martini M, Pondrelli M, Bertaccini A (1998) X-disease-related phytoplasmas in ornamental trees and shrubs with witches' broom and malformation symptoms. *Journal of Plant Pathology* **80**, 261.

Perilla-Henao LM, Franco-Lara L (2013) Especies arbóreas de las familias Euphorbiaceae, Pittosporaceae y Salicaceae son infectadas por '*Ca.* Phytoplasma fraxini' y '*Ca.* Phytoplasma asteris' en infecciones mixtas en Bogotá, Colombia. *Revista Facultad de Ciencias Básicas* **9**, 248–65.

Perilla-Henao LM, Dickinson M, Franco-Lara L (2012) First report of '*Candidatus* Phytoplasma asteris' affecting woody hosts (*Fraxinus udhei*, *Populus nigra*, *Pittosporum undulatum*, and *Croton* spp.) in Colombia. *Plant Disease* **96**, 1372.

Perilla-Henao L, Wilson MR, Franco-Lara L (2016) Leafhoppers *Exitianus atratus* and *Amplicephalus funzaensis* transmit phytoplasmas of groups 16SrI and 16SrVII in Colombia. *Plant Pathology* **65,** 1200–1209.

Pisi A, Marani F, Bertaccini A (1981) Mycoplasma-like organisms associated with elm witches' broom symptoms. *Phytopathologia Mediterranea* **20**, 189–191.

Posada D, Crandall KA (2001) Intraspecific gene genealogies: trees grafting into networks. *Trends in Ecology and Evolution* **16**, 37–45.

Rangaswami S, Griffith AL (1941) Demonstration of *Jassus indicus* (Walk.) as a vector of the spike disease of sandal (*Santalum album* Linn.) *The Indian Forester* **67**, 387–394.

Rao MS, Muniyappa V (1988) Epidemiology of sandal spike-disease. In: Tree Mycoplasmas and Mycoplasma Diseases. Ed Hiruki C. University of Alberta Press, Edmonton, Alberta, Canada, 57–58 pp.

Raychaudhuri SP, Chenulu VV, Ghosh SK, Varma A, Rap PS, Srimathi RA, Nag KC (1972) Chemical control of spike disease of sandal. *Current Science* **41**, 72–73.

Reinert W, Maixner M (1997) Epidemiological studies on a new grapevine yellows in Germany. 12th ICVG Meeting, Lisbon, Portugal, 65–66.

Rosa C, McCarthy E, Duong K, Hoover G, Moorman G (2014) First report of the spittlebug *Lepyronia quadrangularis* and the leafhopper *Latalus* sp. as vectors of the elm yellows associated phytoplasma, '*Candidatus* Phytoplasma' ulmi in North America. *Plant Disease* **98**, 154.

Sato M, Mitsuhashi W, Watanabe K, Kawakita H (1996) PCR detection of mulberry dwarf disease-phytoplasmas in mulberry tissues, phloem sap collected by laser stylectomy and insect vector *Hishimonus sellatus*. *Journal of Sericulture Science of Japan* **65**, 352–358.

Schneider B, Torres E, Martín MP, Schröder M, Behnke H-D, Seemüller E (2005) '*Candidatus* Phytoplasma pini', a novel taxon from *Pinus silvestris* and *Pinus halepensis*. *International Journal of Systematic and Evolutionary Microbiology* **55**, 303–307.

Seemüller E, Lederer W (1988) MLO-associated decline of *Alnus glutinosa*, *Populus tremula* and *Crataegus monogyna*. *Journal of Phytopathology* **121**, 33–39.

Seemüller E, Marcone C, Lauer U, Ragozzino A, Göschl M (1998) Current status of molecular classification of the phytoplasmas. *Journal of Plant Pathology* **80**, 3–26.

Šeruga M, Škorić D, Botti S, Paltrinieri S, Juretić N, Bertaccini A (2003) Molecular characterization of a phytoplasma from the aster yellows (16SrI) group naturally infecting *Populus nigra* L. Italica trees in Croatia. *Forest Pathology* **33**, 113–125.

Sfalanga A, Martini M, Surico G, Bertaccini A (2002) Involvement of phytoplasmas in a decline of *Ulmus chenmoui* in central Italy. *Forest Pathology* **32**, 265–275.

Sharma CD, Cousin MT (1986) Mycoplasmalike organisms (MLOs) associated with the witches' broom disease of poplar. *Journal of Phytopathology* **117**, 349–356.

Sinclair WA (1981) Elm yellows. In: Compendium of Elm Diseases. Eds Stipes RJ, Campana RJ. The American Phytopathological Society, St. Paul, Minnesota, USA, 25–31 pp.

Sinclair WA (2000) Elm yellows in North America. In: The Elms: Breeding, Conservation, and Disease Management. Ed Dunn CP. Kluwer Academic Publisher, Boston, Massachusetts, USA, 121–136.

Sinclair WA, Griffiths HM (1995) Epidemiology of a slow-decline phytoplasmal disease: ash yellows on old-fields sites in New York State. *Phytopathology* **85**, 123–128.

Sinclair WA, Griffiths HM, Davis RE (1996) Ash yellows and lilac witches' broom: phytoplasmal diseases of concern in forestry and horticulture. *Plant Disease* **80**, 468–475.

Sinclair WA, Townsend AM, Griffiths HM, Whitlow TH (2000a). Responses of six Eurasian *Ulmus* cultivars to a North American elm yellows phytoplasma. *Plant Disease* **84**, 1266–1270.

Sinclair WA, Gleason ML, Griffiths HM, Iles JK, Zriba N, Charlson DV, Batzer JC, Whitlow TH (2000b) Responses of 11 *Fraxinus* cultivars to ash yellows phytoplasma strains of differing aggressiveness. *Plant Disease* **84**, 725–730.

Śliwa H, Kaminska M, Korszun S, Adler P (2008) Detection of '*Candidatus* Phytoplasma pini' in *Pinus sylvestris* trees in Poland. *Journal of Phytopathology* **156**, 88–92.

Thomas S, Balasundaran M (1999) Detection of sandal spike phytoplasma by polymerase chain reaction. *Current Science* **76**, 1574–1575.

Valiunas D, Jomantiene R, Ivanauskas A, Sneideris D, Staniulis J, Davis RE (2010) A possible threat to the timber industry: '*Candidatus* Phytoplasma pini' in Scots pine (*Pinus sylvestris* L.) in Lithuania. In: Current Status and Perspectives of Phytoplasma Disease Research and Management. Eds Bertaccini A, Laviña A, Torres E. COST Action FA0807. Sitges, Spain, 38.

Valiunas D, Jomantiene R, Ivanauskas A, Urbonaite I, Sneideris D, Davis RE (2015) Molecular identification of phytoplasmas infecting diseased pine trees in the UNESCO-protected Curonian Spit of Lithuania. *Forests* **6**, 2469–2483.

van der Meer FA (1980) Witches' broom in poplars. *Populier* **17,** 42–43.

Weintraub PG, Beanland L (2006) Insect vectors of phytoplasmas. *Annual Review of Entomology* **51,** 91–111.

Win NKK, Lee SY, Bertaccini A, Namba S, Jung HY (2013) '*Candidatus* Phytoplasma balanitae' associated with witches' broom disease of *Balanites triflora. International Journal of Systematic and Evolutionary Microbiology* **63**, 636–640.

Yue HN, Wu YF, Shi YZ, Wu KK, Li YR (2008) First report of paulownia witches' broom phytoplasma in China. *Plant Disease* **92**, 1134.

Zhao Y, Davis RE (2016) Criteria for phytoplasma 16Sr group/subgroup delineation and the need of a platform for proper registration of new groups and subgroups. *International Journal of Systematic and Evolutionary Microbiology* **66**, 2121–2123

Zhao YY, Contaldo N, Paltrinieri S, Satta E, Bertaccini A (2016) Multigene characterization of phytoplasmas infecting jujube and paulownia in China. *Phytopathogenic Mollicutes* **6**, 63–68.

Zhu TS, Pan YZ, Cui TT, Gao R, Li X-D, Zhu SF (2008) Molecular detection and identification of the phytoplasma associated with elm yellows in China. *Acta Phytopathologica Sinica* **38**, 401–406.

Chapter 11
Phytoplasmas in Weeds and Wild Plants

Bojan Duduk, Jelena Stepanović, Amit Yadav, and Govind Pratap Rao

Abstract Weeds and wild plants as hosts of phytoplasmas play an important role in the epidemiology and emergence of phytoplasma diseases of economically important crops. In this chapter phytoplasmas detected in weeds and wild plants, their geographic origins, symptoms, identification, and their role in natural dissemination of phytoplasmas are described.

Keywords Reservoir · Overwintering · Epidemiology · Source of inoculum · Plant diseases

11.1 Introduction

To date, phytoplasmas have been associated with diseases in several hundred plant species in which they induce symptoms such as virescence, phyllody, sterility of flowers, witches' broom growth, elongation of internodes, overall stunting, discoloration of leaves/shoots, leaf curling, and plant decline. Phytoplasmas are transmitted from plant to plant mainly by sap-sucking insects, and they may overwinter in perennial plants which can act as their reservoirs for spreading in the following spring. In many important crops all over the world, phytoplasmas induce diseases that sometimes lead to severe economic losses in agronomically relevant species such as carrot, corn, potato, rice, grapevine, and palms. Therefore, throughout the world, different weeds and wild plants, with and without symptoms, have been tested to identify possible reservoir plants for phytoplasmas (Schneider et al. 1997;

B. Duduk (✉) · J. Stepanović
Institute of Pesticides and Environmental Protection, Belgrade, Serbia
e-mail: bojan.duduk@pesting.org.rs

A. Yadav
National Centre for Microbial Research, (Formerly, Microbial Culture Collection), National Centre for Cell Science, Pune, India

G. P. Rao
Division of Plant Pathology, Advance Center for Plant Virology, Indian Agricultural Research Institute, New Delhi 110012, India

G. P. Rao et al. (eds.), *Phytoplasmas: Plant Pathogenic Bacteria - I*,
https://doi.org/10.1007/978-981-13-0119-3_11

Mall et al. 2010; Win et al. 2013; Rao et al. 2017b). At the beginning of phytoplasma research, phytoplasmas were detected by characteristic symptoms and by observation of round or filamentous bodies in sieve tubes of diseased plants by transmission electron microscopy (TEM). Over the years, as molecular techniques evolved, introduction of PCR assays for detection and identification enabled further studies of the ecology and genomic diversity of phytoplasmas as well as the epidemiology and physiology of phytoplasma-associated diseases (Seemüller et al. 1994; Lee et al. 2000).

Some weeds or wild plants in which phytoplasma presence was recorded by observation only and therefore without proper identification are listed thereafter. In Korea, phytoplasma bodies were observed in the phloem tissues of *Cnidium officinale*, *Bupleurum falcatum*, and *Plantago asiatica* by electron microscopy (Choi et al. 1985). In Jamaica, Dabek (1983) used electron microscopy to confirm the presence of phytoplasmas in *Rhynchosia minima* with a disease called *Rhynchosia* little leaf (RLL) and managed to transmit the disease agent to *R. minima* test plants by the insect vector *Ollarianus balli* (van Duzee 1907). In India, the association of phytoplasma bodies with white leaf disease was observed in Bermuda grass – *Cynodon dactylon* (Singh et al. 1978). The symptoms associated with phytoplasma presence, which lead to the rice yellow dwarf disease, were observed in the common grass weed, *Echinochloa colonum* (Reddy and Jeyarajan 1988). Pleomorphic phytoplasma bodies were observed in symptomatic *C. dactylon* plants and yellowing diseased *Urochloa panicoides* in South India (Muniyappa et al. 1982). Rao and Singh (1990) observed grassy shoot and white leaf symptoms on *Imperata arundinacea* (Poaceae) growing in the vicinity of sugarcane fields and reported that the symptoms were associated with phytoplasma. *I. arundinacea* was then reported as a new alternative host species of the phytoplasma associated with sugarcane grassy shoot disease. In India, in *Phyllanthus amarus* with overall retarded growth symptoms, phytoplasma presence was confirmed by TEM (Samad et al. 2004).

Besides these reports, phytoplasmas identified in weeds all over the globe mainly belong to the 16SrI, 16SrII, 16SrXI, 16SrXII, and 16SrXIV groups, but some members belonging to the 16SrIII, 16SrIV, 16SrV, 16SrVI, 16SrVII, 16SrIX, 16SrX, and 16SrXXIX groups were also detected. A list of phytoplasmas detected in weeds and wild plants and their geographic origins is provided in Table 11.1.

11.2 Phytoplasmas in 16SrI Group (Aster Yellows)

In Italy, pot marigold (*Calendula officinalis*) collected inside apricot and plum orchards near vegetable crops affected by aster yellows (AY) were infected with phytoplasmas belonging to the 16SrI-B subgroup (Marcone et al. 1997b). AY phytoplasma was detected in *Portulaca oleracea* (purslane) collected from apricot orchards in Italy, in *Cardaria draba* (hoary cress) and *Bunias orientalis* (hill mustard) collected from an agricultural area, and in *Stellaria media* (common chickweed) and *Trifolium repens* (white clover) collected in or around apple/stone fruit orchards in Germany (Schneider et al. 1997). In the United Kingdom, AY

Table 11.1 Summary of phytoplasmas reported in weeds and wild plant species

Genus and species	Common name	Ribosomal group/ subgroup	Country	Reference
Acalypha indica	Indian nettle	16SrI	India	Tiwari et al. (2017)
Achyranthes aspera	Devil's horsewhip	16SrI	India	Raj et al. (2009a)
		16SrII	Oman	Moghal et al. (1998)
Aegilops squarrosa	Goat grass	16SrI	China	Wu et al. (2010)
Aeschynomene americana	American jointvetch	16SrII	Australia	Wilson et al. (2001)
Aeschynomene indica	Indian jointvetch	16SrII	Australia	Schneider et al. (1999)
Ageratum conyzoides	Goat weed	16SrI	India	Tiwari et al. (2012)
Alysicarpus rugosus	Rough chainweed	16SrII	Australia	Davis et al. (1997)
Alysicarpus vaginalis	Alyce clover	16SrII	Australia	Wilson et al. (2001)
Amaranthus retroflexus	Redroot pigweed	16SrI	China	Wu et al. (2010)
		16SrV-B		Yang et al. (2011)
		16SrXII-A	Italy	Credi et al. (2006)
			Czech Republic	Fialová et al. (2009)
Amaranthus sp.		16SrII	India	Arocha et al. (2008)
Aphyllodium sp.		16SrII	Australia	Schneider et al. (1999)
Arachis pintoi	Pinto's peanut	16SrII	Australia	Schneider et al. (1999)
Artemisia vulgaris	Common wormwood	16SrXII-A	Italy	Credi et al. (2006)
Arundo donax	Giant reed	16SrXIV	Saudi Arabia	Omar (2016)
Avena fatua	Wild oat	16SrI	China	Wu et al. (2010)
Axonopus compressus	Broadleaf carpet grass	16SrXIV	Thailand	Sunpapao (2016)
			Singapore	Koh et al. (2008)
Bidens alba	Shepherd's needles	16SrIX	Iran	Hemmati et al. (2017)
Bonamia pannosa		16SrII	Australia	Schneider et al. (1999)
Brachiaria brizantha	Signal grass	16SrXI	Africa	Asudi et al. (2016)
		16SrXIV		
Brachiaria distachya	Brachiaria grass	16SrXIV	Thailand	Seemüller et al. (1998)
Brugmansia candida	Angel's trumpet	16SrII	Australia	Davis et al. (1997)
Bunias orientalis	Hill mustard	16SrI	Germany	Schneider et al. (1997)

(continued)

Table 11.1 (continued)

Genus and species	Common name	Ribosomal group/ subgroup	Country	Reference
Bupleurum falcatum	Chinese thorowax	UDG	Korea	Choi et al. (1985)
Cajanus marmoratus		16SrII	Australia	Schneider et al. (1999)
Calendula arvensis	Field marigold	16SrII-E	Italy	Tolu et al. (2006)
Calendula officinalis	Pot marigold	16SrI	Italy	Marcone et al. (1997b)
		16SrII	Iran	Esmailzadeh Hosseini et al. (2011a)
Calotropis gigantea	Crown flower	16SrVI	India	Madupriya et al. (2010)
Calystegia sepium	Hedge bindweed	16SrXII-A	Italy	Credi et al. (2006)
Cannabis sativa	Hemp	16SrI	India	Mall et al. (2015)
				Raj et al. (2008b)
				Nabi et al. (2015a)
Cardaria draba	Hoary cress	16SrI	Germany	Schneider et al. 1997
		16SrII	Iran	Esmailzadeh Hosseini et al. (2011b)
Cassia italica	Italian senna	16SrXXIX-A	Oman	Al-Saady et al. (2008)
Cenchrus ciliaris	Buffel grass	16SrII	Australia	Tran-Nguyen et al. (2000)
Cenchrus setiger	Birdwood grass	UDG	Australia	Tran-Nguyen et al. (2000)
Centrosema pascuorum	Cavalcade	16SrII	Australia	Wilson et al. (2001)
Chenopodium album	Lamb's quarter	16SrXII-A	Italy	Credi et al. (2006)
Chenopodium ambrosioides	Epazote	16SrI	China	Li et al. (2012)
Chenopodium murale	Nettle-leaved goosefoot	16SrII	Saudi Arabia	Alhudaib et al. (2009)
Chenopodium sp.		16SrII	Italy	Tolu et al. (2006)
Chloris gayana	Rhodes grass	16SrXI	East Africa	Asudi et al. (2016)
Chloris inflata	Purpletop Rhodes grass	16SrXI	Australia	Blanche et al. (2003)
Chrysopogon aciculatus	Golden beard grass	16SrXIV	Myanmar	Win and Jung (2012)

(continued)

Table 11.1 (continued)

Genus and species	Common name	Ribosomal group/ subgroup	Country	Reference
Cirsium arvense	Canada thistle	16SrXI-E	Germany	Schneider et al. (1997)
			Czech Republic	Šafářová et al. (2016)
		16SrXII-A	Italy	Credi et al. (2006)
			Czech Republic	Fialová et al. (2009)
		16SrIII	Serbia	Rančić et al. (2005)
Cirsium sp.		16SrIII	Hungary	Palermo et al. (2004)
Cleome viscosa	Tick weed	16SrII	Australia	Schneider et al. (1999)
			India	Thorat et al. (2016)
Cnidium officinale		UDG	Korea	Choi et al. (1985)
Coix lacryma-jobi	"Otiro"	16SrXI	East Africa	Asudi et al. (2016)
Convolvulus arvensis	Field bindweed	16SrII	Saudi Arabia	Alhudaib et al. (2009)
		16SrIII	Hungary	Palermo et al. (2004)
		16SrX	Germany	Schneider et al. (1997)
		16SrXII-A	Europe, Iran, Israel	Batlle et al. (2000), Berger et al. (2009)
		16SrXII-H	Italy, Serbia, Bosnia and Herzegovina, Germany	Martini et al. (2012)
			Austria	Aryan et al. (2014)
		16SrXXIX-B	Iran	Esmailzadeh Hosseini et al. (2016a)
Conyza canadensis	Canadian horseweed	16SrIII	USA	Schneider et al. (1997)
		16SrVI	Iran	Zibadoost and Rastgou (2016)
Crepis setosa	Hawksbeard	16SrXI	Italy	Marcone et al. (1997b), Schneider et al. (1997)
Crotalaria brevis		16SrII	Australia	Schneider et al. (1999)
Crotalaria crispata		16SrII	Australia	Schneider et al. (1999)
Crotalaria goreensis	Blunt bird flower/ Gambia pea	16SrII	Australia	Davis et al. (1997), Wilson et al. (2001)
Crotalaria novae-hollandiae	New Holland rattlepod	16SrII	Australia	Davis et al. (1997)
Crotalaria pallida	Smooth rattlepods	16SrII	India	Yadav et al. (2016)

(continued)

Table 11.1 (continued)

Genus and species	Common name	Ribosomal group/ subgroup	Country	Reference
Crotalaria sp.		16SrII	Australia	Davis et al. (1997)
Crotalaria spectabilis	Showy rattlebox	16SrI	India	Kumar et al. (2010)
		16SrII	Australia	Schneider et al. (1999)
Crotalaria tetragona		16SrI	India	Baiswar et al. (2010)
Cyanthillium cinereum	Little ironweed	16SrII	Australia	Schneider et al. (1999)
Cynodon dactylon	Bermuda grass/ couch	16SrII	Australia	Tran-Nguyen et al. (2000)
		16SrXI	East Africa	Asudi et al. (2016)
		16SrXIV	Europe, Asia, Africa, Cuba	Chen et al. (1972)
				Mitrović et al. (2015)
				Khanna et al. (2015)
Cyperus rotundus	Coco grass	16SrII	Cuba	Zamora et al. (2015)
Dactyloctenium aegyptium	Crowfoot grass	16SrXI	Australia	Blanche et al. (2003)
		16SrXIV		
Dactyloctenium radulans	Button grass	16SrXI	Australia	Blanche et al. (2003)
Dahlia sp.		16SrXI-E	Czech Republic	Šafářová et al. (2016)
Datura inoxia	Downy thorn-apple	16SrVI	India	Raj et al. (2009b)
Datura stramonium	Jimsonweed	16SrVI	India	Singh et al. (2012)
				Mall et al. (2015)
		16SrXII-A	Italy	Credi et al. (2006)
			Czech Republic	Fialová et al. (2009)
Delphinium sp.		16SrIII	England	Harju et al. (2008)
Descurainia sophia	Flixweed-tansy mustard	16SrI	China	Wu et al. (2010)
Desmodium intortum	Greenleaf tick trefoil	16SrII	Australia	Schneider et al. (1999)
Dichanthium annulatum	Marvel grass	16SrXIV	India	Rao et al. (2009)
				Mall et al. (2015)
Digitaria ciliaris	Southern crabgrass	16SrXIV	India	Mall et al. (2015)
Digitaria sanguinalis	Hairy crabgrass	16SrXIV	India	Rao et al. (2010)
				Mall et al. (2015)
Digitaria scalarum	Couch grass	16SrXI	East Africa	Asudi et al. (2016)

(continued)

Table 11.1 (continued)

Genus and species	Common name	Ribosomal group/ subgroup	Country	Reference
Diplacrum capitatum		16SrXXII-B	Côte D'Ivoire	Arocha Rosete et al. (2016)
Dodonaea angustifolia	Sand olive shrub	16SrXIV	Saudi Arabia	Omar (2016)
Echinochloa colonum	Jungle rice		India	Reddy and Jeyarajan (1988)
Echium vulgare	Blueweed	16SrXI	Italy	Marcone et al. (1997b)
		16SrXII-A		Berger et al. (2009)
Eleusine indica	Goosegrass	16SrXI	Myanmar	Win et al. (2013)
			Africa	Asudi et al. (2016)
		16SrXIV	India	Mall et al. (2015)
			Africa	Asudi et al. (2016)
Emilia fosbergii	Florida tasselflower	16SrIV	Jamaica	Brown et al. (2008a)
Emilia sonchifolia	Lilac tasselflower	16SrII-D	Australia	Schneider et al. (1999)
Enteropogon macrostachyus	Bush rye	16SrXI	East Africa	Asudi et al. (2016)
Eragrostis cilianensis	Stink grass	16SrI	China	Wu et al. (2010)
Eragrostis falcata	Sickle lovegrass	16SrII	Australia	Tran-Nguyen et al. (2000)
Eriachne obtusa		16SrII	Australia	Tran-Nguyen et al. (2000)
Erigeron sp.		16SrVII-B	Brazil	Barros et al. (2002)
Erigeron bonariensis	Flax-leaved fleabane	16SrVII-B	Brazil	Montano et al. (2014)
		16SrVII-D	Brazil	Flôres et al. (2015)
Erysimum cheiranthoides	Wormseed mustard	16SrI	China	Wu et al. (2010)
Euphorbia milii	Crown of thorns	16SrII-D	Australia	Davis et al. (1997)
Festuca arundinacea	Tall fescue	16SrI	Lithuania	Valiūnas et al. (2007)
Galactia tenuiflora		UDG	Australia	Schneider et al. (1999)
Gerbera sp.		16SrII-D	Australia	Davis et al. (1997)
Goodenia sp.		16SrII-D	Australia	Schneider et al. (1999)
Guizotia abyssinica	Niga	16SrII-D	Australia	Davis et al. (1997)
Hyparrhenia cymbaria	Boat thatching grass	16SrXI	East Africa	Asudi et al. (2016)

(continued)

Table 11.1 (continued)

Genus and species	Common name	Ribosomal group/ subgroup	Country	Reference
Hyparrhenia rufa	Giant thatching grass	16SrXI	Africa	Obura et al. (2011)
				Asudi et al. (2016)
		16SrXIV		
Indigofera colutea	Rusty indigo	16SrII	Australia	Schneider et al. (1999)
Ipomoea plebeia	Bell vine	16SrII-D	Australia	Davis et al. (1997)
Knautia arvensis	Field scabious	16SrXI	Italy	Marcone et al. (1997b)
Lavandula officinalis	Lavender	16SrXII-A	Spain	Batlle et al. (2000)
Lavandula angustifolia	Lavender	16SrXII-A	France	Gaudin et al. (2011)
Lithospermum arvense	Corn gromwell	16SrI	China	Wu et al. (2010)
Macroptilium atropurpureum	Purple bean	16SrII-D	Australia	Davis et al. (1997)
Macroptilium bracteatum	Burgundy bean	16SrII	Australia	Schneider et al. (1999)
Macroptilium gracile		16SrII	Australia	Schneider et al. (1999)
Macroptilium lathyroides	Phasey bean	16SrII-D	Australia	Davis et al. (1997)
Malva sylvestris	Common mallow	16SrXII-A	Italy	Credi et al. (2006)
Medicago sativa	Lucerne, alfalfa	16SrII	Australia	Wilson et al. (2001)
		16SrXII-A	Italy	Credi et al. (2006)
			Iran	Esmailzadeh Hosseini et al. (2016c)
		16SrII-D	Oman	Khan et al. (2002)
		16SrII-C/D	Iran	Esmailzadeh Hosseini et al. (2016b)
		16SrVI-A	Iran	Esmailzadeh Hosseini et al. (2016c)
Melochia corchorifolia	Chocolateweed	16SrI	China	Chen et al. (2017)
Mentha arvensis	Wild mint	16SrXII-A	Italy	Credi et al. (2006)
Mikania sp.		16SrI	Bangladesh	Kelly et al. (2009)
Mimosa pudica	Sensitive plant	16SrI	Indonesia	Boa et al. (2010)
Mitracarpus hirtus	Tropical girdlepod	16SrII	Australia	Wilson et al. (2001)
Oplismenus burmannii	Burmann's basketgrass	16SrII	India	Mall et al. (2015)
		16SrXIV	India	Rao et al. (2010)

(continued)

Table 11.1 (continued)

Genus and species	Common name	Ribosomal group/ subgroup	Country	Reference
Parthenium hysterophorus	Santa Maria feverfew	16SrI	India	Raj et al. (2008a)
		16SrII	Ethiopia	Bekele et al. (2011)
			India	Mall et al. (2015)
				Thorat et al. (2016)
			China	Cai et al. (2016)
				Li et al. (2011)
Paspalum conjugatum	Buffalo grass	16SrXIV	Singapore	Koh et al. (2008)
Paspalum vaginatum	Seashore paspalum/biscuit grass	16SrXXII-B	Côte D'Ivoire	Arocha Rosete et al. (2016)
Pennisetum pedicillatum	Desho grass	16SrXXII-B	Côte D'Ivoire	Arocha Rosete et al. (2016)
Phalaris minor	Little seed canary grass	16SrI	India	Mall et al. (2015)
Phlox sp.	Perennial phlox	16SrII-D	Australia	Davis et al. (1997)
Phragmites australis	Common reed	16SrV	China	Li et al. (2013)
Phyllanthus amarus	Shatterstone		India	Samad et al. (2004)
Phyllanthus maderaspanatus		16SrII	Australia	Schneider et al. (1999)
Phyllanthus niruri	Gale of the wind	16SrI	India	Chaube et al. (2015)
Physalis minima	Wild gooseberry	16SrII-D	Australia	Davis et al. (1997)
Phyllanthus muellerianus		16SrXXII-B	Côte D'Ivoire	Arocha Rosete et al. (2016)
Picris echioides	Bristly oxtongue	16SrII-E	Italy	Marcone et al. (1997b)
		16SrIX-C	Italy	Schneider et al. (1997)
		16SrXII-A	Italy	Credi et al. (2006)
Plantago asiatica	Chinese plantain		Korea	Choi et al. (1985)
Plantago lanceolata	Narrowleaf plantain	16SrII	Saudi Arabia	Alhudaib et al. (2009)
		16SrXII-A	Italy	Credi et al. (2006)
			Spain	Batlle et al. 2000
Poa annua	Annual blue grass	16SrXIV-C	Italy	Lee et al. (1997)
Poa pratensis	Common meadow grass	16SrI	Lithuania	Valiūnas et al. (2007)
Polygala mascatense		16SrII	Oman	Livingston et al. (2006)
Polygonum aviculare	Common knotgrass	16SrXII-A	Italy	Berger et al. (2009)

(continued)

Table 11.1 (continued)

Genus and species	Common name	Ribosomal group/ subgroup	Country	Reference
Polygonum convolvulus	Wild buckwheat	16SrXII-A	Spain	Batlle et al. (2000)
Portulaca grandiflora	Moss rose	16SrXIV	India	Ajaykumar et al. (2007)
Portulaca oleracea	Purslane	16SrI	Italy	Schneider et al. (1997)
Potentilla reptans	Creeping cinquefoil	16SrXII-A	Italy	Credi et al. (2006)
Prosopis farcta	Syrian mesquite	16SrII	Iran	Esmailzadeh Hosseini et al. (2011b)
Pterocaulon sp.		16SrII	Australia	Wilson et al. (2001)
Ptilotus distans		16SrII-D	Australia	Schneider et al. (1999)
Ranunculus sceleratus	Cursed buttercup	16SrXIV	India	Singh et al. (2013)
Rhynchosia minima	Rhynchosia	16SrII-D	Australia	Davis et al. (1997)
				Wilson et al. (2001)
			Jamaica	Dabek (1983)
Rubia tinctorum	Common madder	16SrVI	Iran	Zibadoost and Rastgou (2016)
Scaevola taccada	Beach naupaka	16SrII	Oman	Al-Zadjali et al. (2012)
Sclerocarpus africanus	African bonebract	16SrI	India	Nabi et al. (2015b)
Scoparia dulcis	Licorice weed	16SrXXII-B	Côte D'Ivoire	Arocha Rosete et al. (2016)
Senecio jacobaea	Common ragwort	16SrI	United Kingdom	Reeder and Arocha (2008)
Senna obtusifolia	Sicklepod	16SrII	Australia	Schneider et al. (1999)
Setaria viridis	Green foxtail	16SrXII-A	Italy	Credi et al. (2006)
Sida cordifolia	Flannel weed	16SrII-D	Australia	Davis et al. (1997)
Silene alba	White campion	16SrXII-A	Italy	Credi et al. (2006)
Silene vulgaris	Bladder campion	16SrXII-A	Italy	Berger et al. (2009)
Silene niceensis		16SrI-B	Italy	Cozza et al. (2008)
Solanum nigrum	Black nightshade	16SrII-E	Italy	Tolu et al. (2006)
		16SrXII-A	Spain	Batlle et al. (2000)
Sonchus oleraceus	Common sow thistle	16SrXII-A	Italy	Credi et al. (2006)
Sophora alopecuroides		16SrXII-A	Iran	Allahverdi et al. (2014)
Sorghum halepense	Johnson grass	16SrVI	Iran	Zibadoost and Rastgou (2016)

(continued)

Table 11.1 (continued)

Genus and species	Common name	Ribosomal group/ subgroup	Country	Reference
Sorghum stipoideum		16SrXI	Australia	Tran-Nguyen et al. (2000)
Sorghum versicolor	Wild sorghum	16SrXI	Africa	Asudi et al. (2016)
Spermacocci sp.		16SrII	Australia	Schneider et al. (1999)
Sporobolus pyramidalis	Drop-seed grass	16SrXI	Africa	Asudi et al. (2016)
Stachytarpheta indica		16SrXXII-B	Côte D'Ivoire	Arocha Rosete et al. (2016)
Stachytarpheta jamaicensis	Blue snakeweed		India	Pallavi et al. (2011)
Stellaria media	Common chickweed	16SrI	Germany	Schneider et al. (1997)
Stylosanthes hamata	Cheesytoes	16SrII	Australia	Schneider et al. (1999)
Stylosanthes scabra	Scabrous stylo	16SrII	Australia	Davis et al. (1997) Schneider et al. (1999) Schneider et al. (1999)
Synedrella nodiflora	Nodeweed	16SrIV	Jamaica	Brown et al. (2008a)
Taraxacum officinale	Dandelion	16SrXII-A	Italy	Credi et al. (2006)
				Berger et al. (2009)
Tephrosia purpurea	Wild indigo	16SrII	India	Yadav et al. (2014)
				Thorat et al. (2016)
Trichodesma zeylanicum	Cattle bush	16SrII	India	Thorat et al. (2016)
Trifolium repens	White clover	16SrI-C	Canada	Lee et al. (1992)
			Germany	Schneider et al. (1997)
			China	Wu et al. (2010)
Urtica dioica	Stinging nettle	16SrXII-A	Italy	Credi et al. (2006)
				Berger et al. (2009)
		16SrXII-H	Austria	Aryan et al. (2014)
Urtica urens	Dwarf nettle	16SrXII-A	Italy	Berger et al. (2009)
Vernonia cinerea		16SrIV	Jamaica	Brown et al. (2008b)
Veronica didyma	Veronica	16SrI	China	Wu et al. (2010)
Vigna lanceolata	Maloga bean		Australia	Schneider et al. (1999)
Vigna luteola	Dalrymple vigna	16SrII-D	Australia	Davis et al. (1997)
Vigna trilobata	African gram	16SrII-D	Australia	Davis et al. (1997)

(continued)

Table 11.1 (continued)

Genus and species	Common name	Ribosomal group/ subgroup	Country	Reference
Waltheria indica	Sleepy morning	16SrII	Australia	Schneider et al. (1999)
Whiteochloa biciliata	Mauve sandgrass	16SrXI	Australia	Blanche et al. (2003)
Whiteochloa capillipes		16SrXI	Australia	Schneider et al. (1999)
Whiteochloa cymbiformis		16SrXI	Australia	Blanche et al. (2003)
				Tran-Nguyen et al. (2000)

phytoplasma was identified in *Senecio jacobaea* (common ragwort) with little leaf, chlorosis, and proliferation of axillary shoots symptoms (Reeder and Arocha 2008).

In Lithuania, poa stunt (PoaS) phytoplasma and festuca yellow (FesY) phytoplasma were detected in *Poa pratensis* (common meadow grass) and *Festuca arundinacea* (tall fescue) and identified as members of subgroup 16SrI-C (Valiūnas et al. 2007).

In India, AY phytoplasma was detected in *Ageratum conyzoides* (goat weed) collected near sugarcane fields showing little leaf symptoms and yellowing of leaf lamina, *Phalaris minor*, *Cannabis sativa* (Fig. 11.1b), *Parthenium hysterophorus* with virescence and witches' broom (Fig. 11.1f), *Crotalaria tetragona* with witches' broom, *C. spectabilis*, and *Achyranthes aspera* (Raj et al. 2008a, b, 2009a; Baiswar et al. 2010; Kumar et al. 2010; Tiwari et al. 2012; Mall et al. 2015; Nabi et al. 2015a; Rao et al. 2017b). *C. spectabilis* (showy rattlebox) is used as a green manure crop to improve soil properties in India where it is a native plant, like in Malay Peninsula. It has been introduced into other areas, such as the USA and the Pacific Islands, where the plant grows like a weed and invades cultivated fields. Nabi et al. (2015b) and Rao et al. (2017a) determined that the weeds, *Sclerocarpus africanus* and *Ocimum canum* (Fig. 11.1a), showing little leaf and witches' broom symptoms collected in Kushinagar and Gorakhpur, India, are alternative natural hosts for sesame phyllody phytoplasma, subgroup 16SrI-B. Also for the first time in India, the typical phytoplasma symptoms of little leaf, yellowing, chlorosis, witches' broom, and stunted growth were observed on the commonly occurring weed *Acalypha indica* (Tiwari et al. 2017). Based on the 16S rRNA gene sequence and virtual RFLP, the *A. indica* phytoplasma was identified as '*Ca.* P. asteris', 16SrI-B subgroup. In *Phyllanthus niruri*, a common weed with medicinal uses in India, symptoms such as yellowing, little leaf, proliferation of axillary shoots, rosetting, and stunted growth were observed, and phytoplasma bodies were first detected using transmission electron microscopy by Samad et al. (2004) and later by sequence analysis of the 16S rRNA gene where '*Ca.* P. asteris' was identified (Chaube et al. 2015). A '*Ca.* P. asteris'-related strain was reported affecting *Mikania* sp. from Bangladesh (Kelly et al. 2009).

Fig. 11.1 Little leaf and witches' broom of *Ocimum canum* (**a**); witches' broom in *Cannabis sativa* (**b**); typical witches' broom symptoms in *Tephrosia purpurea* (**c**); little leaf disease symptoms in *Trichodesma zeylanicum* (cattle bush) (**d**); phyllody and witches' broom in *Parthenium hysterophorus* (**e**); white leaf in Bermuda grass (**f**)

In China, several weeds were identified as hosts of the wheat blue dwarf phytoplasma (WBD), 16SrI-C subgroup, found near wheat fields. These weeds were redroot amaranth (*Amaranthus retroflexus*), corn gromwell (*Lithospermum arvense*), flixweed-tansy mustard (*Descurainia sophia*), wormseed mustard (*Erysimum chei-*

ranthoides), goat grass (*Aegilops squarrosa*), wild oat (*Avena fatua*), stink grass (*Eragrostis cilianensis*), volunteer wheat seedlings (*Triticum aestivum*), white clover (*Trifolium repens*), and veronica (*Veronica didyma*) (Wu et al. 2010). The invasive weed "epazote" (*Chenopodium ambrosioides*) exhibiting small leaves and fasciation was found in a pepper field in Qijiang County (China), and in it a phytoplasma related to the 16SrI-B group was identified (Li et al. 2012). Also 16SrI-B-related phytoplasmas were found in *Melochia corchorifolia*, a common invasive weed in China, with witches' broom, virescence, and phyllody symptoms (Chen et al. 2017). *Mimosa pudica* is a perennial, widespread serious weed in cultivated grasslands and plantation crops such as coffee, tea, and oil palm, and '*Ca.* P. asteris' was detected in plants with leaf yellowing, little leaf, and proliferation of axillary shoot symptoms in Indonesia (Boa et al. 2010).

In Canada Lee et al. (1992) reported a phytoplasma infecting *Trifolium* sp. (Fabaceae) with clover phyllody symptoms and identified the agent as a 16SrI-C aster yellows group member.

11.3 Phytoplasmas in 16SrII Group (Peanut Witches' Broom)

In Italy, a 16SrII-A subgroup phytoplasma was detected in *Picris echioides* (bristly oxtongue) sampled inside commercial vineyards affected by grapevine yellows (Marcone et al. 1997b).

Tolu et al. (2006) surveyed 14 different chlorotic and stunted weed species growing within a 10-year-old vineyard affected by "bois noir" disease in Italy and identified phytoplasmas belonging to the 16SrII-E subgroup in three *Calendula arvensis*, one *Solanum nigrum* and in one *Chenopodium* sp. samples.

In Saudi Arabia, around 25% of lime trees were declining in 2007, and a survey detected phytoplasmas belonging to the 16SrII group in lime trees; in the weeds, *Chenopodium murale*, *Plantago lanceolate*, and *Convolvulus arvensis*; and in the insect, *Empoasca decipiens* (Alhudaib et al. 2009). In Oman, in *Polygala mascatense* with stunted small leaves, bushy growth, and phyllody symptoms, and in *Scaevola taccada* (beach naupaka) showing witches' broom symptoms, a member of the 16SrII group was detected (Livingston et al. 2006; Al-Zadjali et al. 2012). In *Achyranthes aspera* (an annual herb that grows wild in India), the agent of lime witches' broom disease was detected in the Sultanate of Oman (Moghal et al. 1998). In Iran, peanut witches' broom-related phytoplasmas (16SrII) were detected in *Calendula officinalis* (pot marigold) with phyllody symptoms; in *Prosopis farcta* with small leaves, shortened internodes, proliferation of axillary buds, and bushy growth habit; and in *Cardaria draba* with dwarfing, virescence, phyllody, and infertile flowers (Esmailzadeh Hosseini et al. 2011a, b).

In India, phytoplasmas belonging to group 16SrII were detected in *Amaranthus* sp. with yellowing symptoms, in *Parthenium hysterophorus*, and in *Oplismenus burmannii* (Arocha et al. 2008; Mall et al. 2015). The 16SrII subgroups C and D phytoplasma strains were discovered in all symptomatic *P. hysterophorus* samples and in the previously reported insect vector, *Orosius albicinctus* (Cicadellidae), and other collected Hemipteran insects collected from the same sampling site (Yadav et al. 2015). The 16SrII group phytoplasmas were also found associated with *Crotalaria pallida*, commonly used as green manure in India (Yadav et al. 2016).

In China, a phytoplasma belonging to the 16SrII-A subgroup was detected for the first time in *P. hysterophorus* by Li et al. (2011). This finding was also confirmed by Cai et al. (2016) who identified the same phytoplasma in this well-known invasive weed as well as in symptomatic plants of cowpea, sword bean, string bean, tomato, lettuce, and water spinach which were extensively invaded by *P. hysterophorus*. A new phytoplasma strain classified as a member of subgroup 16SrII-M was detected in *Tephrosia purpurea* (wild indigo), a common weed throughout the Indian subcontinent, collected from Maharashtra, India (Fig. 11.1d). The delineation to subgroup level was achieved using 16S rRNA gene sequencing followed by RFLP analyses (Yadav et al. 2014). During a field survey in India, symptoms of little leaf, phyllody, stunting, and branch proliferation were observed on the common invasive weeds, *Cleome viscosa* (tick weed), *Trichodesma zeylanicum* (cattle bush) (Fig. 11.1e), and *Tephrosia purpurea* (wild indigo), from the same or adjacent fields where symptomatic *Sesamum indicum* (sesame), *Vigna unguiculata* (cow pea), *Phaseolus vulgaris* (French bean), *Dendrocalamus strictus* (bamboo), and *Carica papaya* (papaya) plants were found positive for peanut witches' broom-related phytoplasmas (16SrII). On the basis of 16S rRNA gene sequences, *T. zeylanicum* and *T. purpurea* were infected with phytoplasmas belonging to the 16SrII-C subgroup, while in *C. viscosa* a phytoplasma of the 16SrII-D subgroup was identified (Thorat et al. 2016).

In Australia, tomato big bud, sweet potato little leaf, pigeon pea little leaf, *Waltheria* little leaf, and *Bonamia* little leaf phytoplasmas (all members of 16SrII-D subgroup differentiated by the 16S rRNA spacer region) were detected in more than 40 different plant species. In particular tomato big bud phytoplasma (TBB) was identified in *Gerbera* sp., *Guizotia abyssinica* (niga), *Euphorbia milii*, *Alysicarpus rugosus* (rough chainweed), *Crotalaria novae-hollandiae* (new Holland rattlepod), *Crotalaria* sp., *Macroptilium atropurpureum* (purple bean), *M. lathyroides* (phasey bean), *Rhynchosia minima* (rhynchosia), *Stylosanthes scabra* (scabrous stylo), *Trifolium repens* (white clover), *Vigna luteola* (dalrymple vigna), *V. trilobata*, *Sida cordifolia* (flannel weed), *Phlox sp.* (perennial phlox), *Brugmansia candida* (angel's trumpet), *Physalis minima* (wild gooseberry), *Cynodon dactylon* (Bermuda grass), *Cenchrus ciliaris*, *Eragrostis falcata*, *Ptilotus distans*, *Emilia sonchifolia*, *Macroptilium bracteatum*, *Stylosanthes scabra*, *Goodenia* sp., *Ipomoea plebeia* (bell vine), *Crotalaria goreensis* (blunt bird flower), and *Eriachne obtusa* (Davis et al. 1997; Schneider et al. 1999; Tran-Nguyen et al. 2000). Sweet potato little leaf phytoplasma (variant grafted on vinca, SPLL-V4) was detected in *Cyanthillium cinereum*, *Cleome viscosa*, *Senna obtusifolia*, *Phyllanthus maderaspanatus*, *Aeschynomene indica*, *Aphyllodium* sp., *Arachis pintoi*, *Cajanus marmoratus*, *Crotalaria brevis*, *C. crispata*, *Desmodium intortum*, *Indigofera colutea*,

Macroptilium gracile, *Stylosanthes hamata*, and *S. scabra*. Pigeon pea little leaf phytoplasma (PLL) was detected in *Crotalaria spectabilis*, *Arachis pintoi*, *Macroptilium bracteatum*, and *Stylosanthes scabra*. Waltheria little leaf (WaLL) phytoplasma was detected in *Spermacocci* sp. and *Waltheria indica*. *Bonamia* little leaf (BoLL) phytoplasma, a phytoplasma belonging to group 16SrII with a unique RFLP profile compared to other members of this ribosomal group, was detected in *Bonamia pannosa* (Schneider et al. 1999).

Also in Australia, Wilson et al. (2001) tested non-crop species, associated with sesame (*Sesamum indicum*), mung bean (*Vigna radiata*), and peanut (*Arachis hypogaea*) crops, for phytoplasma presence. SPLL-V4 phytoplasma was identified in *Aeschynomene americana* (American jointvetch), *Alysicarpus vaginalis* (alyce clover), *Centrosema pascuorum* (cavalcade), *Crotalaria goreensis* (gambia pea), *Medicago sativa* (lucerne), *Rhynchosia minima* (rhynchosia), and *Mitracarpus hirtus* (with symptoms of little leaf). PLL phytoplasma was identified in *Mitracarpus hirtus* with symptoms of bunching/little leaf, while WaLL phytoplasma was identified in *Pterocaulon* sp. with symptoms of yellowing/rosette.

In Ethiopia, Bekele et al. (2011) identified a phytoplasma belonging to the 16SrII group in *Parthenium hysterophorus*.

11.4 Phytoplasmas in 16SrIII Group (X-Disease)

In the United Kingdom, 16SrIII phytoplasmas were identified in *Delphinium* sp. with severe phyllody, virescence, and proliferation symptoms (Harju et al. 2008). In the USA, a phytoplasma assigned to the western X-disease group was identified in *Conyza* (*Erigeron*) *canadensis* (horseweed) collected next to an apple orchard (Schneider et al. 1997). Palermo et al. (2004) found 16SrIII phytoplasma in *Cirsium* spp. and *Convolvulus arvensis* in Hungarian vineyards, while Rančić et al. (2005) found the same phytoplasma in *Cirsium arvense* in Serbia. In Australia, poinsettia branching-induced phytoplasma (PoiBI, member of the 16SrIII-H subgroup) was detected in wild *Euphorbia pulcherrima* plants (Schneider et al. 1999).

11.5 Phytoplasmas in Groups 16SrIV (Coconut Lethal Yellowing), 16SrV (Elm Yellows), 16SrVI (Clover Proliferation), 16SrVII (Ash Yellows), 16SrIX (Pigeon Pea Witches' Broom), and 16SrX (Apple Proliferation)

Brown et al. (2008b) sampled *Vernonia cinerea* (L.) (Asteraceae) plants, a prevalent dicotyledonous weed inside coconut farms in Jamaica, and even though the plants showed no symptoms, 44.9% (53 out of 118 tested) of them tested positive for phytoplasma. RFLP analysis identified the detected phytoplasmas as coconut lethal yellowing phytoplasma, from ribosomal group 16SrIV. The same authors also found this phytoplasma in the weeds, *Emilia fosbergii* and *Synedrella nodiflora* (Brown et al. 2008a).

In China, a phytoplasma belonging to the 16SrV-B ribosomal subgroup was detected in amaranth (*Amaranthus retroflexus* L.) and in *Phragmites australis* (Poaceae, a widely distributed weed species in China), both with typical witches' broom symptoms, (Yang et al. 2011; Li et al. 2013). A witches' broom disease on *Cannabis* sp. was earlier found to be associated with a phytoplasma of elm yellows group (16SrV) in China (Zhao et al. 2007). In Iran, phytoplasmas belonging to ribosomal group 16SrVI were detected in *Sorghum halepense* (Johnson grass), *Conyza canadensis* (Canadian horseweed), and *Rubia tinctorum* (common madder) with symptoms of yellowing, little leaf, and witches' broom (Zibadoost and Rastgou 2016). In India, a phytoplasma designated as a member of the ribosomal group 16SrVI was detected in *Datura stramonium* with symptoms of witches' broom and little leaf, in *D. inoxia* with proliferation of branches, shortened internodes, and smaller leaves and in *Calotropis gigantea* (crown flower) with symptoms of leaf yellowing (Raj et al. 2009b; Madupriya et al. 2010; Singh et al. 2012; Mall et al. 2015). In Brazil, *Erigeron* sp. with symptoms of witches' broom and chlorosis were found to be infected with new phytoplasma subgroups B and D that fall within group 16SrVII (Barros et al. 2002; Flôres et al. 2015). Symptoms of a phytoplasma disease including phyllody, virescence, witches' broom, and little leaf were observed on *Bidens alba* growing like a weed in citrus orchards of Hormozgan Province, Iran, and after analyses, a phytoplasma related to '*Ca.* P. phoenicium' (16SrIX group) was detected (Hemmati et al. 2017). In Germany, apple proliferation phytoplasma was detected in a single symptomatic *Convolvulus arvensis* (field bindweed) plant out of 25 collected in or around apple/stone fruit orchards (Schneider et al. 1997).

11.6 Phytoplasmas in 16SrXI Group (Rice Yellow Dwarf)

The 16SrXI or rice yellow dwarf group consists of subgroup A, which includes rice yellow dwarf phytoplasma (RYD) and napier grass stunt phytoplasma (NGS); subgroup B, which includes sugarcane white leaf phytoplasma (SCWL) and sugarcane grassy shoot phytoplasma (SCGS); and a leafhopper-borne (BVK) phytoplasma included in subgroup C (Lee et al. 2000; Jones et al. 2004). In Italy, phytoplasmas belonging to the sugarcane white leaf group (16SrXI-B) were detected in *Picris echioides* (bristly oxtongue) collected in an apple orchard, *Crepis setosa* (hawksbeard) collected in alfalfa fields, *Knautia arvensis* (field scabious) collected in brushwood areas, and *Echium vulgare* (blueweed) collected in vineyards affected by grapevine yellows (Schneider et al. 1997; Marcone et al. 1997b). In Myanmar, goosegrass white leaf (GGWL) phytoplasma was detected in *Eleusine indica* (goosegrass). This phytoplasma is closely related to SGS phytoplasma (Win et al. 2013).

In Australia, sorghum grassy shoot (SGS) phytoplasma was detected and identified for the first time in *Sorghum stipoideum* and *Whiteochloa capillipes* by Schneider et al. (1999) that according to tentative classification by *i*Phyclassifier is a member of the 16SrXI-C ribosomal subgroup (Zhao et al. 2009). Tran-Nguyen et al. (2000) also found SGS phytoplasma in *S. stipoideum* and *W. cymbiformis*. Later, during a generic survey of grasses in Australia, Blanche et al. (2003) detected a SGS-related phytoplasma in *W. cymbiformis*, *W. biciliata*, *Dactyloctenium aegyptium*, *D. radulans*, and *Chloris inflata*. They also tried to associate symptoms with the phytoplasmas identified, but it wasn't possible due to a number of symptomless plants testing positive for phytoplasma.

In Africa, Obura et al. (2011) detected a phytoplasma in *Hyparrhenia rufa* (thatching grass which is common in the tropics) which were stunted and appeared bushy, with small white leaves, and identified it as a member of the 16SrXI ribosomal group. Later in East Africa, Asudi et al. (2016) tested plants from 33 grass species collected from fields bordering farms of napier grass (*Pennisetum purpureum*), an important fodder for livestock. Besides '*Ca*. P. cynodontis', they identified a phytoplasma related to NGS (16SrXI-A) in the following 11 grass species: *Coix lacryma-jobi* (otiro), *Chloris gayana* (rhodes grass), *Digitaria scalarum* (couch grass), *Enteropogon macrostachyus* (bush rye), *Eleusine indica* (goosegrass), *Hyparrhenia cymbaria* (thatch grass), *H. rufa* (thatch grass), *Sorghum versicolor* (wild sorghum), *Sporobolus pyramidalis* (drop-seed grass), *Cynodon dactylon* (Bermuda grass) and *Brachiaria brizantha* (signal grass), and GGWL phytoplasma (16SrXI-C) in two wild grass species (*B. brizantha* and *S. pyramidalis*).

In Germany, *Cirsium arvense* (Canada thistle) collected in or around apple/stone fruit orchards were found to be infected with cirsium phyllody (CIRP) phytoplasma, a phytoplasma closely related to members of the SCWL group, however sharing only 96.9 and 96.7% 16S rRNA sequence identity to both SCWL and BVK phytoplasmas, respectively (Schneider et al. 1997). A new taxon has therefore been introduced, '*Ca*. P. cirsii', comprising the phytoplasma found in *C. arvense* and *Dahlia* sp. that induces symptoms of yellowing, stunting, inflorescence, and proliferation in samples collected from the Czech Republic. Phytoplasmas belonging to this taxon are members of subgroup 16SrXI-E and appear to only infect dicotyledonous plants (Šafářová et al. 2016).

11.7 Phytoplasmas in 16SrXII-A Group ("Stolbur" Group)

"Stolbur" phytoplasma in grapevine induces a disease called "bois noir" (BN) that is one of the most investigated phytoplasma diseases in Europe. In order for BN to spread, herbaceous host plants, which serves as a phytoplasma reservoir, and insect vectors need to be present. Stinging nettle (*Urtica dioica*) and bindweed (*Convolvulus arvensis*) were in most cases found to be the main phytoplasma source. In Slovenia, Mehle et al. (2011) detected and identified "stolbur" in 43% of tested bindweed samples. Marcone et al. (1997b) tested six weed species from Italy that had yellowing of the leaves and among other things identified a new "stolbur" group in field bindweed, while Palermo et al. (2004) detected "stolbur" on stinging nettle in Hungarian vineyards.

According to the sequence and RFLP profile of the *tuf* gene (elongation factor Tu), Langer and Maixner (2004) assigned "stolbur" phytoplasma to two main genetic types, tuf type I (tuf-type a) and tuf type II (tuf-type b) that were involved in different natural epidemic cycles. Strains belonging to tuf-type a are predominately spread via *U. dioica* in Germany, while tuf-type b strains were less specific and were found in *C. arvensis*, *C. sepium*, *Prunus spinosa*, and *Solanum nigrum*. A third type, tuf-type III (tuf-type c) has only been detected in *C. sepium* in the Mosel area in Germany. Fialová et al. (2009) found tuf-type b strains also to be present in other weedy plants such as *Amaranthus retroflexus*, *Cirsium arvense*, and *Datura stramonium*, as well as in *U. dioica* collected in intensive vegetable crop fields and in two vineyards in the Czech Republic. In Austria, between 2003 and 2008, only tuf-type b strains were found to be present in *C. arvensis* and grapevine, while infections of *U. dioica* were rare (Riedle-Bauer et al. 2006, 2008; Tiefenbrunner et al. 2007). Aryan et al. (2014) found an intermediate tuf-type, on the basis of the sequence of *tuf* gene, called tuf-type b2 and discovered that all "stolbur" phytoplasmas from nettle in the studied area belonged to the new tuf-type b2.

Berger et al. (2009) surveyed, among other things, 516 herbaceous plants of 41 potential host species belonging to 21 families, in 15 BN-affected commercial vineyards from South Tyrol, Northern Italy, over 4 years as part of a monitoring study. The "stolbur" phytoplasma was detected in seven species belonging to six families: *C. arvensis* (Convolvulaceae), *Echium vulgare* (Boraginaceae), *Polygonum aviculare* (Polygonaceae), *Silene vulgaris* (Caryophyllaceae), *Taraxacum officinale* (Asteraceae), and the two Urticaceae species *U. dioica* and *U. urens*. For *C. arvensis*, 25.1% (45 out of 179) tested positive for "stolbur" phytoplasma, as well as 4.5% (5 out of 111) of stinging nettle samples and the single *U. urens* (dwarf nettle) sample. Furthermore, positive samples of *C. arvensis*, *E. vulgare*, *P. aviculare*, *S. vulgaris*, and *T. officinale* were assigned to tuf-type b, while positive samples from both *Urtica* plants were assigned to tuf-type a.

Credi et al. (2006) surveyed 162 non-crop native plant samples, consisting of 30 plant species, in vineyards in the region of Emilia-Romagna, Italy, and found that 48.1% samples tested positive for "stolbur" phytoplasma. The 18 positive weed species belonged to the following 13 families: *Amaranthus retroflexus* (Amaranthaceae), *Silene alba* (Caryophyllaceae), *Chenopodium album* (Chenopodiaceae), *Artemisia vulgaris*, *Cirsium arvense*, *Picris echioides*, *Sonchus oleraceus*, *Taraxacum officinale* (Compositae), *Calystegia sepium*, *Convolvulus arvensis* (Convolvulaceae), *Mentha arvensis* (Labiateae), *Medicago sativa* (Leguminosae), *Malva sylvestris* (Malvaceae), *Plantago lanceolata* (Plantaginaceae), *Setaria viridis* (Poaceae), *Potentilla reptans* (Rosaceae), *Datura stramonium* (Solanaceae), and *Urtica dioica* (Urticaceae). These infected weeds included 5 annual, 1 biennial, and 12 perennial species which represents a huge phytoplasma reservoir. Plant symptoms consisted of stunting, rosetting, chlorosis, leaf malformation, little leaf, leaf yellowing, reddening, and necrosis while some species, *A. retroflexus* (redroot pigweed), *C. album* (lambsquarter), and *U. dioica* (stinging nettle), were symptomless. Batlle et al. (2000) also found "stolbur"-positive *C. arvensis*, *Lavandula officinalis*, *Polygonum convolvulus*, and *Solanum nigrum* in three regions of Northeast Spain. They also found *Plantago lanceolata* being sporadically infected. Allahverdi et al. (2014) reported '*Ca.* P. solani' (16SrXII-A group, "stolbur") affecting *Sophora alopecuroides* in Iran where it is considered an invasive weed.

In *C. arvensis* (bindweed), a well-known host of "stolbur" phytoplasma, a phytoplasma that could not be assigned to any previously reported group or subgroup from time to time could be found in Italy. The symptoms observed on the diseased bindweed were undersized leaves, shoot proliferation, and yellowing (Marcone et al. 1997b; Martini et al. 2008). After phylogenetic analyses of the amplified 16S rRNA gene and 16S–23S rRNA spacer region of strains from Italy, Serbia, Bosnia and Herzegovina, and Germany, this phytoplasma was classified into a new subgroup inside the 16SrXII group, subgroup 16SrXII-H. This phytoplasma shares 97.2% similarity of its 16S rRNA gene sequence with "stolbur" phytoplasma (16SrXII-A) and 97.1% with '*Ca.* P. fragariae' (16SrXII-E). RFLP patterns of R16F2n/R16R2 amplicons are most similar to those of phytoplasmas belonging to subgroups 16SrI-C and 16SrXII-A, but RFLP analyses using *Alu*I, *Hae*III, and *Tru*I restriction enzymes could clearly distinguish it (Martini et al. 2012). Aryan et al. (2014) identified '*Ca.* P. convolvuli' (16SrXII-H) in some stinging nettles, and almost all bindweed samples tested positive for phytoplasma in a survey for the presence of BN in Austrian vineyards.

11.8 Phytoplasmas in 16SrXIV Group (Bermuda Grass White Leaf)

Bermuda grass white leaf (BGWL) phytoplasma is the agent of a white leaf disease in *Cynodon dactylon* L. (Bermuda grass) (Fig. 11.1d), and it was first reported in Taiwan by Chen et al. (1972). So far, BGWL phytoplasma has been reported in Italy (Marcone et al. 1997a), Serbia, Albania (Mitrović et al. 2015), Turkey (Çağlar et al. 2013a), Saudi Arabia (Omar 2016), Iran (Salehi et al. 2009), Pakistan (Zahoor et al. 1995), India (Rao et al. 2007; Snehi et al. 2008; Kumar et al. 2015; Mall et al. 2015), Myanmar (Win et al. 2013), Thailand (Sarindu and Clark 1993; Wongkaew et al. 1997; Sdoodee et al. 1999), Malaysia (Nejat et al. 2009a, b), Singapore (Koh et al. 2008), Australia (Schneider et al. 1999; Tran-Nguyen et al. 2000, Blanche et al. 2003), Cuba (Arocha et al. 2005), and Africa (Dafalla and Cousin 1988; Obura et al. 2010; Asudi et al. 2016).

The most frequent symptom of BGWL phytoplasma is extensive chlorosis, but other symptoms such as proliferation of axillary shoots, bushy growing habit, small leaves, shortened stolons and rhizomes, stunting, and in the end death of the host plant can be present (Marcone et al. 2004). The phytoplasma associated with this disease is a member of the 16SrXIV-A subgroup, together with the phytoplasmas of white leaf diseases of other gramineous plants such as *Brachiaria distachya* (brachiaria grass), *Poa annua* (annual blue grass), and *Dactyloctenium aegyptum* (crowfoot grass) (Lee et al. 1997, 1998a, 2000; Seemüller et al. 1998; Sdoodee et al. 1999). Another phytoplasma closely related or identical to BGWL is the agent of Australian cynodon white leaf (CWL) disease (Schneider et al. 1999; Tran-Nguyen et al. 2000). A phytoplasma associated with carpet grass white leaf (CGWL) on *Axonopus compressus* is also considered to be closely related to BGWL (Schneider et al. 1999).

Agents of monocot diseases like sugarcane white leaf (SCWL), sugarcane grassy shoot (SCGS), rice yellow dwarf (RYD), and sorghum (*Sorghum stipoideum*) grassy shoot (SGS) belonging to 16SrXI group ('*Ca*. P. oryzae') are distantly related to this group (they share 98.2–98.5% 16S rRNA identity) (Firrao et al. 2005). All these phytoplasmas form a so-called SCWL branch inside the phytoplasma clade, and in earlier years they were classified as members of the 16SrXI group, with BGWL and closely related strains being separated in subgroup 16SrXI-C (Lee et al. 1997). Marcone et al. (2004) performed the taxonomic study and showed that according to the 16S rDNA gene and 16S–23S rDNA spacer region sequences, serological comparisons, vector transmission, and host-range specificity, BGWL phytoplasma is a discrete taxon at the putative species level and proposed the name '*Ca*. P. cynodontis' for it. They selected BGWL-C1 strain from Italy as the reference strain (GenBank accession number AJ550984). Omar (2016) showed that on the basis of 16S rDNA sequence, the '*Ca*. P. cynodontis' clade was regionally divided into four subclades – two subclades consisting only of strains from Saudi Arabia/Serbia, one of the strains from Italy and Albania, and one of the strains from Myanmar, China, and India what was in concordance with previous work of Salehi et al. (2009) and Mitrović et al. (2015) who classified the strain from Iran into subgroup 16SrXIV-B and strains from Serbia into subgroup 16SrXIV-C.

'*Ca*. P. cynodontis' was associated with many weeds and plant species such as *Dodonaea angustifolia* (sand olive shrub) and *Arundo donax* (giant reed) in Saudi Arabia (Omar 2016); *Dichanthium annulatum* (marvel grass), *Ranunculus sceleratus* with little leaf disease, *Oplismenus burmannii*, *Digitaria sanguinalis* and *D. ciliaris* in India, and *Eleusine indica* (goosegrass) in India and Africa (Rao et al. 2009, 2010, 2011; Singh et al. 2013; Mall et al. 2015; Asudi et al. 2016); *Chrysopogon aciculatus* (golden beard grass) in Myanmar (Win and Jung 2012); *Axonopus compressus* in Singapore and Thailand (Koh et al. 2008; Sunpapao 2016), *Paspalum conjugatum* in Singapore (Koh et al. 2008); and *Brachiaria brizantha* (signal grass) and *Hyparrhenia rufa* (thatch grass) in Africa (Asudi et al. 2016).

In Iran, *Exitianus capicola* was reported as a natural and experimental vector of BGWL agent (Salehi et al. 2009). During a survey in India for potential insect vectors of BGWL phytoplasma, Kumar et al. (2015) found that the leafhopper, *Exitianus indicus*, could be a putative vector since the phytoplasma carried by this insect shared 99% identity of the 16S rRNA gene sequence with BGWL from India and Thailand. To assess the importance of BGWL phytoplasma to agricultural crops such as sugarcane, Khankahdani and Ghasemi (2011) performed serological testing.

According to their results, there was no serological relationship between BGWL and SCWL (sugarcane white leaf), LWB (lime witches' broom), AWB (almond witches' broom), and PY (periwinkle yellowing) phytoplasma. Also, it has been shown that BGWL is not transmitted by the vector of SCWL, *Matsumuratettis hiroglyphicus* (Firrao et al. 2005). On the other hand, Çağlar et al. (2013b) managed to transmit BGWL phytoplasma to wheat plants (*Triticum* spp.) by a root-bridge modality with 30% successful transmission. Even though members of the 16SrXIV group are identical, or nearly identical, on the basis of their 16S rRNA gene sequences and on the basis of their ecological and genetic features, insufficient evidence exists for their relationship. Therefore, Firrao et al. (2005) stated that they should be considered as members of different '*Candidatus* Phytoplasma' species.

In Bermuda grass *Clavibacter xyli* subsp. *cynodontis* can also be detected, and it is thought that this bacterium causes only stunting symptoms. However, when it was detected together with BGWL phytoplasma, the plants showed more severe disease symptoms leading to early death of the plants (Davis et al. 1983). Also spiroplasmas could be detected in Bermuda grass with witches' broom, but they were not apparently associated with this symptom (Chen et al. 1977; Raju and Chen 1980).

11.9 Phytoplasmas in 16SrXXII Group

Very recently a survey was carried out in Grand-Lahou in Côte d'Ivoire where coconut palms are severely affected by a lethal yellowing disease (CILY) associated with the group 16SrXXII-B, '*Ca.* P. palmicola'-related strains. Plant species from the families Poaceae (*Paspalum vaginatum*, *Pennisetum pedicillatum*), Verbenaceae (*Stachytarpheta indica*), Plantaginaceae (*Scoparia dulcis*), Phyllanthaceae (*Phyllanthus muellerianus*), and Cyperaceae (*Diplacrum capitatum*) were positive for the presence of the CILY phytoplasma, suggesting they may have epidemiological implications for disease spread in coconut plants (Arocha Rosete et al. 2016).

11.10 Phytoplasmas in 16SrXXIX Group

Al-Saady et al. (2008) reported '*Ca.* P. omanense' in Italian senna (*Cassia italica*, fam. *Fabaceae*), a native plant from Africa, commonly found throughout the Arabian Peninsula. *C. italica* plants showing witches' broom symptoms were collected in Oman. In Iran, *Convolvulus arvensis* growing in alfalfa fields were found to be infected with phytoplasmas that shared 99% identity with '*Ca.* P. omanense' but were differentiated from it by specific RFLP analyses and were assigned to subgroup 16SrXXIX-B (Esmailzadeh Hosseini et al. 2016a).

11.11 Phytoplasmas in Undesignated Groups

In India, *Stachytarpheta jamaicensis* plants with witches' broom symptoms were confirmed to be infected with phytoplasmas by nested PCR employing universal phytoplasma primers, but they were not identified (Pallavi et al. 2011). In Australia, a new phytoplasma was detected in *Cenchrus setiger* and was named cenchrus bunchy shoot (CBS), as well as detection of *Stylosanthes* little leaf (StLL) in *Stylosanthes scabra* and *Arachis pintoi*, *Galactia* little leaf (GaLL) in *Galactia tenuiflora*, and vigna little leaf (ViLL) in *Vigna lanceolata*. According to *i*Phyclassifier this latter phytoplasma shares 98.2% 16S rDNA sequence identity with '*Ca.* P. omanense', a member of the 16SrXXIX ribosomal group (Schneider et al. 1999; Tran-Nguyen et al. 2000; Zhao et al. 2009).

11.12 Geographic Distribution

In Africa and Oceania (including Australia), phytoplasmas affiliated to a small number of ribosomal groups (three to four) have been detected so far, while in Europe and Asia, numerous phytoplasmas, belonging to more than seven ribosomal groups, were detected (Table 11.2). The wide host range of weeds (18 families in Europe and 22 families in Asia) described on these two continents might be a result of sampling bias, as the two continents have the most detailed record for phytoplasmas on weeds. Phytoplasmas affiliated to ribosomal group 16SrII were detected in Oceania in weeds and wild plants belonging to 13 different families and in Asia belonging to 11 different families. Likewise, "stolbur" phytoplasma (16SrXII-A) was detected in hosts from 14 different families in Europe. These two phytoplasmas have the widest host range among phytoplasmas in weeds. On the other hand, phytoplasmas affiliated to ribosomal groups 16SrXI and 16SrXIV could be found with some exceptions only in family Poaceae regardless of continent where they were detected.

11.13 Conclusion

Wild plants, as natural phytoplasma hosts, are sometimes symptomless, probably due to long coevolution between the host and pathogen. If crop plants are grown in the same environment, this natural epidemiological cycle can branch to cultivated plants as dead-end hosts to form a crop-specific epidemic system. In this way, new diseases of economic importance are emerging (Lee et al. 1998b). In the case that cultivated plants represent a dead-end host, transmission of phytoplasma depends on the presence of wild hosts as a reservoir (source of inoculum). Such diseases can be of high economic impact, and one example is the "stolbur" phytoplasma

Table 11.2 Summary of phytoplasma geographic distribution in weeds and wild plants

Continent	Ribosomal group	Host family	Continent	Ribosomal group	Host family
Europe	16SrI	Asteraceae	Asia	16SrI	Amaranthaceae
		Brassicaceae			Asteraceae
		Caryophyllaceae			Boraginaceae
		Fabaceae			Brasicaceae
		Poaceae			Canabaceae
		Portulacaceae			Euphorbiaceae
	16SrII	Amaranthaceae			Fabaceae
		Asteraceae			Malvaceae
		Solanaceae			Phyllanthaceae
	16SrIII	Asteraceae			Plantaginaceae
		Convolvulaceae			Poaceae
		Ranunculaceae		16SrII	Amaranthaceae
	16SrX	Convolvulaceae			Asteraceae
	16SrXI	Asteraceae			Boraginaceae
		Boraginaceae			Brassicaceae
		Caprifoliaceae			Cleomaceae
	16SrXI-E	Asteraceae			Convolvulaceae
	16SrXII-A	Amaranthaceae			Fabaceae
		Asteraceae			Goodeniaceae
		Boraginaceae			Plantaginaceae
		Caryophyllaceae			Poaceae
		Convolvulaceae			Polygalaceae
		Fabaceae		16SrV	Amaranthaceae
		Lamiaceae			Poaceae
		Malvaceae		16SrVI	Apocynaceae
		Plantaginaceae			Asteraceae
		Poaceae			Poaceae
		Polygonaceae			Rubiaceae
		Rosaceae			Solanaceae
		Solanaceae		16SrIX	Asteraceae
		Urticaceae		16SrXI	Poaceae
	16SrXII-H	Convolvulaceae		16SrXII-A	Fabaceae
	16SrXIV-C	Poaceae		16SrXIV	Poaceae

(continued)

Table 11.2 (continued)

Continent	Ribosomal group	Host family	Continent	Ribosomal group	Host family
Oceania	16SrII	Amaranthaceae			Ranunculaceae
		Asteraceae			Sapindaceae
		Cleomaceae		16SrXXIX	Convolvulaceae
		Convolvulaceae			Fabaceae
		Euphorbiaceae			Apiaceae
		Fabaceae			Phyllanthaceae
		Goodeniaceae			Plantaginaceae
		Malvaceae			Verbenaceae
		Phyllanthaceae	Africa	16SrII	Asteraceae
		Poaceae		16SrXI	Poaceae
		Polemoniaceae		16SrXIV	Poaceae
		Rubiaceae	North America	16SrI	Fabaceae
		Solanaceae		16SrIII	Asteraceae
	16SrIII	Euphorbiaceae		16SrIV	Asteraceae
	16SrXI	Poaceae		16SrXIV	Poaceae
	16SrXIV	Poaceae			Fabaceae
	UDG	Fabaceae	South America	16SrVII	Asteraceae
		Poaceae			

(16SrXII-A, '*Ca.* P. solani'). Natural hosts of "stolbur" are *U. dioica* and *C. arvensis*, while the cultivated hosts are solanaceous crops, grapevine, and corn. "Stolbur"-infected *U. dioica* is usually symptomless and therefore represents an even greater threat to crops. In such cases, weed management by herbicide treatment is an effective method of reducing disease incidence. It is achieved through elimination of the source of inoculum and lowering the density of the vector population (Maixner 2009). Crop-specific epidemic systems are often new epidemiological cycles that evolved from the dead-end host system in the presence of a potentially competent vector. An example of such a system is "flavescence dorée" (16SrV-C/-D), whose epidemiological cycle became independent from its original, natural host. The origin of "flavescence dorée" can now be deduced by analyses of DNA (Arnaud et al. 2007). Although there is no clear recognized role of weeds in this kind of epidemic systems, this underlines the importance of weeds in the emergence of new crop diseases.

References

Ajaykumar PV, Samad A, Shasany A K, Gupta M K, Alam M, Rastogi S (2007) First record of a '*Candidatus* Phytoplasma' associated with little leaf disease of *Portulaca grandiflora*. *Australasian Plant Disease Notes* **2**, 67–69.

Alhudaib K, Arocha Y, Wilson M, Jones P (2009) Molecular identification, potential vectors and alternative hosts of the phytoplasma associated with a lime decline disease in Saudi Arabia. *Crop Protection* **28**, 13–18.

Allahverdi T, Rahimian H, Babaeizad V (2014) The first report of a '*Candidatus* Phytoplasma solani' isolate affecting *Sophora alopecuroides* in Iran. *New Disease Reports* **30**, 22.

Al-Saady NA, Khan AJ, Calari A, Al-Subhi AM, Bertaccini A (2008) '*Candidatus* Phytoplasma omanense,' associated with witches' broom of *Cassia italica* (Mill.) Spreng. in Oman. *International Journal of Systematic and Evolutionary Microbiology* **58**, 461–466.

Al-Zadjali AD, Al-Sadi AM, Deadman ML, Okuda S, Natsuki T, Al-Zadjali TS (2012) Detection, identification and molecular characterization of aphytoplasma associated with beach naupaka witches' broom. *Journal of Plant Pathology* **94**, 379–385.

Arnaud G, Malembic-Maher S, Salar P, Bonnet P, Maixner M, Marcone C, Boudon- Padieu E, Foissac X (2007) Multilocus sequence typing confirms the close genetic interrelatedness of three distinct "flavescence dorée" phytoplasma strain clusters and group 16SrV phytoplasmas infecting grapevine and alder in Europe. *Applied and Environmental Microbiology* **73**, 4001–4010.

Arocha Y, Horta D, Piñol B, Palenzuela I, Picornell S, Almeida R, Jones P (2005) First report of a phytoplasma associated with Bermuda grass white leaf disease in Cuba. *Plant Pathology* **54,** 233.

Arocha Y, Singh A, Pandey M, Tripathi AN, Chandra B, Shukla SK, Singh Y, Kumar A, Srivastava RK, Zaidi NW, Arif M, Narwal S, Tewari AK, Gupta MK, Nath PD, Rabindran R, Khirbat SK, Byadgi AS, Singh G, Boa E (2008) New plant hosts for group 16SrII, '*Candidatus* Phytoplasma aurantifolia', in India. *Plant Pathology* **17**, 36.

Arocha Rosete Y, Diallo HA, Konan Konan JL, Assiri PK, Séka K, Daniel KK, Toualy MN, Koffi EK, Daramcoum MP, Beugré NI, Ouattara WM, Allou K, Fursy-Rodelec ND, Doudjo ON, Yankey N, Dery S, Maharaj A, Saleh M, Summerbell R, Contaldo N, Paltrinieri S, Bertaccini A, Scott J (2016) Detection and identification of the coconut lethal yellowing phytoplasma in weeds growing in coconut farms in Côte d'Ivoire. *Canadian Journal of Plant Pathology* **38**, 164–173.

Aryan A, Brader G, Mörtel J, Pastar M, Riedle-Bauer M (2014) An abundant '*Candidatus* Phytoplasma solani' tuf b strain is associated with grapevine, stinging nettle and *Hyalesthes obsoletus*. *European Journal of Plant Pathology* **140,** 213–227.

Asudi GO, Van den Berg J, Midega CAO, Schneider B, Seemüller E, Pickett JA, Khan ZR (2016) Detection, identification, and significance of phytoplasmas in wild grasses in East Africa. *Plant Disease* **100**, 108–115.

Baiswar P, Arocha Y, Chandra S, Ngachan SV (2010) First report of '*Candidatus* Phytoplasma asteris' associated with witches' broom of *Crotalaria tetragona* in India. *Plant Pathology* **59**, 397.

Barros TSL, Davis RE, Resende RO, Dally EL (2002) Erigeron witches' broom phytoplasma in Brazil represents new subgroup VII-B in 16S rRNA gene group VII, the ash yellows phytoplasma group. *Plant Disease* **86**, 1142–1148.

Batlle A, Martínez MA, Laviña A (2000) Occurrence, distribution and epidemiology of Grapevine Yellows in Spain. *European Journal of Plant Pathology* **106**, 811–816.

Bekele B, Hodgetts J, Tomlinson J, Boonham N, Nikolić P, Swarbrick P, Dickinson M (2011) Use of a real-time LAMP isothermal assay for detecting 16SrII and XII phytoplasmas in fruit and weeds of the Ethiopian Rift Valley. *Plant Pathology* **60**, 345–355.

Berger J, Schweigkofler W, Kerschbamer C, Roschatt C, Dalla Via J, Baric S (2009) Occurrence of Stolbur phytoplasma in the vector *Hyalesthes obsoletus*, herbaceous host plants and grapevine in South Tyrol (Northern Italy). *Vitis* **48**, 185–192.

Blanche KR, Tran-Nguyen LTT, Gibb KS (2003) Detection, identification and significance of phytoplasmas in grasses in northern Australia. *Plant Pathology* **52**, 505–512.

Boa E, Arocha Y, Harling R, Tobing C, Kelly P, Reeder R (2010) First report of group 16SrI, '*Candidatus* Phytoplasma asteris' associated with *Mimosa pudica* yellows in Indonesia. *Plant Pathology* **59**, 397.

Brown SE, Been BO, McLaughlin WA (2008a) First report of the presence of the lethal yellowing group (16SrIV) of phytoplasmas in the weeds *Emilia fosbergii* and *Synedrella nodiflora* in Jamaica. *Plant Pathology* **57**, 770.

Brown SE, Been BO, McLaughlin WA (2008b) First report of lethal yellowing group (16SrIV) of phytoplasmas in *Vernonia cinerea* in Jamaica. *Plant Disease* **92**, 1132.

Cai H, Wang LC, Yang ZX, Wan QL, Wei W, Davis RE, Zhao Y (2016) Evidence for the role of an invasive weed in widespread occurrence of phytoplasma diseases in diverse vegetable crops: implications from lineage-specific molecular markers. *Crop Protection* **89**, 193–201.

Chaube S, Mall S, Dubey DK, Rao GP (2015) *Phyllanthus niruri* L.: a new host of '*Candidatus* Phytoplasma asteris' in India. *Phytopathogenic Mollicutes* **5**(1-Supplement), S87–S88.

Chen TA, Su HJ, Raju BC, Huang WC (1977) A new spiroplasma isolated from Bermuda grass *Cynodon dactylon* L. Per. *Proceedings of The American Phytopathological Society* **4**, 171.

Chen TC, Lee CS, Chen MJ (1972) Mycoplasmalike organisms in *Cynodon dactylon* and *Brachiaria distachya* affected by white leaf disease. *Report of Taiwan Sugar Experiment Station* **56,** 49–55.

Chen W, Li Y, Liu F, Zeng LY, Ren L, Xu L, Chen RK, Fang XP (2017) *Melochia corchorifolia*, a new host of 16SrI-B phytoplasma in China. *Journal of Plant Pathology* **99**, 291.

Choi YM, Lee SH, Lee EK, Kim JS (1985) An investigation of underscribed witches' broom symptom caused by MLO on *Bupleurum falcatum, Cnidium officinale* and *Plantago asiatica* in Korea. *Korean Journal of Mycology* **13**, 49–51.

Credi R, Terlizzi F, Milanesi L, Bondavalli R, Cavallini G, Montermini A, Dradi D (2006) Wild host plants of "stolbur" phytoplasma and its vector, *Hyalesthes obsoletus*, at sites of grapevine "bois noir" occurrence in Emilia-Romagna, Italy. 15th Meeting ICVG, Stellenbosch, South Africa, 182–183.

Çağlar BK, Satar S, Elbeaino T (2013a) Detection and molecular characterization of Bermuda grass (*Cynodon dactylon*) white leaf phytoplasma from Turkey. *International Journal of Agriculture and Biology* **15**, 90–94.

Çağlar BK, Satar S, Elbeaino T (2013b) Experimentally transmission of Bermuda grass [*Cynodon dactylon* (L.) Pers.] white leaf phytoplasma to wheat plants by root-bridge modality. *Journal of Turkish Phytopathology* **40**, 51–56.

Cozza R, Bernardo L, Calari A, Silvestro G, Duduk B, Bertaccini A (2008) Molecular identification of '*Candidatus* Phytoplasma asteris' inducing histological anomalies in Silene nicaeensis. *Phytoparasitica* **36**, 290–293.

Dabek AJ (1983) Leafhopper transmission of *Rhynchosia* little leaf, a disease associated with mycoplasma-like organisms in Jamaica. *Annals of Applied Biology* **103**, 431–438.

Dafalla GA, Cousin MT (1988) Fluorescence and electron microscopy of *Cynodon dactylon* affected with a white leaf disease in Sudan. *Journal of Phytopathology* **122**, 25–34.

Davis RI, Schneider B, Gibb KS (1997) Detection and differentiation of phytoplasmas in Australia. *Australian Journal of Agricultural Research* **48**, 535–544.

Davis MJ, Lawson RH, Gillaspie AG (Jr), Harris RW (1983) Properties and relationships of two xylem-limited bacteria and a mycoplasmalike organism infecting Bermuda grass. *Phytopathology* **73**, 341–346.

Esmailzadeh Hosseini SA, Salehi M, Khanchezar A, Shamszadeh M (2011a) The first report of a phytoplasma associated with pot marigold phyllody in Iran. *Bulletin of Insectology* **64**(Supplement): S109–S110.

Esmailzadeh Hosseini SA, Salehi M, Mirzaie A (2011b) Alternate hosts of alfalfa witches' broom phytoplasma and winter hosts of its vector *Orosius albicinctus* in Yazd-Iran. *Bulletin of Insectology* **64**(Supplement): S247–S248.

Esmailzadeh Hosseini SA, Salehi M, Mirchenari SM, Contaldo N, Paltrinieri S, Bertaccini A (2016a) Occurrence of a '*Candidatus* Phytoplasma omanense'-related strain in bindweed witches' broom disease in Iran. *Phytopathogenic Mollicutes* **6**, 87–92.

Esmailzadeh Hosseini SA, Khodakaramian G, Salehi M, Bertaccini A (2016b). Characterization of 16SrII group phytoplasmas associated with alfalfa (*Medicago sativa*) witches' broom disease in diverse areas of Iran. *Journal of Crop Protection* **5**, 581–590.

Esmailzadeh Hosseini SA, Khodakaramian G, Salehi M, Bertaccini A (2016c) First report of 16SrVI-A and 16SrXII-A phytoplasmas associated with alfalfa witches' broom disease in Iran. *Journal of Plant Pathology* **98**, 369–377.

Fialová R, Válová P, Balakishiyeva G, Danet J-L, Šafářová D, Foissac X, Navrátil M (2009) Genetic variability of "stolbur" phytoplasma in annual crop and wild plant species in South Moravia. *Journal of Plant Pathology* **91**, 411–416.

Firrao G, Gibb K, Streten C (2005) Short taxonomic guide to the genus '*Candidatus* Phytoplasma'. *Journal of Plant Pathology* **87**, 249–263.

Flôres D, de Oliveira AP Amaral Mello AP, Camargo Pereira TB, Marques Rezende JA, Bedendo IP (2015) A novel subgroup 16SrVII-D phytoplasma identified in association with erigeron witches' broom. *International Journal of Systematic and Evolutionary Microbiology* **65,** 2761–2765.

Gaudin J, Semetey O, Foissac X, Eveillard S (2011) Phytoplasma titer in diseased lavender is not correlated to lavender tolerance to "stolbur" phytoplasma. *Bulletin of Insectology* **64**(Supplement), S179–S180.

Harju VA, Skelton AL, Monger WA, Jarvis B, Mumford RA (2008) Identification of an X-disease (16SrIII) group phytoplasma ('*Candidatus* Phytoplasma pruni') infecting delphiniums in the UK. *Plant Pathology* **57**, 769.

Hemmati C, Nikooei M, Bagheri A, Faghihi MM (2017) First report of a '*Candidatus* Phytoplasma phoenicium'-related strain associated with *Bidens alba* phyllody in Iran. *New Disease Reports* **35**, 8.

Jones P, Devonshire BJ, Holman TJ, Ajanga S (2004) Napier grass stunt: a new disease associated with a 16SrXI group phytoplasma in Kenya. *Plant Pathology* **53**, 519.

Khan AJ, Botti S, Al-Subhi AM, Gundersen-Rindal DE, BertacciniA (2002) Molecular identification of a new phytoplasma strain associated with alfalfa witches' broom in Oman. *Phytopathology* **92**, 1038–1047.

Khanna S, Singh J, Singh R, Kumar P, Rani T, Baranwal VK, Sirohi A, Bertaccini A (2015) Evidence of association of a '*Candidatus* Phytoplasma cynodontis' with Bermuda grass (*Cynodon dactylon*) and '*Candidatus* Phytoplasma asteris' with periwinkle (*Catharanthus roseus*) from Western Uttar Pradesh, India. *Crop Protection* **74**, 138–144.

Kelly PL, Arocha Y, Dider SZ (2009) First report of a 16SrI, '*Candidatus* Phytoplasma asteris' isolate affecting eggplant and *Mikania* sp. in Bangladesh. *Plant Pathology* **58**, 789.

Khankahdani RB, Ghasemi S (2011) Serological aspects of phytoplasma associated with Bermuda grass white leaf (BGWL) disease. International Conference on Asia Agriculture and Animal IPCBEE, Singapore **13**, 106–110.

Koh LH, Yap ML, Yik CP (2008) First report of phytoplasma infection of grasses in Singapore. *Plant Disease* **92**, 317.

Kumar S, Singh V, Lakhanpaul S (2010) First report of *Crotalaria spectabilis* fasciation associated with '*Candidatus* Phytoplasma asteris' in India. *Plant Disease* **94**, 1265.

Kumar S, Jadon V, Tiwari AK, Rao GP (2015) *Exitianus indicus* (Distant): a putative vector for '*Candidatus* Phytoplasma cynodontis' in India. *Phytopathogenic Mollicutes* **5**(1-Supplement), S51–S52.

Langer M, Maixner M (2004) Molecular characterisation of grapevine yellows associated phytoplasmas of the "stolbur" group based on RFLP-analysis of non-ribosomal DNA. *Vitis* **43**, 191–199.

Lee I-M, Davis RE, Chen TA, Chiykowski LN, Fletcher J, Hiruki C, Schaff DA (1992) A genotype-based system for identification and classification of mycoplasmalike organisms (MLOs) in the aster yellows MLO strain cluster. *Phytopathology* **82**, 977–986.

Lee I-M, Pastore M, Vibio M, Danielli A, Attathorn S, Davis RE, Bertaccini A (1997) Detection and characterization of a phytoplasma associated with annual blue grass (*Poa annua*) white leaf disease in southern Italy. *European Journal of Plant Pathology* **103**, 251–254.

Lee I-M, Gundersen DE, Davis RE, Bartoszyk I-M (1998a) Revised classification scheme of phytoplasmas based on RFLP analysis of 16S rRNA and ribosomal protein gene sequences. *International Journal of Systematic Bacteriology* **48**, 1153–1169.

Lee I-M, Gundersen-Rindal DE, Bertaccini A (1998b) Phytoplasma: ecology and genomic diversity. *Phytopathology* **88**, 1359–1366.

Lee I-M, Davis RE, Gundersen-Rindal DE (2000) Phytoplasma: phytopathogenic mollicutes. *Annual Review of Microbiology* **54**, 221–255.

Li CL, Du YJ, Xiang BC, Zhang P (2013) First report of the association of a '*Candidatus* Phytoplasma ulmi' isolate with a witches' broom disease of reed in China. *New Disease Reports* **28**, 13–13.

Li TT, Sun XC, Wu GX (2012) First report of a group 16SrI phytoplasma infecting epazote in China. *Journal of Plant Pathology* **94**, S4.89.

Li Z, Zhang L, Che H, Liu H, Chi M, Luo D, Li Y, Chen W, Wu Y (2011) A disease associated with phytoplasma in *Parthenium hysterophorus*. *Phytoparasitica* **39**, 407–410.

Livingston S, Al-Azri MO, Al-Saady NA, Al-Subhi AM, Khan AJ (2006) First report of 16S rDNA II group phytoplasma on *Polygala mascatense*, a weed in Oman. *Plant Disease* **90**, 248.

Madupriya, Chaturvedi Y, Rao GP, Raj SK (2010) First report of phytoplasma '*Candidatus* Phytoplasma trifolii' (16SrVI) group associated with leaf yellows of *Calotropis gigantea* in India. *New Disease Reports* **22**, 2044.

Maixner M (2009) Phytoplasma epidemiological systems with multiple plant hosts. In: Phytoplasmas: genomes, plant hosts and vectors. Eds Weintraub PG, Jones P, CABI United Kingdom, 213–232 pp.

Mall S, Rao GP, Marcone C (2010). Phytoplasma diseases of weeds: detection, taxonomy and diversity. IN: Recent trends in biotechnology and microbiology. Nova Science Publishers, Inc. Hauppauge, New York, USA, 87–108 pp.

Mall S, Kumar S, Jadon V, Rao GP (2015) Identification of phytoplasmas associated with weed species in India. *Indian Phytopathology* **68**, 449–453.

Marcone C, Ragozzino A, Seemüller E (1997a) Detection of Bermuda grass white leaf disease in Italy and characterization of the associated phytoplasma by RFLP analysis. *Plant Disease* **81**, 862–866.

Marcone M, Ragozzino A, Seemüller E (1997b) Detection and identification of phytoplasmas in yellows-diseased weeds in Italy. *Plant Pathology* **46**, 530–537.

Marcone C, Schneider B, Seemüller E (2004) '*Candidatus* Phytoplasma cynodontis', the phytoplasma associated with Bermuda grass white leaf disease. *International Journal of Systematic and Evolutionary Microbiology* **54**, 1077–1082.

Martini M, Carraro L, Marcone C, Maixner M, Delic D, Myrta A (2008) Caratterizzazione molecolare di fitoplasmi associati a giallume in convolvolo. *Petria* **18**, 341–344.

Martini M, Marcone C, Mitrović J, Maixner M, Delić D, Myrta A, Ermacora P, Bertaccini A, Duduk B (2012) '*Candidatus* Phytoplasma convolvuli', a new phytoplasma taxon associated with bindweed yellows in four European countries. *International Journal of Systematic and Evolutionary Microbiology* **62**, 2910–2915.

Mehle N, Ravnikar M, Seljak G, Knapiè V, Dermastia M (2011) The most widespread phytoplasmas, vectors and measures for disease control in Slovenia. *Phytopathogenic Mollicutes* **1**, 65–76.

Mitrović J, Smiljković M, Seemüller E, Reinhardt R, Huttel B, Bertaccini A, Kube M, Duduk B (2015) Differentiation of '*Candidatus* Phytoplasma cynodontis' based on 16S rRNA and *groEl* genes and identification of a new subgroup, 16SrXIV-C. *Plant Disease* **99**, 1578–1583.

Moghal SM, Zidgali AD, Moustafa SS (1998) Natural host range and reactions of citrus species to witches' broom disease of lime (WBDL) in Oman. Proceedings of the IPM Conference, Sultan Qaboos University, Muscat, Oman, 143–152 pp.

Montano HG, Bertaccini A, Pimentel JP, Paltrinieri S, Contaldo N (2014). *Erigeron (Conyza) bonariensis*, a host of '*Candidatus* Phytoplasma fraxini'-related strain in Brazil. *Phytopathogenic Mollicutes* **4**, 72–76.

Muniyappa V, Rao MS, Govindu HC (1982) Yellowing disease of *Urochloa panicoides* Beauv. *Current Science* **51**, 427–428.

Nabi S, Madhupriya DD, Rao GP (2015a) Identification of *Cannabis sativa* L. ssp. sativa as putative alternate host of sesame phyllody phytoplasma belongs to a 16SrI group in India. *Medicinal Plants* **7**, 68–70.

Nabi S, Dubey DK, Rao GP, Baranwal VK, Sharma P (2015b) Molecular characterization of '*Candidatus* Phytoplasma asteris' subgroup I-B associated with sesame phyllody disease and identification of its natural vector and weed reservoir in India. *Australasian Plant Pathology* **44**,289–297.

Nejat N, Sijam K, Abdullah SNA, Vadamalai G, Dickinson M (2009a) First report of a 16Sr XIV '*Candidatus* Phytoplasma cynodontis' group phytoplasma associated with coconut yellow decline in Malaysia. *Plant Pathology* **58**, 389.

Nejat N, Sijam K, Abdullah SNA, Vadamalai G, Dickinson M (2009b) Phytoplasmas associated with disease of coconut in Malaysia: phylogenetic groups and host plant species. *Plant Pathology* **58**, 1152–1160.

Obura E, Masiga D, Midega CAO, Wachira F, Pickett JA, Deng AL, Khan ZR (2010) First report of a phytoplasma associated with Bermuda grass white leaf disease in Kenya. *New Disease Reports* **21**, 23.

Obura E, Masiga D, Midega CAO, Otim M, Wachira F, Pickett J, Khan ZR (2011) Hyparrhenia grass white leaf disease, associated with a 16SrXI phytoplasma, newly reported in Kenya. *New Disease Reports* **24**, 17.

Omar AF (2016) Association of '*Candidatus* Phytoplasma cynodontis' with Bermuda grass white leaf disease and its new hosts in Qassim province, Saudi Arabia. *Journal of Plant Interactions* **11**, 101–107.

Palermo S, Elekes M, Botti S, Ember I, Alma A, Orosz A, Bertaccini A, Kölber M (2004) Presence of stolbur phytoplasma in Cixiidae in Hungarian vineyards. *Vitis* **43**, 201–203.

Pallavi MS, Rangaswamy KT, Naik RH, Divya BL (2011) Molecular detection of a phytoplasma associated with witches' broom in South India. *Indian Journal of Plant Protection* **39**, 308–313.

Raj SK, Khan MS, Snehi SK, Kumar S, Mall S, Rao GP (2008a) First report of phytoplasma '*Candidatus* phytoplasma asteris' (16SrI) from *Parthenium hysterophorus* L. showing symptoms of virescence and witches' broom in India. *Australasian Plant Disease Notes* **3**, 44–45.

Raj SK, Snehi SK, Khan MS, Kumar S (2008b) '*Candidatus* Phytoplasma asteris' (group 16SrI) associated with a witches' broom disease of *Cannabis sativa* in India. *Plant Pathology* **57**, 1173.

Raj SK, Snehi SK, Kumar S, Pratap D, Khan MS (2009a) Association of '*Candidatus* Phytoplasma asteris' (16SrI group) with yellows of *Achyranthes aspera* in India. *Plant Pathology* **58**, 390.

Raj SK, Snehi SK, Kumar S, Khan MS (2009b) First finding of '*Candidatus* Phytoplasma trifolii' (16SrVI group) associated with little leaf disease of *Datura inoxia* in India. *Plant Pathology* **58**, 791.

Raju BC, Chen TA (1980) Growth, morphology and ultrastructural studies of a spiroplasma associated with Bermuda grass showing witches' broom symptoms. *Zeitschrift Pflanzenkrankheit Pflanzenschutz* **87**, 37–45.

Rančić D, Paltrinieri S, Toševski I, Petanović R, Stevanović B, Bertaccini A (2005) First report of multiple inflorescence disease of *Cirsium arvense* and its association with a 16SrIII-B subgroup phytoplasma in Serbia. *Plant Pathology* **54**, 561–561.

Rao GP, Singh HN (1990) Occurrence of grassy shoot diseases pathogen (MLO) of sugarcane on *Imperata arundinacea* in India. *National Academy Science Letters* **13**, 403.

Rao GP, Raj SK, Snehi SK, Mall S, Singh M, Marcone C (2007) Molecular evidence for the presence of '*Candidatus* Phytoplasma cynodontis', the Bermuda grass white leaf agent, in India. *Bulletin of Insectology* **60**, 145–146.

Rao GP, Mall S, Singh M, Marcone C (2009) First report of a '*Candidatus* Phytoplasma cynodontis'-related strain (group 16SrXIV) associated with white leaf disease of *Dichanthium annulatum* in India. *Australasian Plant Disease Notes* **4**, 56–58.

Rao GP, Mall S, Marcone C (2010) '*Candidatus* Phytoplasma cynodontis' (16SrXIV group) affecting *Oplismenus burmannii* (Retz.) P. Beauv. and *Digitaria sanguinalis* (L.) Scop. in India. *Australasian Plant Disease Notes* **5**, 93–95.

Rao GP, Mall S, Raj SK, Snehi SK (2011) Phytoplasma diseases affecting various plant species in India. *Acta Phytopathologica et Entomologica Hungarica* **46**, 59–99.

Rao GP, Madhupriya, Tiwari AK, Dubey DK (2017a) Identification of *Ocimum canum* as possible alternative host of sesame phyllody phytoplasma strain 16SrI-B in India. *Phytopathogenic Mollicutes* **7**, 62–65.

Rao G, Madhupriya, Thorat V, Manimekalai R, Tiwari A, Yadav A (2017b) Review: a century progress of research on phytoplasma diseases in India. *Phytopathogenic Mollicutes* **7**, 1–38.

Reddy AV, Jeyarajan R (1988) A new weed host for rice yellow dwarf (RYD) pathogen. *International Rice Research Newsletter* **13**, 35.

Reeder R, Arocha Y (2008) '*Candidatus* Phytoplasma asteris' identified in *Senecio jacobaea* in the United Kingdom. *Plant Pathology* **57**, 769.

Riedle-Bauer M, Tiefenbrunner W, Otreba J, Hanak K, Schildberger B, Regner F (2006) Epidemiological observations on "bois noir" in Austrian vineyards. *Mitteilungen Klosterneuburg* **56**, 177–181.

Riedle-Bauer M, Sára A, Regner F (2008) Transmission of a "stolbur" phytoplasma by the Agalliinae leafhopper *Anaceratagallia ribauti* (Hemiptera, Auchenorrhyncha, Cicadellidae). *Journal of Phytopathology* **156**, 687–690.

Salehi M, Izadpanah K, Siampour M, Taghizadeh M (2009) Molecular characterization and transmission of Bermuda grass white leaf phytoplasma in Iran. *Journal of Plant Pathology* **91**, 655–661.

Samad A, Ajayakmar PV, Zaim M, Sattar A, Khaliq A (2004) Little leaf of "bhumyamalaki" (*Phyllanthus amarus*) – a new disease in India. *Acta Phytopathologica et Entomologica Hungarica* **39**, 49–54.

Sarindu N, Clark MF (1993) Antibody production and identity of MLOs associated with sugarcane white leaf disease and Bermuda grass white leaf disease from Thailand. *Plant Pathology* **42**, 396–402.

Schneider B, Marcone C, Kampmann M, Ragozzino A, Lederer W, Cousin M-T, Seemüller E (1997) Characterization and classification of phytoplasmas from wild and cultivated plants by RFLP and sequence analysis of ribosomal DNA. *European Journal of Plant Pathology* **103**, 675–686.

Schneider B, Padovan A, De La Rue S, Eichner R, Davis R, Bernuetz A, Gibb K (1999) Detection and differentiation of phytoplasmas in Australia: an update. *Australian Journal of Agricultural Research* **50**, 333–342.

Sdoodee R, Schneider B, Padovan A, Gibb K (1999) Detection and genetic relatedness of phytoplasmas associated with plant disease in Thailand. *Journal of Biochemistry, Molecular Biology and Biophysics* **3**, 133–140.

Seemüller E, Schneider B, Mäurer R, Ahrens U, Daire X, Kison H, Lorenz KH, Firrao G, Avinent L, Sears BB, Stackebrandt E (1994) Phylogenetic classification of phytopathogenic mollicutes by sequence analysis of 16S ribosomal DNA. *Internatonal Journal of Systematic Bacteriology* **44**, 440–446.

Seemüller E, Marcone C, Lauer U, Ragozzino A, Göschl M (1998) Current status of molecular classification of the phytoplasmas. *Journal of Plant Pathology* **80**, 3–26.

Singh UP, Sakai A, Singh AK (1978) White leaf disease of *Cynodon dactylon* Pers., a mycoplasmal disease in India. *Cellular and Molecular Life Sciences* **34**, 1447–1448.

Singh N, Madhupria, Rao GP, Upadhyaya PP (2012) '*Candidatus* Phytoplasma trifoli' associated with little leaf and witches' broom disease of *Datura stramonium* L. in India. *Phytopathogenic Mollicutes* **2**, 69–71.

Singh N, Mall S, Upadhaya PP, Rao GP (2013) *Ranunculus sceleratus*: a new host of '*Candidatus* Phytoplasma cynodontis' in India. *Phytopathogenic Mollicutes* **3**, 82–86.

Snehi SK, Khan MS, Raj SK, Mall S, Singh M, Rao GP (2008) Molecular identification of '*Candidatus* Phytoplasma cynodontis' associated with Bermuda grass white leaf disease in India. *Plant Pathology* **57**, 770.

Sunpapao A (2016) Preliminary molecular identification of phytoplasma associated with *Axonopus compressus* white leaf disease. *Songklanakarin Journal of Plant Science* **3**, 40–43.

Šafářová D, Zemánek T, Válová P, Navrátil M (2016) '*Candidatus* Phytoplasma cirsii', a novel taxon from creeping thistle [*Cirsium arvense* (L.) Scop.]. *International Journal of Systematic and Evolutionary Microbiology* **66**, 1745–1753.

Thorat V, Bhale U, Sawant V, More V, Jadhav P, Mane SS, Nandanwar RS, Tripathi S, Yadav A (2016) Alternative weed hosts harbor 16SrII group phytoplasmas associated with little leaf and witches' broom diseases of various crops in India. *Phytopathogenic Mollicutes* **6**, 50–55.

Tiefenbrunner W, Leitner G, Gangl H, Riedle-Bauer M (2007) Epidemische ausbreitung der schwarzholzkrankheit ("stolbur" phytoplasma) in südostösterreichischen weingärten. *Mitteilungen Klosterneuburg* **57**, 185–189.

Tiwari AK, Vishwakarma SK, Singh SP, Kumar P, Khan MS, Chun SC, Rao GP (2012) First report of a '*Candidatus* Phytoplasma asteris' associated with little leaf disease of *Ageratum conyzoides* in India. *New Disease Reports* **26**, 18.

Tiwari AK, Khan MS, Kumar P, Tiwari A (2017) Molecular characterization of phytoplasma of 16SrI-B group association with *Acalypha indica* in India. *3 Biotech* **7**, 49.

Tolu G, Botti S, Garau R, Prota VA, Sechi A, Prota U, Bertaccini A (2006) Identification of a 16SrII-E phytoplasma in *Calendula arvensis*, *Solanum nigrum*, and *Chenopodium* spp. *Plant Disease* **90**, 325–330.

Tran-Nguyen L, Blanche KR, Egan B, Gibb KS (2000) Diversity of phytoplasmas in northern Australian sugarcane and other grasses. *Plant Pathology* **49**, 666–679.

Valiūnas D, Urbanavičienė L, Jomantienė R, Davis RE (2007) Molecular detection, classification and phylogenetic analysis of subgroup 16SrI-C phytoplasmas detected in diseased *Poa* and *Festuca* in Lithuania. *Biologija* **53**, 36–39.

Wilson D, Blanche KR, Gibb KS (2001) Phytoplasmas and disease symptoms of crops and weeds in the semi-arid tropics of the Northern Territory, Australia. *Australasian Plant Pathology* **30**, 159–163.

Win NKK, Jung H-Y (2012) '*Candidatus* Phytoplasma cynodontis' associates with white leaf disease of golden beard grass (*Chrysopogon aciculatus*). *Tropical Plant Pathology* **37**, 76–79.

Win NKK, Kim Y-H, Jung H-Y, Ohga S (2013) Molecular characterization of white leaf phytoplasma associated with the Graminae in Myanmar. *Journal of the Faculty of Agriculture, Kyushu University* **58**, 225–229.

Wongkaew P, Hanboonsong Y, Sirithorn P, Choosai C, Boonkrong S, Tinnangwattana T, Kitchareonpanya R, Damak S (1997) Differentiation of phytoplasmas associated with sugarcane and gramineous weed white leaf disease and sugarcane grassy shoot disease by RFLP and sequencing. *Theoretical and Applied Genetics* **95**, 660–663.

Wu Y, Hao X, Li Z, Gu P, An F, Xiang J, Wang H, Luo Z, Liu J, Xiang Y (2010) Identification of the phytoplasma associated with wheat blue dwarf disease in China. *Plant Disease* **94**, 977–985.

Yadav A, Bhale U, Thorat V, Shouche Y (2014) First report of a new subgroup 16SrII-M '*Candidatus* Phytoplasma aurantifolia' associated with witches' broom disease of *Tephrosia purpurea* in India. *Plant Disease* **98**, 990.

Yadav A, Thorat V, Bhale U, Shouche Y (2015) Association of 16SrII-C and 16SrII-D subgroup phytoplasma strains with witches' broom disease of *Parthenium hysterophorus* and insect vector *Orosius albicinctus* in India. *Australian Plant Disease Notes* **10**, 31.

Yadav V, Thorat V, Madevkumar S, Janardana G, Yadav A (2016) First report on the association of 16SrII-D group phytoplasma with little leaf disease of *Crotalaria pallida* from Karnataka, India. *Plant Disease* **100**, 2523.

Yang Y, Zhao W, Li Z, Zhu S (2011) Molecular Identification of a '*Candidatus* Phytoplasma ziziphi'-related strain infecting amaranth (*Amaranthus retroflexus* L.) in China. *Journal of Phytopathology* **159**, 635–637.

Zahoor A, Bashir M, Nakashima K, Mitsueda T, Murata N (1995) Bermuda grass white leaf caused by phytoplasmas in Pakistan. *Pakistan Journal of Botany* **27**, 251–252.

Zamora L, Contaldo N, Paltrinieri S, Quiñones ML, Piñol B, Acosta KI, Bertaccini A (2015) *Cyperus rotundus* L. a new host species for '*Candidatus* Phytoplasma aurantifolia'-related phytoplasmas in Cuba. *Phytopathogenic Mollicutes* **5**, 42–46.

Zhao Y, Sun Q, Davis RE, Lee IM, Liu Q (2007) First report of witches' broom disease in a *Cannabis* spp. in China and its association with a phytoplasma of elm yellows group (16SrV). *Plant Disease* **91**, 227–227.

Zhao Y, Wei W, Lee I-M, Shao J, Suo X, Davis RE (2009) Construction of an interactive online phytoplasma classification tool, *i*PhyClassifier, and its application in analysis of the peach X-disease phytoplasma group (16SrIII). *International Journal of Systematic and Evolutionary Microbiology* **59**, 2582–2593.

Zibadoost S, Rastgou M (2016) Molecular identification of phytoplasmas associated with some weeds in West Azerbaijan province, Iran. *Acta Agriculturae Slovenica* **107**, 129–136.

The manufacturer's authorised representative in the EU is Springer Nature Customer Service Centre GmbH, Europaplatz 3, 69115 Heidelberg, Germany. If you have any concerns regarding our products, please contact ProductSafety@springernature.com

Printed and bound by CPI Group (UK) Ltd, Croydon, CR0 4YY
25/09/2026
02222891-0011